Circum-Pacific Council
for Energy and Mineral Resources
Earth Science Series

Volume 14

Geology and Offshore Mineral Resources of the Central Pacific Basin

Circum-Pacific Council
for Energy and Mineral Resources
Earth Science Series

H. Gary Greene, Editor-in-Chief
Florence L. Wong, Executive Editor

Volume 1
Tectonostratigraphic Terranes of the Circum-Pacific Region
Edited by D.G. Howell

Volume 2
Geology and Offshore Resources of Pacific Island Arcs—Tonga Region
Edited and Compiled by D.W. Scholl and T.L. Vallier

Volume 3
Investigations of the Northern Melanesian Borderland
Edited by T.M. Brocher

Volume 4
Geology and Offshore Resources of Pacific Island Arcs—Central and Western Solomon Islands
Edited by J.G. Vedder, K.S. Pound, and S.Q. Boundy

Volume 5A
The Antarctic Continental Margin: Geology and Geophysics of Offshore Wilkes Land
Edited by S.L. Eittreim and M.A. Hampton

Volume 5B
The Antarctic Continental Margin: Geology and Geophysics of the Western Ross Sea
Edited by A.K. Cooper and F.J. Davey

Volume 6
Geology and Resource Potential of the Continental Margin of Western North America and Adjacent Ocean Basins—Beaufort Sea to Baja California
Edited by D.W. Scholl, A. Grantz, and J.G. Vedder

Volume 7
Marine Geology, Geophysics, and Geochemistry of the Woodlark Basin—Solomon Islands
Edited by B. Taylor and N.F. Exon

Volume 8
Geology and Offshore Resources of Pacific Island Arcs—Vanuatu Region
Edited by H.G. Greene and F.L. Wong

Volume 9
Geology and Offshore Resources of Pacific Island Arcs—New Ireland and Manus Region, Papua New Guinea
Edited by M.S. Marlow, S.V. Dadisman, and N.F. Exon

Volume 10
Petroleum Resources of China and Related Subjects
Edited by H.C. Wagner, L.C. Wagner, F.F.H. Wang, and F.L. Wong

Volume 11
Geology of the Andes and Its Relation to Hydrocarbon and Mineral Resources
Edited by G.E. Ericksen, M.T. Cañas Pinochet, and J.A. Reinemund

Volume 12
Geology and Offshore Resources of Pacific Island Arcs—Solomon Islands and Bougainville, Papua New Guinea Regions
Edited by J.G. Vedder and T.R. Bruns

Volume 13
Terrane Analysis of China and the Pacific Rim
Edited by T.J. Wiley, D.G. Howell, and F.L. Wong

Volume 14
Geology and Offshore Mineral Resources of the Central Pacific Basin
Edited by B.H. Keating and B.R. Bolton

3, and 5A through 13 available from AAPG, P.O. Box 979, Tulsa, OK 74101-0979.

Barbara H. Keating
Barrie R. Bolton
Editors

Geology and Offshore Mineral Resources of the Central Pacific Basin

Circum-Pacific Council
for Energy and Mineral Resources
Earth Science Series, Volume 14

With 173 Illustrations

Springer-Verlag
New York Berlin Heidelberg London Paris
Tokyo Hong Kong Barcelona Budapest

1992

Barbara H. Keating
Hawaii Institute of Geophysics
University of Hawaii
Honolulu, HI 96822-3188, USA

Barrie R. Bolton
BHP-International Inc.
Herndon, VA 22070-5200, USA

Cover Illustration: This three-dimensional image of Machias Seamount merges processed SeaMARC II sidescan sonar imagery with smoothed SeaMARC II bathymetry. Software developed by Thomas B. Reed, IV, allowed correction, mosaicking, and smoothing of the bathymetry. The sidescan imagery was then geometrically stretched to co-register with the bathymetry. The bathymetry and sidescan were merged to create a composite image depicting depth by color hue and sidescan texture by intensity (black to white). The composite image was draped over the smoothed digital elevation model of the seamount at 2 : 1 vertical exaggeration and viewed from the southeast. Bright areas are regions of low reflectivity, typically sediments. Shadowed areas of the image are a result of highly reflective materials such as basalts. The summit is at 800 m, and the surrounding seafloor at left in the image is at 7000 m. Machias Seamount is in Western Samoa's Exclusive Economic Zone (EEZ). Machias Seamount is located at the boundary of the Pacific plate where it is currently being subducted into the Tonga Trench. As subduction takes place, the western half of the seamount is broken into thin slivers by normal faulting. A series of faults resembling steps can be seen. The eastern margin of the seamount has not yet been fractured and broken into segments; instead it shows a more typical distribution of debris slides that are the product of mass wasting. The geologic history of Machias Seamount is discussed in the chapter "The Geology of the Samoan Islands" by Keating (this volume, p. 127–178.)

Library of Congress Cataloging-in-Publication Data
Geology and offshore mineral resources of the Central Pacific Basin /
edited by Barbara H. Keating and Barrie R. Bolton.
p. cm. – (Earth science series / Circum-Pacific Council for
Energy and Mineral Resources ; v. 14)
Includes bibliographical references and index.
ISBN 0-387-97771-6. – ISBN 3-540-97771-6
1. Mines and mineral resources – Central Pacific Basin.
2. Submarine geology – Central Pacific Basin. I. Keating, Barbara
H. II. Bolton, Barrie R. III. Series: Earth science series
(Houston, Tex.) ; v. 14.
TN123.G46 1991
551.46′08′0949 – dc20 91-43873

Printed on acid-free paper

Production managed by Ellen Seham; manufacturing supervised by Robert Paella.
Camera-ready copy provided by the editors.
Printed and bound by Edwards Brothers, Inc., Ann Arbor, MI.
Printed in the United States of America.

9 8 7 6 5 4 3 2 1

ISBN 0-387-97771-6 Springer-Verlag New York Berlin Heidelberg
ISBN 3-540-97771-6 Springer-Verlag Berlin Heidelberg New York

CIRCUM-PACIFIC COUNCIL FOR ENERGY AND MINERAL RESOURCES

Foreword

The Earth Science Series of the Circum-Pacific Council for Energy and Mineral Resources (CPCEMR) is designed to convey the results of geologic research in and around the Pacific Basin. Topics of interest include framework geology, petroleum geology, hard minerals, geothermal energy, environmental geology, volcanology, oceanography, tectonics, geophysics, geochemistry, and applications of renewable energy. The CPCEMR supports and publishes results of scientific research that will advance the knowledge of energy and mineral resource potential in the circum-Pacific region. The Earth Science Series is specifically designed to publish papers that include new data and new maps, report on CPCEMR-sponsored symposia and workshops, and describe the results of onshore and marine geological and geophysical explorations.

This volume reports the results of one of fourteen internationally sponsored surveys to investigate the energy and mineral resources in the Southwest Pacific. The 1982, 1984 and 1986-1987 surveys were fostered by the Australia-New Zealand-United States Tripartite Agreement. Geophysical and geological data were collected aboard the U.S. Geological Survey's (USGS) R/V *S.P. Lee* and the Hawaii Institute of Geophysics' (HIG) R/V *Kana Keoki* and R/V *Moana Wave* in the waters of Fiji, Kiribati, Papua New Guinea, Solomon Islands, Tonga, Western Samoa, and Vanuatu.

Funding for ship time was made available through the U.S. Agency for International Development, the USGS, the U.S. Office of Naval Research (for HIG's 1982 work), the Australian Development Assistance Bureau, the Australian Bureau of Mineral Resources (BMR), the New Zealand Ministry of Foreign Affairs, the New Zealand Department of Scientific and Industrial Research (DSIR), the New Zealand Geological Survey, and the New Zealand Oceanographic Institute (NZOI). Coordination of the program was provided by the U.S. Department of State and the South Pacific Applied Geoscience Commission (SOPAC, formerly the United Nations-sponsored Committee for the Coordination of Joint Prospecting for Mineral Resources in South Pacific Offshore Areas CCOP/SOPAC) in Fiji. Over 150 scientists and technicians participated in the cruises and represented the South Pacific island nations and funding countries mentioned above, as well as SOPAC, New Caledonia, the United Kingdom and France.

Michel T. Halbouty

Chairman and President

SOPAC

Acknowledgment

The research recorded here is a major contribution by the member countries of the South Pacific Applied Geoscience Commission (SOPAC) and by the wider international community to the understanding of the natural resource potential and the geological origin and significance of the seabed of national Exclusive Economic Zones (EEZ) and the oceanic areas of the South Pacific. At the same time, it is an outstanding example of international partnership and co-operation, with mutual benefits to all concerned.

SOPAC warmly welcomes the success of this collaborative enterprise, and, on behalf of the member governments and its Technical Secretariat, I wish to place on record our sincere appreciation for the very generous contribution by the Tripartite Partners—the governments of Australia, New Zealand, and the United States—and by the Circum-Pacific Council for Energy and Mineral Resources.

Jioji Kotobalavu

Director, SOPAC Technical Secretariat

Preface

The results of two scientific cruises are summarized in this volume: The initial sampling cruise (MW-86-02, Chief Scientists Barbara H. Keating and Barrie R. Bolton.) departed from Ponape in the Federated States of Micronesia in February 1986, and ended in Honolulu. The SeaMARC II sidescan mapping and sampling cruise (MW-87-02, Chief Scientists William Coulbourn and Peter Hill) departed Pago Pago in American Samoa a year later. In addition to the shipboard scientists (see List of Participants), numerous other scientists have been involved in the land-based laboratory studies. This volume presents the results of both shipboard and laboratory studies resulting from these cruises, as well as from similar resource-related cruises in the central Pacific Basin that have been sponsored by other international institutions during recent years. The contributions from other scientists working in the region provide a regional understanding of the geology of the central Pacific Basin. The results of the SeaMARC II mapping conducted during the second cruise are being reproduced separately as the Pacific Seafloor Atlas.

The sampling and mapping cruises (MW-86-02 and MW-87-02) were conducted under the terms of the Tripartite II Agreement between Australia, New Zealand, and the United States in association with the United Nations-sponsored Committee for Coordination of Joint Prospecting for Mineral Resources in South Pacific Offshore Areas (CCOP/SOPAC). [In 1989, the name was changed to South Pacific Applied Geoscience Commission (SOPAC)]. Major funding was provided by the U.S. Agency for International Development and by the government of Australia. We wish to thank the governments of Kiribati, Samoa, and the Cook Islands for encouraging these scientific studies within their waters and for providing personnel to assist in the scientific endeavors.

The names of Pacific Islands often change; some islands have had more than 20 names during the last 200 years. In the chapters by Keating the names generally used in the literature have been chosen. Readers should also be aware that several of the island names of the Line Islands (part of the Kiribati nation) have recently been changed. The older names have been retained in this publication in order to be consistent with the published literature. Place names within the Samoan group have also been utilized without apostrophes; that punctuation is not commonly shown in scientific work, and we have chosen to be consistent with other scientific publications about Samoa.

The shipboard work was carried out aboard the Hawaii Institute of Geophysics research vessel *Moana Wave* under Captain R. Hayes. We wish to thank the captain, ship's crew, and scientific technicians aboard the ship for their support, which greatly contributed to the success of these cruises. We also thank the support personnel at the University Marine Center in Honolulu for their efforts in fabricating equipment for these cruises. We acknowledge support from the Hawaii Institute of Geophysics for the production of this volume, in particular the work of the Publication Services group. We are grateful to those who critically reviewed the chapters in this volume for their timely assistance and responses.

Barbara H. Keating
Hawaii Institute of Geophysics
University of Hawaii

Barrie R. Bolton
University of Adelaide
(now at BHP-Utah International, Inc., Herndon, VA)

Participants

Shipboard Scientists, Leg MW-86-02

Barbara H. Keating
Chief Scientist
Hawaii Institute of Geophysics

Co-Chief Scientist
Barrie R. Bolton
LaTrobe University
now at BHP-Utah International Inc.

Jules Bogi
New South Wales Institute of Technology

Bryan Baker
Florida State University

Terry Hamilton
University of Melbourne

Loren Kroenke
Hawaii Institute of Geophysics

Edward Murphy
U.S. Geological Survey

Patricia Pennywell
Hawaii Institute of Geophysics

Peter Humphrey
Hawaii Institute of Geophysics

Jim Kellogg
Hawaii Institute of Geophysics
now at University of South Carolina

Bruce Wedgeworth
Hawaii Institute of Geophysics

Dr. A.M.V. Raghava Rao
Geological Survey of India

Anote Tong
University of the South Pacific

Being Yee-Ting
Kiribati, Ministry of Natural
Resources and Development

Shipboard Scientists, Leg MW-87-02

Co-Chief Scientist
William T. Coulbourn
(deceased June 1990)
Hawaii Institute of Geophysics

Co-Chief Scientist
Peter J. Hill
Bureau of Mineral Resources, Australia

Jules Bogi
New South Wales Institute of Technology

Eric H. DeCarlo
Hawaii Institute of Geophysics

Barbara H. Keating
Hawaii Institute of Geophysics

Douglas Bergersen
Hawaii Institute of Geophysics

Patricia Pennywell
Hawaii Institute of Geophysics

Steven Kamu
Apia Observatory, Samoa

Being Yee-Ting
Kiribati, Ministry of Natural
Resources and Development

Contributors

Nicolas Baudry
Seafloor Imaging, Inc.
New Caledonia
BP 8039, Noumea-Sud

Jules Bogi
Faculty of Physical Science
University of Technology
Sydney, P.O. Box 123
Broadway, N.S.W. 2007, Australia

Barrie R. Bolton
BHP-Utah International Inc.
200 Fairbrook Drive
Herndon, VA 22070

William C. Burnett
Department of Oceanography
Florida State University
Tallahassee, FL 32306

David S. Cronan
Department of Geology
Imperial College of Science and Technology
London, SW7, England

Eric H. De Carlo
Department of Oceanography
University of Hawaii
Honolulu, HI 96822

Charles M. Fraley
Hawaii Institute of Geophysics
University of Hawaii
Honolulu, HI 96822

Jeffrey T. Freymueller
Department of Geophysics
Stanford University
Stanford, CA 94305

Masahide Furukawa
Department of Earth Sciences
Graduate School of Science and Technolgoy
Kobe University
Nada, Kobe 657, Japan

Lisa M. Gein
U.S. Geological Survey
345 Middlefield Road
Mail Stop 999
Menlo Park, CA 94025

James R. Hein
U.S. Geological Survey
345 Middlefield Road
Mail Stop 999
Menlo Park, CA 94025

Peter J. Hill
Division of Marine Geosciences
and Petroleum Geology
Bureau of Mineral Resources,
Geology and Geophysics
GPO Box 378
Canberra, ACT 2601, Australia

Barbara H. Keating
Hawaii Institute of Geophysics
University of Hawaii
Honolulu, HI 96822

James N. Kellogg
Department of Geological Science
University of South Carolina
Columbia, SC 29208

Masaaki Kimura
Department of Marine Sciences
University of the Ryukyus
Okinawa 903-01, Japan

Atsuyuki Mizuno
Department of Earth Sciences
Ehime University
Matsuyama 790, Japan

Tomoyuki Moritani
Marine Geology Department
Geological Survey of Japan
1-1-3 Higashi, Tsukuba City
Ibaraki 305, Japan

Akira Nishimura
Marine Geology Department
Geological Survey of Japan
1-1-3 Higashi, Tsukuba City
Ibaraki 305, Japan

V. Purnachandra Rao
Department of Oceanography
Florida State University
Tallahassee, FL 32306-3048
On leave from
National Institute of Oceanography
Dona Paula
Goa-403004, India

Bruce M. Richmond
U.S. Geological Survey
345 Middlefield Road
Mail Stop 999
Menlo Park, CA

William W. Sager
Departments of Oeanography, Geophysics, and
Geodynamics Research Institute
Texas A&M University
College Station, TX 77843-3146

Marjorie S. Schulz
U.S. Geological Survey
345 Middlefield Road
Mail Stop 999
Menlo Park, CA

David R. Stoddart
Department of Geography
Earth Sciences Building
University of California
Berkeley, CA 94720

Kensaku Tamaki
Ocean Research Institute
University of Tokyo
Tokyo 164, Japan

Akira Usui
Marine Geology Department
Geological Survey of Japan
1-1-3 Higashi, Tsukuba City
Ibaraki 305, Japan

Contents

Keating, B.H., and Bolton, B.R., editors, 1992, Geology and offshore mineral resources of the central Pacific basin, Circum-Pacific Council for Energy and Mineral Resources Earth Science Series, V. 14. New York, Springer-Verlag.

CONTRIBUTIONS OF THE 1838-1842 U.S. EXPLORING EXPEDITION

Barbara H. Keating
Hawaii Institute of Geophysics, University of Hawaii, Honolulu, HI, 96822

ABSTRACT

The years 1988–1992 mark the 150th anniversary of the U.S. Exploring Expedition, one of the first scientific missions undertaken by the United States to explore the oceans of the world. It circumnavigated the world and mapped large portions of the coast of Antarctica, the islands of the central Pacific Ocean, and the northwest Pacific coast of North America. The explorers were the first to sight land on the Antarctic continent. The eastern Antarctic coastline was mapped by the expedition and later named Wilkesland in honor of the commander of the U.S. Exploring Expedition. The expedition mapped most of the island groups in the central Pacific Ocean, eliminating the confusion over the number and position of islands in this region. The important legacies from the expedition include the establishment of the U.S. National Museum (Smithsonian Museum) to house the collections of the cruise and the establishment of the United States as an internationally recognized scientific leader in the area of botanical, zoological, and marine studies, and as a leader and contributor to international ocean-going expeditions.

INTRODUCTION

One hundred and fifty years ago, 490 sailors and officers in six ships departed the United States in an expedition that would take four years to circumnavigate the world. The expedition left Hampton Roads, Virginia in August, 1838, and returned to New York in June, 1842. The motivation for the expedition was in large part to find new whaling grounds for already wealthy New England merchants. The expedition was also supported by the Navy since it provided a means of responding to attacks against American whalers in distant ports. Scientifically, the expedition was to examine the theory of the "holes in the poles" suggested by John Symmes (Viola and Margolis, 1985). Regardless of the reasoning behind the expedition, it succeeded and produced results that far exceeded the expectations of the then Secretary of the Navy, whose responsibility it was to see that the expedition took place.

The expedition was the first to claim definitive sighting of an Antarctic continent, and it mapped large portions of the eastern Antarctic coastline, most of the island groups of the Pacific, and the Pacific northwest coast of North America (Fig. 1). It established the precedent for the United States government's charting both national and foreign waters for the safety of commercial and naval interests, and founded American interests in scientific studies of the Antarctic. The charts produced by the expedition eliminated the confusion over the number and position of islands within the central Pacific. These charts produced for the navy as a consequence of this cruise provided accurate maritime information for seafarers and increased the safety of ocean passages. The expedition's work led to the founding of the U.S. Hydrographic Office and provided many of the natural history collections of the Smithsonian Institution National Museum. The expedition also established the United States' role in international scientific research and enhanced the nation's prestige in the international scientific community.

Later scientific expeditions to the Southern and Pacific Ocean (including those discussed in this volume) benefited from the precise locations and detailed hydrographic maps produced by the U.S. Exploring Expedition. Though plagued with difficulties, America's first naval exploring expedition resulted in a proud heritage that is a tribute to its participants.

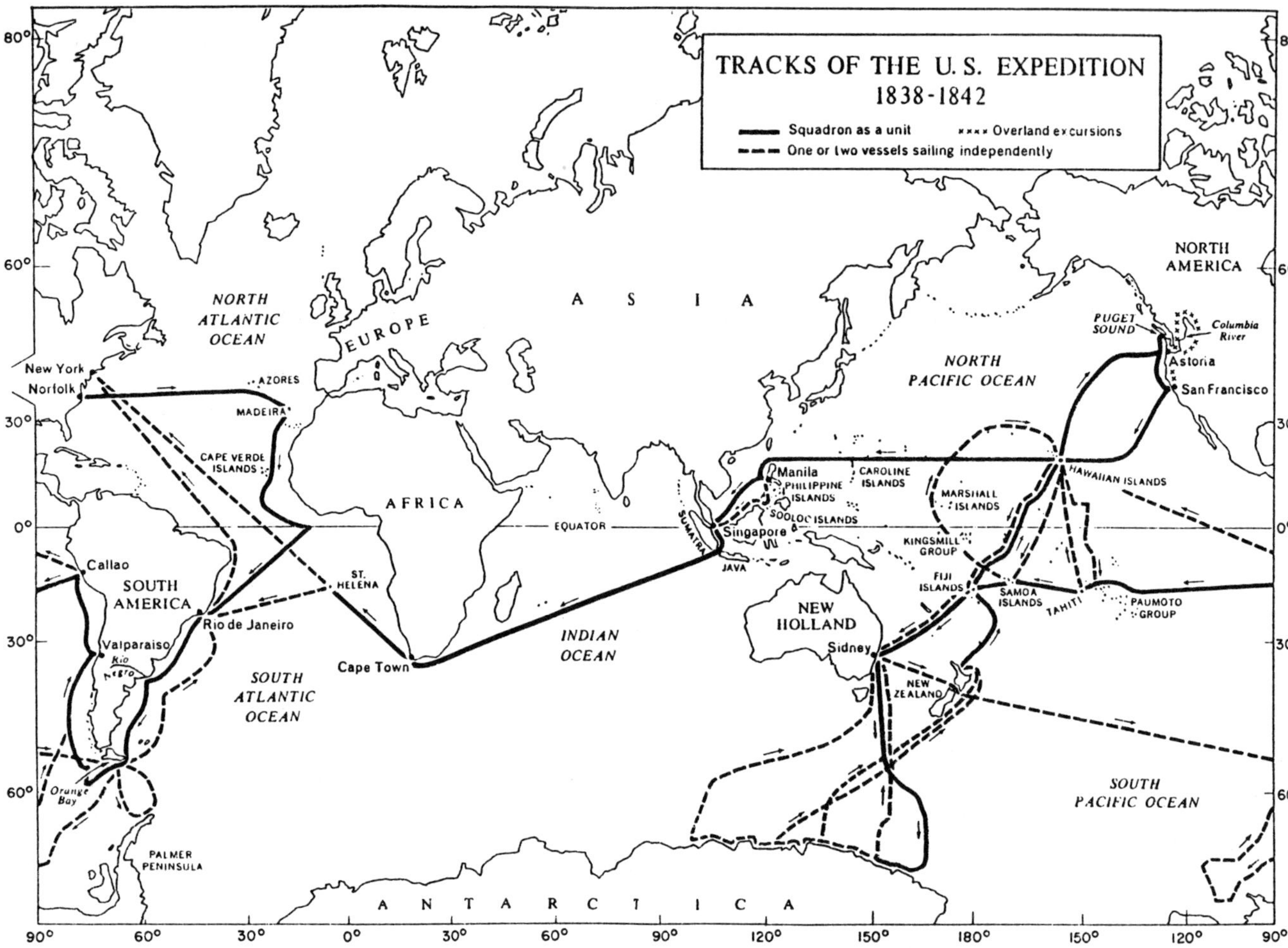

Figure 1. Chart of the world showing the ship tracks of the U.S. Exploring Expedition (redrawn from Tyler, 1968).

BACKGROUND

The U.S. Exploring Expedition began with the support of New England merchants who were eager to find new sealing and whaling grounds in the South Seas. They allied themselves with John Symmes, Jr. and friends in order to lobby Congress to sponsor a South Seas Exploring Expedition. Symmes had hypothesized that the world was hollow and the entrances to the interior would be found by sailing to the poles. His theory, known as the "Holes in the Poles Theory," was not widely accepted even then.

The U.S. Navy welcomed the opportunity to learn more about this distant part of the world and thus supported the expedition. Naval interests wished to display the American flag in the Pacific islands where American property and lives had been taken.

With the support of merchants, the scientific community, and the navy, President John Quincy Adams encouraged the U.S. Congress to support the expedition. In 1828, Congress passed the resolution authorizing the president to send one public ship to the Pacific to examine coasts, islands, harbors, shoals, and reefs – if it could be done without a specific appropriation. With such unenthusiastic support from Congress it is not surprising that ten years elapsed before the expedition sailed. There were endless squabbles over the number and types of ships required, the commander and the authority he would be given, the size, nature and background of the scientific crew.

Officially the expedition was known as the South Seas Exploring Expedition; however, it became better known as the U.S. Exploring Expedition or the Wilkes Expedition (after its strong-willed commander Charles Wilkes). It was also known as the "Deplorable Expedition" as a result of the seemingly endless delays and controversies which surrounded it (Viola and Margolis, 1985).

Lieutenant Wilkes

Charles Wilkes was made commander of the expedition because of his skills in naval surveying, familiarity with the navigation instruments placed on board the vessels, and because the job had already been turned down by many senior officers. Of 40 lieutenants in the service at the time of his appointment, 38 of them had more experience than Wilkes, including his second-in-command, but Wilkes was appointed nevertheless. Prior to the cruise, Wilkes was denied a promotion to the acting rank of captain, which he deserved and expected. It was not until after the expedition, in 1843, that Lieutenant Wilkes was promoted to the rank of Commander.

Wilkes was described as aloof, stern, and resolute. Most scientists who took part in the expedition disliked Wilkes. They suspected that Wilkes did not want their achievements to overshadow the expeditions' naval accomplishments. He zealously promoted the interests of the Navy and took personal charge of the studies in the physical sciences. He insisted that the surveying duty was "a service second to none in the Expedition" (Wilkes, 1845).

One of the few participants in the Expedition who had praise for Wilkes was James Dana, the expedition geologist. Dana states that Wilkes drove himself harder than he drove his ships and his men. Wilkes was determined to find the southern continent. His single-mindedness alienated his crew. One of his crew wrote of his "twenty years before the mast," telling of his desire to murder Wilkes during the voyage (Erskine, 1890). Jaffe (1976) convincingly argues that the menacing figure of Captain Ahab described in the book *Moby Dick* by Melville (1851) was Lt. Wilkes and the great white whale that Ahab searched for represented the great white continent Wilkes so desperately sought. Wilkes and Melville grew up in the same neighborhood and there were many family acquaintances. Melville also knew the expedition geologist James Dana.

Wilkes, a strict disciplinarian, often ordered more than the allowable 12 lashes in disciplining his crew, sometimes as many as 40. Immediately after the completion of the cruise several officers filed charges against Wilkes on 11 counts. A court martial upheld one charge of illegal punishment and sentenced Wilkes to a public reprimand by the Secretary of the Navy. In the 1840's and 50's Wilkes was on special duty with the navy, supervising the publication of the expedition reports. In 1843, however, he was promoted to the rank of Commander and in 1855 to Captain.

His real fame in life came during the civil war, when he was in command of the U.S.S. *San Jacinto*. In 1861, he took two confederate commissioners from the British mail packet *Trent*, in international waters. Wilkes was hailed as a hero of the Union navy, since he stopped the Confederate commissioners from requesting badly needed money and weapons support from Europe. Britain demanded the release of the prisoners and a suitable apology, which was forthcoming from the embarrassed American government.

Wilkes then served his country on the James and Potomac rivers and was given an acting appointment of Rear Admiral. He soon was in trouble again for blatantly disobeying and ignoring orders and was recalled. He received notification of his promotion to Commodore in August 1862. However, a mix-up with his records was found, and the rank was rescinded. In March, 1863, he had the rank restored. Unfortunately, his frustration became known when a letter from Wilkes to Secretary of the Navy Welles was printed in a newspaper. A court of inquiry found that Wilkes had foreknowledge of the publication. He again was court-martialled in the spring of 1864. He was found guilty of five counts of disobedience of orders, insubordinate conduct, disrespectful language, disobedience to a general order or regulation, and conduct unbecoming an officer. He was sentenced to a public reprimand and three years suspension from the Navy. In December, 1864, President Lincoln reduced the suspension to one year. In 1866, Wilkes was promoted to Rear Admiral on the retired list. In 1871, Wilkes began his autobiography but was unable to find a publisher for the volume in his lifetime. It was not until a hundred years later, in 1978, that the book was printed by the Naval Historical Center.

The Ships and Crews

The U.S. Exploring Expedition squadron consisted of six sailing ships. The flagship was the U.S.S. *Vincennes*, a sloop-of war 127 feet long, under the command of Lieutenant Wilkes, with a crew of 190 aboard. The U.S.S. *Peacock*, under the second-in-command, Lt. William Hudson, was also a sloop-of-war. It was 118 feet long and carried a crew of 130. This ship did not complete the expedition but ran aground on a sand bar at the mouth of the Columbia River off Oregon and sank. All the members of the crew were saved but the valuable collections on board were lost. The ship was replaced with a merchant brig, the *Oregon*, which was 85 feet long.

The next largest ship was a brigantine, *Porpoise*, which was 88 feet long and carried a crew of 65. The supply ship for the fleet was the *Relief*, which was 109 feet long and carried a crew of 75. The other ships of the squadron were two almost identical schooners, the *Flying Fish* and the *Seagull*. The former was 70 feet long and carried a crew of fifteen; the later was 73 feet long, and carried a similar crew. The *Flying Fish* did not complete the voyage around the world. Wilkes determined that it was not seaworthy and sold it in Singapore. The *Seagull* was lost at sea.

Before departing the U.S., the ships were modified to include accommodations for scientists and to strengthen the hulls for work in polar waters. Much additional work was required in Australia. The work in polar waters was very difficult. The crews were very poorly dressed for polar work, and illness and many deaths resulted.

Wilkes, however, was determined to find the southern continent. Despite the demands of the ship's surgeon and the crew's pleading to turn the ship around and head back to Australia, Wilkes pushed the ships and men to their absolute limits. One ship sank with all of the crew lost at sea. Another was badly damaged and nearly crushed by ice.

The Scientific Crew

Civilian scientists of the expedition were referred to as "scientifics." Twenty-five were named to the cruise originally, but in the end only nine participated. Some became frustrated over the delays and arguments associated with the expedition and dropped out. Several were excused by Wilkes in an effort to reorganize the expedition. Those who did participate established an important precedent for American scientists and naval personnel: combining resources for peaceful scientific endeavors. That tradition continues today.

In all, nine civilians took part in the expedition. Geologist James Dwight Dana is the best known of the civilian scientists who took part in the expedition. He made major contributions to the geology of Australia and Hawaii and the understanding of coral reefs as a result of his research on the expedition. One volume of the exploring expedition publications describes his geologic studies. He distinguished the linear trends of islands in the Pacific and substantiated Darwin's theory of island subsidence. His map is strikingly similar to modern tectonic maps showing the hot spot traces that mark many Pacific Island chains (e.g., Jarrard and Clague, 1977). Dana is said to have been "the most influential American geologist of the nineteenth century and remains a towering figure in the history of Geology" (Viola and Margolis, 1985). Dana devised a system of mineral identification so universally accepted it remains the prime method of mineral identification used today. His text, entitled *A System of Mineralogy*, originally published in 1837, is now in the 20th edition. Many historians suggest Dana's contributions to geology are comparable to Charles Darwin's contributions to biology. Other participants in the expedition included Horatio Hale (philologist), Titian R. Peale (artist and naturalist), Charles Pickering (naturalist), William Brackenridge (botanist), William Rich (botanist), Joseph P. Couthouy (conchologist), Alfred Agate (artist), and Joseph Drayton (artist).

SCIENTIFIC RESULTS OF THE EXPEDITION

By far the biggest collection acquired during the expedition was pressed plants. Charles Pickering was originally suggested for the expedition as an ichthyologist and herpetologist, later as chief zoologist. Asa Gray at age twenty had already been the author of a botany textbook. William Rich was chosen as "plantsman". He was evidently well-liked by the Secretary of the Navy for organizing Washington's garden shows. The Secretary of the Navy originally planned to make Rich, an unknown hobbyist, equal in position to Gray. Protests were lodged, however, and Rich was made an assistant, much to his own liking. At the last moment, Gray was offered a job which forced him to withdraw and appeal to Rich to be chief botanist. An assistant, William Brackenridge, was hired as an aide.

Rich unfortunately proved incapable of writing the botanical results of the expedition. Gray eventually agreed to take over the writing. Brackenridge wrote the fern report (Volume 16; 1856). Pickering lost interest in the herpetology and ichthyology collections, and instead wrote *Races of Man and Their Geographical Distribution* (Expedition Volume 9; 1848), and *Geographical Distribution of Animals and Plants* (Volume 15; 1848). Wilkes was forced to approach Spencer Baird to continue Pickering's work. Baird and an assistant, Charles Girard, published preliminary reports of the reptiles collected during the expedition and published them in American journals. Girard wrote the final report, published as Volume 20 of the Expedition volumes (1858).

Wilkes hired Louis Agassiz to prepare the scientific volume on the fish collection with artist J.H. Richard. Agassiz wrote a 2,000-page manuscript; however, because of the Civil War, funds were not available for publication of the volume, and it was never published. The collection of invertebrates was the responsibility of J. Couthouy; that of insects was Titian Peale's. Wilkes resented Couthouy, who had been appointed to the expedition at the intervention of President Andrew Jackson. After numerous confrontations on the expedition, Wilkes relieved Couthouy of his responsibilities and ordered him to return to the U.S.; Agate and Dana then continued the collecting of invertebrates. Dana wrote two volumes on the marine fauna (of 1618 pages, 1855) and edited another; A.A. Gould prepared one on Mollusca (1852). The color drawings by Drayton were invaluable in naming and describing the specimens collected.

The expedition's zoological collections were vast. Where possible they were crated and shipped back to the U.S. from ports of opportunity. Poor preparation of specimens for exhibition, rudimentary labeling, careless data handling, recording and bungling delayed publication of results. The artist Peale collected the majority of the birds and mammals that were described by Cassin (1858).

SURVEYS AND CHARTS

Prior to the Wilkes Expedition, the only American charts available to the nation's seafarers were "loose accounts from whalers, who were careless in some instances and forgetful in others" according to Reynolds (1935, 1941). With the invention of the Hadley octant and reliable clocks, seamen could finally make accurate determinations of longitude. Wilkes, who was trained and experienced in the fundamentals of triangulation and hydrographic survey, was chosen to go to Europe and purchase the needed theodolites and mapping equipment for the Navy. His experience in mapping the dangerous Georges Bank area off the coast of New England, gave him the crucial background needed for the U.S. Exploring Expedition. Wilkes was perhaps the logical choice for commander of the expedition, despite his lack of naval seniority.

Mapping by Triangulation

In order to construct bathymetric maps of coastal regions two different types of information are required, accurate positions and depths. In order to map a coastline, a framework of triangles over the region is established as shown in Figures 2 and 3. This network of triangles provides a base for the area being sounded. The construction of the triangulation maps requires that measurements be made along triangle legs by determining distance by sound. The angles of the triangles are determined using horizontal sextant (or theodolite) angles generally taken from a ship, which are connected to coastal or island observations made with carefully leveled theodolites. This gridwork of triangles forms a regular trigonometric network over the whole region being charted.

Crew members used gunfire to determine "baseline" distances between the ships and horizontal angular measurements by sextant to the fixed stations on shore. The process involved firing shots alternately in quick succession from each vessel, "noting the elapsed time between the flash and the report." During the firing, officers measured the angles between the ships and the shore. The operations continued along the coastline and around the island, with the ships changing positions, until the entire circuit was completed and a network of triangles was established. At the same time observers were placed on prominent geographic features on land, where they determined precise geodetic positions of latitude and longitude by astronomical measurements.

After delineating the island and establishing points of known location along its perimeter, the officers deployed survey boats in sounding lines radiating from the shores. Using sextants, two officers in each boat would simultaneously measure the horizontal angles to the known points on land (and anchored ships offshore) and measure water depths with lead weighted lines. The angular measurements produced a left and a right angle

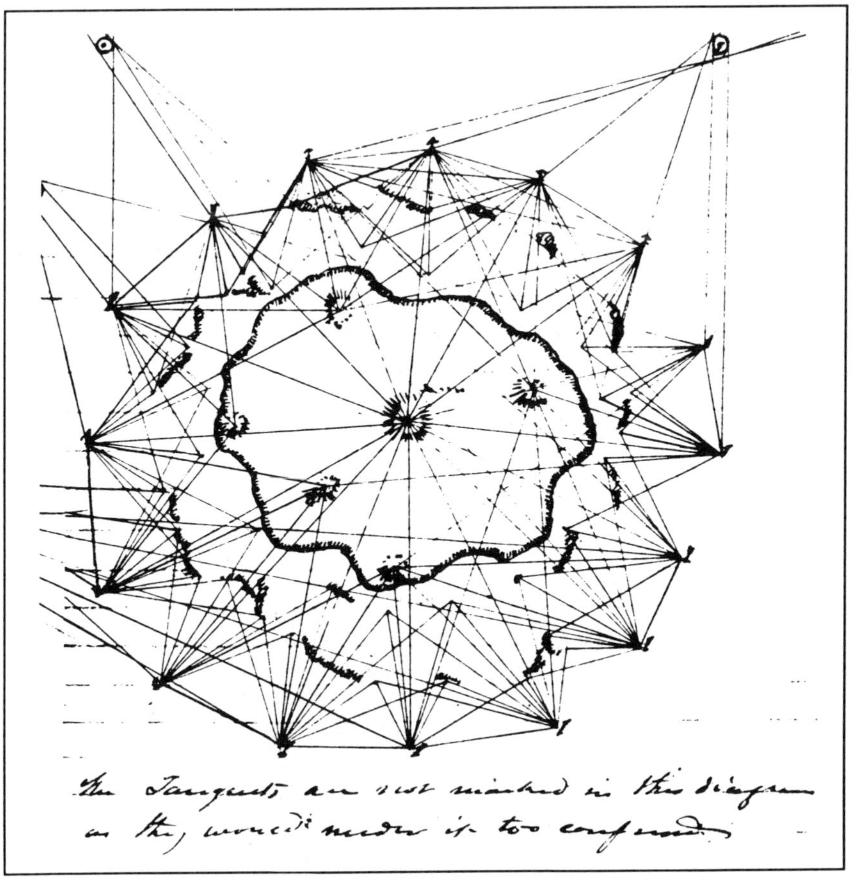

Figure 2. Drawing from Charles Wilkes' notes of the U.S. Exploring Expedition illustrating the network of triangulations utilized to construct maps of Central Pacific islands (Courtesy Manuscript Division, The Library of Congress; reproduced from Viola and Margolis, 1985, p. 164).

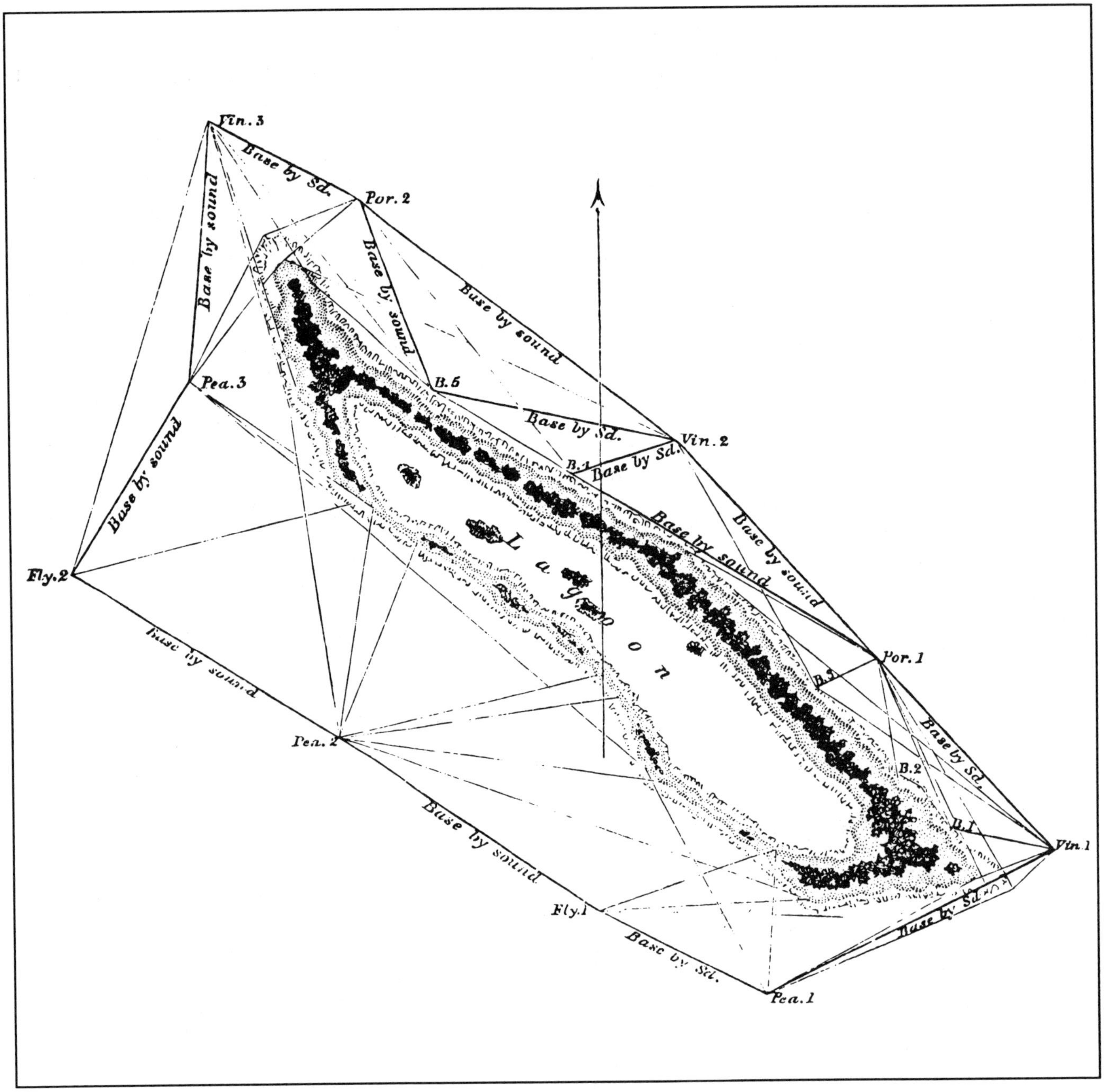

Figure 3. U.S. Exploring Expedition chart of a Pacific atoll, published with the Narrative Atlas volume.

from which the survey boats' position could be accurately located. These location "fixes" or stations were normally taken at convenient time intervals along the sounding lines so as to coincide with lead line measurements of depth, as well as to mark major changes in course.

The survey notebooks of the data were then compiled on the ship by the officers, including Wilkes. A total of 241 separate charts were made. The earliest printed maps appeared in the five-volume Narrative as well as a five-sheet atlas. Wilkes also published a two-volume folio, "Atlas of Charts", in 1850 and 1858. The maps produced by the expedition were numerous and accurate. Single-sheet maps cost thirteen cents, and double-page maps cost twenty-five cents when sold to the public. Their value was immediately recognized and over fifty thousand charts were sold. The Secretary of the Navy was so pleased with the charts of the Pacific Northwest Coast that he reported these alone were worth the full cost of the expedition.

The Charts

The surveys conducted in the Central Pacific were carried out by the three prime sailing vessels. The *Vincennes* surveyed in the Hawaiian, Fijian, Samoan, and Marquesas Islands. The *Porpoise* surveyed in the Fiji, Samoa, Tuamotus, Tahiti, Penrhyn and Flint Islands. The *Peacock* returned and completed work in Samoa and then sailed northward to chart Ellice and Kingsmill islands. Figure 4 is a chart of Charlotte Island (now Abaiang in the Gilbert Islands).

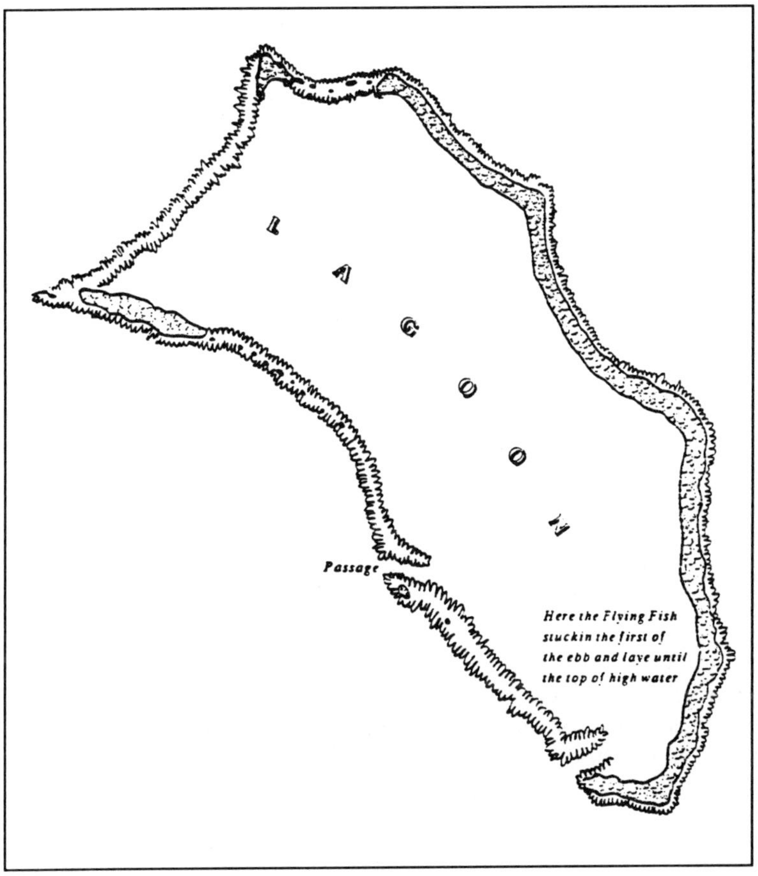

Figure 4. Chart of Charlotte Island, redrawn from Wilkes' Narrative Atlas.

IMPORTANCE OF THE EXPEDITION

The U.S. Exploring Expedition set important precedents for American scientific efforts during the next 150 years. Military and civilian personnel evolved a relationship that allowed them to work together on peaceful scientific pursuits. The federal government's responsibility for charting U.S. and foreign waters for the safety of commercial and naval interests dates from the expedition's work. The U.S. Hydrographic Office was founded in 1871 and the Smithsonian as the National Museum in 1858. The completed surveys established American interests in Antarctica and its interests in pursuing scientific studies there. The charts produced by the expedition led to great time savings in ocean passages and increased safety. The charts also originated the tradition of the U.S. Navy providing accurate maritime information to seafaring nations around the world. The expedition established the American scientists with international reputations in numerous scientific fields. The publication of findings from the expedition established a protocol for publishing cruise results under government auspices as soon as possible after scientific cruises.

A modern-day equivalent of the first U.S. Exploring Expedition is the Ocean Drilling Program, formerly called the Deep Sea Drilling Program, an American-sponsored program supported by the international scientific community. The program supports a modern drill ship with a crew of 25 which circumnavigates the world drilling for scientific objectives. More than 100 volumes of the Initial Reports of the Deep Sea Drilling Program have been published. This successful international scientific program is a model for scientific cooperative studies.

MAPPING IN THE CENTRAL PACIFIC BASIN

In 1859, the New York Daily Tribune published a list of 48 guano bearing islands claimed by American companies and individuals. Strommel (1984) states that the locations of most islands were only poorly known, and that some bore several names. Strommel states the existence of 21 of these islands was considered doubtful at the time of publication of the list.). The exploration and mapping completed by the U.S. Exploring Expedition provided the first accurate map of the Pacific Ocean showing the true positions and locations of the island groups. As a result of this mapping Dana (1849) was able to distinguish the linear trends of islands in the Pacific (Fig. 5) as well as important trends of subsidence of islands.

Scientific exploration of the Central Pacific Basin has been aided by the charts and maps of the U.S. Exploring Expedition for the last 150 years. Thus it seems fitting that this volume on the Geology and Offshore Mineral Resources of the Central Pacific Basin commemorates the 150th anniversary of the sailing expedition.

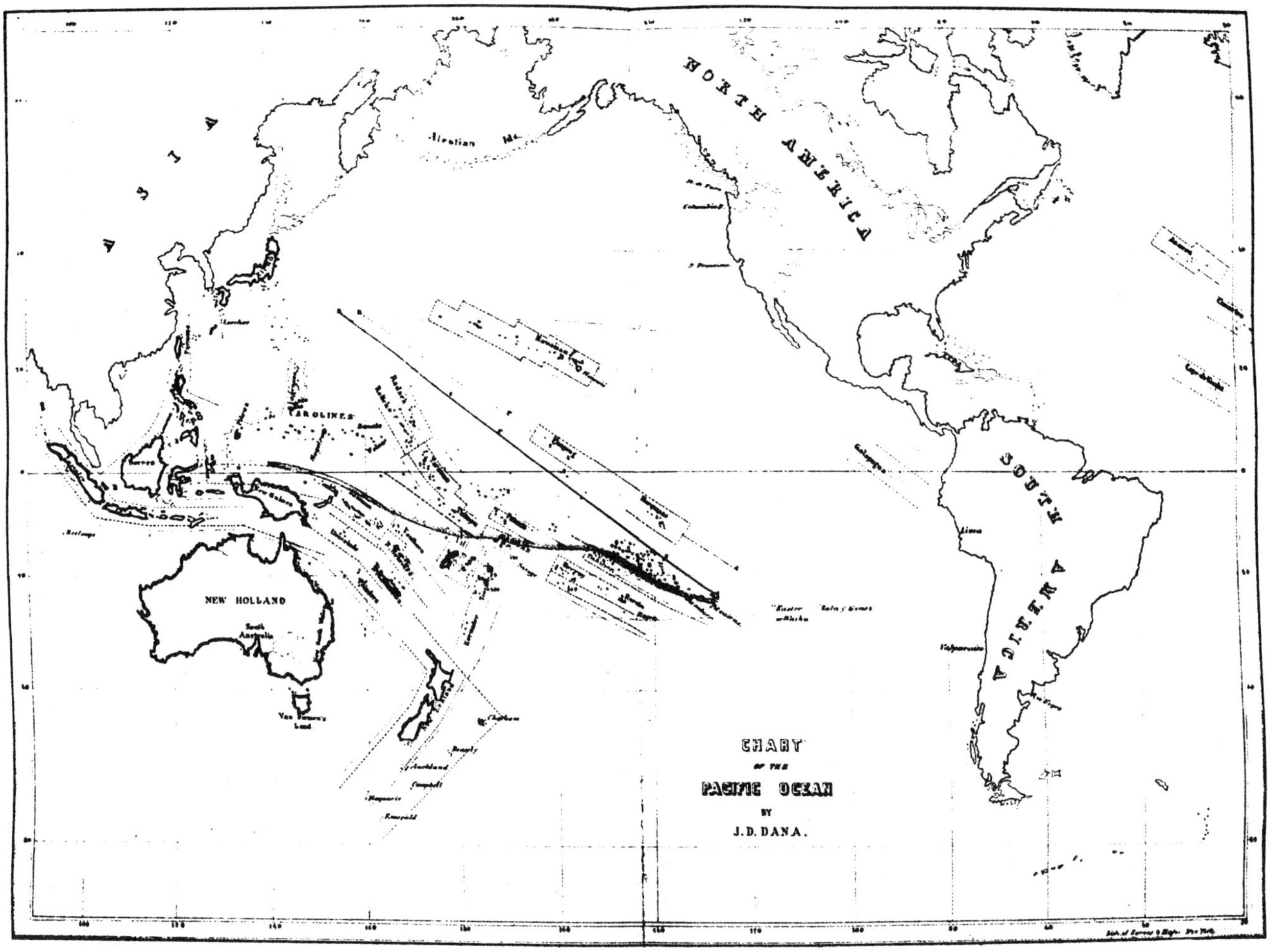

Figure 5. Chart of the Pacific Ocean showing the location of islands and atolls. Dotted line AB is the axis of greatest subsidence from positions of island chains, determined from study of coral islands. Solid lines show limits of coral reef formation. Dashed lines are island chains; dotted lines are trends of island chains. (From Dana, *Geology,* after p. 352; Courtesy Smithsonian Institution Libraries; Viola and Margolis, 1985, p. 94.)

REFERENCES

Brackenridge, W.D., 1856, Botany. Cryotogamia. Filices, including Lycopodiacaea and Hydropterides. Vol. 16, U.S. Exploring Expedition: Philadelphia, C. Sherman and Son, 357 pp.

Cassin, J., 1858, Mammalogy and Ornithology, U.S. Exploring Expedition: Philadelphia, C. Sherman and Son, 466 pp.

Dana, J.D., 1837, A System of Mineralogy: New Haven, Durrie and Peck and Herrick and Noyes, 354 pp.

Dana, J.D., 1849, Geology, Vol. 10, U.S. Exploring Expedition: Philadelphia, C. Sherman and Son, with atlas, 756 pp.

Dana, J.D., 1855, Crustacea, Vols. 13 and 14, U.S. Exploring Expedition: Philadelphia, C. Sherman and Son, 96 pp.

Fremont, J.C., 1845, Report of the exploring expedition to the Rocky Mountains in the year 1842: Washington, Blair and Rives, 588 pp.

Girard, C., 1858, Herpetology. U.S. Exploring Expedition: Philadelphia, C. Sherman and Son, 492 pp.

Gould, A.A., 1852, Mollusca and Shells, Vol. 12, U.S. Exploring Expedition: Philadelphia, C. Sherman and Son, 492 pp.

Jarrard, R.D., and Clague, D.A., 1977, Implications of Pacific island and seamount ages for the origin of volcanic chains: Reviews of Geophysics and Space Physics, v. 15, p. 57-76.

Pickering, C., 1848, The Races of Man and their Geographical Distribution, Vol. 9, U.S. Exploring Expedition: Philadelphia, C. Sherman and Son, 447 pp.

Pickering, C., 1848, The Geographical Distribution and Animals and Plants, Vol. 15, U.S. Exploring Expedition: Boston, Little Brown.

Reynolds, J.N., 1935, Pacific Ocean and South Seas., 23d Cong., 2d sess., House Document 105, Serial 273.

Reynolds, J.N., 1941, Pacific and Indian Oceans: The South Sea Surveying and Exploring Expedition: its inception, progress, and objects: New York, Harper and Brothers, 516 pp.

Strommel, H., 1984, Lost Islands The Story of Islands That Have Vanished from Nautical Charts: University of British Columbia Press, Vancouver, p. 36-50.

Viola, H.J., and Margolis, C., 1985, Magnificent Voyagers: The U.S. Exploring Expedition, 1838-1842: Washington, DC, Smithsonian Institution Press, 303 pp.
Wilkes, C., 1842, Synopsis of the Cruise of the U.S. Exploring Expedition: Washington, DC, Peter Force, 56 pp.
Wilkes, C., 1845, Narrative of the United States Exploring Expedition During the Years 1838, 1839, 1840, 1841, 1842: 5 vols. + folio atlases. Philadelphia, C. Sherman and Son, 5 vols + folio atlases, 372 pp.
Wilkes, C., 1978, Autobiography of Rear Admiral Charles Wilkes, U.S. Navy, 1798-1877, ed. W.J. Morgan et al.: Washington, DC, U.S. Government Printing Office, 944 pp.

Keating, B.H., and Bolton, B.R., editors, 1992, Geology and offshore mineral resources of the central Pacific basin, Circum-Pacific Council for Energy and Mineral Resources Earth Science Series, V. 14. New York, Springer-Verlag.

THE FOUNDATIONS OF ATOLLS: FIRST EXPLORATIONS

D.R. Stoddart
Department of Geography, University of Cambridge, Cambridge, England

ABSTRACT

Apart from a boring, about which little is known, in the Maldive Islands during 1834-1837 by R. Moresby, the two earliest attempts to drill an atoll were made in the Tuamotus in 1840-1841 by E. Belcher and R. E. Johnson. The prevailing belief was that atoll reefs were superficial veneers on the rims of volcanic craters. Specific knowledge of ocean depths surrounding atolls was lacking. The boring technology then available (for drilling water wells) appeared adequate for the anticipated thickness of the reefs, but all the bores made were entirely in sandy material and none reached a reef foundation. After Darwin published his theory of subsidence in 1842, with its deduced consequence of the existence of great thicknesses of shallow-water reef limestones beneath atoll rims, and after deep ocean soundings became available in the tropical Pacific in the 1870s, the perceived technical requirements for a successful boring to reef foundations were enormously increased. The first successful deep boring took place at Funafuti Atoll during 1896-1898, though it failed to reach reef foundations. Darwin's theory was not demonstrated to be correct until more than 110 years after those first tentative attempts by Moresby, Belcher, and Johnson, when basalt was reached beneath shallow-water reef limestones at a depth of 1.2 km at Enewetak Atoll.

INTRODUCTION

Cook and his naturalists were the first European observers to speculate scientifically on the nature and origin of 'lagoon islands' or atolls, ring-shaped coral reefs isolated in the open ocean, during that remarkable decade of Pacific exploration that began in 1769. Cook was aware that 'Philosophers have attempted to account for the formation of Low isles which are in this sea' (Journal, June 1774; Beaglehole, ed., 1961, 438). Forster, on Cook's second voyage, drew a distinction between volcanic and coral islands, but his explanation that corals formed atolls 'by instinct, to shelter themselves the better against the Impetuosity & constancy of the SW winds' (Journal, 15 August 1773; Hoare, ed., 1982, 324; also Forster, 1778, 148-151) seemed inadequate even to himself: 'He whosoever can say something more clever can do it, if he is able to support his opinion by proofs.' Cook's lieutenant, William Anderson, on board *Resolution* during the last voyage, was pessimistic about the possibility of any general explanation for the existence of a mid-ocean reef: 'the reason of its rising in any particular part of the ocean or the method of its formation will probably never be determin'd' (Journal, 17 April 1777: Beaglehole, ed., 1967, 2, 855).

During Kotzebue's voyage to the Marshall Islands, Eschscholtz (in Chamisso, 1821) suggested that the form of atoll reefs 'probably depends on the size of the submarine mountain tops, on which their basis is founded' (Sluiter, 1892, 334). Beechey, who explored extensively in the Tuamotus, generalised the argument: atolls 'have their foundations upon submarine mountains, or upon extinguished volcanoes, which are not more than four or five hundred feet immersed in the ocean; and . . .their shape depends on the figure of the base whence they spring' (Beechy, 1831, 1, 192).

The link between atoll and submarine volcano had first been inferred from the crateriform nature of the enclosed lagoon, and from the association of atolls in linear chains reminiscent of terrestrial mountains (Eschscholtz in Chamisso, 1821). As such it had clear structural and stratigraphic implications. These were given greater point by the discovery, made by the French naturalists

Quoy and Gaimard, during Louis de Freycinet's voyage with the *Uranie* in 1817-1820, that corals were strictly limited in the depths at which they could grow, perhaps as little as 25-30 ft. (Quoy and Gaimard, 1824). Thus when Lyell published his "Principles of Geology" he adopted the volcanic missing crater theory completely (Lyell, 1832, 283-301). If atoll reefs did indeed consist of thin veneers of coral limestone underlain by volcanic rocks, boring through an atoll rim should readily reveal the predicted stratigraphy at depths corresponding to the lower limit of coral growth. Such a test was indeed positively required, if only because of the many obvious problems the volcanic crater theory raised, which included the great size of some atolls, their occasional intricacy of outline, and the constitution of some atoll rims by individual faros (Stoddart, 1976).

It is important to realise that the ocean basins in which atolls were located were technically unfathomable at the time the reef problem was being formulated. Cook made few soundings, except in shoal waters, and at sea he routinely failed to find bottom with 100-250 fathoms (183-457 m) of line. Careful soundings by Beechey at Whitsunday Island (Pinaki Atoll) in the Tuamotus, by Moresby and Powell in the Maldives and Chagos, and by FitzRoy at Cocos-Keeling Atoll, indicated simply that, as Banks had described under alarming circumstances on the Great Barrier Reef in August 1770, a reef is 'a wall of Coral rock rising almost perpendicularly out of the unfathomable ocean' (Beaglehole, ed., 1962, 2, 105). At Cocos-Keeling FitzRoy described a gradual seaward slope to 25 fathoms (46 m), then a steepening to 45°, and finally no bottom with 7200 ft (2195 m) of line at a distance of 6600 ft (2012 m) from the breakers (see Darwin, 1842, 21-23; Jukes, 1863, 71-72).

Sounding with hemp rope was slow, tedious and inaccurate, and it was not until the 1840s that Sir James Clark Ross, using twine rather than rope, began regularly to make soundings of more than 2400 fathoms (4389 m). His methods were used in the tropical Pacific by Sir Edward Belcher in H.M.S. *Samarang* and by Captain Owen Stanley in H.M.S. *Rattlesnake* (Deacon, 1971, 282-283). The first contoured map of the ocean floor was not produced by Maury until 1855, and then for the north Atlantic only, and it was not until the 1870s that deep-sea soundings by H.M.S. *Challenger* and U.S.S. *Tuscarora* began to appear on charts of the tropical Pacific (Dana, 1885, 95).

Those speculating on the foundations of atolls from the time of Cook to that of Darwin thus had no specific idea of the dimensions of the features they were attempting to explain, except that most of them rose in water depths greater than 250 fathoms (460 m), a figure which underestimates the true height of most atolls by approximately an order of magnitude.

Moresby in the Maldives

The first documented attempt to determine the foundations of a coral reef by boring was made by Captain Sir Edward Belcher in the Tuamotu Archipelago in 1840. After mentioning Belcher's attempt, however, Darwin (1842, 73) records that 'On one of the Maldiva atolls, Capt. Moresby bored to a depth of twenty-six feet, when his auger also broke: he has had the kindness to give me the matter brought up; it is perfectly white, and like finely triturated coral-rock'. Captain Robert Moresby, first of the Bombay Marine and then the Indian Navy, surveyed through the Maldives on board *Benares*, supported by Lieutenant T. Powell, also Indian Navy, in *Royal Tiger* during 1834-1837. Unfortunately, while Darwin in his book (1842) makes many references to Moresby's kindness in supplying information about the Maldives and Chagos atolls, his surviving papers in Cambridge University Library contain no material relevant to any boring, and no correspondence between Darwin and Moresby has survived (Burkhardt and Smith, 1983). Moresby (1835, 1844) briefly mentions the superficial stratigraphy of reef islands in the northern Maldives, but only to a depth of 2 m and evidently based on an inspection of wells. Neither his "Nautical Directions" for the Maldives (1840) nor the manuscript on which it is based (in the India Office Library, London) mention the boring, and I have located no other reference to it. It would, however, be worth searching the Bombay Archives, Elphinstone College, Bombay, India, for possible Moresby material. This I have not been able to do.

Belcher at Hao

Hao, 56 km long, is one of the largest Polynesian atolls. It had been discovered by Bougainville in March 1768 (Bougainville, 1772, 206-208). During his first voyage it was seen again by Cook on 6 April 1769, when he named it Bow Island because of its shape (Beaglehole, ed., 1955, 70-71; Beaglehole, ed., 1962, 1, 245-246). Neither Bougainville nor Cook effected a landing. Captain F. W. Beechey, H.M.S. *Blossom*, entered the lagoon in February 1826, spent four days there, and made a running survey which remains the basis of published British Admirably charts (Beechey, 1831, 1, 227-248). His observations at Hao and numerous other Tuamotuan atolls formed the basis for Beechey's extended discussion of the nature and origin of atolls (Beechey, 1831, 1, 261-262).

Belcher, then Lieutenant, played an active part in the *Blossom* expedition, in his post as Assistant Surveyor to Beechey. He had joined the Royal Navy at the age of thirteen in 1812, and was commissioned Lieutenant in 1818. By the time of the Pacific voyage he already had experience in the West Indies, West Africa and Canada. During the *Blossom* voyage he twice almost lost his life at Oeno Atoll when his boat capsized. On the first occasion the gunwale pinned him to the seabed by his neck, and on the second, he was washed into the sea where 'his trowsers . . . (were) entangled in the coral at the bottom of a deep chasm. Fortunately they gave way, and he rose to the surface' and survived.

A second expedition to the eastern Pacific was launched in H.M.S. *Sulphur* in 1835 and Beechey was again appointed to command it. He was forced to retire because of illness, however, and was succeeded by Belcher, by then promoted Captain. Thus it was Belcher who was responsible for the first documented atoll boring. Two things distinguished the new expedition. The first was that, with Belcher in command, the ship was under the control of a ruthlessly ambitious, personally difficult, and professionally highly competent surveyor. 'Perhaps no officer of equal ability has ever succeeded in inspiring so much personal dislike' (Dawson, 1885, 142). Admiral Cockburn complained in 1834 of Belcher's 'violent and overbearing conduct towards his officers' (Friendly, 1977, 320). Captain Bethune observed in 1845: 'How unfortunate it is that such a capital fellow for work should be such a devil incarnate with his officers' (Day, 1967, 65). Life on every ship he commanded was 'a living hell', and he routinely brought his officers back under arrest (Friendly, 1977, 319). But as a hydrographic surveyor he was a perfectionist, and was already responsible for the standard "Treatise on Nautical Surveying" published in the year the *Sulphur* sailed.

The second factor was the appointment of Francis Beaufort, a former commander of H.M.S. *Blossom*, as Hydrographer of the Navy in 1829. Beaufort, best-known for his wind scale, was a nautical surveyor of wide-ranging scientific curiosity. He had not had any personal experience of coral reefs during his active years at sea, though he had made pioneer observations of beach-rock in the eastern Mediterranean (Goudie, 1969), but he was well aware of the issues involved. The "Hydrographic Instructions" he wrote for his surveyors, such as Captain FitzRoy and Captain Wickham in H.M.S. *Beagle* in 1831 and 1837, regularly directed their attention to the problem of the origin of reefs.

Darwin did not announce his theory of the subsidence of reef foundations, the upgrowth of coral formations on them, and the consequent transformation of fringing reefs into barrier reefs and then into atolls, until 31 May 1837, in a lecture to the Geological Society of London (Darwin, 1838). Hence Beaufort's theoretical point of departure in his "Instructions" remained that argued by Lyell in his "Principles" in 1832, that reefs were thin veneers on volcanic craters slightly submerged below the surface of the sea (Lyell, 1832, 290).

Beaufort issued his "Instructions" for the voyage of H.M.S. *Sulphur* on 19 December 1835. Under the heading of 'Coral Islands' he wrote as follows:

'It has been suggested by some geologists that the coral insect, instead of raising its super-structure directly from the bottom of the sea, works only on the summits of submarine mountains, which have been projected upwards by volcanic action. They account, therefore, for the basin-like form so generally observed in coral islands, by supposing that they insist on the circular lip of extinct volcanic craters.'

'In order, by a satisfactory experiment, to bring this question to a direct issue, their Lordships have ordered you to be supplied with a complete set of the boring apparatus used by miners; leaving it to your own judgment to select any coral island which may be well adapted to the purpose, and which will lead you as little as possible from the line of your survey. They wish you to fix upon a convenient spot of the island where the operations cannot be disturbed by the surf, and there to bore perpendicularly, so as to perforate the whole thickness of the coral, and to enter the tool sufficiently deep in the rock on which it is based to furnish specimens for future analysis. You will of course keep a register of the contents of the auger every time it is withdrawn, and if the structure or density of the coral appear to change, it will be desirable to have a series of such specimens also preserved, and tallied with their corresponding depths.

'Immediately that the bore hole arrives at its greatest depths, provided no water has been allowed to enter, it will be well to contrive some method of sending down a registering thermometer, so as to ascertain the temperature of the bottom of the hole' (Belcher, 1843, xxvii-xxviii; original manuscript in Hydrographic Department archives, Taunton).

After his earlier experience with Beechey, Belcher decided to drill at Hao. The site chosen was 12 km from the northern entrance to the lagoon, at the village of Otepe, 20 m inland from the lagoon shore, at a point where the lagoonal fringing reef is 140 m wide. The equipment used was 'similar to that used for boring for water in England, ... comprising augers of various sizes; iron tubes of three inches internal bore; and twenty-feet bars connected by strong joints with male and female screws'. Belcher described the operation: 'To work these augers it was necessary to raise a strong scaffold more than twenty feet above our working level; and during the interval employed upon this duty, a party sank a well of six feet square and six feet in depth, which was secured from tumbling in by strong piles, plank, and wattling of the cocoa-nut and pandanus leaves. In the centre of this area a forty-gallon cask was sunk, forming a well, into which one of the lift pumps was introduced. By this arrangement we gained a depth of seven feet ... before entering the tools' (Belcher, 1843, 1, 366-367).

Forty five feet (13.7 m) of pipe were entered, until on the 35th day of the operation the auger jammed and could not be moved. With fifty men, exerting a force estimated at 20 tons, the one-inch iron bar sheared and the first boring had to be abandoned. A second attempt near the seaward shore ended when the tool jammed at 9 feet (2.7 m).

Not surprisingly Belcher found the whole exercise 'very tedious' (letter, 3 April 1840, to Beaufort: Hydrographic Department archives, Taunton) and 'extremely harassing' even to those of 'sanguine temperament' (Belcher, 1843, 1 368-369), a class which undoubtedly did not include himself. The lagoon-shore boring went through water-logged fine sand, with almost a creamy consistency, exerting lateral pressure 'so strong as to force up the fine sandy mixture faster than we could get the tube down' (letter, 12 April 1840, to Beaufort: Hydrographic Department archives, Taunton). Belcher considered that similar materials extended deeper, at least to the maximum depth of 100 feet (30 m) he could have reached with the equipment supplied (letter, 12 April 1840, to Beaufort: Hydrographic Department archives, Taunton), but the results achieved clearly said nothing on the origin of atolls.

'With naval men', Belcher commented, 'the word "impossible" only turns up when that expression can be clearly demonstrated': success would require 'an express borer by profession, with his smithy force', and the proper tools (Belcher, 1843, 1, 368). He anticipated the criticism that 'those unaccustomed to labour themselves will probably think that we ought to have bored better & deeper' (letter, 12 April 1840, to Beaufort: Hydrographic Department archives, Taunton).

But if the boring failed, Belcher produced the first bathymetric maps of the entrance channel and of the north-eastern part of the lagoon (original manuscripts HO L3211, L3143 and L3636: Hydrographic Department archives, Taunton), and published charts, based on these surveys, still refer to the area as "Boring Bay". He also carried out magnetic observations round the atoll rim, and made the world's first sub-surface temperature measurements within an atoll reef, at 6 and 23 feet (1.8 and 7 m) depth in the lagoon borehole, on 20 and 21 March 1840 (measurements in Hydrographic Department archives, Taunton). When he sailed from Hao he recorded the attempt on a copper plate fixed to a coconut tree, but of this there is no subsequent record; nor have I been able to trace the borehole log, last recorded in the Hydrographic Department archives fifty years ago and not seen since, nor the sediment samples he collected from the auger.

The Wilkes Expedition

Meanwhile the United States Exploring Expedition began its four-year circumnavigation in August 1838 under the command of Lieutenant Charles Wilkes (Viola and Margolis, 1985). A year later the expedition reached the Tuamotus, worked westwards through Tahiti and Samoa, charting atolls, and then on to Sydney, New South Wales. On board was the geologist and zoologist James Dwight Dana, subsequently to become the leading American authority on coral reefs and their development. Dana later recalled that during the visit to Sydney, in December 1839, he read a newspaper account of Darwin's new theory of subsidence, which 'threw a flood of light over the subject' (Dana, 1875, x). I have failed to find any such reference in Australian newspapers published during 1839 and 1840: the only reference I have found to Darwin is a note on petrified trees in Chile in the "Australasian Chronicle", 24 December 1839 (Australian newspaper collection, Mitchell Library, Sydney). It is, of course, possible that Dana learned of Darwin's views from an overseas rather than Australian newspaper. Six months later Dana used the new theory to interpret the coral reefs of Fiji.

It was in Fiji that Wilkes and the United States Exploring Expedition fell in with Belcher, who had abandoned his own boring attempt at Hao four months previously. Wilkes, whose own character was scarcely beyond reproach, emphasized 'the madcap tyrrany' that Belcher practised on board his ship (Wilkes, 1978, 463; Wilkes here wrongly identifies Belcher's ship as H.M.S. *Samarang*, it was still *Sulphur*, since Belcher did not translate to *Samarang* until 1843). Wilkes met Belcher again at the Cape of Good Hope: 'He was indeed what might be termed a brute. ... One and all of his officers detested him and his crew deserted whenever they had an opportunity to effect their escape' (Wilkes, 1978, 516). Wilkes himself, it may be recalled, was in no position to cast such aspersions. When in Hawaii many of his crew decided to leave the expedition on expiration of contract, Wilkes chose three, put them in irons and had them slowly flogged, personally enquiring between every stroke whether they would now do their duty. Eight strokes of the cat was sufficient to make them sign on once more, and Wilkes was subsequently court-martialled for it (Stanton, 1975, 220). Wilkes himself was generally known as violent, overbearing, insulting and offensive (Stanton, 1975, 284). Belcher's own account of his meeting with Wilkes in Fiji is much more generous (Belcher, 1843, 2, 46-48), but whether the two men discussed the practicalities of reef boring is unrecorded.

After Fiji the ships of the American squadron separated: between November 1840 and March 1841 the 27 m brig *Porpoise*, under the command of Lieutenant Cadwalader Ringgold, worked in the Tuamotu Archipelago. It

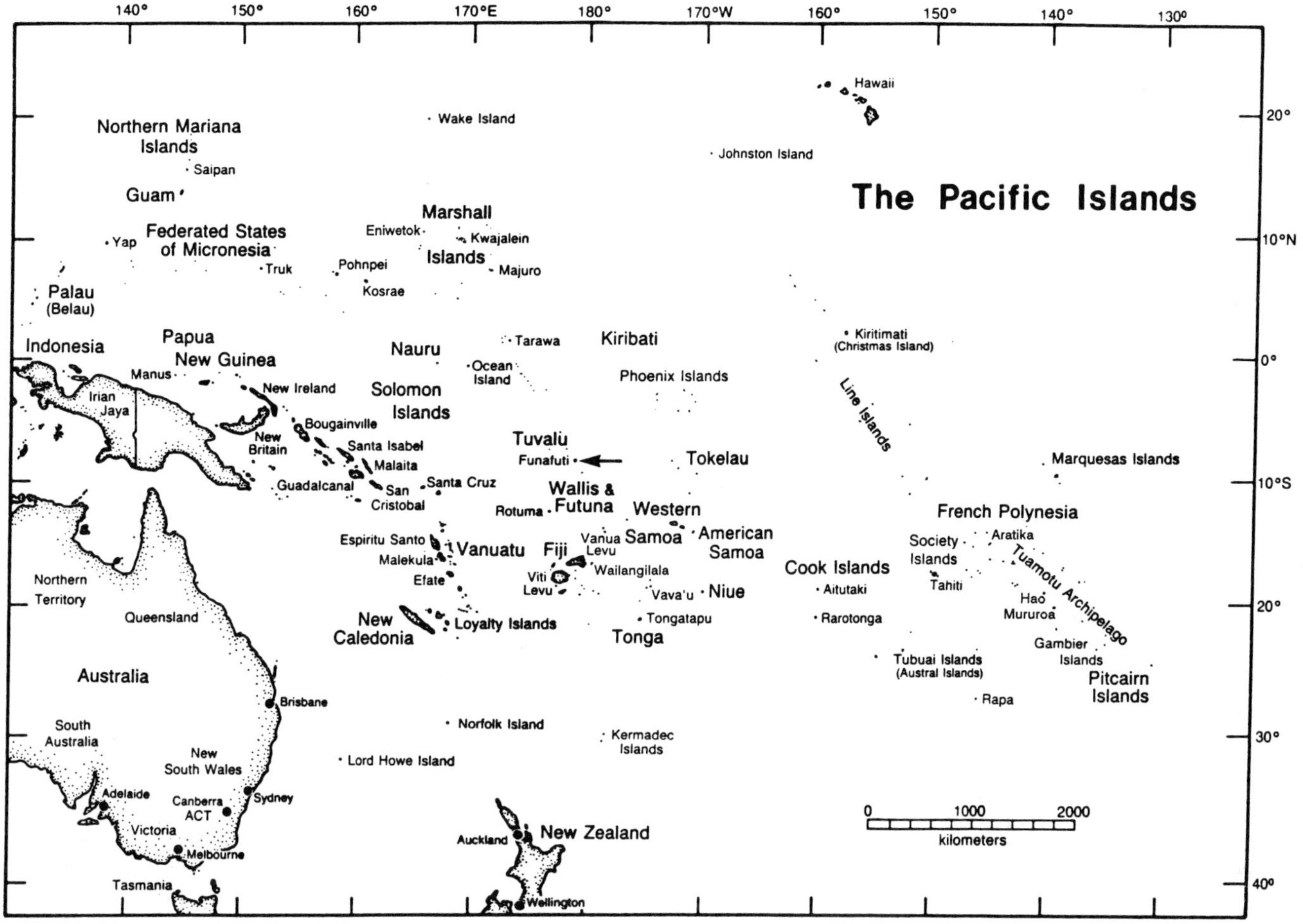

Figure 1. Map of the southwestern Pacific showing locations of major island groups. Arrow points to Funafuti.

was here, 550 km northwest of Hao, that the second documented boring into a coral reef was attempted, at Aratika (or Carlshoff) Atoll, which the expedition had briefly visited the previous year.

When this boring was planned, or on whose initiative, I have been unable to determine. The Expedition's "Instructions" from the Secretary of the Navy, dated 11 August 1838, simply stated that 'No special directions are thought necessary, as to the mode of conducting the scientific researches and experiments which you are enjoined to prosecute' (Wilkes, 1856, 1, 365). Dana's remit at the outset was mineralogical. Joseph Couthouy, also interested in reef development, had left the expedition long before *Porpoise* arrived at Aratika. Presumably the equipment was shipped either in the United States or in Sydney, but if the latter it is strange that Dana played no part in the exercise.

Aratika, south-southwest of Takapoto, had been discovered in 1824 by Otto von Kotzebue during his second voyage (Sharp, 1960, 206-207). Lieutenant Ringgold landed a party of 13 to effect a boring on 15 December 1840; they were ashore until 19 January 1841. The operation was under the command of Lieutenant Robert E. Johnson, with orders to bore day and night for 30-40 days (Tyler, 1968, 223). Lieutenant Johnson's log is extant in the U.S. National Archives, Washington, D.C. (National Archives microfilm M-75, roll 15). Unfortunately under the date of 16 December 1840 he writes: 'Of my stay and proceedings at Carlshoff (Aratika) I shall have here nothing to say, inasmuch as every circumstance connected therewith was duly recorded in a diary which I have delivered by order to Capt. Ringgold'. Johnson's log resume after his departure from the atoll.

Lieutenant Johnson's diary of the boring has unfortunately not been located, in spite of extensive searches in the National Archives, Washington, D.C., and also in the Exploring Expedition papers at Yale University, by Mr.

Myron M. Weinstein. The shore party also included, however, Acting Master George T. Sinclair, and his journal is preserved in the National Archives (microfilm M-75, roll 21). Under the date of 18 January 1840 he gives this summary account of the boring:

'The experiments with the drill had not been attended with much success only having reached the depth (sic) of 22 feet. The failure is to be attributed to the nature of the soil which being sand caved in as they bored and instead of making a small hole, they had formed a pit 8 feet in diameter. In sinking the pipes large blocks of coral would change their direction from the perpendicular and the heavy drill falling and fetching up on the side of the pipes broke a number of them and it required considerable labour to replace them. They had advanced further in the last few days than they had done in several weeks before. Samples of the nature of the sand at different depths (sic) were preserved as also a number of shells &c. It is all of the same coral formation. The water which came from the pit was fresh and good. It is a riddle of me where it comes from for the Island is only ten feet above the sea and the pit was only a few hundred feet from the margin of the ocean. The bottom of the pit was therefore twelve feet below the surface of the ocean and yet the water obtained was fresh. I cannot imagine how this is unless the water in passing through the sand is freed of the particles of salt'.

The shore party was no more enamoured of its task than was Belcher's, more particularly when they discovered that of the few 'tolerably pretty' girls in the local population, the one who was unattached was as impenetrable as the reef (Stanton, 1975, 227).

The only formal account of the Aratika boring is that provided by Wilkes himself (Wilkes, 1856, 1, 267-268). 'I was aware of the difficulty of the task that would be imposed upon the officer who directed the operation', he writes. 'The undertaking proved fully as laborious as I had anticipated, and Lieutenant Johnson's exertions were worthy of better success'. Wilkes then repeats the observations given in Sinclair's log, but adds: 'The greatest depth to which he succeeded in reaching was twenty-one feet: ten to eleven feet were generally accomplished without much difficulty; but after that depth was arrived at, they frequently did not succeed in getting down beyond one foot per day. The coral shelf, composed of conglomerates and compact coral rock, seems to have afforded an impediment to further progress'. In spite of the 'Breaking of pipes and augers, and the occurrence of various other accidents', Wilkes professed himself 'well satisfied that there is no insuperable difficulty in boring into coral islands'.

No detailed account of the Aratika boring was published, and in the absence of Johnson's boring log they cannot be reconstructed. Wilkes left the details to the Expedition's Geological Report, to be written by Dana. Dana, however, was not present during the Aratika boring, and makes no mention of it in his report (Dana, 1849). Wilkes also notes that 'agreeably to my instructions, a specimen of each foot reached was preserved': I have seen no subsequent reference to this material, the location of which is unknown. The only contemporary commentary on the significance of the bore was that by Wilkes himself, who was 'unable to believe that these great formations (atolls) are or can possibly be the work of zoophytes', a proposition he found 'almost absurd' and 'certainly inconsistent with the fact' (Wilkes, 1856, 1, 268, 270).

Dana referred to the Aratika boring only in a footnote to his final analysis of Darwin's theory (Dana, 1885, 104), and it is clear from this that his information is derived only from Wilkes's narrative. He recommends that deep borings on atolls would provide 'a sure method of obtaining a final decision of the coral island question and should be tried'.

The End of the Century

Darwin's book on "The Structure and Distribution of Coral Reefs" with its theory of the upward growth of coral reefs during subsidence, which was published in 1842, changed in terms of the debate. No longer was the issue simply to demonstrate the existence of a volcanic or other foundation at or close to the limiting depth of coral growth; it was to determine how deeply shallow-water reef facies extended beneath the surface, and thus to calibrate the subsidence which had occurred. Whereas the technology of shallow drilling for water had seemed adequate to test the accepted theoretical conjectures of earlier times, vastly improved methods were needed to test the new ones. No further attempts at boring were made for sixty years.

Alexander Agassiz began his studies of reefs in the East Pacific and southwestern United States in the 1800s. He was not disposed to accept that subsidence was a necessary condition of reef formation. He sent his results to Darwin, who wrote to him on 5 May 1881: 'I wish that some doubly rich millionaire would take it into his head to have borings made in some of the Pacific and Indian atolls, and bring home cores for slicing from a depth of 500 or 600 feet' (F. Darwin, ed., 1888, 3, 184; F. Darwin and Seward, eds., 1903, 2, 198).

Agassiz himself, with his vast copper mining interests, met Darwin's criterion, but he failed adequately to carry out the test. In November 1987 he sent a drilling team to the island of Wailangilala, northeast Fiji, where boring operations occupied three months. The upper 50 ft (15 m) of the record was in coral or coralline sand; thence to 85 ft (26 m) in limestone equivalent to the widespread elevated Tertiary limestones of the Fiji Islands (Agassiz, 1899, 47). Agassiz discontinued the operation as 'foolish' on the grounds that

Figure 2. Atoll drilling in Funafuti lagoon, from the *Porpoise*, (from *The Atoll of Funafuti*, Royal Society of London, 1904)

such limestones could readily be studied at outcrop (G.R. Agassiz, 1913, 331).

By then, of course, his efforts, at least in a theoretical sense, had largely been made redundant through the work of a committee established by the British Association for the Advancement of Science in 1891, and subsequently developed by the Royal Society of London, to undertake a definitive boring on an open-ocean atoll.

Funafuti in the Ellice Islands (Tuvalu) was selected, and three successive expeditions were mounted. The first was led by W. J. Sollas, with J.S. Gardiner and C. Hedley, 21 May-30 June 1896. It succeeded in putting down holes to 22 m on the seaward side of an island and to 32 m on the lagoon side. The following year a party under T.W. Edgeworth David spent 19 July-30 November on the atoll, and reached 213 m depth, through sand and coral, on the seaward side. Finally, in June 1898, during the third expedition led by A.E. Finckh, a boring reached 340 m, entirely in calcareous material, and borings of 34 and 44 m were also made in the lagoon floor in water depths of about 30 m. The Royal Society's report on the drillings (1904) remained a standard work on atoll structure, as well as establishing new standards in carbonate petrography and the study of reef morphology, until U.S. deep drillings at Enewetak in the Marshall Islands in 1947 (Ladd et al., 1948) and 1951–1952 (Ladd et al., 1953; also MacLeod, 1985).

CONCLUSION

It is clear that the first attempts to determine the nature of the foundations of atolls had no hope of success. There were two reasons for this. One was simply inadequate technology. Both Belcher and Johnson had equipment capable of drilling only some 30 m, and neither in fact achieved more than 15 m. Although there have been no subsequent investigations of the depth of reef basement at either Hao or Aratika, the deep drillings on the comparable atoll of Mururoa, approximately 420 km south of Hao, in 1964-1965 revealed a thickness of 415-438 m of limestone overlying basalt (Deneufbourg, 1969). The second reason was conceptual. Under Lyell's influence, reinforced by Beechey's own conclusions from his own investigations in the Pacific, atolls were seen as thin veneers of coral formed on volcanic craters; the true dimension of the difficulty of boring to reef foundations could not be apprehended before the publication of Darwin's theory. Even then, the technical problems remained formidable, as the repeated expeditions to Funafuti demonstrated. Although drilling on that atoll reached 340 m, later geophysical studies have indicated that 550-760 m of reef limestone overlies volcanic rock (Gaskell and Swallow, 1953), suggesting a shortfall of at least 40% in the length of the drill. It unfortunately required the military necessity of nuclear weapons testing, and the national sense of purpose and commitment of resources this entailed, at both Enewetak and Mururoa, before the nature of the foundations of atolls could at last be directly confirmed.

ACKNOWLEDGMENTS

I am extremely grateful to Mr. Myron M. Weinstein, formerly of the Library of Congress, Washington, D.C., for pursuing the Wilkes and other U.S. Exploring Expedition papers through the National Archives and Yale University for me, and for correcting my decipherment of the Sinclair log. I thank also the Hydrographer of the Navy for access to Sir Edward Belcher's papers housed in the Ministry of Defence (Navy), Taunton, England, and the Superintendent of Sailing Directions, Captain R.J. Campbell, R.N., for his assistance. This paper was written at Churchill College, Cambridge, where H.W. Menard was an Overseas Fellow.

REFERENCES

Agassiz, A., The islands and coral reefs of Fiji, Bull. Mus. Comp. Zool. Harvard Univ., 33, 1-167, 1899.

Agassiz, G. R., ed., Letters and recollections of Alexander Agassiz with a sketch of his life and work, Houghton Mifflin Company, Boston, 1913.

Beaglehole, J., ed., The Journals of Captain James Cook on his voyages of discovery: the voyage of the *Endeavour* 1768-1771, Cambridge University Press, Cambridge, 1955.

Beaglehole, J., ed., The Journals of Captain James Cook on his voyages of discovery: the voyage of the *Resolution* and *Adventure* 1772-1775, Cambridge University Press, Cambridge, 1961.

Beaglehole, J., ed., The *Endeavour* Journal of Joseph Banks 1768-1771, 2 vols., Public Library of New South Wales, Sydney, 1962.

Beaglehole, J., ed., The Journals of Captain James Cook on his voyages of discovery: the voyage of the *Resolution* and *Discovery* 1776-1780, 2 vols., Cambridge University Press, Cambridge, 1967.

Beechey, F. W., Narrative of a voyage to the Pacific and Beering's Strait, to co-operate with the polar expeditions, 2 vols., Colburn and Bentley, London, 1831.

Belcher, E., Narrative of a voyage round the world performed in Her Majesty's Ship *Sulphur*, during the years 1836-1842, 2 vols., Colburn, London, 1843.

Bougainville, L. A. de, A voyage round the world, Nourse, London, 1772.

Burkhardt, F. and Smith, S., eds., A calendar of the correspondence of Charles Darwin, 1821-1882, Garland Publishers, New York, 1983.

Chamisso, A. von, Remarks and opinions of the naturalist of the expedition, *in* O. von Kotzebue, A voyage of discovery into the South Sea and Beering's Straits for the purpose of exploring a north-east passage, undertaken in the years 1815-1818, 2, 349-433, 3, 1-318, Longman, London, 1821.

Dana, J. D., United States Exploring Expedition. During the years 1838, 1839, 1840, 1841, 1842. X. Geology. Sherman, Philadelphia, 1849.

Dana, J. D., Corals and coral islands, Sampson Low, London, 1875.

Dana, J. D., Origin of coral reefs and islands, Amer. J. Sci., (3) 30, 89-105, 169-191, 1885.

Darwin, C. R., On certain areas of elevation and subsidence in the Pacific and Indian Oceans, as deduced from the study of coral formations, Proc. Geol. Soc. Lond., 2, 552-554, 1838.

Darwin, C. R., The structure and distribution of coral reefs, Smith, Elder, London, 1842.

Darwin, F., ed., The life and letters of Charles Darwin, including an autobiographical chapter, Murray, London, 3 vols., 1887.

Darwin, F., and Seward, A. C., eds., More letters of Charles Darwin: a record of his work in a series of hitherto unpublished letters, Murray, London, 2 vols, 1903.

Dawson, L. S., Memoirs of hydrography including brief biographies of the principal officers who have served in H.M. Naval Surveying Service between the years 1750 and 1885, Henry Keay, Eastbourne, 2 vols, 1885.

Day, A., The Admiralty Hydrographic Service 1795-1919, Her Majesty's Stationery Office, London, 1967.

Deacon, M., Scientists and the sea 1650-1900: a study of marine science, Academic Press, New York, 1971.

Deneufbourg, G., Les forages de Mururoa, Cahiers du Pacifique, 13, 47-58, 1969.

Forster, J. R., Observations made during a voyage round the world, on physical geography, natural history, and ethic philosophy, Robinson, London, 1778.

Friendly, A., Beaufort of the Admiralty: the life of Sir Francis Beaufort 1774-1857, Hutchinson, London, 1977.

Gaskell, T. F., and Swallow, J. C., Seismic experiments on two Pacific atolls, Occas. Papers Challenger Soc., 3, 1-8, 1953.

Goudie, A., A note on Mediterranean beachrock: its history, Atoll Res. Bull., 126, 11-14, 1969.

Hoare, M. E., ed., The *Resolution* Journal of Johann Reinhold Forster 1772-1775, Cambridge University Press for the Hakluyt Society, Cambridge, 4 vols., 1982.

Jukes, J. B., The school manual of geology, Black, Edinburgh, 1863.

Ladd, H. S., Tracey, J. I., Jr., and Lill, G. G., Drilling on Bikini Atoll, Marshall Islands, Science, 107, 51-55, 1948.

Ladd, H. S., Ingerson, E., Townsend, R. C., Russell, M., and Stephenson, H. K., Drilling on Enewetok Atoll, Marshall Islands, Bull. Amer. Ass. Petrol. Geol., 37, 2257-2280, 1953.

Lyell, C., Corals and coral reefs, Principles of Geology, 2, 283-301, Murray, London, 1832.

MacLeod, R., Imperial reflections in the southern seas: the Funafuti expeditions, 1896-1904, Abstr. XVII Int. Congr. Hist. Sci. (Berkeley, 1985), 8.2, 1985.

Moresby, R., Extracts from Commander Moresby's report on the northern atolls of the Maldivas, J. Roy. Geogr. Soc., 5, 398-404, 1835.

Moresby, R., Nautical directions for the Maldive Islands and Chagos Archipelago, East India Company, London, 1840.

Moresby, R., Reports on the Maldivas, Trans. Bombay Geogr. Soc., 1, 102-108, 1844.

Quoy, J. R., and Gaimard, J. P., Memoire sur l'accroissement des polypes lithophytes considere geologiquement, Voyage autour du monde entrepris par ordre du Roi par M. Louis de Freycinet: Zoologie, 658-671, Pillet Aine, Paris, 1824.

Royal Society of London, 1904, The atoll of Funafuti; borings into a coral reef and the results, being the report of the Coral Reef Committee of the Royal Society: London, The Royal Society of London, 420 pp..

Sharp, A., The discovery of the Pacific islands, Clarendon Press, Oxford, 1960.

Sluiter, C. P., Die Korallentheorie von Eschscholtz, Zool. Anz., 400, 326-327, 1892.

Stanton, W., The great United States Exploring Expedition of 1838-1842, University of California Press, Berkeley, 1975.

Stoddart, D. R., Darwin, Lyell, and the geological significance of coral reefs, Br. J. Hist. Sci., 9, 199-218, 1976.

Tyler, D. B., The Wilkes Expedition: the first United States Exploring Expedition (1838-1842), Mem. Amer. Philos. Soc. Philadephia, 73, 1-435, 1968.

Viola, H. J., and Margolis, C., eds., Magnificent voyagers: the U.S. Exploring Expedition, 1838-1842, Smithsonian Institution Press, Washington, D.C., 1985.
Wilkes, C., Narrative of the United States Exploring Expedition during the years 1838, 1839, 1840, 1841, 1842, G. P. Putnam, New York, 5 vols., 1856 (originally published by Lea and Blanchard, Philadelphia, 1845).
Wilkes, C., Autobiography of Rear Admiral Charles Wilkes, U.S. Navy, 1798-1897, Department of the Navy, Washington, D.C., 1978.

Keating, B.H., and Bolton, B.R., editors, 1992, Geology and offshore mineral resources of the central Pacific basin, Circum-Pacific Council for Energy and Mineral Resources Earth Science Series, V. 14. New York, Springer-Verlag.

SEAMOUNT AGE ESTIMATES FROM PALEOMAGNETISM AND THEIR IMPLICATIONS FOR THE HISTORY OF VOLCANISM ON THE PACIFIC PLATE

William W. Sager
Departments of Oceanography, Geophysics, and Geodynamics Research Institute
Texas A&M University, College Station, TX 77843 USA

ABSTRACT

Of 100 reliable Pacific seamount paleomagnetic poles, only 36 have ages that also seem to be reliably determined. Nevertheless, it is possible to make age estimates for most of the undated seamounts by comparing their paleomagnetic poles to the Pacific apparent polar wander path. According to their locations along the polar path, seamount paleopoles were divided into four groups: (1) early Tertiary, about 36–45 Ma; (2) Late Cretaceous, 64–79 Ma; (3) mid-Cretaceous, 80 Ma to about 120 Ma, and (4) Early Cretaceous-Jurassic, older than about 120 Ma. The greatest number of paleopoles lie along the mid- to Late Cretaceous portion of the polar wander path, implying a heavy concentration of volcanism of this age in the central and western Pacific. Though there appears to be a significant fraction of seamounts with ages within about 20 Ma of the underlying seafloor throughout this area, many of the seamounts between the western Pacific trenches and the Line Islands in the central Pacific are intraplate volcanoes, formed long after the underlying lithosphere. Moreover, these off-ridge seamounts generally show a trend of decreasing paleomagnetic age toward the southeast, consistent with hypotheses that explain the intraplate volcanism by the northwestward drift of the Pacific plate over mantle plumes presently located in the southeast Pacific.

INTRODUCTION

One of the most notable features of the Pacific plate is the abundance of seamounts and guyots, particularly in the central and western Pacific. The origins of most Pacific seamounts are unclear. It appears that there are at least two basic types: those formed near a spreading ridge and those formed in the interior of a plate. The source of the former is evidently magma from the ridge-crest magma chamber (Batiza, 1977). The latter, intraplate volcanoes, are some of the best and least understood. Perhaps the best understood are some linear chains, such as the Hawaiian-Emperor, Louisville, and several Gulf of Alaska chains (e.g., Clague and Dalrymple, 1987; Lonsdale, 1988; Dalrymple et al., 1987). These display convincing age progressions implying that each was formed by the motion of the Pacific plate over a persistent source of magma, often called a "hotspot", more-or-less fixed relative to the mantle. Indeed, some authors believe that most intraplate seamounts form by this mechanism (e.g., Epp, 1984).

Other seamount groups have had complex tectonic histories that defy such simple explanations. Examples are the Line Islands (Schlanger et al., 1984) and Austral-Cook (Jarrard and Clague, 1977) chains, both of which contain volcanoes that fit a simple linear age progression as well as those that do not. Most western Pacific seamounts, however, are enigmas simply because of the scarcity of geologic data.

Perhaps the most important datum for any seamount is its age. With such information, it is possible to discover progressions, overlaps, and areal extents of volcanic episodes – all important observations in distinguishing among various hypotheses of seamount formation. Additionally, age data are critical in assessing the occurrence and abundance of mineral resources such as ferromanganese deposits. Excepting the well-known hotspot chains, however, only a few dozen Pacific seamounts have been reliably dated.

In this study, paleomagnetic data were used to make age estimates for a large number of Pacific seamounts. The procedure is an extension of a method used by many geoscientists to assess the age of a rock unit whose age is unknown but for which there exists a reliable paleomagnetic pole. The pole is compared to the paleomagnetic apparent polar wander path (APWP) for the plate upon which the site is located. If the pole falls within the error limits of the APWP or its error ellipse overlaps the APWP, then the age of that segment of APWP is used as an estimate of the pole's age. Making use of this technique, ages were estimated for 64 central and western Pacific seamounts that had not been reliably dated previously.

Seamount Paleomagnetism

Seamount paleomagnetism is a method of obtaining paleomagnetic data by remote sensing. It has been applied extensively to Pacific seamounts because of the importance of the Pacific plate's tectonic history and because this plate is almost entirely submerged, making it difficult to obtain paleomagnetic data by traditional means.

Most seamounts are made up mainly of basalt, a rock which typically retains a strong remanent magnetization, usually acquired during cooling. As a result, the seamount often has a significant magnetic field that perturbs the local geomagnetic field, causing a magnetic anomaly. The shape of the anomaly depends on the shape of the seamount and the direction of magnetization (Fig. 1). It is positive where the seamount field has a component parallel to the geomagnetic field and negative where it has a component in the opposite direction. As the geomagnetic field direction, seamount shape, and magnetic anomaly are easily measured, the magnetization direction can be derived using a three-dimensional magnetic model to reproduce the anomaly.

To find the magnetization precisely, the problem is usually treated as an inversion. Two main types of inversion have been applied in seamount anomaly studies: (1) least-squares and (2) seminorm minimization. The major difference between the two methods is that the former assumes that the magnetization of the seamount is homogeneous but the latter does not. In the least-squares method, the constant magnetization assumption is made to allow the total field anomaly to be written as a linear combination of the unknown magnetization vector components and volume integrals describing the seamount's shape. The least-squares solution is that which minimizes the sum of the squared residuals (observed minus modeled anomaly). Three similar least-squares methods have been widely used for seamount paleomagnetic inversion (Vacquier, 1962; Talwani, 1965; Plouff, 1976), differing primarily in the approximations of the volume integrals.

Noting that homogeneously magnetized seamounts are probably rare, Parker et al. (1987) developed an inversion method that specifically addresses inhomogeneities. Their method divides the magnetization into uniform and nonuniform parts. Using a Hilbert space of magnetization functions, it minimizes the squared norm (seminorm) of the nonuniform part of the magnetization model, producing a model that has the maximum uniform part consistent with the observed magnetic anomaly (Parker et al., 1987; Hildebrand and Parker, 1987). Subsequently, Parker (1988) formulated a statistical model that propagates random fluctuations of the magnetic anomaly into an estimate of confidence limits for paleomagnetic poles calculated by the

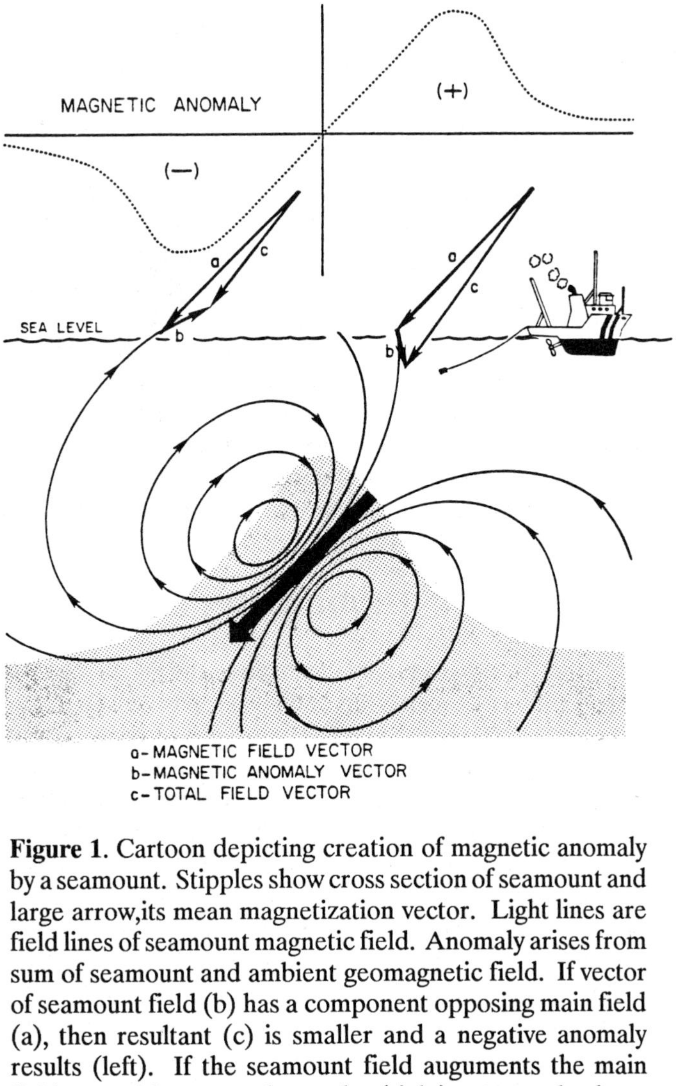

Figure 1. Cartoon depicting creation of magnetic anomaly by a seamount. Stipples show cross section of seamount and large arrow,its mean magnetization vector. Light lines are field lines of seamount magnetic field. Anomaly arises from sum of seamount and ambient geomagnetic field. If vector of seamount field (b) has a component opposing main field (a), then resultant (c) is smaller and a negative anomaly results (left). If the seamount field auguments the main field, a positive anomaly results (right). Anomaly shape depends on shape of seamount and direction of mean magnetization vector.

seminorm minimization method. In contrast, the least-squares method produces no such error estimates.

The final result of these analyses is a paleomagnetic pole, calculated from the declination and inclination of the mean magnetization vector. Making the usual paleomagnetic assumptions that the average geomagnetic field has the form of a geocentric axial dipole, that the seamount basalts record the field direction existing at the time they were erupted, and that the seamount erupted over a long enough period to average out secular variation, the paleomagnetic pole is interpreted as the position of the geographic pole relative to the seamount.

PALEOMAGNETIC DATA

Two types of paleomagnetic information were needed for this study, seamount paleomagnetic poles and an APWP with which to compare them. An updated list of reliable paleomagnetic poles from seamounts older than Neogene was prepared (Appendix 1). Younger seamounts were not considered because there are few reliable paleomagnetic poles for these seamounts and because the amount of apparent polar wander that they display is small. Most of the seamounts that have been studied paleomagnetically are restricted in location to the central and western Pacific (Fig. 2), reflecting the abundance of Cretaceous seamounts located in this region.

The appendix contains paleomagnetic results from 100 seamounts: 67 with least-squares models, 11 with seminorm models, and 22 with both. Twenty-five new poles are included. Ten are seminorm method results from seamounts with previously published least-squares models (seamounts 15, 46, 50, 51, 53, 54, 59, 66, 75, 90); the remaining fifteen are from seamounts whose magnetism had not been analyzed previously. Of these, four are least-squares models using data gleaned from the National Geophysical Data Center (NGDC) archives (seamounts 72, 91, 92, 97); and eleven are seminorm models of seamounts surveyed recently (seamounts 16-19, 35, 93-96, 99-100).

The new survey data, mainly from the Roundabout Leg 10 cruise (RNDB10WT) of the R/V *Thomas Washington* (Scripps Institution of Oceanography), were the basis for the analysis of eleven seamounts (16-19, 35, 93-97, 99-100). Additionally, data from cruise KK81062602 of the R/V

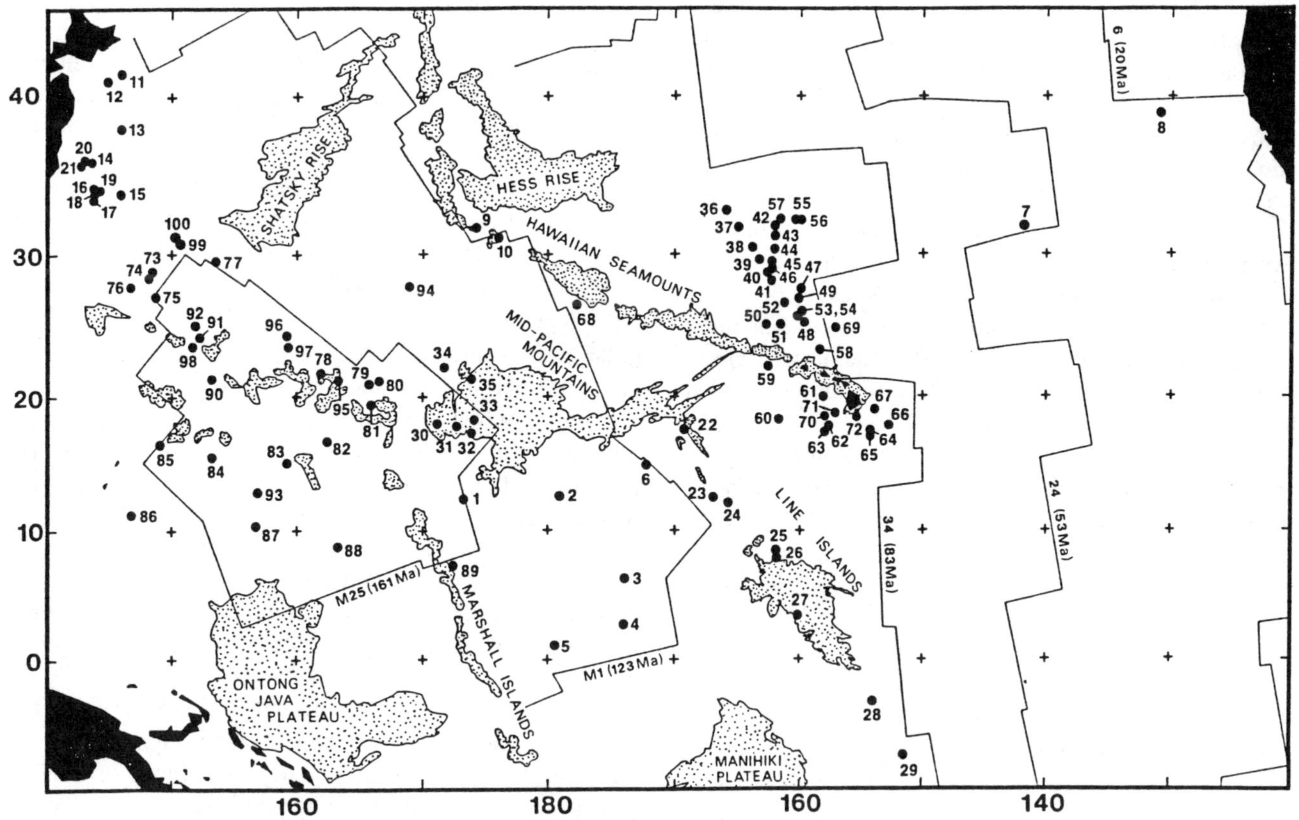

Figure 2. Locations of Pacific seamounts from which reliable paleomagnetic poles have been determined. Numbers correspond to Appendix 1. Land shown in black; major bathymetric features are stippled; light lines show representative magnetic reversal isochrons and their ages from geomagnetic polarity timescale of Harland et al. (1982)

Kana Keoki (Hawaii Institute of Geophysics) were used to model Ita Mai Tai Guyot (seamount 93). Ship tracks from the NGDC collection were used to augument the newer data in most of the models.

For the new least-squares models, I followed established procedures, which have been described at length elsewhere (e.g., Harrison et al., 1975; Sager, 1984; Sager and Pringle, 1987; 1988). Some explanation of the seminorm modeling is desirable, however, because this technique is new. For these models, I followed the steps outlined by Parker et al. (1987) and Hildebrand and Parker (1987). Using the seminorm technique, it is possible to derive a magnetization model that produces an exact fit to the observed anomaly data. However, an exact fit is usually unreasonable because of ionospheric noise and survey errors in the data set. Consequently, the method makes use of a precision parameter, λ, that allows the nonuniform part of the model to be adjusted from an exact fit of the observed anomaly ($\lambda = 0$) to no compensation for inhomogeneities at all ($\lambda = \infty$, equivalent to the least-squares model).

Typically, as λ is varied from large to small values, the calculated paleomagnetic pole moves rapidly away from the least-squares position and gradually approaches the exact fit position (see Figure 10 in Parker et al., 1987). As there is no criterion for judging which pole is best, Parker and Hildebrand (1987) recommend adjusting λ so that the rms misfit between the observed and calculated anomalies is approximately 30 nT, a value they consider reasonable for well-navigated surveys. For the new seminorm models whose poles are given in Appendix 1, λ was lowered until either the pole was within a few degrees of the exact fit pole, at which point subsequent lowering of λ had little effect on the pole, or an rms misfit of 30-40 nT was achieved.

Figure 3 shows the distribution of poles from Appendix 1. Most are clustered in the north Atlantic in a cloud extending from Spitzbergen at 80°N down the Greenwich Meridian to the British Isles at about 55°N, there turning westward and stretching south of Iceland towards Laborador. No poles with locations significantly displaced from the cluster are found in the Atlantic south of 40°N or in Asia east of about 45°E. However, two are located in southern Europe (14 and 57) and about 16 are found in Greenland and northern North America. The two in southern Europe are probably a result of inaccurate pole determinations, but those found in Greenland and North America may represent the Early Cretaceous-Jurassic part of the APWP, as explained below.

Sources of Error

Before attempting to interpret seamount paleomagnetic poles in terms of age or apparent polar wander, it is prudent to consider the possible errors in determining a pole. For the purpose of discussion, these errors are divided into two classes, random and systematic.

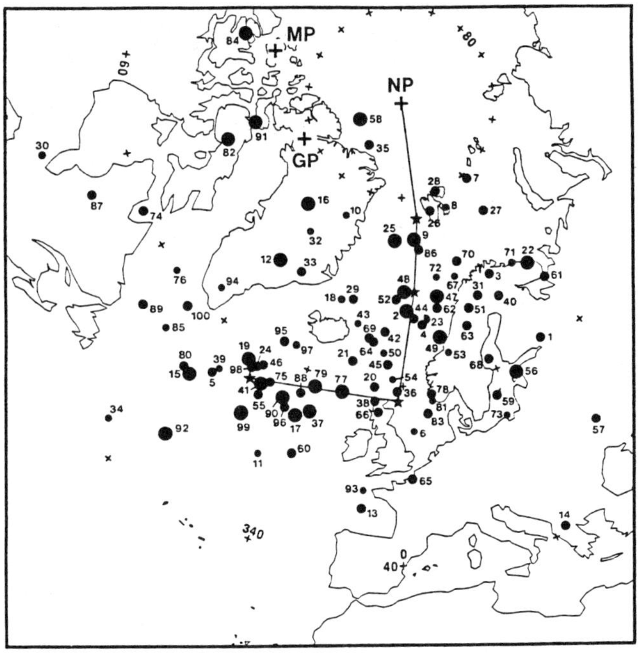

Figure 3. Reliable paleomagnetic poles from seamounts older than about 35 Ma. Symbol sizes indicate reliability of poles: largest, most reliable, GFR > 4.5 or major semiaxis length of 95% confidence ellipse < 6°; medium, reliable, 4.5 > GFR > 2.5, or semiaxis length between 6-12°; smallest, least reliable, GFR < 2.5 or semiaxis length > 12°. Numbers correspond to Appendix 1. Thin line is apparent polar wander path of Sager and Pringle (1988). NP is north geographic pole; MP is north magnetic pole; GP is north geomagnetic pole. Land outlines shown for reference.

Random scatter will arise from many sources including uncorrected ionospheric noise in the magnetic field recordings, navigation errors, incorrect or incomplete regional field corrections, and the complexities and variations of a seamount's magnetic properties. In my experience, geomagnetic variations and survey errors rarely seem to be at fault in causing large errors in pole position. The other two problems, regional removal and magnetization complexities, are another matter.

If a large magnetic anomaly interferes with the seamount anomaly being analyzed, perhaps because of magnetic lineations in the underlying seafloor or another nearby seamount, a misleading result can occur. This is probably the case for pole 14, calculated from a seamount with a small anomaly superimposed on relatively large amplitude seafloor anomalies of the Japanese lineations. However, pole 14 is clearly recognizable as discordant.

More troublesome is a pole that is located on the wrong part of the APWP because of regional removal problems.

Erimo seamount is a particularly bad example. An initial inversion yielded a pole at 56.4°N, 359.4°E (Uyeda and Richards, 1966) consistent with poles with ages of 80-84 Ma (Sager and Pringle, 1988). Furthermore, K-Ar analysis of rocks dredged from the seamount yielded an age of 80 Ma (Ozima et al., 1983). Because of this apparent agreement Erimo's pole was one of nine used to determine the location of the 82 Ma mean Pacific paleomagnetic pole (Sager and Pringle, 1988). However, in a recent reanalysis of this seamount's magnetic anomaly, the effect of underlying magnetic lineations was removed and a new pole position of 69.0°N, 321.0°N derived (Yamazaki, 1988). This pole position is among those of seamounts that are probably Jurassic to Early Cretaceous in age. What is more, new age constraints (Table 1) also imply that the seamount is older than 80 Ma. Usually the effects of inadequate removal of regional magnetic gradients are not as severe as this. Part of the reason is that many of the modeled seamounts are located on seafloor of the Cretaceous Quiet Zone or the Jurassic 'Quiet' Zone where background anomalies are usually small.

Scatter may also result from the way in which regional gradients are removed. Some authors simultaneously invert for the magnetization vector and a planar regional, but Sager (1983) shows that this creates a greater scatter in pole positions. McNutt (1986) suggests that the greater errors

Table 1. Age constraints for Jurassic–Early Cretaceous seamounts.

ID	Name	Age (Ma)	Constraints
34	Darwin	118–157	Reversed polarity indicates age of M0 (118 Ma) or older; capped by Albian-Aptian coral (Ladd et al., 1974); seafloor age M23 (157 Ma).
12	Erimo	104–130	^{40}Ar-^{39}Ar age of 104±9 Ma, but gravity study gives 2 km elastic plate thickness implying seamount is same age as seafloor, M8 (130 Ma) (Yamazaki, 1988).
90	Golden Dragon	95	^{40}Ar-^{39}Ar age, no error limits given (Ozima et al., 1983).
93	Ita Mai Tai	110–115	Nearby DSDP Site 585 yielded late Aptian-Early Albian volcaniclastics shed by guyot during edifice building stage (Baltuck et al., 1986).
77	Makarov	94±1	^{40}Ar-^{39}Ar age (Ozima et al., 1983).
19	Seiko	102	^{40}Ar-^{39}Ar age no error limits given (Ozima et al., 1983).
16	Takuyo-Daini	98–144	Guyot topped by Albian-Aptian corals (Matthews et al., 1974); seafloor age, M16 (144 Ma).
32	Thomas	≈162	Thin elastic plate thickness implies age close to that of seafloor (Watts et al., 1980); seafloor age approximately M25 (Winterer and Metzler, 1984).
75	Z43	93.7±4.2	^{40}Ar-^{39}Ar age (Ozima et al., 1983).
99	Isakov	98–122	Albian-Aptian rudists dredged from summit (Matthews et al., 1974); seafloor age M23 (157 Ma); flexure indicates >35 Ma younger than seafloor (Watts et al., 1980).
Seamounts with reversed polarity on Jurassic-Early Cretaceous seafloor			
85	Campbell	118–161	Seafloor age M25 (161 Ma)
34	Darwin	118–158	See above
94	Guadaloupe	118–139	Seafloor age, M13 (139 Ma)
92	Unnamed	118–167	Seafloor slightly older than M29 (165 Ma)
96	UCSD	≥118?	On flank of Scripps Guyot, 98±3 Ma (Ozima et al., 1983)

Note: Magnetic anomaly ages from timescale of Harland et al. (1982)

are a result of the inversion routine confusing part of the seamount anomaly with the regional. Consequently, seamount paleomagnetic results in which the regional has been removed prior to the inversion calculations are probably somewhat more trustworthy.

Complex magnetizations are probably the most troublesome problem. They might result from large variations in rock type or compostion within a seamount, eruption of a seamount during more than one polarity interval, or the resurgence of volcanism on a seamount after a long quiescence. Recognizing inaccurate results caused by these problems is not always straightforward.

A least-squares inversion of a seamount with a grossly nonuniform magnetization can produce a pole that is 30-40° or more in error (Lumb et al., 1973; Blakely and Christiansen, 1978; Parker et al., 1987). Often a pole this far off will be recognized as discordant or its inversion rejected because the match of the model and observed magnetic anomalies is poor. The goodness-of-fit ratio (GFR), the mean of the observed anomaly values divided by the mean of the residual values, has frequently been used as a rejection criterion with most authors discarding results from inversions with GFR less than 2. Lower GFR often result from complex magnetic anomalies caused by nonuniform magnetizations. Though application of this criterion has undoubtedly rejected many unreliable data, it is arbitrary and probably does not completely screen out erroneous poles.

The seminorm inversion technique should be less susceptible to this type of error because it was designed to account for nonuniform magnetizations (Parker et al., 1987) Moreover, it calculates a confidence ellipse for the paleomagnetic pole based on the distribution of magnetization heterogeneities (Parker, 1988). Consequently, when available, seminorm results should usually be favored. However, even this technique has its limitations.

For example, a seamount consisting of two nearly equal blocks of opposing polarity would be unlikely to yield a useful paleomagnetic pole (Parker et al., 1987). The same can probably be said of seamounts whose magnetizations are so heterogeneous as to have little or no coherent underlying uniform direction as well as those that acquired their magnetizations during two or more volcanic events widely separated in time. In such instances, the estimated error ellipse of the pole may not accurately reflect the true degree of error because the assumptions of the statistical model used in the calculation are violated (Parker, 1988).

Applying an empirical approach to the estimation of least-squares paleopole errors, Sager and Pringle (1988) analyzed the scatter of poles from dated Pacific seamounts and found standard deviations of 4–7°. Similar values for random errors are implied by the major semi-axis lengths of the 95% confidence ellipses calculated for seamount paleopoles using Parker's (1988) method. These are generally 4–12° for all but the most inhomogeneously magnetized seamounts (Appendix 1).

Harder to evaluate is the effect of systematic error, of which two main types may cause problems in Pacific seamount paleopole analyses. One is a bias resulting from an induced or viscous magnetization overprint and the other is errors in tectonic reconstructions.

The complication of tectonic reconstruction errors is the possibility of misinterpreting paleopoles that are from seamounts on lithosphere that was not always a part of the Pacific plate. For example, Sager and Pringle (1987) suggest that some volcanoes with radiometric ages between 83-90 Ma in the northern Musicians and southern Hawaiian seamount groups yield poles consistent with younger Cretaceous data because of the rotations of microplates that existed briefly at the Pacific-Farallon ridge during the Late Cretaceous. Moreover, the complex tectonics of the Pacific during the Cretaceous Quiet Period and the prevalence of microplates during its history (Sager et al., 1988; Tamaki and Larson, 1988; Mammerickx and Sharman, 1988; Mammerickx et al., 1988) imply that this problem could be potentially troublesome; however, Pacific paleopoles display a remarkable degree of consistency (Sager, 1983; Gordon, 1983; Sager and Pringle, 1988) indicating otherwise.

The roles of induced or viscous magnetization overprints on seamount paleomagnetic data have received much attention in the recent literature (e.g., Williams et al., 1983; Moberly and Campbell, 1984; Merrill, 1985; Verhoef et al., 1985; Hildebrand and Staudigel, 1986). The effect of an induced magnetization or a Brunhes normal polarity overprint will be to cause poles of normally polarized seamounts to be displaced towards the geomagnetic pole. Poles of reversely polarized seamounts will be shifted in the opposite direction. This error is about 3–6° or less for seamounts with Koenigsberger ratios (induced divided by remanent magnetization) of 5 or greater, typical of oceanic basalts; however, it can be 15–20° for ratios as low as 1 (Sager, 1983).

Determining the amount of overprint is difficult because detailed studies of individual seamounts suggest that they can contain a wide range of intrusive and extrusive rocks whose magnetizations and Koenigsberger ratios depend not only upon lithology, but also on metamorphic and alteration state (Gee et al., 1988; 1989). In a practical sense, it is impossible to infer the magnetizations of most seamounts from samples because these are usually either scarce or nonexistent. However, general estimates have been made in several ways.

Comparing the paleopoles of normally and reversely polarized Pacific seamounts of Late Cretaceous age or younger, Sager (1987) found little systematic difference and concluded the effect of overprinting on these seamounts is negligible. In contrast, Hildebrand and Staudigel (1986) compared anomaly amplitudes of a large number of normally and reversely polarized Pacific seamounts and estimated that the overprint comprised an average of 20% of the anomalies. Similarly, Gee et al. (1989) used a knowlege of the average magnetizations of various seamount rocks and a structural model of Mauna Loa volcano to estimate that about 1/6 of that edifice's magnetization should be an induced or viscous overprint. Although an overprint of this size should only bias a seamount's paleopole by a few degrees, there is likely a variation in individual seamounts with some displaying less overprint and some more.

The Pacific Apparent Polar Wander Path

To make an age estimate based upon a paleomagnetic pole position, a well-determined APWP is needed. Until recently, the Pacific APWP was poorly constrained because of the scarcity of paleomagnetic data. New data have allowed the refinement of the path to the point where it is relatively precise, at least from mid-Cretaceous to early Tertiary time. Sager and Pringle (1988) review the available data and determine two APWP, one with four mean paleopoles and the other with six (Figures 4a and 4b), using 39 data including 22 seamount paleomagnetic poles. The difference between these two polar paths is simply that the data used for calculating the 72 and 88 Ma poles were each subdivided into two portions with smaller age ranges.

The Pacific APWP has a "hook" shape, noted by a number of investigators, including Jarrard and Sasajima (1980), Gordon (1983), and Cox and Gordon (1984). Since 82 Ma, apparent polar wander has been north-south, but before that it was east-west. Additionally, the rate has been uneven. Rapid wandering occurred between about 76 and 85 Ma making this 9 Ma stretch of the APWP longer than the subsequent 66 Ma portion (Fig. 4). Conversely, little or no apparent polar wandering is evident for the data used to calculate the 66, 76 and 85, 94 Ma poles, encompassing about 60-78 Ma and 84-95 Ma.

The APWP outlined by Sager and Pringle (1988) covers only about half of the lifetime of the Pacific plate. They did not define the Early Cretaceous-Late Jurassic portion of the APWP because few Pacific paleomagnetic data of this age have both reliable ages and pole estimates.

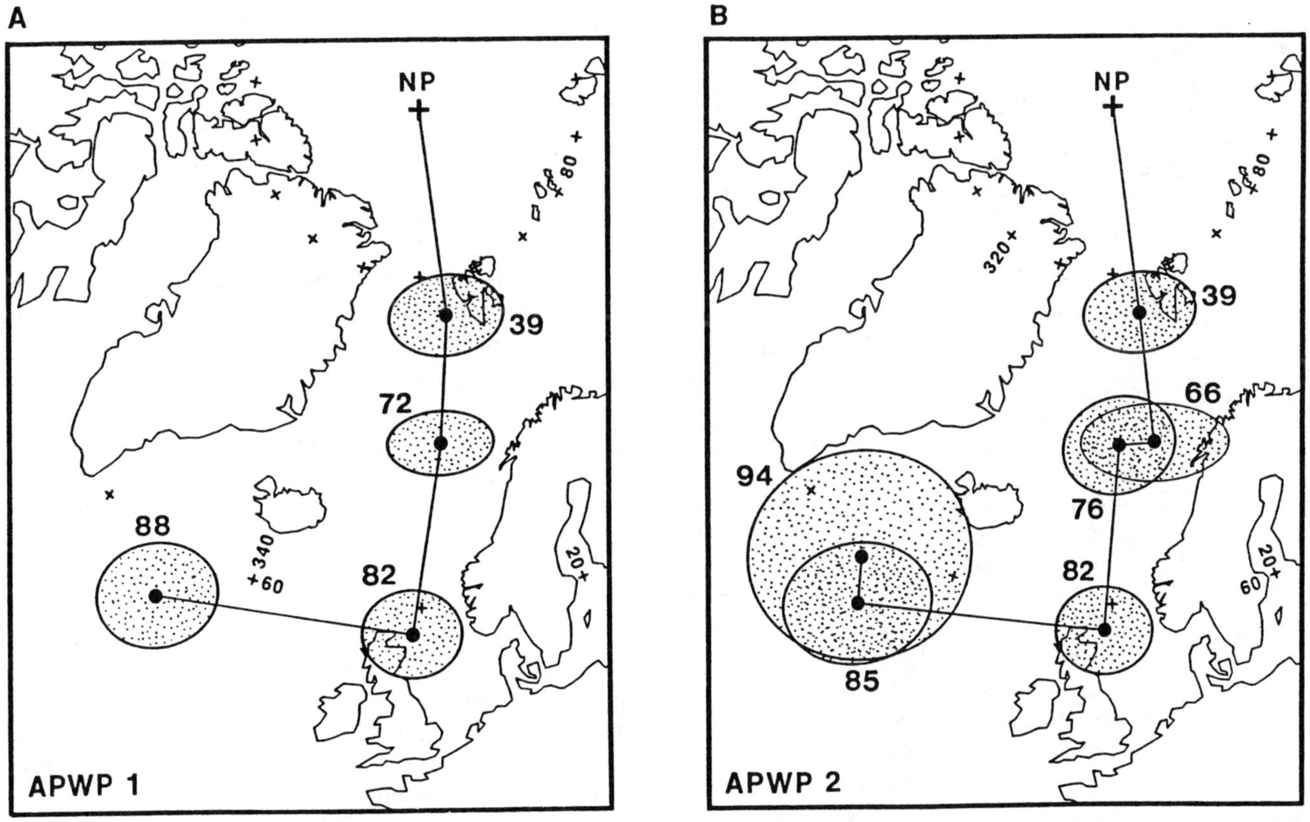

Figure 4. Pacific apparent polar wander. *A* and *B*, apparent polar wander paths calculated by Sager and Pringle (1988). Bold numbers are mean ages of poles in Ma. Path 2 differs from path 1 in that the data were subdivided into more nearly equal time windows to show slower polar wander around times of 66, 76 and 85, 94 Ma poles. Stippled ellipses are 95% confidence regions of mean pole positions.

Several authors suggest that the "tail" of the APWP turns northward and ends in the vicinity of central Greenland (Larson and Lowrie, 1975; Jarrard and Sasajima, 1980; Cox and Gordon, 1984; Hildebrand and Parker, 1987). Unfortunately, this supposition is uncertain because reliably dated Pacific paleomagnetic data of this age are scarce. Jarrard and Sasajima (1980) and Hildebrand and Parker (1987) calculated mean paleomagnetic poles in this region using undated seamounts from Jurassic or Early Cretaceous seafloor in the western Pacific. Furthermore, Jarrard and Sasajima (1980) also presented an undocumented Early Cretaceous pole from an analysis of marine magnetic lineation skewness (Fig. 5). Larson and Lowrie (1975) and Cox and Gordon (1984) based a similar interpretation on a single paleomagnetic datum, an azimuthally unoriented

C

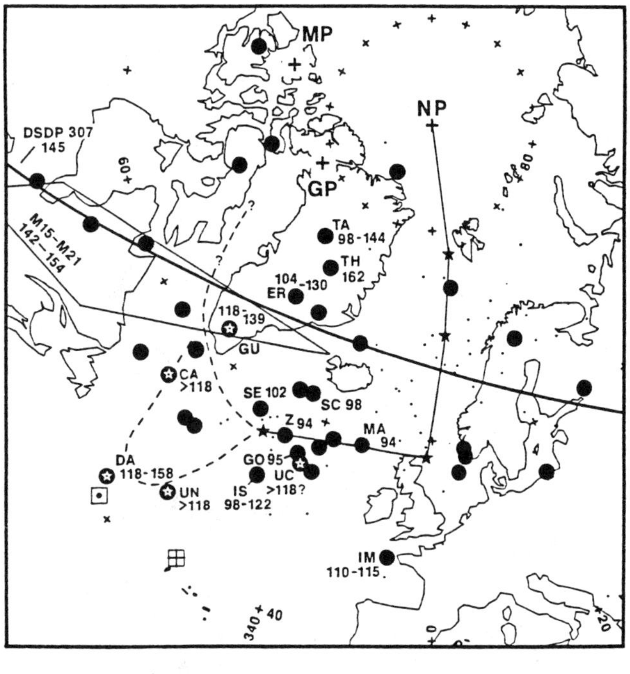

Figure 5. Polar wander trends for the Jurassic and Early Cretaceous. Large filled circles are poles of seamounts with ages of 94 Ma and older or are located on Jurassic age seafloor. Circles with embedded stars have reversed polarities. Seamounts with age constraints (Table 1) are labeled with abbreviated name and age. Small dots are other seamount poles. Thick arc is locus of poles constrained by paleolatitude from DSDP Site 307 (Larson and Lowrie, 1975). Quadrilateral is pole location prediced from magnetic lineations M15-M21 (Jarrard and Sasajima, 1980). Small box with dot and small box with cross are Early Cretaceous paleomagnetic poles determined from DSDP core data by Sager (1983) and Cox and Gordon (1984), respectively. Dashed lines are two possible trends of pre-mid-Cretaceous polar wander (see text).

paleolatitude from DSDP Site 307 (Fig. 5), drilled on magnetic chron M21 (Larson and Lowrie, 1975). This datum is of uncertain reliability because it only sampled approximately three independent flow units (Cox and Gordon, 1984).

An additional indication of the APWP trend prior to the mid-Cretaceous is obtained by plotting poles from seamounts known to have ages of about 95 Ma or greater along with poles from seamounts on seafloor of Jurassic age (Fig. 5). These include all but two of the poles located to the west of the main cluster of poles and most of those located at the west end of the APWP. The critical question is whether these poles accurately represent the APWP or reflect systematic errors.

The few existing age data for these seamounts are summarized in Table 1. Only five seamounts have Ar^{40}-Ar^{39} radiometric ages, and in one instance (Erimo, 12) these data conflict with other geophysical data. The lack of additional reliable radiometric ages is a problem, for the remainder of the seamounts have only loose age constraints. Ages for another five seamounts are implied by reversed polarities; they formed either before or after the Cretaceous Quiet Period(118–83 Ma). One, Darwin Guyot (34), is capped by mid-Cretaceous corals, so it formed at or before 118 Ma. Three, 85, 92, and 94, have poles inconsistent with Late Cretaceous pole positions (Fig. 5), so they likely did the same. The other, UCSD (96), has a pole position too close to the Late Cretaceous portion of the APWP to make such an inference. The remaining seamounts in Table 1 are dated by coral caps (16, 99), fossils in nearby sediments (93), or by lithospheric flexure (32).

If the pole positions of these older seamounts are simply averaged, a mean near the 88-94 Ma end of the APWP is obtained; however, this result is probably misleading. There are several poles located near the Darwin Guyot pole (34) southwest of the end of the APWP (Fig. 5). These poles may indicate a southwestward trend of the APWP. Additionally, the poles located in Greenland and North America imply that the APWP must bend back to the north. Consequently, two possible APWP trends are shown in Figure 5. One assumes that the Darwin Guyot pole lies near the bend of the APWP. Consistent with this hypothesis are mean paleomagnetic poles located at 41.4°N, 317.5°E and 41°N, 329°E calculated by Sager (1983) and Cox and Gordon (1984), respectively. The former was found using the Darwin pole and four paleolatitudes measured from DSDP sediment cores, whereas the latter was determined from basalts drilled at four other DSDP sites.

The other possible APWP ignores the poles in this region and trends northward from the end of the known APWP. This trend may be more accurate if induced magnetism contributes significantly to the magnetizations of some of these seamounts. Note that the five reversely polarized seamounts from Table 1 have poles that are farther from the magnetic and geomagnetic poles than their normally polarized counterparts. What is more, a significant fraction of the poles of normally polarized seamounts on Jurassic seafloor are located in the vicinity of the magnetic and geomagnetic poles. Such a distribution is expected if the seamounts are contemporaneous and a significant part of their magnetizations are induced or have a present-day normal polarity overprint.

A satisfactory solution of this problem requires more reliably dated paleomagnetic data. Despite this uncertainty, the data strongly suggest that the Jurassic portion of the APWP trends northwestward into the region of Greenland and perhaps northeastern North America. If the paleomagnetic pole of Darwin Guyot and other nearby poles are reliable, then the APWP may dip southwestward before bending to the north.

PALEOMAGNETIC AGE ESTIMATES

To make paleomagnetic age estimates, the seamount paleomagnetic poles were assembled into four groups according to their position along the APWP (Fig. 6, Appendix 1). In doing so, several factors were considered. The most important was that the groups were divided to be consistent with the ages of reliably dated seamounts. The groupings also take into account changes in polar wander direction and gaps in the distribution of poles. Additionally, poles near the boundaries of the age groups were sorted so as to maximize the consistency of the ages of neighboring seamounts. In order of increasing age, the groups are (1) early Tertiary; (2) Late Cretaceous; (3) mid-Cretaceous; and (4) Early Cretaceous to Jurassic.

The easiest group to delineate was the early Tertiary because many of these seamounts have been dated and there is a natural gap between these poles and those of Late Cretaceous age. The reason for the gap is that no paleomagnetic studies have been done on seamounts with early Eocene or Paleocene ages. The data used to define the early Tertiary mean pole ranged in age from 36-45 Ma (Sager, 1987), so the seamount paleomagnetic poles located along this section of the APWP are probably Eocene in age.

The main cluster of poles, distributed along the mid to Late Cretaceous part of the APWP was divided into two groups, one representing the north-south section and the other the east-west section. The division between these two groups, near the APWP bend, is the most arbitrary of all; however, it was drawn to produce the maximum age consistency as stated above. Following the APWP analysis of Sager and Pringle (1988), the Late Cretaceous group spans an age range of about 64-79 Ma. The mid-Cretaceous group ends at about 80 Ma, but its beginning is less clear. The oldest dated seamount in this group is Seiko (19) with an age of 102 Ma (Ozima et al., 1983); however, as these seamounts are all normally polarized, they may have ages as old as the beginning of the Cretaceous Quiet Period, about 118 Ma.

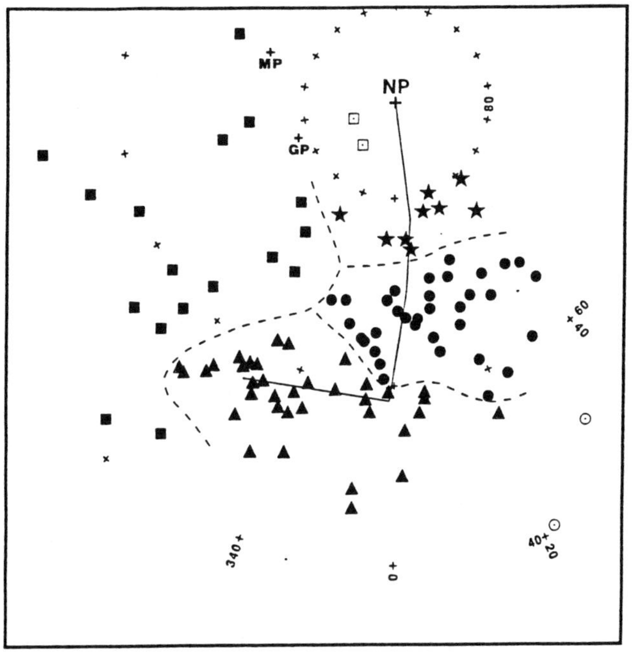

Figure 6. "Paleomagnetic age" estimated by grouping seamount paleopoles along the Pacific apparent polar wander path. Stars, early Tertiary (36-45 Ma); filled circles, Late Cretaceous (64-79 Ma); triangles, mid-Cretaceous (80-120 Ma); squares, Early Cretaceous-Jurassic (older than 120 Ma); open circles, undifferentiated Cretaceous; open squares, not interpreted. Thin line represents early Tertiary to mid-Cretaceous APWP.

Constituting the oldest group are poles located in North America and Greenland as well as those from seamounts with reversed polarity believed to have been formed prior to the Cretaceous Quiet Period. Because of the uncertainty surrounding the beginning of the Pacific APWP, the age resolution and accuracy of this group is the worst of all. These seamounts may range in age from as young as the beginning of the Cretaceous Quiet Period to as ancient as the oldest Pacific crust, about mid-Jurassic in age. Consequently, the paleomagnetic ages assigned to this group are simply Early Cretaceous-Jurassic. Because it is unclear whether the poles in northern North America and Greenland are reliable or biased by induced or viscous magnetism, those to the south of the geomagnetic and magnetic poles were included with the Early Cretaceous-Jurassic group. The two nearest the geographic pole (35 and 58) were not given ages.

Two other "anomalous" poles, located in southern Europe (14 and 57) also are not easily fit into the groups outlined above. Their distance from the main cluster brings their reliability into question. Nevertheless, they show 30–40° of displacement from the geographic pole, implying that these seamounts were probably formed during the Cretaceous. Consequently, they were assigned undifferentiated Cretaceous paleomagnetic ages.

A

B

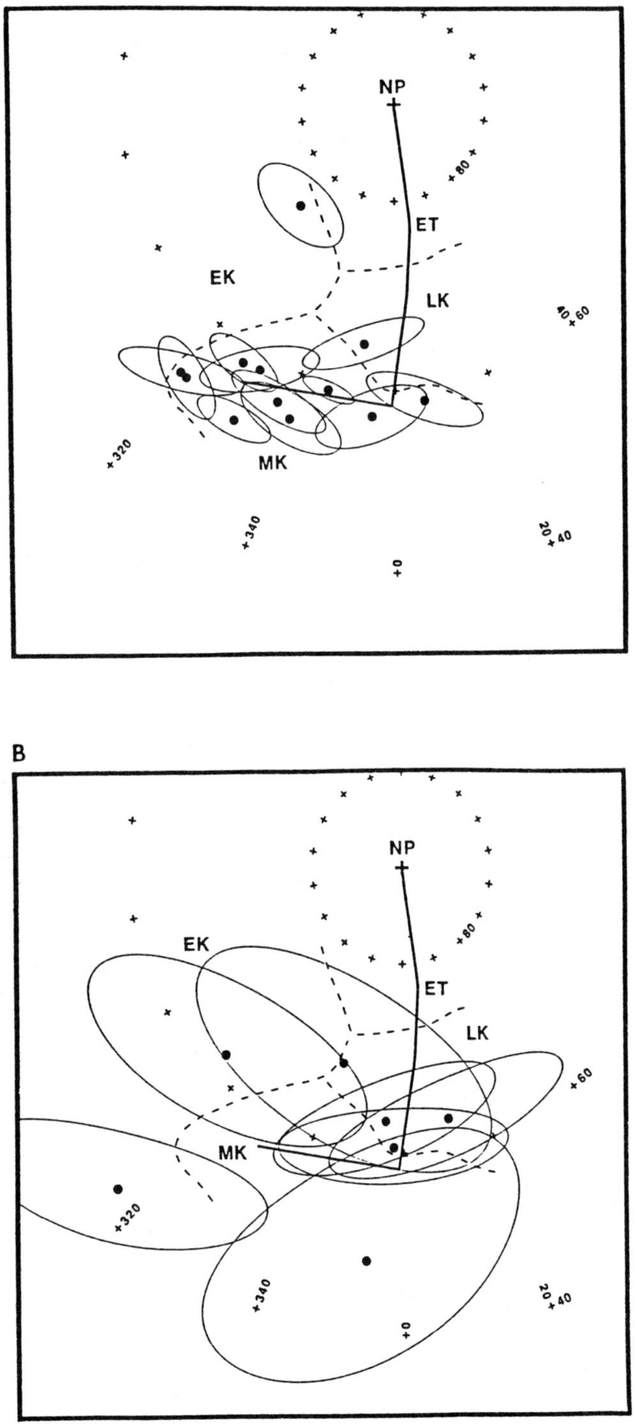

DISCUSSION

Because the distribution of paleomagnetic poles is more or less continuous, their division into age groups was necessarily somewhat aribitrary, ignoring the errors associated with each of the poles. As seen in Figure 7, many of the poles have 95% confidence ellipses that extend into two or more age groups. Hence, there is a chance that they have been assigned misleading paleomagnetic ages. A more correct method would be to estimate a range of ages for a seamount according to the portion of the APWP covered by the 95% confidence oval of its pole. However, this technique has several drawbacks. One problem is that about 70% of the existing seamount paleopoles have no estimated error bounds. Another is that the Jurassic-Early Cretaceous portion of the APWP is ill-defined, so many seamounts with poles along the end of the known APWP would be assigned very large age ranges, defeating much of the purpose of this analysis. Finally, an aim of this study is to use the paleomagnetic ages to examine the distribution of Pacific volcanism, a task made more difficult by trying to use variable age ranges. However, because of the expected errors in age assignments that result from this simple approach, one must not place too much weight on a single paleomagnetic age, but instead observe the age patterns and trends.

Distribution of Pacific Volcanism

The arrangement of paleomagnetic ages (Fig. 8) provides some interesting clues about the history of volcanism in the central and western Pacific. Early Cretaceous seamounts appear restricted to the western Pacific whereas Late Cretaceous edifices are mainly in the central Pacific. Mid-Cretaceous seamounts are abundant and scattered over most parts of the Pacific plate that existed at that time. In contrast, Tertiary poles are few in the western Pacific.

If most of the seamounts had erupted near the time of formation of the seafloor upon which they reside, then the ages would be arranged in concentric bands, younging outward and centered on the old western Pacific litho-

Figure 7. Comparison of 95% confidence ellipses of selected paleopoles derived from seminorm inversion (Parker, 1988) with the paleomagnetic age groups. Solid dots are pole positions and ovals are 95% confidence ellipses. (A) Paleopoles with major semi-axis lengths ≤6°. (B) Paleopoles with major semi-axis lengths ≥12°. Many of the larger ellipses overlap more than one age group, indicating the possibility of misleading paleomagnetic age assignments.

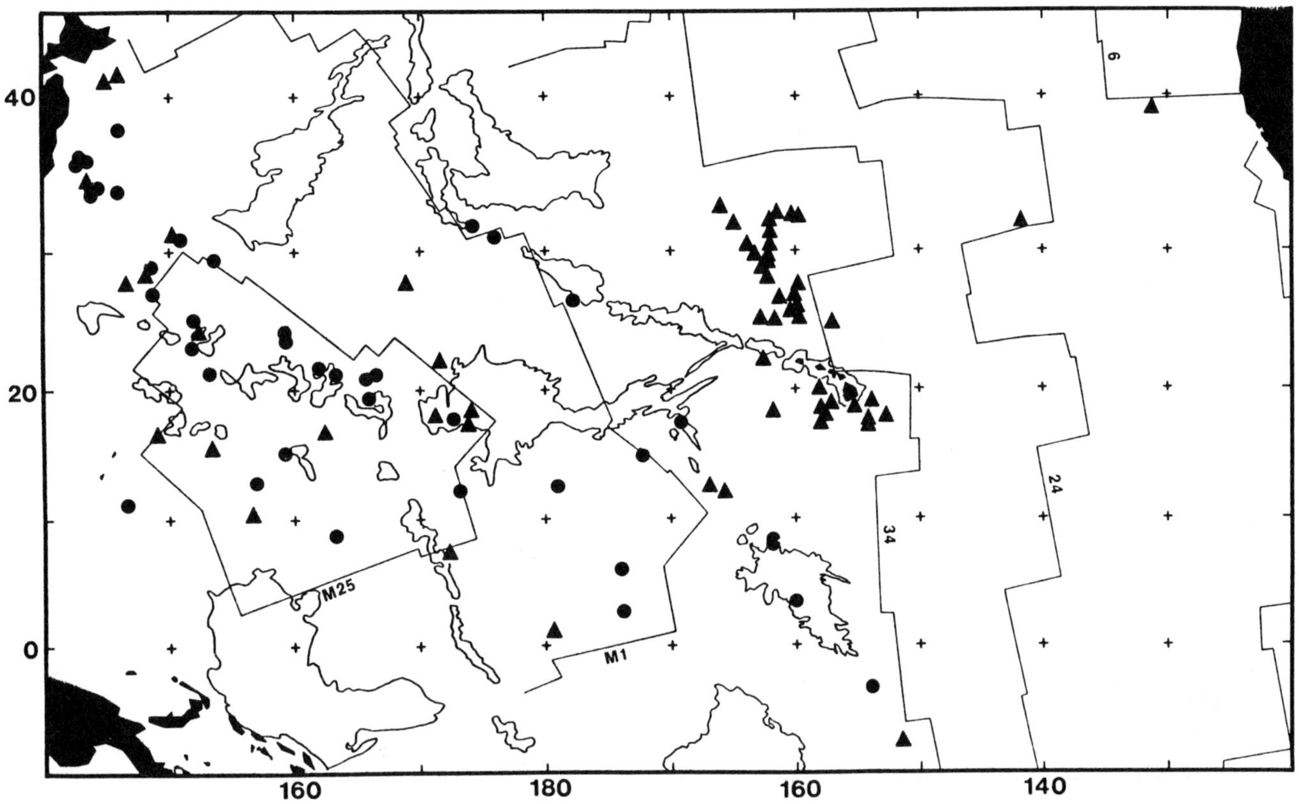

Figure 8. Distribution of "paleomagnetic ages". Symbols represent same age groups as in Figure 6: stars, early Tertiary; filled circles, Late Cretaceous; triangles, mid-Cretaceous; squares, Early Cretaceous-Jurassic; open circles, undifferentiated Cretaceous; open squares, not interpreted.

sphere. This is clearly not the case (Fig. 8). Though the Early Cretaceous-Jurassic seamounts are found on the oldest lithosphere and many of the Late Cretaceous seamounts are located on lithosphere of similar age, seamounts of the latter age are also found on older seafloor in the vicinity of 180° longitude. Moreover, mid-Cretaceous volcanoes are found in greater abundance on the old lithosphere of the western Pacific than on lithosphere of like age. These observations imply that although some seamounts are near the age of the underlying seafloor, a substantial portion are intraplate volcanoes, formed long after the lithosphere.

This outcome is not surprising, for Watts et al. (1980) obtained similar results from a study of the gravity fields of Pacific seamounts. Using lithospheric flexure, they were able to determine whether a seamount formed on thin elastic lithosphere near the ridge or on thick lithosphere away from the ridge. They found about equal numbers of on-ridge and off-ridge seamounts. The former were most abundant around the plateaus and large uplifts such as the Shatsky Rise, Hess Rise, and Line Islands, whereas the latter were found in greatest number in the known hotspot chains and the Magellan and Wake seamounts of the western Pacific.

Using paleomagnetic ages compared to the age of the underlying seafloor, a plot of on-ridge and off-ridge seamounts was constructed (Fig. 9). This assessment is not as easily made or as direct as that arising from the gravity analysis because it relies on an accurate determination of the age of the seafloor from magnetic isochrons along with an accurate paleopole. Nevertheless, the results are very similar to those reported by Watts et al. (1980). In constructing Figure 9, a relatively broad definition of on-ridge was used. To be labeled "off-ridge" seamounts had to be clearly older than the underlying seafloor, by a span of perhaps 20 Ma or more. As a consequence, the Musicians and southern Hawaiian seamounts were all deemed to be on-ridge, even though volcanism in that province lasted about 20-30 Ma (Sager and Pringle, 1987). However, in most cases the distinction was relatively clear.

The paleomagnetic data indicate that many of the central and western Pacific seamounts formed off-ridge. If the densely surveyed, on-ridge Musicians and southern Hawaiian seamounts are ignored, nearly 75% of the remain-

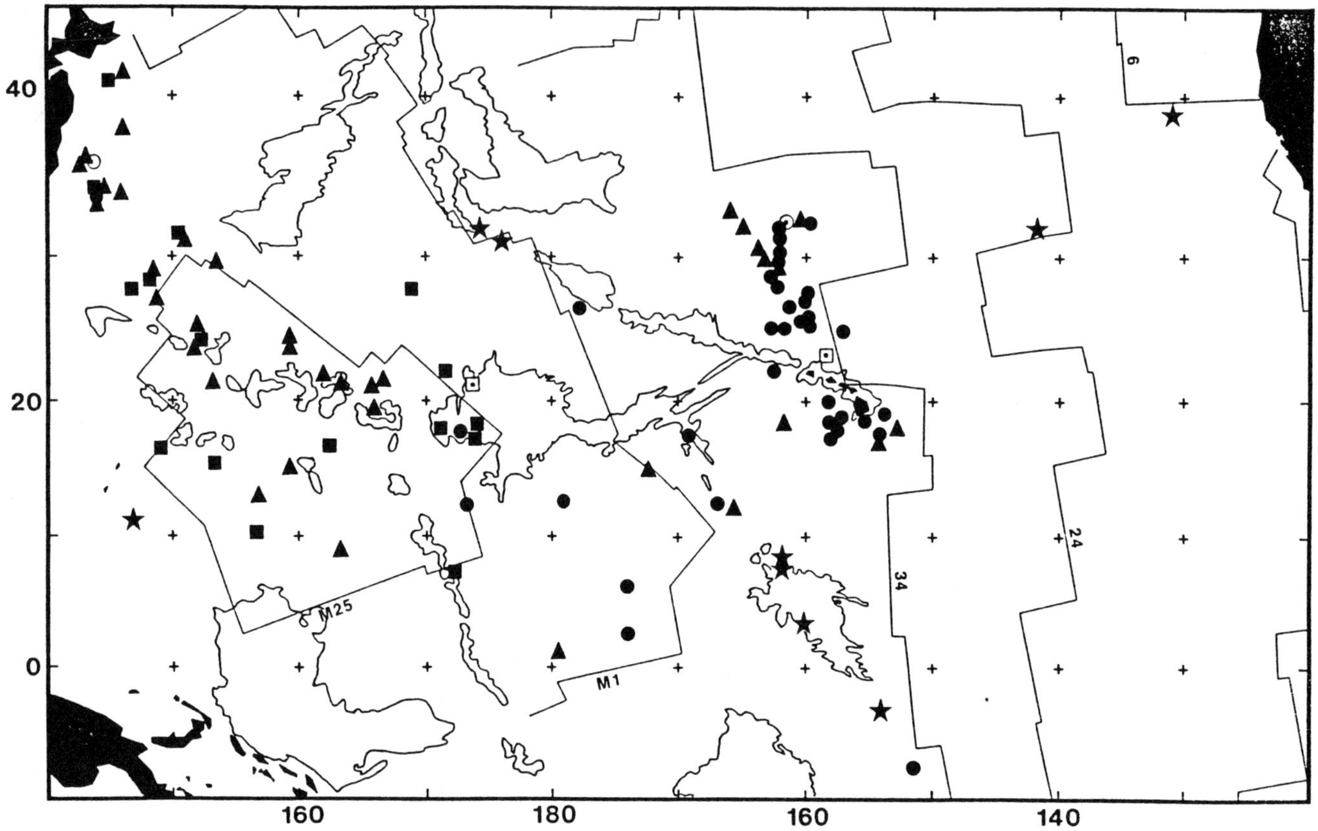

Figure 9. Occurrence of on-ridge versus off-ridge volcanism interpreted from "paleomagnetic ages" and age of underlying seafloor. On-ridge seamounts formed within about 20 Ma of seafloor age. Triangles represent on-ridge volcanoes; filled circles denote off-ridge seamounts.

ing volcanoes fall into the off-ridge category. Furthermore, the off-ridge seamounts are less restricted to the westernmost Pacific than indicated by the gravity study of Watts et al. (1980). Specifically, the paleomagnetic data suggest that there are also off-ridge seamounts in the central Pacific basin and Line Islands. Indeed, the Line Islands results show the greatest difference between the gravity and paleomagnetic studies. The gravity implied that these seamounts were formed almost entirely on-ridge (Watts et al., 1980), whereas the paleomagnetic data suggest the opposite (Fig. 9).

The paleomagnetic data are generally consistent with two classes of hypotheses that attempt to explain the origin of the intraplate volcanism in the western Pacific. One idea, outlined by Henderson and Gordon (1981), presumes that these volcanoes formed as the western Pacific passed over a congregation of mantle plumes, specifically the Easter, Macdonald, Marquesas, Pitcairn, and Tahiti hotspots, all currently located in a relatively small portion of the southeast Pacific. McNutt and Fischer (1987) expanded upon this idea and proposed that the hotspot group owes its coherence to a broad mantle upwelling, which they named the "superswell", that causes the lithosphere to be more vulnerable to hotspot magma penetration.

Another hypothesis, advanced by a number of investigators, attempts to explain the origin of the intraplate seamounts by one or more widespread pulses of voluminous volcanism. One of the first such conjectures was Menard's "Darwin Rise" hypothesis which suggested that a huge, ephemeral mantle upwelling occurred, caused widespread volcanism, and then subsided (Menard, 1964; 1984). Other investigators found evidence implying that widespread volcanic episodes occurred during the mid-Cretaceous, Late Cretaceous, and Eocene (Schlanger et al., 1981; Schlanger and Premoli-Silva, 1981; Haggerty et al., 1982; Rea and Vallier, 1983).

Though the paleomagnetic data probably do not have sufficient resolution to disprove or confirm either hypothesis, they do show a trend that appears to be more consistent with the superswell model. Comparing Figures 8 and 9, the off-ridge seamounts appear to become younger to the southeast. In the western Pacific they are mid-

Cretaceous in age; in the Central Pacific Basin, mostly Late Cretaceous; and, in the Line Islands there is a significant early Tertiary component.

Other research results also favor this hypothesis. Comparisons of the isotopic geochemistry and thermal subsidence histories of seamounts on the superswell today and those in the western Pacific show significant similarities (Smith et al., 1989; McNutt et al., 1990). Additionally, the superswell hypothesis is attractive because it offers a simple mechanism and a present-day counterpart to the western Pacific volcanism, rather than calling upon a unique event.

ACKNOWLEDGMENTS

I thank Barbara Keating and two anonymous reviewers for constructive reviews. B. Dayharsh assisted with seamount paleomagnetic analyses. J. Hildebrand and R. Parker provided a copy of their seminorm magnetic modeling program. E. Winterer, J. Natland, and M. McNutt obliged with useful information and discussions on cruise Roundabout Leg 10 on the R/V *Thomas Washington*. Support was provided by Office of Naval Research grant N0001487K0271 and National Science Foundation grant OCE-8715733. Texas A&M Geodynamics Research Institute Contribution No. 81.

REFERENCES

Baltuck, M., R. Moberly, and S. O. Schlanger, 1986, Introduction and explanatory notes: Initial Reports of the Deep Sea Drilling Project, v. 89, p. 5-28.

Batiza, R., 1977, Age, volume, compositional and spatial relations of small isolated oceanic central volcanoes: Marine Geology, v. 24, p. 169-183.

Blakely, R. J. and R. L. Christiansen, 1978, The magnetization of Mount Shasta and implications for virtual geomagnetic poles determined from seamounts: Journal of Geophysical Research, v. 83, p. 5971-5878.

Clague, D. A. and G. B. Dalrymple, 1987, The Hawaiian-Emperor volcanic chain, part I. Geologic evolution, *in* R. W. Decker, T. L. Wright, and P. H. Stauffer, eds., Volcanism in Hawaii: USGS Professional Paper 1350, p. 5-54.

Cox, A. and R. G. Gordon, 1984, Paleolatitudes determined from paleomagnetic data from vertical cores: Reviews of Geophysics and Space Physics, v. 22, p. 47-72.

Dalrymple, G. B., D. A. Clague, T. L. Vallier, and H. W. Menard, 1987, $^{40}Ar/^{39}Ar$ age, petrology, and tectonic significance of some seamounts in the gulf of Alaska, *in* B. H. Keating, P. Fryer, R. Batiza, and G. W. Boehlert, eds., Seamounts, Islands, and Atolls: American Geophysical Union Monograph 43, p. 297-315.

Dymond, J. and H. L. Windom, 1968, Cretaceous K/Ar ages from Pacific ocean seamounts: Earth and Planetary Science Letters, v. 4, p. 47-52.

Epp, D., 1984, Possible perturbations to hotspot traces and implications for the origin and structure of the Line Islands: Journal of Geophysical Research, v. 89, p. 11,273-11,286.

Duncan, R. A. and D. A. Clague, 1984, The earliest volcanism on the Hawaiian Ridge (abstract): EOS, Transactions of the American Geophysical Union, v. 65, p. 1076.

Francheteau, J., C. G. A. Harrison, J. G. Sclater, and M. L. Richards, 1970, Magnetization of Pacific seamounts: a preliminary polar wander curve for the northeastern Pacific: Journal of Geophysical Research, v. 75, p. 2,035-2,061.

Francheteau, J., J. G. Sclater, and H. Craig, 1969, Magnetization of a recently discovered Pacific seamount in the central Pacific: Geophysics, v. 34, p. 645-651.

Garcia, M. O., D. G. Grooms, and J. J. Naughton, 1987, Petrology and geochronology of volcanic rocks from seamounts along or near the Hawaiian Ridge: Lithos, v. 20, p. 323-336.

Gee, J., H. Staudigel, and L. Tauxe, 1989, Contribution of induced magnetization to magnetization of seamounts: Nature., v. 342, p. 170-173.

Gee, J., L. Tauxe, J. A. Hildebrand, H. Staudigel, and P. Lonsdale, 1988, Nonuniform magnetization of Jasper Seamount: J. Geophys. Res., v. 93, p. 12,159-12,175.

Gordon, R. G., 1983, Late Cretaceous apparent polar wander path of the Pacific plate: evidence for a rapid shift of the Pacific hotspots with respect to the paleomagnetic axis: Geophysical Research Letters, v. 10, p. 709-712.

Haggerty, J. A., S. O. Schlanger, and I. Premoli-Silva, 1982, Late Cretaceous and Eocene volcanism in the southern Line Islands and implications for hotspot theory: Geology, v. 10, p. 433-437.

Harland, W. B., A. V. Cox, P. G. Llewellyn, C. A. G. Pickton, A. G. Smith, and R. Walters, 1982, A geologic time scale: Cambridge, Cambridge University Press, 131 p.

Harrison, C. G. A., R. D. Jarrard, V. Vacquier, and R. L. Larson, 1975, Palaeomagnetism of Cretaceous Pacific seamounts: Geophysical Journal of the Royal Astronomical Society, v. 42, p. 859-882.

Henderson, L. J. and R. G. Gordon, 1981, Oceanic plateaus and the motion of the Pacific plate with respect to the hotspots (abstract): EOS, Transactions of the American Geophysical Union, v. 62, p. 1028.

Hildebrand, J. A. and R. L. Parker, 1987, Paleomagnetism of Cretaceous Pacific Seamounts revisited: Journal of Geophysical Research, v. 92, p. 12,695-12,712.

Hildebrand, J. A. and H. Staudigel, 1986, Seamount magnetic polarity and Cretaceous volcanism of the Pacific basin: Geology, v. 14, p. 456-458.

Jarrard, R. D. and D. A. Clague, 1977, Implications of Pacific island and seamount ages for the origin of volcanic chains: Reviews of Geophysics and Space Physics, v. 15, p. 57-76.

Jarrard, R. D., and S. Sasajima, 1980, Paleomagnetic synthesis for southeast Asia: constraints on plate motions, *in* D. E. Hayes, ed., The Tectonic and Geologic Evolution of Southeast Asian Seas and Islands: American Geophysical Union Geophysical Monograph 23, p. 293-317.

Keating, B. and W. Sager, 1980, Watkins Seamount: preliminary paleomagnetic results: Journal of Geophysical Research, v. 85, p. 3,567-3,571.

Ladd, H. S., W. A. Newman, and N. F. Sohl, 1974, Darwin Guyot, the Pacific's oldest atoll: Proceedings of the 2nd International Symposium on Coral Reefs, p. 513-522.

Larson, R. L. and W. Lowrie, 1975, Paleomagnetic evidence for the motion of the Pacific plate from Leg 32 basalts and magnetic anomalies: Initial Reports of the Deep Sea Drilling Project, v. 32, p. 571-577.

Lonsdale, P., 1988, Geography and history of the Louisville hotspot chain in the southwest Pacific: Journal of Geophysical Research, v. 93, p. 3,078-3,104.

Lumb, J. T., M. P. Hochstein, and D. J. Woodward, 1973, Interpretation of magnetic measurements in the Cook Islands, southwest Pacific Ocean, *in* P. J. Coleman, ed., The Western Pacific Island Arcs, Marginal Seas, Geochemistry: New York, Crane, Russak, and Company, p. 79-101.

Mammerickx, J. and G. F. Sharman, 1988, Tectonic evolution of the north Pacific during the Cretaceous Quiet Period: Journal of Geophysical Research, v. 93, p. 3,009-3,024.

Mammerickx, J., D. F. Naar, and R. L. Tyce, 1988, The Mathematician paleoplate: Journal of Geophysical Research, v. 93, p. 3,025-3,040.

Matthews, J. L., B. C. Heezen, R. Catalano, A. Coogan, M. Tharp, J. Natland, and M. Rawson, 1974, Cretaceous drowning of reefs on Mid-Pacific and Japanese guyots: Science, v. 184, p. 462-464.

McNutt, M., 1986, Nonuniform magnetization of seamounts: a least-squares approach: Journal of Geophysical Research, v. 91, p. 3,686-3,700.

McNutt, M. K. and K. M. Fischer, 1987, The south Pacific superswell, *in* B. H. Keating, P. Fryer, R. Batiza, and G. W. Boehlert, eds., Seamounts islands, and atolls: American Geophysical Union Geophysical Monograph 43, p. 25-34.

McNutt, M. K., E. L. Winterer, W. W. Sager, J. H. Natland, and G. Ito, 1990, The Darwin Rise: a Cretaceous superswell?, Geophysical Research Letters, in press.

Menard, H. W., 1964, Marine Geology of the Pacific: New York, McGraw-Hill, 271 p.

Menard, H. W., 1984, Darwin reprise: Journal of Geophysical Research, v. 89, p. 9,960-9,968.

Merrill, R. T., 1985, Correlating magnetic field polarity changes with geologic pheonomena: Geology, v. 13, p. 487-490.

Moberly, R. and J. F. Campbell, 1984, Hawaiian hotspot volcanism mainly during geomagnetic normal intervals: Geology, v. 12, p. 459-463.

Ozima, M., I. Kaneoka, K. Saito, M. Honda, Y. Yanagisawa, and Y. Takigami, 1983, Summary of geochronological studies of submarine rocks from the western Pacific Ocean, *in* T. W. C. Hilde and S. Uyeda, eds., Geodynamics of the Western Pacific-Indonesian Region: Amerian Geophyical Union Geodynamics Series v. 11, p. 137-142.

Parker, R. L., 1988, A statistical theory of seamount magnetism: Journal of Geophysical Research, v. 93, p. 3,105-3,115.

Parker, R. L., L. Shure, and J. A. Hildebrand, 1987, The application of inverse theory to seamount magnetism: Reviews of Geophysics, v. 25, p. 17-40.

Plouff, D., 1976, Gravity and magnetic fields of polygonal prisms and application to magnetic terrain corrections: Geophysics, v. 41, p. 717-741.

Rea, D. K. and T. L. Vallier, 1983, Two Cretaceous volcanic episodes in the western Pacfic Ocean: Geological Society of America Bulletin, v. 94, p. 1,430-1,437.

Richards, M. L., V. Vacquier, and G. D. Van Voorhis, 1967, Calculation of the magnetization of uplifts from combining topographic and magnetic surveys: Geophysics, v. 32, 678-707.

Sager, W. W., 1983, Seamount paleomagnetism and Pacific plate tectonics: Ph. D. dissertation, University of Hawaii, Honolulu, 472 p.

Sager, W. W., 1984, Paleomagnetism of Abbott Seamount and implications for the latitudinal drift of the Hawaiian hotspot: Journal of Geophysical Research, v. 89, p. 6,271-6,284.

Sager, W. W., 1987, Late Eocene and Maastrichtian paleomagnetic poles for the Pacific plate: implications for the validity of seamount paleomagnetic data: Tectonophysics, v. 144, 301-314.

Sager, W. W., G. T. Davis, B. H. Keating, and J. A. Philpotts, 1982, A geophysical and geologic study of Nagata Seamount, northern Line Islands: Journal of Geomagnetism and Geoelectricity, v. 34, p. 283-305.

Sager, W. W., D. W. Handschumacher, T. W. C. Hilde, and D. R. Bracey, 1988, Tectonic evolution of the northern Pacific plate and Pacific-Farallon-Izanagi triple junction in the Late Jurassic and Early Cretaceous (M21-M10): Tectonophysics, v. 155, p. 345-364.

Sager, W. W., and B. H. Keating, 1984, Paleomagnetism of Line Islands seamounts: evidence for Late Cretaceous and early Tertiary volcanism: Journal of Geophysical Research, v. 89, p. 11,135-11,151.

Sager, W. W., and M. S. Pringle, 1987, Paleomagnetic constraints on the origin and evolution of the Musicians and South Hawaiian seamounts, central Pacific Ocean, *in* B. H. Keating, P. Fryer, R. Batiza, and G. W. Boehlert, eds., Seamounts, Islands, and Atolls: American Geophysical Union Geophysical Monograph 43, p. 133-162.

Sager, W. W. and M. S. Pringle, 1988, Mid-Cretaceous to Early Tertiary apparent polar wander path of the Pacific plate: Journal of Geophysical Research, v. 93, p. 11,753-11,771.

Saito, K. and M. Ozima, 1977, ^{40}Ar-^{39}Ar geochronological studies on submarine rocks from the western Pacific area: Earth and Planetary Science Letters, v. 33, p. 353-369.

Schimke, G. R. and C. G. Bufe, 1968, Geophysical description of a Pacific Ocean seamount: Journal of Geophysical Research, v. 73, p. 559-569.

Schlanger, S. O., M. O. Garcia, B. H. Keating, J. J. Naughton, W. W. Sager, J. A. Haggerty, and J. A. Philpotts, 1984, Geology and geochronology of the Line Islands, Journal of Geophysical Research, v. 89, p. 11,261-11,272.

Schlanger, S. O., H. C. Jenkyns, and I. Premoli-Silva, 1981, Volcanism and vertical tectonics in the Pacific basin related to global Cretaceous transgressions: Earth and Planetary Science Letters, v. 52, p. 435-449.

Schlanger, S. O. and I. Premoli-Silva, 1981, Tectonic, volcanic, and paleogeographic implications of redeposited reef faunas of Late Cretaceous and Tertiary age from the Nauru Basin and Line Islands: Initial Reports of the Deep Sea Drilling Project, v. 61, p. 817-827.

Smith, W. H. F., H. Staudigel, A. B. Watts, and M. S. Pringle, 1989, The Magellan Seamounts: Early Cretaceous record of the south Pacific isotopic and thermal anomaly: Journal of Geophyisical Research, v. 94, p. 10,501-10,523.

Talwani, M., 1965, Computation with the help of a digital computer of magnetic anomalies caused by bodies of arbitrary shape: Geophysics, v. 30, p. 797-817.

Tamaki, K. and R. L. Larson, 1988, The Mesozoic tectonic history of the Magellan microplate in the western central Pacific: Journal of Geophysical Research, v. 93, p. 2857-2874.

Ueda, Y., 1985, Geomagnetic study on seamounts Daiiti-Kasima and Katori with special reference to a subduction process of Daiiti-kasima: Journal of Geomagnetism and Geoelectricity, v. 37, p. 601-625.

Ueda, Y., 1988, Geophysical study of two seamounts near Minami-Tori Sima (Marcus) Island, western Pacific Ocean: Journal of Geomagnetism and Geoelectricity, v. 49, p. 1,481-1,501.

Uyeda, S. and M. L. Richards, 1966, Magnetization of four Pacific seamounts near the Japanese islands: Bulletin of the Earthquake Research Institute, Tokyo, v. 44, p. 179-213.

Vacquier, V., 1962, A machine method for computing the magnitude and direction of magnetization of a uniformly magnetized body from its shape and a magnetic survey, *in* T. Nagata, ed., Benedum Earth Magnetism Symposium: Pittsburgh, Univerity of Pittsburgh Press, p. 123-127.

Vacquier, V and S. Uyeda, 1967, Paleomagnetism of nine seamounts in the western Pacific and of three volcanoes in Japan: Bulletin of the Earthquake Research Insitute, Tokyo, v. 45, p. 815-848.

Verhoef, J., B. J. Collette, and C. A. Williams, 1985, Comment on "Hawaiian hotspot volcanism mainly during geomagnetic normal intervals": Geology, v. 13, p. 314-315.

Watts, A. B., J. H. Bodine, and N. M. Ribe, 1980, Observations of flexure and the geological evolution of the Paficic Ocean Basin: Nature, v. 283, p. 532-537.

Williams, C. A., J. Verhoef, and B. J. Collette, 1983, Magnetic analysis of some large seamounts in the north Atlantic: Earth and Planetary Science Letters, v. 63, p. 399-407.

Winterer, E. L. and C. V. Metzler, 1984, Origin and subsidence of Guyots in the Mid-Pacific Mountains: Journal of Geophysical Research, v. 89, p. 9,969-9,979.

Yamazaki, T., 1988, Magnetization of Erimo Seamount: Journal of Geomagnetism and Geoelectricity, v. 40, p. 715-728.

Appendix 1. Seamount Paleomagnetic Parameters.

ID	Name	Location Lat. (°N)	Location Long. (°E)	Pole Lat. (°N)	Pole Long. (°E)	Inc.	Dec.	Inten. U/NU	GFR	Age Pmag	Age Other	Error Ellipse Maj	Error Ellipse Min	Error Ellipse Az	Type	Reference Pmag	Reference Age
Central Pacific Basin																	
1	Magnet	12.3	173.2	61.0	31.2	-20.9	342.3	9.5	3.9	LK					L	1	
2	Dixon	12.6	180.9	68.0	1.0	-18.9	0.0	6.7	5.6	LK					L	2	
3	L1	6.2	186.0	69.7	27.5	-24.3	352.5	7.1	2.8	LK					L	3	
4	L2	2.7	186.0	66.5	5.5	37.3	180.2†	7.5	3.8	LK					L	3	
5	L3	1.0	180.5	54.9	324.1	-45.1	22.4	10.5	4.3	MK					L	3	
6	Unnamed	14.9	187.7	55.0	1.8	-35.9	3.6	3.8	2.0	MK	83P				L	4	1
East Pacific																	
7	Moonless	31.9	218.2	79.4	41.3*	38.0	359.4	10.2	4.2	ET					L	6	
8	39131	39.0	229.0	78.0	23.0*	47.1	5.9	1.3	2.0	ET					L	7	
Hawaiian Chain																	
9	Abbott	31.8	174.3	75.5	4.6*	-32.2	177.3†	3.1	6.6	ET	39.5A				L	8	2
10	Colahan	31.0	176.0	76.8	333.0*	34.3	5.4	9.2	2.1	ET	38.5A				L	9	2
Japan Area																	
11	Unnamed	41.3	146.0	49.4	337.1	2.4	352.8	15.6	1.8	MK					L	10	
12	Erimo	40.9	144.9	69.0	321.0	36.0	1.0	5.8	5.7	EK	130L				L	11	3
13	Ryofu	38.0	146.0	53.0	352.7	2.2	343.2	9.4	3.4						L	10	
				46.2	353.9	-4.3	341.1	7.4/1.2	13.0	MK		7.0	2.4	113	S	12	
14	Unnamed	36.0	143.5	40.9	21.6	7.0	320.0	8.3	2.7	K					L	3	
15	Maiko	34.0	145.9	50.5	327.1	-10.9	359.3	12.2	4.7						L	3	
				53.1	321.1	-5.5	2.9	13.4/4.2		MK		4.7	2.1	113	S	5	
16	Takuyo-Daini	34.3	143.9	75.4	315.7	35.8	2.2	5.0/2.6		EK	>98F	5.6	3.0	88	S	5	
17	Jenson	33.8	143.8	55.0	340.5	0.9	350.7	19.4/8.3		MK		5.9	2.6	102	S	5	
18	Jenson N	34.1	143.9	68.6	341.7	25.7	353.3	23.1/9.6		LK		18.0	8.5	103	S	5	
19	Seiko	34.2	144.3	58.2	328.2	4.9	357.9	11.1/6.8		MK	102A	4.3	1.9	96	S	5	4
20	Katori	36.1	143.0	59.9	354.1	18.3	344.8	13.9	2.8	MK					L	13	
21	Daiiti-Kashima	35.8	142.7	62.3	348.8	20.1	348.0	10.7	3.1	MK					L	13	
Line Islands																	
22	Watkins	17.5	190.8	68.3	38.7	-3.9	350.0	5.8	5.2	LK					L	14	
23	Nagata	12.5	193.0	65.6	6.1*	-29.0	4.4	3.8	3.7						L	15	
				67.1	6.3	-19.9	2.6	5.4/2.6		LK	83P	9.2	4.4	62	S	12	1
24	Kapsitotwa	12.0	194.2	47.5	333.5*	-36.6	28.0	5.1	3.7						L	3	
				58.1	330.5	-21.8	21.8	6.1/3.5		MK	84.4A	10.1	4.0	47	S	12	5
25	Stanley	8.2	198.1	75.6	356.5*	-10.5	5.3	3.2	4.7	ET	39.3A				L	16	6
26	Willoughby	7.9	198.1	78.2	14.6*	-7.8	0.7	3.5	4.4	ET					L	16	
27	Chapman	3.4	199.9	75.7	37.8	-19.7	355.6	4.7-8.0	4.3	ET					L	16	
28	Clarke	-3.3	206.0	80.0	20.4	-25.3	1.0	3.3	4.0	ET					L	16	
29	Wageman	-7.5	208.5	68.5	345.6*	40.1	195.7†	6.9	3.1	LK	71.9A				L	16	6
Mid-Pacific Mountains																	
30	Sio	18.0	171.2	50.9	277.9	7.7	37.2	3.6	2.0	EK					L	1	
31	Harvey	17.8	172.7	68.1	21.9	-3.0	348.5	4.3	2.7	LK					L	1	
32	Thomas	17.3	173.9	73.3	324.2	5.3	8.2	3.5	2.5	EK	165L				L	1	7
33	Allen	18.3	174.1	69.2	328.6	-1.2	8.8	4.1	3.6	EK					L	1	
34	Darwin	22.1	171.6	39.9	316.8	36.2	207.8†	1.9	2.4						L	3	
				43.4	316.6	31.5	205.9†	2.0/0.9		EK	109F	15.5	5.9	55	S	12	8
35	Unnamed	21.2	173.8	84.4	321.5	30.6	3.1	2.0/1.4		?		11.1	3.3	52	S	5	

(continued)

ID	Name	Location Lat. (°N)	Location Long. (°E)	Pole Lat. (°N)	Pole Long. (°E)	Inc.	Dec.	Inten. U/NU	GFR	Age Pmag	Age Other	Error Ellipse Maj	Error Ellipse Min	Error Ellipse Az	Type	Reference Pmag	Reference Age
Musicians Seamounts																	
36	Berlin	32.9	194.0	59.4	358.8	5.1	7.1	12.1	3.9	MK					L	17	
37	Mahler	31.8	195.0	56.0	342.6*	4.9	17.5	6.0	6.7	MK	87.8A				L	17	1
38	Mussorgski	30.4	196.1	58.2	354.5	0.9	11.2	5.6	3.3	MK					L	3	
39	Rachman-inov	29.6	196.7	55.6	324.6*	11.9	26.6	9.3	2.5	MK	86.2A				L	3	1
40	Paganini	28.7	197.4	67.2	26.9	12.2	356.3	8.3	3.2	LK					L	17	
41	Khatch-aturian	28.1	197.7	56.5	332.5	5.9	23.0	12.8	5.3	MK	81.7A				L	3	9
42	Schubert	31.9	197.9	65.7	355.3	17.9	9.2	6.4	4.3	LK					L	17	
43	Brahms	31.2	197.9	66.3	348.2	19.8	11.6	7.4	2.0	LK	88.9A				L	3	9
44	Debussy	30.3	197.9	67.3	3.1	16.2	5.7	5.9	3.2	LK					L	17	
45	Tchaikov-ski	29.4	197.7	62.4	356.6	6.7	9.6	7.8	3.4	LK					L	17	
46	Liszt	29.0	197.7	59.2	333.8*	10.6	20.9	7.0	5.1						L	17	
				58.5	331.6	11.0	22.3	7.9/2.6		MK	84.4A	6.6	2.2	50	S	5	1
47	Handel	27.5	200.1	69.3	10.2	14.0	3.5	4.2	7.4	LK					L	17	
48	Rimski-Korsakov	25.3	200.3	70.1	0.0	12.8	6.8	2.9	9.3	LK					L	17	
49	Gluck	26.9	199.9	64.9	9.5	4.3	4.4	6.7	5.2	LK					L	17	
50	Mendels-sohn-W	25.1	197.2	57.8	3.1*	-12.4	7.5	9.7	3.3						L	17	
				63.5	355.6	0.9	9.7	9.6/8.2		LK	82.4A	12.9	4.2	65	S	5	1
51	Mendels-sohn-E	25.1	198.3	68.3	14.3*	6.8	1.5	7.3	3.5						L	17	
				67.3	18.1	5.5	356.2	7.4/4.9		LK	78.5A	9.7	3.3	78	S	5	1
52	Haydn	26.6	198.7	69.2	357.9*	13.8	7.3	7.7	3.3	LK	75.1A				L	17	1
53	Schumann-W	25.7	199.8	59.2	349.9*	-3.1	14.8	3.3	2.3						L	17	
				63.3	10.7	-1.4	4.1	3.3/2.3		LK	83.0A	13.8	4.1	78	S	5	1
54	Schumann-E	25.9	200.1	55.6	343.3	-5.3	19.8	4.4	2.3						L	17	
				60.8	357.8	-2.9	10.6	4.7/3.7		LK		12.5	4.0	85	S	5	
55	Donizetti-W	32.2	199.7	55.3	333.2	10.8	23.6	5.6	3.5	LK					L	17	
56	Donizetti-E	32.2	200.0	58.9	23.4	2.3	358.1	3.0	5.0	MK					L	17	
57	Bizet	32.3	198.4	50.1	32.1	-15.0	2.8	4.3	3.6	K?					L	17	
South Hawaiian Area																	
58	Kaluaka-lana	23.3	201.6	85.2	289.9	40.8	5.2	2.0	5.7	?	80.5A				L	17	9
59	Chatauqua	22.2	197.4	60.0	358.7*	12.7	189.3†	4.8							L	18	
				57.3	18.2	20.1	179.6†	4.2/1.2		LK	80P	9.7	3.6	95	S	5	1
60	HD1	18.3	198.2	52.0	342.0*	-24.2	21.8	1.9	4.7						L	7	
				50.8	342.0	-26.4	22.6	2.6/0.2	4.9	MK	89.4K	9.6	4.1	54	S	12	10
61	HD4	20.0	201.8	66.0	40.0	-6.7	350.7	4.3	3.4	LK					L	7	
62	Finch	17.7	202.3	68.0	10.0	-7.7	4.6	8.3	4.4	LK					L	17	
63	Finch-SW	17.4	202.1	65.5	16.6	-13.8	2.3	4.1	3.0	LK					L	17	
64	Kona 4N	17.3	205.8	64.6	352.8	-8.4	13.5	5.8	2.9	LK					L	6	
65	Kona 5S	17.1	205.8	49.8	1.5*	35.3	196.4†	0.9	2.9	MK	81P				L	6	1
66	Show	17.9	207.3	56.5	350.1*	18.0	199.8†	3.8	3.2						L	6	
				57.1	355.3	20.1	197.0†	4.9/1.0		MK	81P	6.2	2.7	63	S	5	1
67	Wini	19.0	206.2	70.9	17.1	0.2	2.9	3.9	2.3	LK					L	6	
68	Unnamed	26.5	182.2	68.0	8.9	9.2	357.5	4.7	3.6						L	9	
				61.3	18.9	-2.4	352.1	3.9/2.0		LK	74.3K	9.2	2.9	104	S	12	11
69	Paumakua	24.9	202.9	67.7	1.8*	-7.8	187.0†	4.0	4.9						L	17	
				64.8	352.1	-5.6	192.6†	4.2/0.4		LK	65.5A	6.8	1.8	63	S	12	9
70	Cross	18.5	202.0	72.4	19.4	1.8	0.8	2.7	4.2	LK	84.6A				L	17	9
71	McCall	18.8	202.8	69.2	35.0	-3.1	355.7	4.5	2.1	LK	82.7A				L	17	9
72	Apuupuu	18.6	204.6	71.2	11.2	0.5	4.3	6.7	2.5	LK					L	5	

(continued)

ID	Name	Location Lat. (°N)	Location Long. (°E)	Pole Lat. (°N)	Pole Long. (°E)	Inc.	Dec.	Inten. U/NU	GFR	Age Pmag	Age Other	Error Elipse Maj	Error Elipse Min	Error Elipse Az	Type	Reference Pmag	Reference Age
Western Pacific																	
73	Z41	28.8	148.4	55.0	19.0	9.0	334.0	1.8	2.3	MK					L	19	
74	Z42	28.4	148.2	53.0	278.0	5.0	28.0	4.2	2.6						L	19	
				59.9	292.0	6.1	17.2	4.1/1.6	4.0	EK		7.9	2.7	59	S	12	
75	Z43	27.1	148.7	55.9	323.0*	-13.7	3.2	9.2	3.3						L	4	
				57.1	334.1	-11.3	357.0	8.4/5.1		MK	93.7K	6.7	2.5	93	S	5	4
76	Z44	27.8	146.7	60.0	306.0	-1.0	11.0	3.0	1.8	EK					L	19	
77	Makarov	29.5	153.5	63.7	331.3*	6.4	1.0	8.4	3.4						L	3	
				59.2	346.2	-1.3	353.5	8.9/2.1		MK	94A	2.9	0.9	101	S	5	4
78	Miami	21.7	161.9	51.7	341.5	-30.8	0.3	7.5	2.6						L	3	
				58.9	5.9	-12.2	348.1	8.1/4.7		MK		6.5	2.1	116	S	12	
79	Birdseye	20.9	165.7	58.6	342.3	-20.4	1.8	13.0	8.1	MK					L	1	
80	Woods Hole	21.2	166.5	55.2	349.5	-25.7	358.2	11.2	3.2						L	3	
				53.2	319.7	-23.2	16.0	8.0/4.4		MK		6.8	2.0	61	S	5	
81	Unnamed	19.4	165.9	58.4	5.9	-20.3	349.5	6.2	2.2	MK					L	3	
82	Unnamed	16.7	162.4	71.0	281.3	13.7	16.7	1.6	5.5	EK					L	1	
83	Seascan	15.1	159.3	57.0	4.8	-27.9	346.0	8.9	4.2	MK					L	1	
84	Unnamed	15.5	153.3	71.7	246.5	26.0	18.8	1.8	9.5	?					L	1	
85	Campbell	16.5	149.0	55.0	312.9	31.8	189.6†	4.8	2.4	EK	>118P				L	1	
86	Unnamed	11.2	146.8	74.5	6.4	-1.7	350.2	7.5	2.7	ET					L	1	
87	Winchester	10.3	156.7	55.8	296.6	-29.8	22.1†	2.8	4.1						L	1	
				55.3	286.0	-23.0	26.8	2.7/0.5		EK		8.6	3.2	38	S	12	
88	Heezen	8.8	163.2	57.4	340.3	-41.3	1.7	5.4	3.4	MK					L	1	
89	Von Valtier	7.3	172.3	63.2	2.2	-34.7	355.3	7.4	4.4		80F				L	4	12
				54.2	307.2	-32.7	25.8	7.6/5.5	13.6	EK		9.5	4.4	43	S	12	
90	Golden	21.3	153.2	57.3	330.1*	-21.9	1.7	4.5	3.7						L	4	
	Dragon			56.2	337.4	-23.8	357.6	3.4/0.8		MK	95A	5.6	1.9	98	S	5	4
91	Unnamed	24.3	152.3	74.2	276.9	27.7	13.4	3.4	5.5	EK	78A				L	5	4
92	Unnamed	25.1	151.8	46.2	323.7	33.7	185.5†	3.9	5.6	MK	>118P				L	5	
93	Ita Mai Tai	12.9	156.9	48.2	353.9	-45.7	347.3	1.0/6.7		MK	>110F	19.1	10.6	48	S	5	13
94	Guadaloupe	27.8	168.8	62.4	314.6	-8.5	195.1†	2.6/4.4		EK	>118P	16.2	7.3	71	S	5	
95	Wilde	21.2	163.3	61.8	333.1	-13.2	4.9	7.3/3.9		MK	86A	7.1	2.1	82	S	5	4
96	UCSD	24.3	159.2	55.3	338.4	20.3	180.5†	6.9/0.6		MK		9.2	3.7	85	S	5	
97	Scripps	23.7	159.3	62.0	335.7	-8.5	1.7	6.2	2.2	MK	98A				L	5	4
98	Smt948	23.7	151.9	57.5	329.5	-17.5	1.3	5.1	2.8	MK					L	20	
99	Isakov	31.5	151.2	52.3	331.5	-12.1	-0.2	10.5/3.5		MK	>98F	4.2	1.7	88	S	5	
100	Isakov-NW	31.8	150.9	58.0	311.9	2.4	10.0	16.1/7.2		EK		9.4	3.4	74	S	5	

† Indicates reversed polarity
* Pole used for apparent polar wander path calculations (Sager and Pringle, 1988)

Table Heading Abbreviations:

Inc., inclination; *Dec.*, declination; *Inten.*, magnetization intensity (*U*, uniform; *NU*, nonuniform) units in amperes/meter; *GFR*, goodness-of-fit ratio; *Error Ellipse*, 95% confidence region; *Maj*, major semi-axis length (in degrees) ; *Min*, minor semi-axis length (in degrees); *Az*, azimuth of major semi-axis (degrees clockwise from north).

Type, inverse technique; L, least squares minimization; S, seminorm minimization; *Ref*, reference number (see below)

Age Abbreviations:

Paleomagnetic ages: ET, Early Tertiary; *LK,* Late Cretaceous; *MK,* mid-Cretaceous; *EK,* Early Cretaceous-Late Jurassic. *Other age types: A,* ^{40}Ar-^{39}Ar; *K,* Potassium-Argon; *F,* Fossil; *L,* Lithospheric flexure; *P,* Magnetic polarity.

Paleomagnetic References:

(1) Sager, 1983; (2) Francheteau et al., 1969; (3) Harrison et al., 1975; (4) Sager and Pringle, 1988; (5) This study; (6) Francheteau et al., 1970; (7) Richards et al., 1967; (8) Sager, 1984; (9) Sager, 1987; (10) Uyeda and Richards, 1966; (11) Yamazaki, 1988; (12) Hildebrand and Parker, 1987; (13) Ueda, 1985; (14) Keating and Sager, 1980; (15) Sager et al., 1982; (16) Sager and Keating, 1984; (17) Sager and Pringle, 1987; (18) Schimke and Bufe, 1968; (19) Vacquier and Uyeda, 1967; (20) Ueda, 1988.

Age References:

(1) Sager and Pringle, 1988; (2) Duncan and Clague, 1984; (3) Yamazaki, 1988; (4) Ozima et al., 1983; (5) Saito and Ozima, 1977; (6) Schlanger et al., 1984; (7) Watts et al., 1980; (8) Harrison et al., 1975; (9) Sager and Pringle, 1987; (10) Dymond and Windom, 1968; (11) Garcia et al., 1987; (12) J. A. Haggerty, personal communication, 1982; (13) Baltuck et al., 1986.

Keating, B.H., and Bolton, B.R., editors, 1992, Geology and offshore mineral resources of the central Pacific basin, Circum-Pacific Council for Energy and Mineral Resources Earth Science Series, V. 14. New York, Springer-Verlag.

ISOSTASY AND TECTONIC ORIGINS OF PACIFIC SEAMOUNTS

Jeffrey T. Freymueller and James N. Kellogg
Department of Geological Sciences University of South Carolina, Columbia, SC 29208

ABSTRACT

Previous studies have shown that the response of the Pacific plate to oceanic island and seamount loads can be used to estimate the distribution of ridge-crest, superswell, and normal mid-plate volcanism. Similarly, estimation of the degree of Airy isostatic compensation can be used to predict the tectonic origin of an oceanic island or seamount. In this study 150 Pacific seamounts, each surveyed with at least three sea surface profiles, are classified as ridge-crest or superswell, mid-plate, or uncertain origin based on bathymetry and on-axis gravity. The results correlate well with those obtained from three-dimensional flexural and isostatic studies. Most of the analyzed seamounts rimming the Central Pacific Basin are locally isostatically compensated, i.e., ridge crest or superswell, including the Marshall Islands (10), the Mid-Pacific Mountains (5), the Line Islands (6), Tuvalu (1), and Phoenix seamounts (1). Sixteen compensated edifices were estimated on 150–170 Ma Jurassic sea floor in the Japanese (2), Magellan (3), Marcus Wake (1), and Marshall seamounts (10), but reef fossil and radiometric age determinations show no evidence of ridge crest eruptions during the Jurassic. The gravity results are therefore compatible with recent flexural studies, geochemical results, seafloor depths, age dating, and plate rotation models for other edifices in these seamount clusters that suggest eruption from a South Pacific isotopic and thermal anomaly or superswell during the Cretaceous (80–120 Ma). Recent radiometric age determinations and geochemistry indicate that four Late Cretaceous Marshall seamounts were erupted from mantle sources that are presently in the South Pacific. Our determination of 10 isostatically compensated edifices suggests that many of the Marshall seamounts may have been erupted from the South Pacific isotopic and thermal anomaly. The passage of Jurassic lithosphere over at least four superswell hot spots may also explain the apparent multiple episodes of volcanism in the Marshall Islands.

INTRODUCTION

Progress has been made in the last 15 years in relating the gravity anomalies of oceanic seamounts to their tectonic origins. Walcott (1970) quantitatively examined the hypothesis that the Hawaiian deep and arch were the product of lithospheric flexure, and computed the flexural rigidity of the lithosphere near Hawaii. To first order, the effective elastic thickness (T_e) of the oceanic lithosphere under mid-plate volcanoes increases linearly with the square root of plate age at the time of loading, in agreement with the predicted relationship based on a cooling plate model (Watts, 1978; Watts et al., 1980). The T_e value is approximately given by the depth to the 550 isotherm (Mc-Nutt, 1984). Watts (1978) interpreted low T_e values for the lithosphere at a seamount as indicative of a ridge-crest tectonic origin for the seamount, and high T_e values as indicative of an off-ridge origin. Under many of the mid-plate volcanic chains of the south-central Pacific, however, the elastic layer is found to be much thinner than predicted by the simple square root of age law (Calmant and Cazenave, 1986 and 1987). This broad regional thermal anomaly or superswell is associated with anomalously shallow seafloor depths (McNutt and Fischer, 1987), a shallow low velocity zone for Love waves (Nishimura and Forsyth, 1985), and volcanoes with a geochemical signature termed the Dupal anomaly (Hart, 1984). Volcanic rocks dredged from two Early Cretaceous volcanic edifices in the Magellan Seamounts (Smith et al., 1989) and four Late Cretaceous volcanic edifices in the Ratak chain of the Marshall Islands

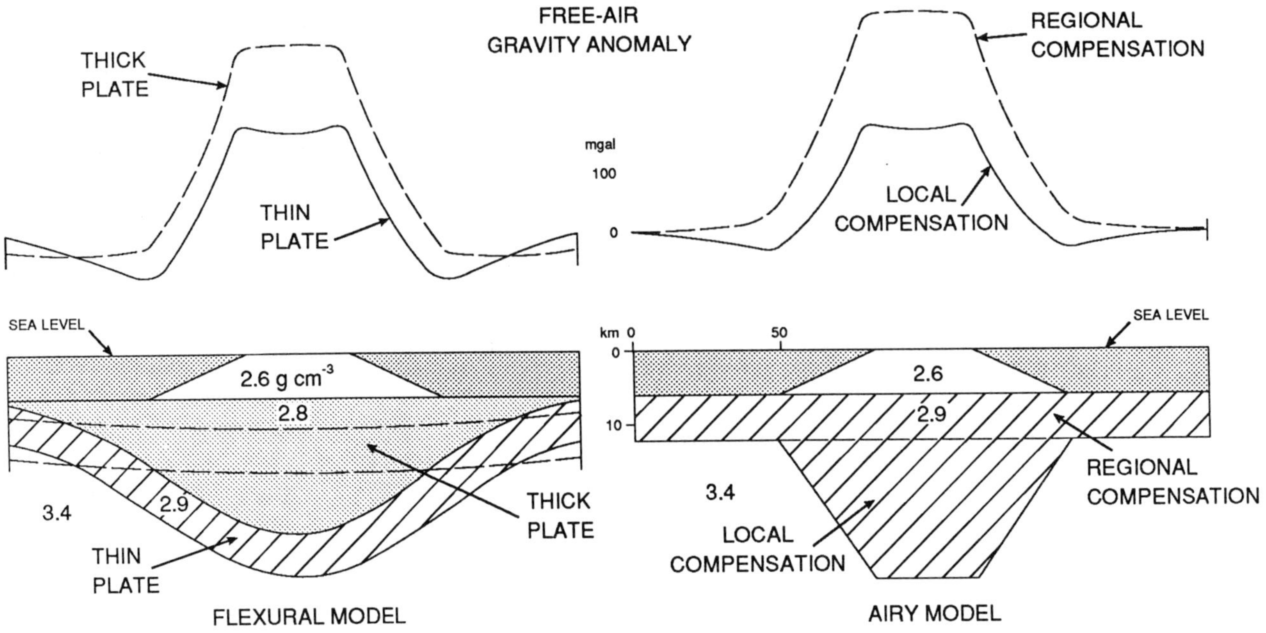

Figure 1. Schematic comparison of flexural and Airy models of seamount isostatic compensation.

(Davis et al., 1989) show the same distinctive geochemical mantle signature. Lithosphere under five mid-Cretaceous guyots surveyed in the Japanese Seamounts exhibits low T_e values (Wolfe and McNutt, submitted). The geochemical data, low T_e values, anomalously shallow Cretaceous seafloor depths, and plate rotation models indicate that these edifices in the Japanese and Magellan seamounts and Marshall Islands formed at the South Pacific isotopic and thermal anomaly or superswell during the Cretaceous (80–120 Ma). Low T_e values for the lithosphere at a seamount may therefore be indicative of either a ridge-crest or superswell tectonic origin for the seamount.

If the effective elastic thickness of the lithosphere is small, then there is a high level of local Airy isostatic compensation, i.e., the excess mass of a seamount is compensated to a large degree by a mass deficiency directly below the seamount (Fig. 1). If the effective elastic thickness of the lithosphere is large, the compensation is regional in nature and the level of local compensation is low. The degree of local compensation (Schwank and Lazarewicz, 1982) can be interpreted in a manner analogous to the interpretation of effective elastic thickness by Watts (1978), which permits a first-order estimate of the tectonic origin of a seamount. In this study, we estimate the degree of local Airy compensation of a seamount from its bathymetry and maximum gravity anomaly. Sea surface data were used to avoid the uncertainties introduced by Seasat's wide track spacing and errors in digital bathymetric data bases. Seamounts were modeled as cones or frustums of cones and the on-axis anomalies were computed (Appendix). The degree of compensation was estimated from the difference between the observed anomaly and the calculated anomaly for the bathymetry alone.

Using only the on-axis anomaly permits rapid determinations of the tectonic origins of a large number of seamounts by minimizing the amount of data and computation required. The on-axis partial Airy compensation method cannot replace a detailed three-dimensional gravity study for a single seamount. However, it gives a valuable first-order approximation of the tectonic origins of the seamount, and as discussed in the results section of this paper, our estimates correlate very well with those obtained from three-dimensional flexural and isostatic studies. In this study 150 Pacific seamounts are classified as ridge-crest or superswell, normal mid-plate, or uncertain origin based on their computed degrees of local compensation. Regional groupings in the degree of local compensation give insight into the tectonic origins of seamount clusters or chains.

DATA ANALYSIS

The degree of local compensation (μ) was determined from the observed on-axis gravity anomaly and the bathymetry. The seamount body and the root were modeled as frustums of cones (Fig. 2) using an exact analytical expression for the on-axis gravity anomaly (Appendix). The density contrast across the crust-mantle interface and uniform densities for the seamount edifice and root were assumed (Table 1). Bouguer corrections with a reduction density of 2.3 g/cm^3 were applied to data from islands, and

Table 1. Input parameters and typical values and uncertainties. The bottom three parameters were held fixed for all seamounts.

Parameter	Symbol	Value (range)	Uncertainty (range)
Top radius (m)	a_0	(0–60000)	(1000–5000)
Bottom radius (m)	a_1	(20000 –80000)	(1000–5000)
Minimum depth (m)	z_0	(0–2500)	200
Maximum depth (m)	z_1	(3500–6000)	(200–300)
Observed gravity (mgal)	g_{obs}	(45–50)	10
Density contrast rock-water (g/cm^3)	$\Delta\rho_{rw}$	1.57	0.15
Density contrast crust*-mantle (g/cm^3)	$\Delta\rho_{cm}$	−0.4	0.15
Thickness of normal crust (m)	z_{crust}	6000	1000

* oceanic crust

the islands were considered to be cut off at sea level. About half of the data set (mostly islands and atolls) was taken from Wedgeworth (1985).

The marine data were mostly from surveys carried out by research vessels of the Hawaii Institute of Geophysics (University of Hawaii), although data from several other institutions were also used. Ship tracks were checked to ensure that a complete survey of the seamount was performed, and the maximum observed free air anomaly measured from a profile was taken to be the on-axis anomaly. It is unlikely that any gravity anomaly will be overestimated by more than 10 mgal, a conservative upper bound on the error of all of the gravity surveys due to navigational uncertainties. However, it is possible that the maximum observed anomaly is significantly smaller than the actual on-axis anomaly, resulting in an overestimate of μ. This could happen if the trackline did not pass over the summit of the seamount. This potential source of error has been minimized by using sea surface data instead of Seasat data and choosing only surveys which made at least three passes over the seamount.

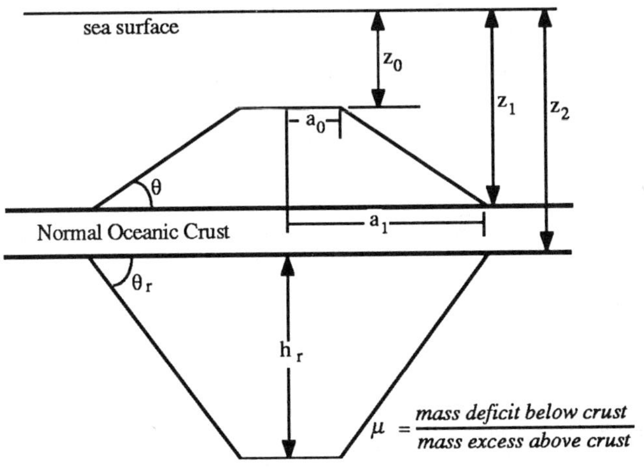

Figure 2. Airy compensation model of a seamount with a low-density root underlying the seamount. The degree of compensation μ is defined to be the mass deficit below the crust divided by the mass excess above the crust (Schwank and Lazarewicz, 1982).

Uncertainties were estimated by calculating numerical derivatives of the degree of compensation with respect to eight parameters, as described in the Appendix. In general, the uncertainties were larger for small seamounts whose low amplitude gravity signatures were difficult to invert.

RESULTS

A complete listing of the estimated Airy compensation values (μ) is given in Table 2. The existence of values outside the range of 0–100% suggests that the approximation of a uniform density conical seamount and root was not always good. However, only 6.6% of the seamounts in this study had estimated values of more than one sigma outside the 0–100% range; none had values of more than two sigma outside that range.

Figure 3 shows a plot of the estimated local compensation values vs. age of loading for eleven well-dated seamounts (Table 3). The solid curve in Figure 3 is based on the deflection of a two-dimensional continuous elastic plate subjected to a line load (Turcotte and Schubert, 1982) and the empirical relationship between effective elastic thick-

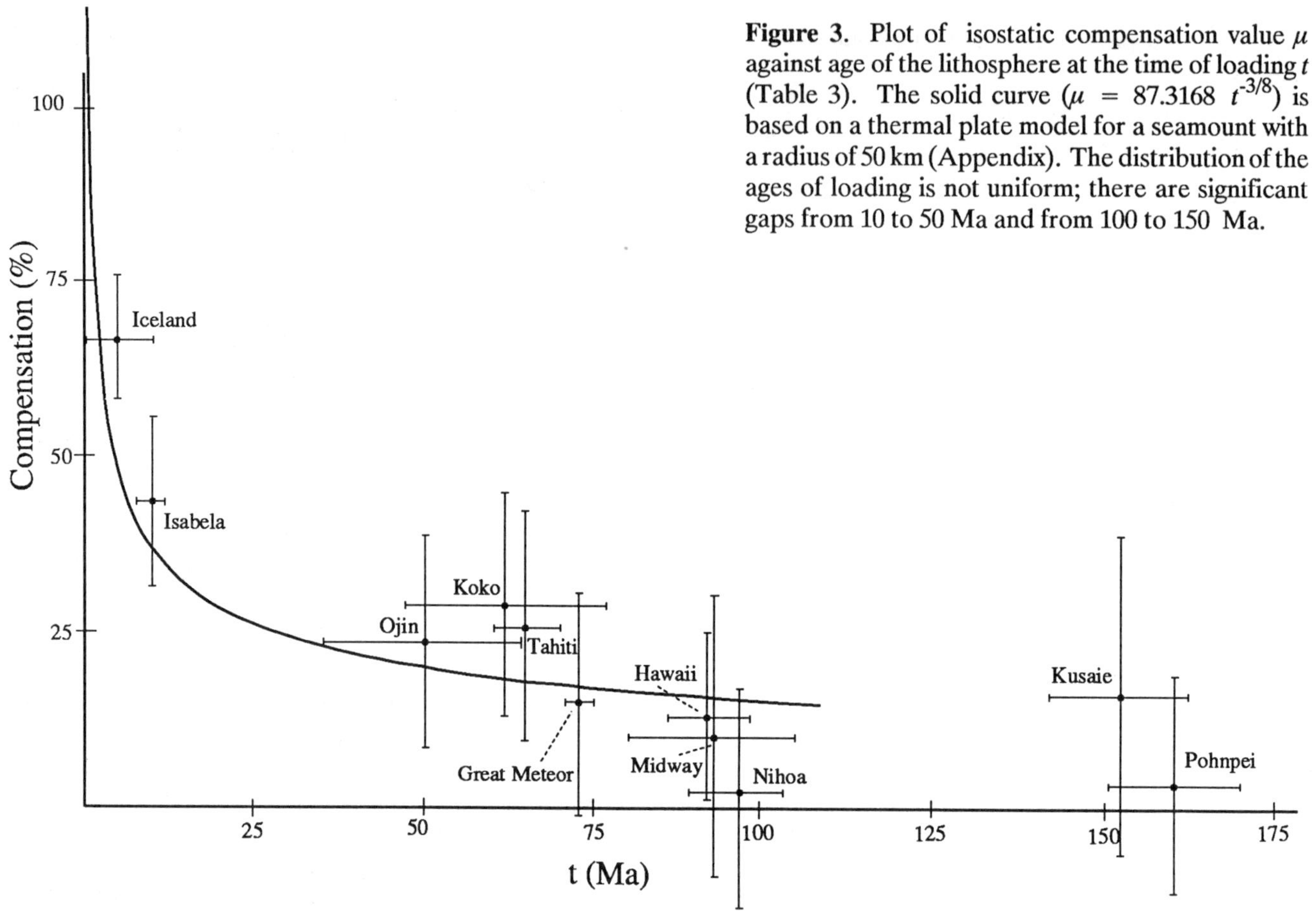

Figure 3. Plot of isostatic compensation value μ against age of the lithosphere at the time of loading t (Table 3). The solid curve ($\mu = 87.3168\ t^{-3/8}$) is based on a thermal plate model for a seamount with a radius of 50 km (Appendix). The distribution of the ages of loading is not uniform; there are significant gaps from 10 to 50 Ma and from 100 to 150 Ma.

ness and age of loading (Watts, 1978; McNutt, 1984), with μ in percent and t in millions of years (Appendix):

$$\mu = 87.3\, t^{-3/8}\ (\%) \tag{1}$$

The estimates are consistent with the theoretical curve predicted by thermal models. There is a consistent trend to lower values of μ as the age of loading increases. Following Watts (1978) and Calmant and Cazenave (1986 and 1987), we associate seamounts with high degrees of local Airy compensation with near-ridge or superswell volcanism, and seamounts with low degrees of local compensation with mid-plate volcanism.

All 150 Pacific seamounts studied were classified as either ridge-crest or superswell, mid-plate, or uncertain origin (Fig. 4), based on their degrees of local compensation. If the estimated degree of compensation was greater than 40%, which corresponds to an age of loading of approximately 8 Ma (Fig. 3), the seamount was classified as ridge-crest or superswell. Seamounts whose compensation estimates were less than 40% were classified as normal mid-plate. If the uncertainty range of the compensation estimate crossed 40%, the seamount was classified as uncertain.

We recognize that although data for a large number of seamounts have been used in this study, the number of estimates (150) is small compared with the total number of oceanic islands and seamounts that exist in the western Pacific. The agreement with previous flexural and isostatic studies and the occurrence of several similar estimates from the same seamount cluster or chain is evidence that the overall distribution of ridge crest or superswell and mid-plate estimates (Fig. 4) is representative.

Normal Mid-Plate Volcanism

The Hawaiian seamount chain has been interpreted as a mid-plate hotspot trace (Fig. 4). Except for Ladd Bank, Pioneer Bank, and two seamounts west of Midway Island with compensation values ranging from 26% to 45%, all seamounts in the chain are 16% locally compensated or less. The compensation estimates for the entire chain are consistent with the formation of the chain by a mid-plate hotspot. Estimates of Moho depths from calculated values of μ are sensitive to assumptions of the crust-mantle density contrast, conical edifice approximations, and homogeneous edifice densities. Comparison of our predictions with those determined by seismic refraction suggest that we have

Table 2. Input data and calculated results for 150 seamounts in the central and western Pacific. Input parameters are defined in Table 1. Gravity and bathymetric data were obtained from cruises from several institutions, as well as from published sources. Sources for bathymetric data are indicated in the table by (*, †, or ‡). If there is no mark beside any bathymetric datum, that datum was obtained directly from the shipboard gravity profile.

Seamount		Latitude	Longitude	FAA	z_0	z_1	a_0	a_1	μ	Source
Japanese-Magellan-Marcus Wake										
1		28.10°N	153.10°E	242	1350	5750	2000†	23000†	−35 ± 27	KK77031705
2	Z4/3	27.05°N	148.53°E	178	1125	5800	0	50000	58 ± 6	V3312
3		25.07°N	151.80°E	52	2400	5400	0	14000	176 ± 102	V3214
4		24.25°N	152.25°E	100	1750	5300	2500†	18500†	149 ± 98	V3214
		23.80°N	148.75°E	210	1000	5500	4400	36000	32 ± 28	UOTKH7104
6		21.55°N	151.61°E	97	2400	5750	2000	34000	94 ± 39	V3312
7		21.50°N	159.50°E	200	1151	5500	17500†	50000†	36 ± 19	Wedgeworth
8		21.30°N	153.25°E	260	1125	5900	8000	44000	7 ± 21	V3214
9		19.70°N	153.35°E	240	1300	5900	7000	56000	16 ± 20	V3214
10		19.50°N	162.33°E	125	1136	5125	5500†	28000†	135 ± 64	Wedgeworth
11		19.50°N	147.70°E	63	2625	6625	12400	25000†	306 ± 164	KK81062603
12		19.25°N	164.85°E	206	1400	5500	16000	46000†	22 ±19	KK81062602
13	Wake	19.25°N	166.67°E	275	0	5300	4000†	42500†	24 ±25	Wedgeworth
14		16.50°N	148.93°E	143	1600	5400	2000	26000	55 ± 43	V3401
15		15.70°N	152.00°E	285	1610	6050	16500	64000	0 ±17	KK81062603
16		15.50°N	153.10°E	71	2125	6000	2000	20000	282 ± 127	V3401
Shatsky Rise										
17	Shatsky Rise	33.00°N	158.00°E	50	2050	6000	42500†	325000	92 ±7	
E. Marianas Basin										
18	Zubov	15.67°N	160.00°E	255	1100	5600	2500†	40000†	−8 ±23	Wedgeworth
19	Seascan	15.33°N	158.75°E	195	1157	5500	3500†	30000†	33 ±33	Wedgeworth
20		15.10°N	159.25°E	250	1100	5880	5600	40000	11 ±23	KK81062602
21		5.25°N	158.10°E	85	1600	3500	6000	22000	35 ±44	LDGO-V2813
22		14.00°N	157.67°E	260	1250	6000	4250†	23500†	22± 25	Wedgeworth
23	Ita Mai Tai	13.00°N	156.80°E	255	1425	5800	18330	47000	1 ±18	KK81062602
24		12.25°N	156.33°E	175	1755	5150	3500†	29500†	3± 29	Wedgeworth
25		11.40°N	156.70°E	203	1375	5600	5000	35000†	22 ± 27	KK81062602
26		11.17°N	156.83°E	205	1550	5900	3000†	46500†	24 ±24	Wedgeworth
27		10.42°N	156.75°E	170	1750	5500	9000†	33000†	32± 27	Wedgeworth
28		10.35°N	158.70°E	168	1550	5500	4000	26800	37 ±36	KK81062602
Caroline Islands										
29		11.13°N	146.70°E	236	2025	6000	2000	21200	61 ± 22	V3611
30	Truk	7.41°N	151.75°E	200	0	4390	27500†	70000	38 ±14	Wedgeworth
31	Minto Reef	8.13°N	154.25°E	250	0	5120	5500†	32500†	37 ±31	Wedgeworth
32	Oroluk	7.50°N	155.33°E	250	0	4950	16500†	50000†	28 ± 18	Wedgeworth
33	Pohnpei	6.83°N	158.25°E	290	0	4750	18330	66900	3 ±16	Wedgeworth
34	Kusaie	5.17°N	163.00°E	250	0	4570	6000†	40000†	16 ± 23	Wedgeworth
Marshall Islands										
35	Ujelang	9.83°N	160.91°E	200	0	4750	4000†	30500†	80 ± 45	Wedgeworth
36	Enewetak	11.50°N	162.33°E	203	0	4575	19500†	47500†	45 ± 19	Wedgeworth
37		16.09°N	163.09°E	100	1373	5300	10000	37500	140 ± 49	Wedgeworth
38		8.80°N	163.16°E	170	1100	5000	10000	38000	46 ± 26	V3401
39	Sylvania	12.00°N	165.00°E	180	1310	5300	13200	66500†	37 ± 18	KK81062604
40	Bikini	11.58°N	165.41°E	175	0	4575	17500†	45000†	69 ± 23	Wedgeworth
41	Ujae	9.09°N	165.58°E	150	0	4950	7500†	30500†	186 ± 93	Wedgeworth
42	Wotho	10.09°N	166.00°E	150	0	4575	7500†	21000†	226 ± 140	Wedgeworth
43		16.70°N	166.50°E	74	2175	6000	1600	17500†	247 ± 122	KK81062604
44	Rongelap	11.25°N	166.83°E	175	0	4400	20000†	45000†	60 ± 20	W edgeworth
45	Kwajalein	9.16°N	167.41°E	216	0	4950	22500†	50000†	46 ± 18	Wedgeworth
46	Namorik	5.67°N	168.17°E	225	0	4575	3500†	25000†	36 ± 39	Wedgeworth
47	Ebon	4.67°N	168.75°E	200	0	4400	7500†	25000	59 ± 42	Wedgeworth
48	Ailinglapalap	7.41°N	168.83°E	200	0	4950	15000†	42500†	65 ± 26	Wedgeworth
49	Taongi	14.58°N	168.91°E	225	0	5500	7500†	42500†	73± 32	Wedgeworth
50	Jaluit	6.00°N	169.17°E	200	0	4400	22000†	35000†	43 ± 20	Wedgeworth

51	Utirik	11.16°N	169.75°E	175	0	4575	7500†	32000†	99 ± 47	Wedgeworth
52	Erikub	9.17°N	170.00°E	240	0	4400	8500†	33000†	16 ± 24	Wedgeworth
53	Bikar	12.25°N	170.09°E	175	0	4950	6000†	30000†	137 ± 69	Wedgeworth
54	Maloelop	8.75°N	171.09°E	200	0	4950	19000†	55000†	57 ± 19	Wedgeworth
55	Mili	6.17°N	172.00°E	197	0	4800	15000†	50000†	59 ± 22	Wedgeworth
56	Van Valtier	7.35°N	172.35°E	145	1250	5200	2000	25500†	75 ±49	KK81062602
57	Harrie	5.60°N	172.40°E	226	800	4800	8000	28000	−1 ± 24	KK81062602
58		6.67°N	172.41°E	70	2260	4700	0†	25500†	79 ± 51	Wedgeworth
59		12.25°N	173.20°E	135	1400	5650	800	28000	97 ± 52	KK81062605
Kiribati										
60	Makin	3.17°N	172.83°E	200	0	4750	10000†	28000†	75 ± 43	Wedgeworth
61	Tarawa	1.50°N	172.91°E	225	0	4400	10500†	31500†	28 ± 25	Wedgeworth
Nauru										
62	Nauru	0.26°S	166.91°E	175	0	4000	3000†	20000†	84 ± 68	Wedgeworth
Mid-Pacific Mountains										
63	Sio	18.40°N	171.20°E	125	1150	5500	30940	147333	69 ± 11	KK81062605
64	Niemayer	18.09°N	173.58°E	100	1400	4200	1000†	32500†	70 ± 39	Wedgeworth
65	Harvey	17.85°N	172.55°E	120	1050	5000	6800	17500†	219 ±137	KK81062605
66	Mac Donald	19.20°N	173.30°E	131	1300	5000	18400	30500†	77 ±30	KK81062605
67	Thomas	17.33°N	173.91°E	140	1400	5200	11000†	37500†	72 ± 30	Wedgeworth
68		18.25°N	174.10°E	101	1400	5000	5600	32000	137 ± 53	KK81062605
Emperor Seamounts										
69	Tench	48.83°N	168.33°E	140	2500	6150	17500†	55000†	52 ± 20	Wedgeworth
70		48.40°N	169.00°E	127	2300	6150	6000	62500†	72 ± 22	KK77031705
71	Suiko	44.55°N	170.28°E	350	1025	6250	41000*	150000*	−6 ±14	Wedgewort
72	Yomei	42.30°N	170.40°E	235	100	0 5500	7000†	47500†	16 ± 21	Wedgeworth
73	Nintoku	41.02°N	170.59°E	335	1002	5500	35000*	80000†	−22 ±16	Wedgeworth
74	Jingu	38.79°N	171.14°E	277	812	6000	9000†	45000†	−2 ±19	Wedgeworth
75	Ojin	37.80°N	170.40°E	225	1033	5500	26000*	70000*	24 ± 15	Wedgeworth
76	Koko	35.25°N	171.58°E	257	293	5500	17000†	67500†	29 ±16	Wedgeworth
Hawaiian Ridge										
77		29.85°N	179.10°E	221	280	5250	1600	24000†	48 ± 48	KK720702
78		28.85° N	178.90°W	255	210	5300	4000	35000†	27 ± 28	KK76080601
79	Midway	28.30°N	177.30°W	306	0	5500	13500†	35000†	10 ±20	Wedgeworth
80	Ladd Bank	28.45°N	176.60° W	250	100	5150	8500†	35000†	32 ±27	KK720702
81	Pearl & Hermes Reef	27.90°N	175.60°W	285	50	5125	17500†	60000†	13 ± 6	Wedgeworth
82	Lisianski	26.00°N	174.80°W	314	0	4950	21000†	50000†	−4 ± 16	Wedgeworth
83		29.10°N	174.00°W	280	700	5250	8000	40000	−10 ±20	KK84042805
84	Pioneer Bank	26.10°N	173.30°W	213	600	5250	12500†	37500†	42 ± 25	KK720702
85	Laysan	25.80°N	171.70°W	290	0	4950	15000†	42500†	5 ± 18	Wedgeworth
86	Gardner Pinnacles	25.00°N	168.50°W	265	250	5000	38800†	88500†	14 ± 14	KK720702
87	French Frigate Shoals	23.75°N	166.16°W	275	0	4750	2500†	50000†	7 ± 21	Wedgeworth
88	Brooks Bank	23.10°N	166.40°W	268	300	5000	12500	45000†	8 ± 19	KK720702
89	La Perouse Pinnacles	24.00°N	165.70°W	292	300	5000	20000	55000†	−1 ± 16	KK720702
90	Necker Island	24.00°N	164.40°W	295	300	5200	24000	62500†	4 ± 15	KK720702
91		23.26°N	162.70°W	250	150	4500	6000	45000†	8 ± 21	KK720702
92		23.10°N	162.10°W	235	325	4500	10000†	62500†	13 ± 18	KK720702
93	Nihoa	23.10°N	161.90°W	285	0	4700	13500†	52500†	2 ± 17	Wedgeworth
94	Niihau	21.90°N	160.20°W	290	0	4575	12500†	55000†	−4 ± 17	Wedgeworth
95	Kauai	20.80°N	159.50°W	340	0	4400	26000†	80000†	−25 ± 16	Wedgeworth
96	Oahu	21.50°N	158.00°W	310	0	4600	30500†	75000†	−8 ± 15	Wedgeworth
97	Molokai	21.20°N	157.00°W	270	0	4575	22500†	62500†	8 ± 16	Wedgeworth
98	Lanai	20.80°N	156.90°W	250	0	4400	12500†	125000†	14 ± 14	Wedgeworth
99	Kahoolawe	20.58°N	156.66°W	250	0	4400	7500†	75000†	13 ± 17	Wedgeworth
100	Maui	20.70°N	156.10°W	280	0	5125	32500†	85000†	17 ± 14	Wedgeworth
101	Hawaii	19.50°N	155.50°W	307	0	5400	59400†	108000†	13 ±12	Kinoshita
Musicians										
102		33.80°N	167.00°W	87	2450	6000	2400	24000†	146 ± 71	KK80041402
103	Berlin	32.90°N	166.00°E	85	2550	6000	1400†	12800†	60 ± 87	KK80041402
104	Hammerstein	32.46°N	165.76°W	85	3600	5850	2500†	18500†	118 ± 84	Wedgeworth
105	Donizetti	32.33°N	160.00°W	40	3900	5900	1500†	15000†	96 ± 97	Wedgeworth
106	Bizet	32.26°N	161.63°W	40	3700	5950	1500†	15000†	137 ± 106	Wedgeworth

107	Schubert	31.90°N	162.10°W	94	2400	6000	8800	22000†	169 ± 91	KK80041402
108	Verdi	31.60°N	162.90°W	103	1950	5950	6000†	25000†	183± 90	Wedgeworth
109	Mahler	31.60°N	165.00°W	68	2450	6000	1600	18000	221± 111	KK80041402
110	Godard	31.48°N	164.56°W	45	2600	5900	7500†	18500†	421 ±201	Wedgeworth
111	Brahms	31.15°N	162.31°W	130	2125	5700	1500†	28000†	47± 41	Wedgeworth
112	Debussy	30.30°N	162.09°W	115	2200	5600	1000†	19000†	30 ±58	Wedgeworth
113	Tchaikovsky	29.39°N	162.09°W	85	2000	5500	2500†	12500†	175± 155	Wedgeworth
114	Liszt	29.00°N	162.00 °W	130	1586	5500	1600	19500†	76 ± 65	KK80041402
115	Handel	27.50°N	159.80°W	71	2500	5550	2800	16000†	172± 106	KK80071500
116	Haydn	26.67°N	161.20°W	50	3600	4950	0†	17500†	34 ±79	Wedgeworth
117	Rapano Ridge	26.67°N	159.80°W	75	2500	5400	2000	44000†	100 ±32	Wedgeworth
118	Bach Ridge	26.50°N	158.40°W	95	2700	5300	3200	18000†	26 ±52	KK80071500
119	Chopin	26.10°N	162.00°W	130	1845	5000	2000	30500†	36 ±34	KK80071500
120	Schumann	25.70°N	160.00°W	105	2200	4920	4800	27500†	51 ±38	KK80071500
121	Rimsky-Korsakov	25.46°N	159.75°W	47	3627	5000	1500†	8500	126 ±06	Wedgeworth
122	Mendelssohn	25.10°N	162.00°W	144	1655	4990	7000	40000†	42± 26	KK80071500
123		24.90°N	157.00°W	105	1900	4950	2400	19000	66 ± 60	KK80071500
124	Paumakua	24.83°N	157.08°W	124	1880	4500	2000†	16500†	29 ±41	Wedgeworth
125	Kaluakalana	23.30°N	158.40°W	60	1850	4800	0	13500†	166 ±109	KK80071500
126		23.12°N	154.45°W	95	1550	4600	1200	21200	101± 65	LDGO-C1203
127		22.67°N	159.75°W	50	50	4575	6500	40500	325 ±140	Wedgeworth
Geologists										
128	Finch	17.53°N	157.58°W	123	1000	5300	2500†	17500†	146 ± 104	KK78080700
Line Islands										
129		17.50°N	169.25°W	81	1800	5100	2000	20000	184 ± 96	KK79102901
130	Johnston Island	16.75°N	169.50°W	251	0	4950	8500†	47500†	29 ± 22	Wedgeworth
131		14.40°N	165.60°W	148	1500	5500	2000	23500	59 ± 49	KK78100300
132	Nagata	12.50°N	167.00°W	87	1559	5303	1000†	14000†	192 ± 130	Wedgeworth
133		10.40°N	165.00°W	185	1175	5500	2400	26500	36 ± 38	KK79102901
134		9.60°N	164.20°W	150	1275	5500	3000	20000	90 ± 69	KK79102901
135	Willoughby	7.80°N	162.00°W	60	2300	4500	1600	10500†	50 ± 108	KK79080801
136		7.10°N	163.00°W	98	1600	4525	4000	38000	83± 33	KK79102901
137		4.60°N	160.10°W	89	1380	4525	8800	37000†	112±38	KK79102901
138		0.60°N	157.80°W	104	1100	4900	2500†	25000†	184 ± 89	KK79080801
139		7.55°S	151.50°W	137	1200	5450	0	38000	89 ± 37	KK79080802
Tuvalu										
140		7.40°S	179.01°E	139	788	5000	2500†	37500†	109 ± 43	Wedgeworth
Phoenix Seamounts										
141		3.82°S	173.31°W	107	1725	5500	1200†	22500†	121 ± 78	Wedgeworth
142		3.00°S	171.40°W	173	1090	5000	3500†	21000†	30 ± 44	Wedgeworth
Central Pacific Basin										
143		2.10°N	164.50°W	220	780	5300	0	26000	7 ± 29	MH70042201
Samoa										
144	Tutuila	14.33°S	170.67°W	290	0	4760	12500†	50000†	1 ± 18	Wedgeworth
145	Ofu and Olosega	14.16°S	169.67°W	310	0	4800	13000†	44000†	9 ± 18	Wedgeworth
Tuamoto										
146		12.90°S	148.15°W	111	1225	5290	3200	28000	172 ± 77	KK79080802
Society Islands										
147	Moorea	17.50°S	149.83°W	215	0	3660	7500†	20000†	3 ± 28	Wedgeworth
148	Tahiti	17.67°S	149.50°W	230	0	4500	17500†	70000†	26 ± 16	Wedgeworth
Tonga Trench										
149	Capricorn	18.65°N	172.30°W	245	725	6500	6250†	35000†	57± 36	LDGO-V1813
W. Pacific										
150	Jasper	30.53°N	122.7°W	98	550	4020	1000†	16000†	262 ± 72	Wedgeworth

Abbreviations for vessels:
Number signs (##) indicate a cruise number.

KK########	R/V *Kana Keoki*, University of Hawaii
V####	R/V *Vema*, Lamont-Doherty Observatory
UOTKH####	R/V *Hakuho-Maru*, University of Tokyo
MH########	R/V *Mahi*, University of Hawaii
LDGO-V####	R/V *Vema*, Lamont-Doherty Observatory
LDGO-C####	R/V *Conrad*, Lamont-Doherty Observatory

Wedgeworth	Wedgeworth (1985)
Kinoshita	Kinoshita et al. (1963)
†	determined from Mammerickx (1982)
‡	determined from Kellogg and Ogujiofor (1985)
*	determined from Smoot (1982)

slightly underestimated the degrees of compensation for the Hawaiian islands. However, our tectonic conclusion that the Hawaiian islands were erupted off-ridge on thick elastic lithosphere and were not locally compensated, i.e., shallow depths to Moho, is in agreement with seismic and flexural studies.

Based on seismic profiles, Hill and Zucca (1987) have estimated that the Moho under Hawaii lies at a depth of only 18 km, despite the island's enormous size. The degree of local compensation for Hawaii, based on the maximum gravity anomaly, is 13%± 12% (Table 2), which, assuming a crust-mantle density contrast of -0.4±0.15 g/cm^3, results in a Moho depth prediction of 14±3.4 km.

Seismic refraction and reflection and gravity profiles have suggested a Moho depth of approximately 18–19 km for the Hawaiian ridge between Oahu and Molokai (Watts et al., 1985). In the model of Watts et al. (1985), the flexed oceanic lithosphere is underlain by a deep crustal sill complex, and possibly a high-density volcanic plug. Interpreting a seismic refraction profile in the same area, Lindwall (1988) suggested a shallower Moho depth of 15–16 km between Oahu and Molokai. Using seismic refraction and gravity modeling with densities almost identical to those used in this study, Suyenaga (1978) estimated a Moho depth similar to that later found by Watts et al. (1985) for the Hawaiian ridge near Oahu. The model of Watts et al. (1985) used a density contrast across the crust-mantle boundary of −0.25 g/cm^3, compared to our adopted value of −0.4 g/cm^3. Table 2 shows that the maximum gravity value for Oahu is actually higher than that for Hawaii. The value for Oahu comes from the eroded Koolau caldera (East Oahu), where dense basalts (2.9–3.0 g/cm^3) are found at the surface (Strange et al., 1965). Although the assumption of a homogeneous edifice density of 2.6 g/cm^3 results in an underestimate of μ for Oahu, our estimate of a mid-plate tectonic origin for Oahu is in agreement with results of flexural models.

The Emperor chain (Fig. 4) is the continuation of the trace of the Hawaiian hotspot. Except for two seamounts at the north end of the chain, all of the Emperors were less than 22% compensated. Using Seasat data, Watts and Ribe (1984) suggested that the Emperors north of 40°N were erupted onto young lithosphere, while the southern Emperors were erupted onto older lithosphere. With two exceptions, our estimates from sea surface data agree with their conclusions. Two of the seamounts north of 40°N, Suiko and Nintoku, have very large (>330 mgal) observed anomalies, are locally uncompensated, and must have been erupted onto old lithosphere rather than near a ridge-crest. Although the ages of these seamount edifices are known, the ages of the underlying lithosphere are not well determined.

Seamounts in the East Marianas Basin are mostly uncompensated. One of these seamounts, Ita Mai Tai, has been studied by three-dimensional gravity modeling (Wedgeworth and Kellogg, 1987). A single high-density conduit was included in this model. The crustal root under Ita Mai Tai was estimated to be only 1.5 km thick, giving a Moho depth directly under the seamount of 13.3 km. In this paper we estimate the degree of compensation to be 1± 18% (Table 2), and the Moho depth to be 12±3.4 km, so the seamount is clearly locally uncompensated. This interpretation is in agreement with the results of three-dimensional gravity modeling and flexural studies (Watts et al., 1980).

Ita Mai Tai is a member of a linear chain of uncompensated seamounts trending NE-SW from about (10°N, 155°E) to (15°N, 160°E). This chain appears as a linear geoid high on the Seasat-derived *Gravity Map of the World's Oceans* (Haxby, 1987), which supports the hypothesis that this chain of uncompensated seamounts is the trace of a Cretaceous(?) mid-plate hotspot.

The crust in the East Marianas Basin is generally considered to be of Jurassic age, but has been interpreted as upper Cretaceous (Hilde et al., 1977). The argument for a Cretaceous age stems from the discovery of Cretaceous basalts in the area by the Deep Sea Drilling Project. If these Cretaceous basalts do not represent oceanic crust, they may be associated with the volcanism that produced the seamounts in the basin. Wedgeworth and Kellogg (1987) assigned a probable Aptian/Albian age to the volcanic edifice of Ita Mai Tai, so the low degree of compensation of that seamount and others in its vicinity argue for Jurassic rather than late Cretaceous crust in the East Marianas basin.

The Caroline Islands appear as an east-west oriented chain of uncompensated seamounts. The low estimated degrees of compensation suggest a mid-plate hotspot origin. This finding supports the interpretation of Keating et al. (1984), who argued for a hotspot origin on the basis of chemistry, paleomagnetics, and age progression.

The islands of Tutuila, Ofu, and Olosega in Samoa are locally uncompensated. These results support three off-ridge estimates for Samoa based on flexural studies (Watts et al., 1980). Historic volcanism is known in Samoa (Richard, 1962), and the islands are probably the result of mid-plate volcanism.

In the Society Islands, Tahiti forms a broad plateau upon which Moorea was erupted. We have treated the structures as being separate, and both are uncompensated. We made no attempt to account for the gravity effect of the thickened plateau on which Moorea lies. Based on seismic refraction modeling, Talandier and Okal (1987) estimated the Moho depth under Tahiti to be 12.3 km. The crust was modeled using a three layer model of sediments (V_P = 1.9 km/s), basalts (V_P = 4.37 km/s), and gabbro (V_P = 7.64 km/s). We estimate the degree of compensation for Tahiti to be 26±16%, which corresponds to a Moho depth of 15±4 km.

Our results are in agreement with those of seismic refraction and flexural models (Watts et al., 1980; Calmant and Cazenave, 1986) for the Society Islands.

Ridge Crest and Superswell Volcanism

Most of the analyzed seamounts rimming the Central Pacific Basin are locally compensated, including the Marshall Islands (10), the Mid-Pacific Mountains (5), the Line Islands (6), Tuvalu (1), and Phoenix seamounts (1). Makin Island in Kiribati also appears to be compensated although the uncertainty is large. These results agree with the predictions of flexural studies for the Mid-Pacific Mountains (Watts et al., 1980; Wolfe and McNutt, submitted) and the Line Island chain (Watts et al., 1980). The authors are not aware of any previous flexural or isostatic gravity studies of

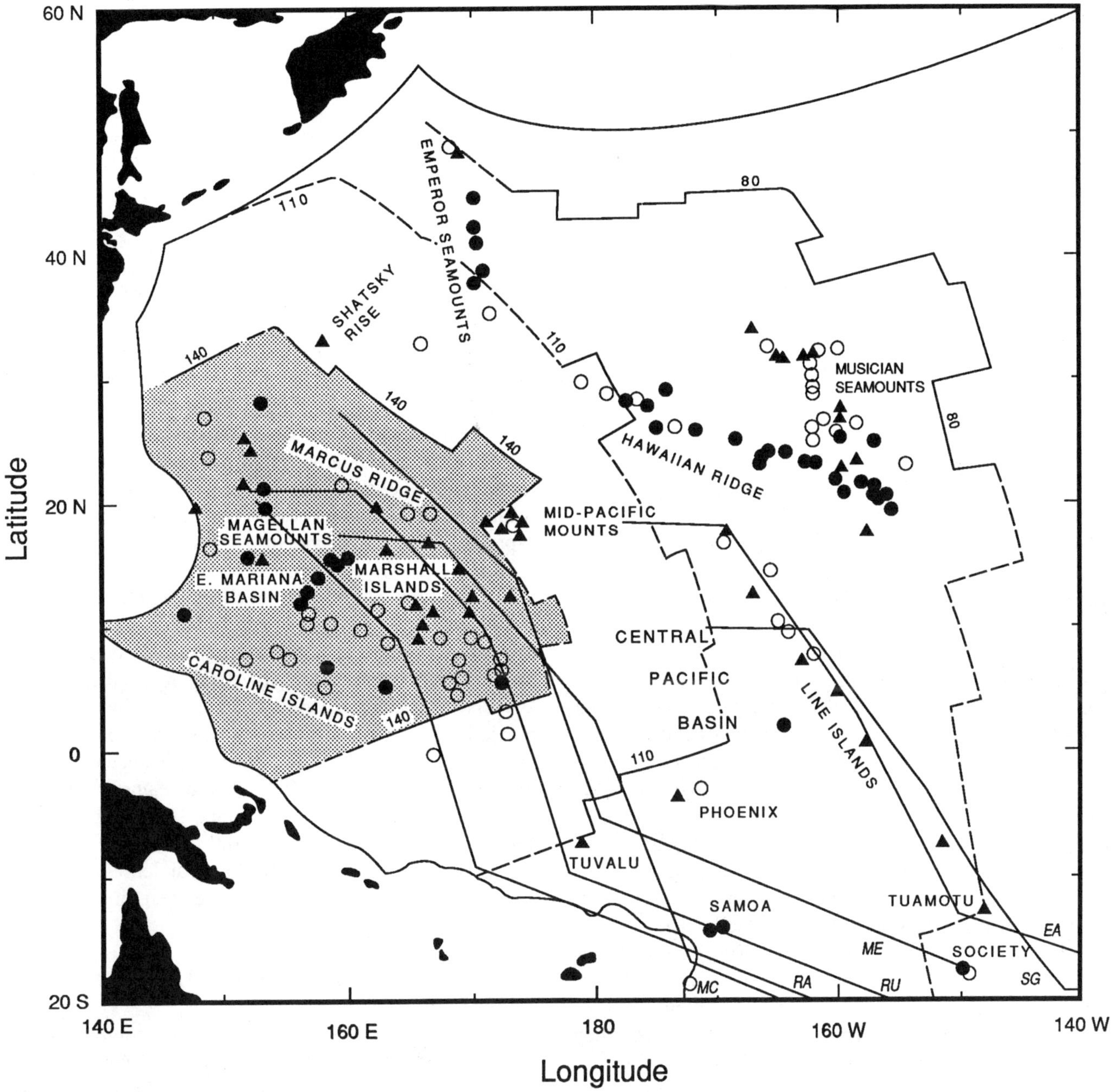

Figure 4. Plot of estimates of ridge crest or superswell (triangles), normal off-ridge (solid circles), and uncertain (open circles) volcanism in the Western Pacific. The sea floor isochrons are based on magnetic lineations (Circum-Pacific Council for Energy and Mineral Resources, 1981). Predicted tracks for South Pacific hot spots: Macdonald (MC), Rarotonga (RA), Rurutu (RU), Mehetia (ME), Easter (EA), and Sala y Gomez (SG) are from Duncan and Clague (1985).

Table 3. Values for compensation μ vs. age of loading t curve (Figure 3).

Seamount	Edifice age (Ma)	Source	Seafloor age (Ma)	Source	t	μ
Iceland	0	Epp84	5 ± 5	Epp84	5 ± 5	67 ± 9
Galapagos (Isabela)	0	Epp84	10 ± 2	Epp84	10 ± 2	44 ±12
Tahiti	1	Epp84	66 ± 5	Epp84	65 ± 5	26 ± 15
Great Meteor	11	V84	84 ± 2	V84	73 ± 2	155 ± 16
Hawaii	0	Epp84	92 ± 5	Epp84	92 ± 5	13 ±12
Pohnpei	5.4	Epp84	165 ± 10	Epp84	160 ± 10	3 ± 1
Nihoa	3.5 ± 5	LVS84	100 ± 5	LVS84	96.5 ± 7	2 ± 17
Midway	27 ± 5	LVS84	120 ± 10	LVS84	93 ± 12	10 ± 20
Kusaie	1.3	Epp84	153 ± 10	Epp84	152 ± 10	16 ± 23
Koko	48	V&S84	110 ± 15	V&S84	62 ± 15	29 ± 16
Ojin	55 ± 1	V&S84	105 ± 15	V&S84	50 ± 15	24 ± 15

Sources of data are: Epp84 = Epp (1984), V84 = Verhoef (1984), LVS84 = Livermore et al. (1984), and V&S84 = Vogt and Smoot (1984).

the Marshall Islands. At least 10 of the seamounts in the group appear to have been erupted onto older lithosphere that was reheated near the time of volcanism (see discussion).

The Mid-Pacific Mountains are a huge submarine mountain range stretching for thousands of kilometers but which have been only sparsely surveyed. The six seamounts surveyed appear to be highly compensated and were probably erupted onto young crust. This interpretation agrees with that of Watts and Ribe (1984) for a sea surface gravity profile across the central Mid-Pacific Mountains and the results of Wolfe and McNutt (submitted) for 17 guyots based on modeling Seasat gravity and deflection of the vertical. Winterer and Metzler (1984) have suggested a mid-Cretaceous hotspot origin for the Mid-Pacific Mountains, followed by a separate phase of Late Cretaceous volcanism, Kroenke et al. (1985) proposed a mid-Cretaceous mid-plate rift, and McNutt et al. (in press) proposed eruption on a shallow Darwin Rise superswell. In all three models the volcanoes of the Mid-Pacific Mountains are erupted onto young oceanic crust, probably less than 10–15 Ma in age at the time of eruption. Natland and Wright (1984) determined that tholeiites of the Mid-Pacific Mountains are fractionated ferrobasalts with moderately enriched trace element characteristics resembling lavas of Iceland and the Galapagos Islands, and concluded that they formed near a spreading ridge in the Early Cretaceous.

Sio Guyot in the Mid-Pacific mountains has been studied by extensive three-dimensional gravity modeling (Kellogg and Ogujiofor, 1985). Two high-density conduits within the seamount were included in the model. The depth to the crust-mantle boundary directly underneath the seamount was estimated to be 22 ± 2 km. We have estimated the degree of compensation to be 68 ± 10%, and the Moho depth to be 23.2 ± 6 km (Table 2), which agrees well with the estimate derived from three-dimensional modeling.

Six additional compensated edifices were found in the Japanese, Magellan, and Marcus Wake seamounts. These results are in agreement with recent flexural studies of the Japanese (Wolfe and McNutt, submitted) and Magellan (Smith et al., 1989) seamounts and may reflect eruption on lithosphere reheated and thinned by a Cretaceous superswell (see discussion). Additional ridge crest or superswell estimates include the Shatsky Rise and Tuamotu as also predicted by flexural studies (Watts et al., 1980).

The Musicians seamounts have been the subject of previous studies of partial Airy compensation (Schwank and Lazarewicz, 1982) and flexure (Dixon et al., 1983; Freedman and Parsons, 1986). These seamounts are very small, with maximum gravity anomalies for some being as small as 40 mgal, and thus it is difficult to invert for their degree of compensation. However, at least nine of the Musicians Seamounts appear to have been formed at ridge crests (Table 2).

DISCUSSION

The relationship between seamount isostatic compensation levels (μ) and age of seafloor loading (Fig. 3) suggests that compensated seamounts ($\mu \geq 40\%$) were erupted within 8 Ma of the age of the underlying sea floor or of a lithosphere reheating event. We tabulated the number of ridge-crest or superswell estimates by underlying Pacific sea floor ages (Fig. 4) in three intervals through the Mesozoic:

16 estimates for the interval 140–170 Ma (Kimmeridgian to Jurassic magnetic quiet interval), 6 estimates for the interval 110–140 Ma (Aptian to Tithonian), and 20 estimates for the interval 80–110 Ma (Cretaceous magnetic quiet interval).

The large number of compensated seamounts in the interval 80–110 Ma is not surprising. The interval is known to be a period of significant volcanism on the Pacific plate (Schlanger and Premoli Silva, 1981; Menard, 1984). Flexural studies include 26 estimates with low T_e values for the 90–120 Ma interval, associated with Hess Rise, Necker Ridge, the Line Islands ridge, Manihiki Plateau, and Robbie Ridge (Watts et al., 1980).

Only six compensated estimates are associated with sea floor in the interval 110–140 Ma. These features include the Shatsky rise (1) and the Mid-Pacific Mountains (5). The Mid-Pacific seamounts may have been erupted near the end of this time interval. Edifice construction in the Mid-Pacific Mountains has been dated from 80–120 Ma (Saito and Ozima, 1977; Thiede et al., 1981; Kroenke et al., 1985). Winterer and Metzler (1984), Kellogg and Ogujiofor (1985), Kroenke et al. (1985), and Wolfe and McNutt (submitted) have proposed models in which the volcanoes of the Mid-Pacific Mountains are erupted onto young oceanic lithosphere, probably less than 10–15 Ma in age at the time of eruption.

The large number of compensated estimates (16) for 140–170 Ma Jurassic sea floor is surprising in view of the limited sea floor of that age remaining in the Pacific (Fig. 3) and the lack of evidence for Jurassic age seamount volcanism. All of these compensated seamounts on 140–170 Ma sea floor are located in the Jurassic magnetic quiet zone (150–170 Ma). Compensated estimates for 150–170 Ma sea floor (Fig. 3) are associated with the Japanese seamounts (2), Magellan seamounts (3), Marcus Wake seamounts (1), and the Marshall Island group (10). Rocks dredged from the Japanese and Marcus Wake seamounts gave ^{40}Ar-^{39}Ar ages of 86–102 Ma (Ozima et al., 1977) and 65–120 Ma Cretaceous reef fossils (Matthews et al., 1974). Paleomagnetic age determinations for the seamounts range from 85-120 Ma (McNutt et al., 1990). Two Magellan seamounts yielded ^{40}Ar-^{39}Ar ages of 100 and 120 Ma (Smith et al., 1989). Volcanics from the Marshall Island group have been dated at 51–88 Ma, and reef fossil ages are as old as 115 Ma (Lincoln, 1990). Since no ages older than 120 Ma have been determined for the Japanese, Magellan, Marcus Wake, and Marshall seamounts, it is improbable that these seamounts were erupted near Jurassic ridge crests.

Figure 4 shows predicted tracks for South Pacific hot spots based on rotation poles for the Pacific plate over a fixed hot spot reference frame (Duncan and Clague, 1985). During the Cretaceous the predicted tracks for the Macdonald, Rurutu, Mehitia, and Rarotonga hot spots passed through 140–170 Ma Jurassic sea floor. Volcanic rocks dredged from two Early Cretaceous volcanic edifices in the Magellan seamounts (Smith et al., 1989) and four Late Cretaceous volcanic edifices in the Ratak chain of the Marshall Islands (Davis et al., 1989) are isotopically similar to Rurutu and Rarotonga. Anomalously shallow Cretaceous seafloor depths and low T_e values similar to those presently found in the South Pacific have been determined for the two Magellan seamounts (Smith et al., 1989) and five mid-Cretaceous Japanese seamounts (Wolfe and McNutt, submitted). Smith et al. (1989), Davis et al. (1989), and Wolfe and McNutt (submitted) have concluded that the seamounts formed at Cretaceous melt sources in the broad thermal anomaly or superswell that is now located in the South Pacific. Our determination of 16 additional isostatically compensated seamounts on 140–170 Ma Jurassic sea floor supports the superswell origin for many of the Japanese, Magellan, Marcus Wake, and Marshall seamounts. Particularly interesting is our discovery of 10 compensated edifices in the Marshall seamounts. Since no reefs or volcanic edifices in the Marshall seamounts have been dated as older than 115 Ma, these compensated seamounts were probably erupted on thin, reheated lithosphere. This agrees with conclusions based on geochemistry (Davis et al., 1989) that a number of Marshall seamounts were erupted from mantle sources that are presently in the South Pacific. The passage of Jurassic lithosphere over at least four superswell hot spots may also explain the apparent multiple episodes of mid-plate volcanism in the Marshall Islands reported by Lincoln (1990).

APPENDIX

Modeling Gravity Anomalies for Conical Seamounts

An exact expression in closed form can be derived for the on-axis free air anomaly for the case of a cone or frustum of a cone with constant density (e.g., Kellogg et al., 1987). If the radius of the cone is given by $a(z) = bz + c$ in the depth interval $[z_0, z_1]$, then the gravitational effect is given by

$$g_z(z_0, z_1, \Delta\rho, b, c) = I(z_1, \Delta\rho, b, c) - I(z_0, \Delta\rho, b, c) \quad \text{(A1a)}$$

$$I(z,\Delta\rho,b,c) = 2\pi G\Delta\rho \Big[z - \frac{1}{\beta^2}\sqrt{\beta^2 z^2 + 2bcz + c^2} + \frac{bc}{b^3}\ln\Big(z + \frac{bc}{\beta^2} + \frac{1}{\beta}\sqrt{z^2 + 2bcz + c^2}\Big)\Big] \quad \text{(A1b)}$$

where $\beta^2 = 1 + b^2$.

The parameters b and c can be determined from the top and bottom radii of the seamount (a_0 and a_1, respectively), its minimum depth (z_0), and the local seafloor depth (z_1) (Fig. 2). These quantities are readily measurable from ship track profiles or bathymetric maps.

$$b = (a_0 - a_1)/(z_0 - z_1) \quad \text{(A2a)}$$
$$c = (z_0 a_1 - z_1 a_0)/(z_0 - z_1) \quad \text{(A2b)}$$

The degree of local Airy compensation μ is defined as the ratio of the mass deficiency below the normal crust directly below the structure to be compensated to the excess mass above the normal crust (Schwank and Lazarewicz, 1982). The crustal root is modeled as an inverted cone or frustum of a cone (Fig. 2).

The thickness of the low-density crustal root is proportional to $\mu : h_r = h_{Airy}$, where h_{Airy} is the root thickness for a 100% locally compensated seamount. The density contrasts at the seamount-water and crust-mantle interfaces are $\Delta\rho_{rw} = \rho_{rock} - \rho_{water}$, and $\Delta\rho_{cm} = \rho_{crust} - \rho_{mantle}$, so

$$h_{Airy} = \frac{\Delta\rho_{rw}}{\Delta\rho_{cm}} \left[z_0 - z_1 \right] \quad \text{(A3)}$$

The free-air gravity anomaly due to the seamount plus the root is $g_{obs} = g_{bathy} + g_{root}(\mu)$, where g_{bathy} is the gravity effect of the bathymetry, and g_{root} is the gravity effect of the root. The first approximation for μ is obtained by ignoring the terms non-linear in μ in the expression for g_{root} (Eq. A1)

$$\mu_{(1)} = \frac{g_{observed} - g_{bathy}}{2\pi G \Delta\rho_{cm} h_{Airy}} \quad \text{(A4)}$$

Successive estimates of μ can be improved by including the non-linear terms using the iterative relation

$$\mu_{(n+1)} = \mu_{(1)} + \frac{1}{h_{Airy}} \left[\frac{1}{b_r} \sqrt{z^2 + \frac{2b_r c_r z}{b_r^2} + \frac{c_r^2}{b_r^2}} - \frac{b_r c_r}{\beta_r^3} \ln\left(z + \frac{b_r c_r}{\beta_r^2} + \sqrt{z^2 + \frac{2b_r c_r z}{\beta_r^2} + \frac{c_r^2}{\beta_r^2}} \right) \right]_{z_2}^{z_2 + \mu_{(n)} h_{Airy}} \quad \text{(A5)}$$

which is derived from Eq. (A1).

The Moho depth directly under the seamount, z_{Moho}, can be computed from the degree of compensation μ :

$$z_{Moho} = z_1 + z_{crust} + h_{Airy} \quad \text{(A6)}$$

Our computed Moho depth should be treated as a very rough approximation, as we model all seamounts as having roots of the same shape. In reality, the shape of the root is largely controlled by the elastic properties of the lithosphere at the seamount.

Uncertainties and Sensitivities

Kellogg et al. (1987) use $\Delta\rho_{rw} = 1.57$ g/cm^3 for the density contrast across the rock-water interface. Other authors use different density contrasts, and model seamounts as multiple rather than single bodies. For this reason, this value must be compared to an approximate volume average density from these more complex models. Watts et al. (1985) use approximately 1.4 g/cm^3 in their gravity models of the Hawaiian ridge, and Hill and Zucca (1987) use approximately 1.7 g/cm^3 in their gravity model of structures on the island of Hawaii. We take as our value $\Delta\rho_{rw} = 1.57 \pm 0.15$ g/cm^3 for the density contrast at the rock-water interface, which corresponds to a seamount edifice density of 2.6 g/cm^3.

Lateral density variations, due to high-density conduits or other internal structures (Kellogg et al., 1987), could be significant causes of error in the estimation of the degree of local compensation. These density variations would have the largest effect on the estimates when the top of the seamount is at or near the sea surface. There is no way to correct for this potential error with our data, so we assign a large uncertainty to the on-axis gravity data.

There is a considerable range of values given in the literature for the value of the density contrast between the crust and the mantle underneath seamounts. Suyenaga (1978) and Kellogg et al. (1987) assumed a density contrast of -0.5 g/cm^3 across the crust-mantle boundary, Hill and Zucca (1987) assumed a density contrast of -0.35 g/cm^3 under Hawaii, and Watts et al. (1985) assumed a density contrast of approximately -0.25 g/cm^3 for this boundary under the Hawaiian Ridge between Oahu and Molokai. The value from Watts et al. (1985) is based on an estimate of the bulk density of the thickened crust, as they modeled the thickened crust under the seamount with multiple bodies. Given this variation, we adopted a value of $\Delta\rho_{cm} = -0.4 \pm 0.15$ g/cm^3.

The uncertainty in the degree of local compensation μ was determined from the sensitivity of μ to each parameter and the uncertainty in each parameter:

$$\sigma_\mu^2 = \sum_j \sigma_j^2 = \sum_j \left[\frac{\partial u}{\partial x_j} \delta x_j \right]^2 \quad \text{(A7)}$$

The eight input parameters and standard uncertainties are given in Table 1. Table 4 and Figure 5 show the sensitivity of the degree of local compensation μ to each of the eight parameters. The vertical axis in Figure 5 is σ_j, the mean contribution of parameter j to the uncertainty in μ, averaged over the entire data set (Eq. A7). Clearly the most important parameter is the seamount edifice-water density contrast. Uncertainties in the minimum and maximum depths and observed gravity also contribute to the uncertainty in μ. The remaining four parameters do not contribute significantly. The uncertainty in the density contrast between the crust and mantle does not contribute to the uncertainty in the degree of local compensation because the dominant term in the expression for μ is independent of this parameter.

Derivation of a theoretical compensation vs. age of loading curve

The theoretical compensation-age of loading curve, Eq. (1), was derived by equating the maximum deflection of the Moho from the partial Airy compensation model with the maximum deflection of the Moho from a continuous flexed elastic plate subject to a line load (Turcotte and Schubert, 1982):

$$\mu h_{Airy} = \frac{V_0 \alpha^3}{8D} \tag{A8}$$

where $V_0 = a_0 h_{Airy} \Delta\rho_{cm} g$ (A9)

$$\alpha = \left(\frac{4D}{\Delta\rho_{cm} g}\right)^{1/4} \tag{A10}$$

$$D = \frac{E T_e^3}{12\,(1 - \nu^2)} \tag{A11}$$

We assume the following values for the constants above: Young's modulus, E = 1×10^{12} dyne/cm^2; Poisson's ratio, ν = 0.25; density contrast at the crust-mantle interface, $\Delta\rho_{cm}$ = 0.4 g/cm^3; acceleration of gravity, g = 980 cm/sec^2. For a typical seamount radius a_0 = 50 km, this gives:

$$\mu = 456\, T_e^{-3/4} \;(\%) \tag{A12}$$

with the effective elastic thickness T_e in meters.

The effective elastic thickness is related to the age of the seafloor at the time of loading by the empiricial relation $T_e = 4200\, t^{1/2}$ meters, with t in Ma (t < 105 Ma) (McNutt, 1984). Substituting this relation into (A12) gives Eq. (1):

$$\mu = 87.3\, t^{-3/8} \;(\%) \tag{A13}$$

Acknowledgments

The authors thank Bruce Wedgeworth for contributing his western Pacific data set, and Terri Duennebier for helping with ship track maps and profiles. Sy Schlanger and Marcia McNutt made several preprints available. Critical reviews by Andy Lazarewicz and two anonymous reviewers and error analyses suggested by Ian Lerche improved this paper. This work was funded by the Office of Naval Research, Code 425GG (University of Hawaii) and the University of South Carolina.

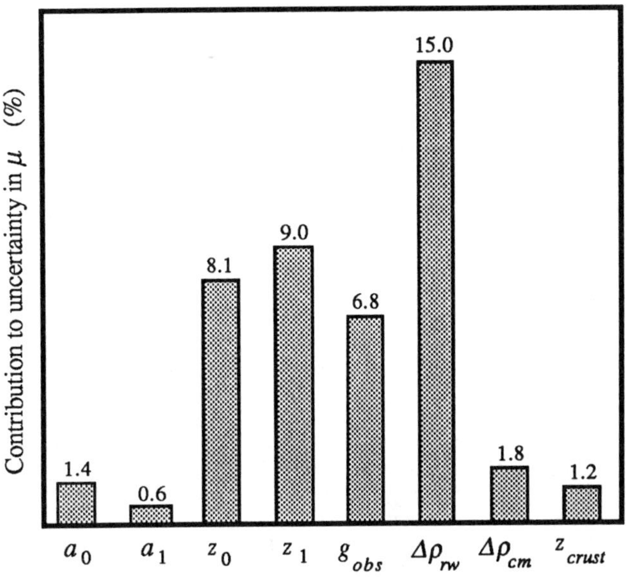

Figure 5. The mean sensitivities for μ (using the entire data set) to the eight parameters considered in this analysis are multiplied by typical values of the parameter uncertainties to produce the uncertainties plotted in this figure. The greatest contribution to uncertainty in μ is uncertainty in the seamount density, followed by seamount and sea floor depths, and the observed gravity.

Table 4. Contributions to the uncertainty in the degree of compensation μ and Moho depth z_{Moho} for all eight parameters. The standard uncertainty in each parameter is listed in the first column (for units see Table 1). The second and third columns show the effects of each parameter's uncertainty on the estimates of μ and z_{Moho}. The values given are the mean perturbations averaged over all seamounts with uncertainties less than 30%, and the rms scatter of the perturbations.

Parameter	δx_j	$\delta\mu$ (%)	δz_{Moho} (m)
a_0	2000	1.4 ± 3.0	240 ± 520
a_1	2000	0.6 ± 0.9	100 ± 150
z_0	200	8.1 ± 2.5	1600 ± 470
z_1	275	9.0 ± 2.2	2100 ±480
g_{obs}	10	6.8 ± 2.2	1200 ± 340
$\Delta\rho_{rw}$	0.15	15.0 ± 3.6	3000 ± 880
$\Delta\rho_{cm}$	0.15	1.8 ± 2.4	2100 ± 1900
z_{crust}	1000	1.3 ± 1.2	1200 ±230

REFERENCES

Calmant, S., and A. Cazenave, 1986, The effective elastic thickness under the Cook-Austral and Society islands: Earth and Planetary Science Letters, v. 77, p. 187-202.

Calmant, S., and A. Cazenave, 1987, Anomalous elastic thickness of the oceanic lithosphere in the south-central Pacific: Nature, v. 328, p. 236-238.

Circum-Pacific Council for Energy and Mineral Resources, 1981, Plate-tectonic map of the Circum-Pacific region, Northwest Quadrant: The American Association of Petroleum Geologists, Tulsa, Oklahoma, Scale 1:10,000,000.

Davis, A. S., M. S. Pringle, L. G. Pickthorn, D. A. Clague, and W. C. Schwab, 1989, Petrology and age of alkalic lava from the Ratak chain of the Marshall Islands: Journal of Geophysical Research, v. 94, p. 5757-5774.

Dixon, T. H., M. Naraghi, M. K. McNutt, and S. M. Smith, 1983, Bathymetric prediction from SEASAT altimeter data: Journal of Geophysical Research, v. 88, p. 1563-1571.

Duncan, R.A., and D.A. Clague, 1985, Pacific plate motion recorded by linear volcanic chains, *in* A.E.M. Nairn, R.G. Stehli, and S. Uyeda, eds.: Plenum, N.Y., Ocean Basins and Margins, v. 7A, p. 89-121.

Epp, D., 1984, Implications of volcano and swell heights for thinning of the lithosphere by hotspots: Journal of Geophysical Research, v. 89, p. 9991-9996.

Freedman, A. P., and B. Parsons, 1986, SEASAT-derived gravity over the Musicians seamounts: Journal of Geophysical Research, v. 91, p. 8325-8340.

Hart, S. R., 1984, A large-scale isotope anomaly in the Southern Hemisphere mantle: Nature, v. 309, p. 753-757.

Haxby, W. F., 1987, Gravity field of the world's oceans; Office of Naval Research: National Geophysical Data Center, Boulder, Colorado, Approx. Scale 1:40,000,000.

Hilde, T. W. C., S. Uyeda, and L. Kroenke, 1977, Evolution of the Western Pacific and its margin: Tectonophysics, v. 38, p. 145-165.

Hill, D. P., and J. J. Zucca, 1987, Constraints on the structure of Kilauea and Mauna Loa volcanoes, Hawaii, and some implications for seismo-magmatic processes: Transactions of Conference on how volcanoes work, Hilo, Hawaii, 1987.

Keating, B. H., D. P. Mattey, C. E. Helsley, J. J. Naughton, D. Epp, A. Lazarewicz, and D. Schwank, 1984, Evidence for a hot spot origin of the Caroline Islands: Journal of Geophysical Research, v. 89, p. 9937-9948.

Kellogg, J. N., and I. J. Ogujiofor, 1985, Gravity field analysis of Sio Guyot: An isostatically compensated seamount in the Mid-Pacific Mountains: Geo-Marine Letters, v. 5, p. 91-97.

Kellogg, J. N., B. S. Wedgeworth, and J. T. Freymueller, 1987, Isostatic compensation levels, and conduit structures of Western Pacific seamounts: Results of three-dimensional gravity modeling, *in* B. H. Keating, P. Fryer, R. Batiza, and G. W. Boehlert, eds., Seamounts, Islands, and Atolls: Washington, D.C., American Geophysical Union, Geophysical Monograph 43, p. 85-96.

Kinoshita, W. K., H. L. Krivoy, D. R. Mabey, and R. R. MacDonald, 1963, Gravity survey of the island of Hawaii: U.S. Geological Survey Professional Paper 475-C, p. C114-116.

Kroenke, L. W., J. N. Kellogg, and K. Nemoto, 1985, Mid-Pacific Mountains revisited: Geo-Marine Letters, v. 5, p. 77-81

Lincoln, J. M., 1990, Regional tectonic history and Cenozoic sea levels deduced from drowned carbonate banks and atoll stratigraphy in the Marshall Islands, West Central Pacific Ocean: Ph.D. Dissertation, Northwestern University, 245 p.

Lindwall, D. A., 1988, A two-dimensional seismic investigation of crustal structure under the Hawaiian Islands near Oahu and Kauai: Journal of Geophysical Research, v. 93, p. 12,107-12,122.

Livermore, R. A., F. J. Vine, and A. G. Smith, 1984, Plate motions and the geomagnetic field - II. Jurassic to Tertiary: Geophysical Journal of the Royal Astronomical Society, v. 79, p. 939-961.

Mammerickx, J., 1982, Bathymetric Atlas of the Western Pacific: Scripps Institution of Oceanography.

Matthews, J. L., B. C. Heezen, R. Catalano, A. Coogan, M. Tharp, J. Natland, and M. Rawson, 1974, Cretaceous drowning of reefs on Mid-Pacific and Japanese guyots: Science, v. 184, p. 462-464.

Menard, H. W., 1984, Darwin reprise: Journal of Geophysical Research, v. 89, p. 9960-9968.

McNutt, M. K., 1984, Lithospheric flexure and thermal anomalies: Journal of Geophysical Research, v. 89, p. 11,180-11,194.

McNutt, M. K., and K. M. Fischer, 1987, The South Pacific superswell, *in* B. H. Keating, P. Fryer, R. Batiza, and G. W. Boehlert, eds., Seamount, Islands, and Atolls: Washington, D.C., American Geophysical Union, Geophysical Monograph 43, p. 25-34.

McNutt, M. K., E. L. Winterer, W. W. Sager, J. H. Natland, and G. Ito, 1990, The Darwin Rise: a Cretaceous Superswell?: Geophysical Research Letters, in press.

Natland, J. H. and E. Wright, 1984, Magmatic lineages and mantle sources of Cretaceous seamounts of the central Pacific [abst.]: Eos, Transactions of the American Geophysical Union, v. 65, p. 1075-1076.

Nishimura, C. E. and D. W. Forsyth, 1985, Anomalous Love-wave phase velocities in the Pacific: sequential pure-path and spherical harmonic inversion: Geophysical Journal of the Royal Astronomical Society, v. 81, p. 389-407.

Ozima, M., H. Masahiko, and K. Saito, 1977, ^{40}Ar-^{39}Ar ages of guyots in the western Pacific and discussion of their evolution: Geophysical Journal of the Royal Astronomical Society, v. 51, p. 475-485.

Richard, J. J., 1962, International Volcanol. Ass., Naples.

Saito, K, and M. Ozima, 1977, 40Ar-39Ar geochronological studies on submarine rocks from the western Pacific area: Earth and Planetary Science Letters, v. 33, p. 353-369.

Schwank. D. C., and A. R. Lazarewicz, 1982, Estimation of seamount compensation using satellite altimetry: Geophysical Research Letters, v. 9, p. 907-910.

Schlanger, S. O., and I. Premoli Silva, 1981, Tectonic, volcanic, and paleogeographic implications of redeposited reef faunas of late Cretaceous and Tertiary age from the Nauru Basin and the Line Islands: Initial Reports of the Deep Sea Drilling Project, v. 61, p. 817-828.

Smith, W. H. F., H. Staudigel, A. B. Watts, and M. S. Pringle, 1989, The Magellan Seamounts: Early Cretaceous record of the South Pacific Isotopic and Thermal Anomaly: Journal of Geophysical Research, v. 94, p.10,501-10,523.

Smoot, N. C., 1982, Guyots of the mid-Emperor chain mapped with multibeam sonar: Marine Geology, v. 47, p. 153-163.

Strange, W. E., G. P. Woollard, and J. C. Rose, 1965, An analysis of the gravity field over the Hawaiian Islands in terms of crustal structure: Pacific Science, v. 19, p. 381-389.

Suyenaga, W., 1978, Isostasy and flexure of the lithosphere under the Hawaiian Islands: Journal of Geophysical Research, v. 84, p. 5599-5604.

Talandier, J., and E. A. Okal, 1987, Crustal structure in the Society and Tuamoto Islands, French Polynesia: Geophysical Journal of the Royal Astronomical Society, v. 88, p. 499-528.

Thiede, J., T. L. Vallier, and others, 1981, Site 463: Western Mid-Pacific Mountains, DSDP Leg 62: Initial Reports of the Deep Sea Drilling Project, v. 62, p. 33-156.

Turcotte, D. L., and G. Schubert, 1982, Geodynamics: Applications of continuum physics to geological problems: John Wiley & Sons, New York.

Verhoef, J., 1984, A Geophysical Study of the Atlantis-Meteor Seamount Complex, Drukkerij Elinkwijk, B. V., Utrecht.

Vogt, P. R., and N. C. Smoot, 1984, The Geisha Guyots: Multibeam bathymetry and morphometric interpretation: Journal of Geophysical Research, v. 89, p. 11,085-11,107.

Walcott, R. I., 1970, Flexure of the lithosphere at Hawaii: Tectonophysics, v. 9, p. 435-446.

Watts, A. B., 1978, An analysis of isostasy in the world's oceans, Part 1. Hawaiian-Emperor Seamount chain: Journal of Geophysical Research, v. 83, p. 5989-6004.

Watts, A. B., J. H. Bodine, and N. M. Ribe, 1980, Observations of flexure and the geological evolution of the Pacific ocean basin: Nature, v. 283, p. 532-537.

Watts, A. B., and N. M. Ribe, 1984, On geoid heights and flexure of the lithosphere at seamounts: Journal of Geophysical Research, v. 89, p. 11,152-11,170.

Watts, A. B., U. S. ten Brink, P. Buhl, and T. M. Brocher, 1985, A multichannel seismic study of lithospheric flexure across the Hawaiian-Emperor seamount chain: Nature, v. 315, p. 105-111.

Wedgeworth, B. S., 1985, Ita Mai Tai Guyot: A comparative geophysical study of Western Pacific seamounts: M.S. Thesis, Univ. of Hawaii.

Wedgeworth, B. S., and J. N. Kellogg, 1987, A 3-D gravity-tectonic study of Ita Mai Tai Guyot, an uncompensated seamount in the East Marianas Basin, *in* B. H. Keating, P. Fryer, R. Batiza, and G. W. Boehlert, eds.,Seamounts, Islands, and Atolls: Washington, D.C., American Geophysical Union, Geophysical Monograph 43, p. 73-84.

Winterer, E. L., and C. V. Metzler, 1984, Origin and subsidence of guyots in the Mid-Pacific Mountains: Journal of Geophysical Research, v. 89, p. 9969-9979.

Keating, B.H., and Bolton, B.R., editors, 1992, Geology and offshore mineral resources of the central Pacific basin, Circum-Pacific Council for Energy and Mineral Resources Earth Science Series, V. 14. New York, Springer-Verlag.

MORPHOLOGY AND GEOLOGY OF THE MAGELLAN TROUGH AREA IN THE CENTRAL PACIFIC

Masahide Furukawa
Graduate School of Science and Technology, Kobe University, Kobe 657, Japan

Kensaku Tamaki
Ocean Research Institute, University of Tokyo, Tokyo 164, Japan

Atsuyuki Mizuno
Department of Earth Sciences, Ehime University, Matsuyama 790, Japan

Masaaki Kimura
Department of Marine Sciences, University of the Ryukyus, Okinawa 903-01, Japan

ABSTRACT

The Magellan trough is a WNW-trending linear depression with flanking ridges that extends more than 300 km in the Central Pacific Basin. Depth of the trough axis exceeds 6600 m. Compilations and analyses of bathymetric, seismic, and geologic data of the trough and its adjacent areas reveal three discriminate acoustic layers: a transparent layer (Unit I), a semi-opaque layer (Unit II), and an opaque layer (acoustic basement) in descending order. In the southern part of the study area (south of the trough), calcareous turbidites (Unit I) whose ages range probably from the Oligocene to the middle Eocene, were derived from the Magellan Rise. This turbidite layer tends to become thinner towards the trough. The southern flanking ridge of the trough, therefore, may have been a barrier to the deposition of the turbidites. The thickness of the transparent layer in Unit I varies from place to place, particularly in the area north of the trough, whereas the thickness of Unit II is rather constant. Some 3.5- kHz profiles show the truncation and abrupt thinning of the transparent layer (Unit I). A hiatus between the late Pliocene and the late Miocene exists in Unit I. These geological findings suggest that the deposition of Unit I has been affected by bottom water currents such as the recent Antarctic bottom water current, which plays a significant role in the deposition of sediment. Major features of the sea floor, namely the surface topography of the acoustic basement and the pattern of faults within the basement around the trough, show trends concordant with those of the trough. The concordant features trend WNW-ESE and converge to the northwest. A detailed bathymetric chart of the study area has been constructed showing left and right lateral offsets of the trough together with the flanking ridges. These bathymetric features of the trough area are concordant with the fan-shaped magnetic anomaly lineations. The bathymetric and magnetic lineations are evidence of sea-floor spreading at the axis of the trough in the Early Cretaceous (M9 to M11). These morphotectonic observations are good indicators of fundamental processes of sea-floor spreading that produced the Magellan system.

INTRODUCTION

The Magellan trough is a WNW-trending linear topographic depression northeast of the Magellan Rise in the Central Pacific Basin (Fig. 1). A bathymetric chart by Winterer, Ewing et al. (1973) showed the general topographic features of the Magellan trough. Larson et al. (1972) reported NW-trending magnetic anomaly lineations around the Magellan trough, north of the Phoenix lineation set, and noted their fan-shaped appearance. Tamaki et al. (1979) identified the symmetric magnetic anomaly lineations, the Magellan set, from M9 (121 Ma) to M11 (126 Ma) of Early Cretaceous age, with a symmetric axis along the Magellan trough and concluded that the trough is a remnant spreading center which ceased opening in the later half of M9 time. Tamaki and Larson (1988) identified new Early Cretaceous magnetic anomaly lineation sets in previously unmapped areas in the western Central Pacific Basin. Their new mapping revealed a pair of opposite-sensed and fanned lineation patterns associated with the fossil Magellan microplate. They concluded that the microplate activity originated at about M14 time and ceased when the microplate became welded to the Pacific plate at M9 time, and that the Magellan rift then failed as a result of a large ridge reorganization at the triple junction of the Pacific, Farallon, and Phoenix plates at M10 time, and leaving a fossil rift, the Magellan trough, at M9 time.

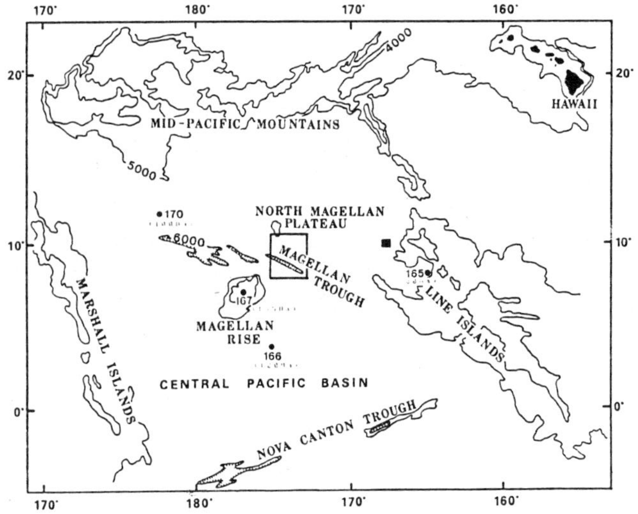

Figure 1. Outline of the topography in the Central Pacific Basin. Contours are in meters (4000 m, 5000 m, and 6000 m). Deep Sea Drilling Project sites are shown as solid circles with the basement ages. The central rectangle denotes the study area, the Magellan basin. A small solid rectangle denotes the GH79-1 detailed survey area. This map is modified from Winterer, Ewing, et al. (1973).

Bathymetric, seismic, geomagnetic, and geologic data of the Magellan trough and its adjacent areas were obtained during four cruises of the R/V *Hakurei-Maru* by the Geological Survey of Japan from 1976 to 1980 (GH76-1, GH77-1, GH80-1, and GH80-5 cruises). These data revealed characteristic features of the Magellan trough and the associated magnetic anomaly lineations (Mizuno and Moritani, 1977; Moritani, 1979; Mizuno and Nakao, 1982; Nakao and Moritani, 1984). In this study, we compiled and interpreted seismic reflection data and 3.5-kHz/12-kHz

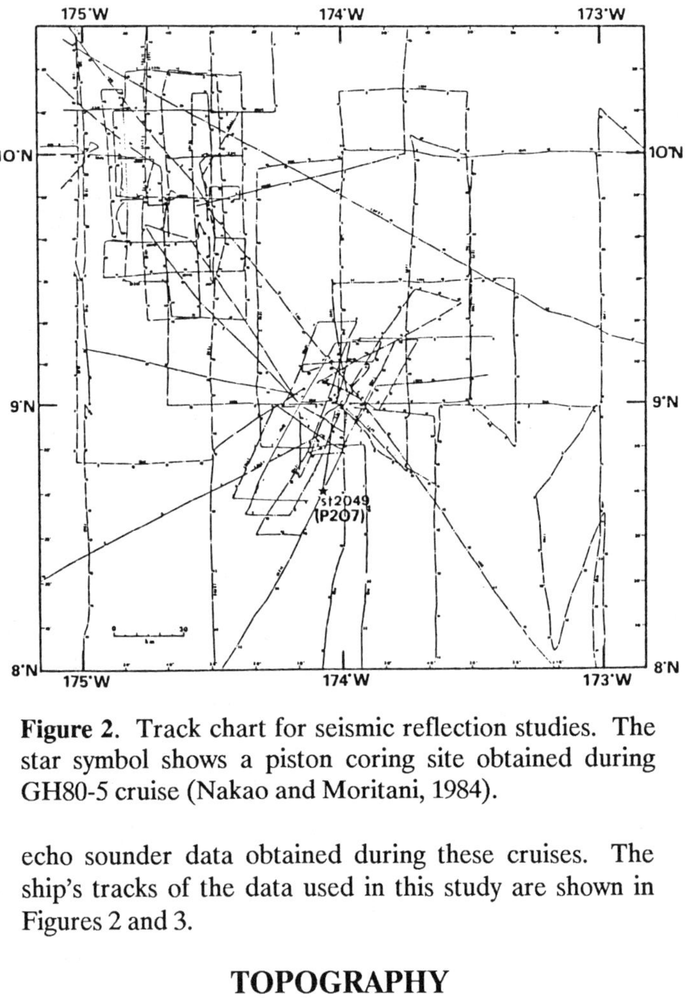

Figure 2. Track chart for seismic reflection studies. The star symbol shows a piston coring site obtained during GH80-5 cruise (Nakao and Moritani, 1984).

echo sounder data obtained during these cruises. The ship's tracks of the data used in this study are shown in Figures 2 and 3.

TOPOGRAPHY

The study area, i.e., the Magellan basin, is located in the Central Pacific Basin about 2000 km southwest of the Hawaiian Islands and northeast of the Magellan Rise (Fig. 1). Bathymetric data were obtained from 3.5- and 12-kHz records along most of the track lines shown in Figure 2. A detailed bathymetric chart of this study area is shown in Figure 3.

The general trend of the topography is WNW to EW, and it converges to the northwest. The Magellan basin is divided into four areas from north to south: the north

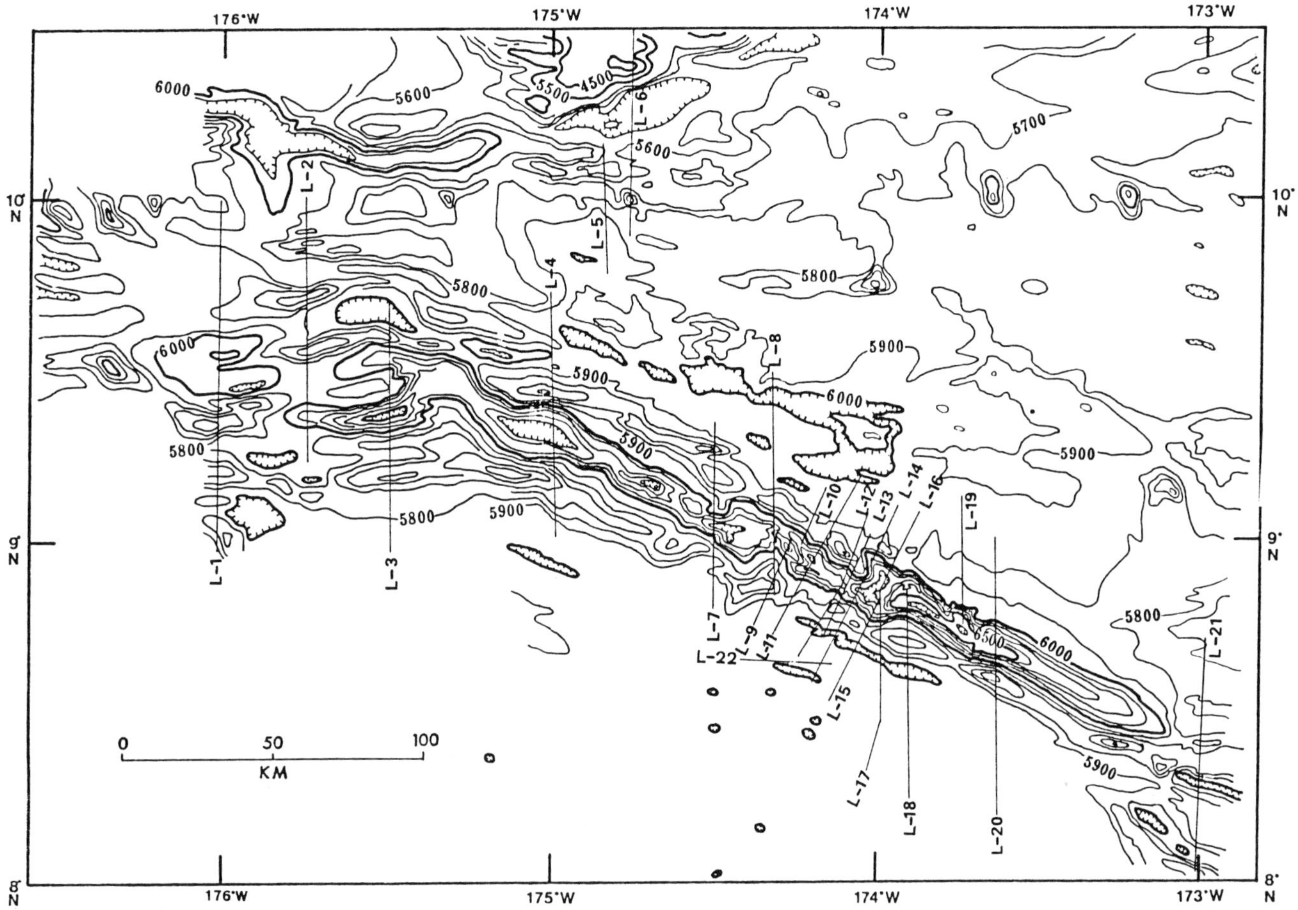

Figure 3. Detailed bathymetric contour chart of the study area. Water depth is shown in meters. Contour interval is 100 meters. Solid lines show the ship's track for the reference of seismic profiles and 3.5-kHz echo sounder profiles that appear in the following figures. Western part of the chart between 175° 10'W and 176° 30 'W was modified from Onodera et al. (1979).

Magellan plateau with minimum water depth of about 3800 m; the north Magellan basin with a distribution of small deep-sea knolls north of the Magellan trough; the Magellan trough, the most prominent feature in the area, with a remarkable lineated depression (Fig. 4); and the south Magellan basin, an abyssal plane about 6000 m deep.

The north Magellan plateau is located 300 km north of the Magellan Rise at about 11°N and 175°W. The study area includes the southern part of this plateau. The Magellan Rise and this plateau face each other symmetrically across the Magellan trough (Fig. 1). This plateau and the rise may be separated parts of a once-continuous feature created about 135 Ma (Mammerickx and Smith, 1979).

The north Magellan basin occupies the northeastern part of the study area, and measures generally 5800 m in depth. The maximum depth of 6000 m occurs in some small depressions trending WNW, almost parallel to the Magellan trough. Small deep-sea knolls in the area are aligned in an EW direction. The relative height of these knolls to the surrounding basin floor is 700–800 m. The basin narrows to the west, then broadens and tends to become flat toward the east.

The Magellan trough is a WNW-trending lineated depression associated on both sides with flanking ridges (Fig. 4) that extend more than 300 km from 173°W to 176°W. The general width of the trough is about 10 km and the width of the entire trough, together with the flanking ridges, is 20–35 km. The axis of the trough is 6200–6600 m in depth, which is about 300–600 m deeper than the surrounding sea floor.

The south Magellan basin in the southwestern part of the study area averages 6000 m in depth. The basin often includes some linear small, WNW-trending, depressions concordant with the general direction of the Magellan trough. This area is flatter and deeper than the north Magellan basin area.

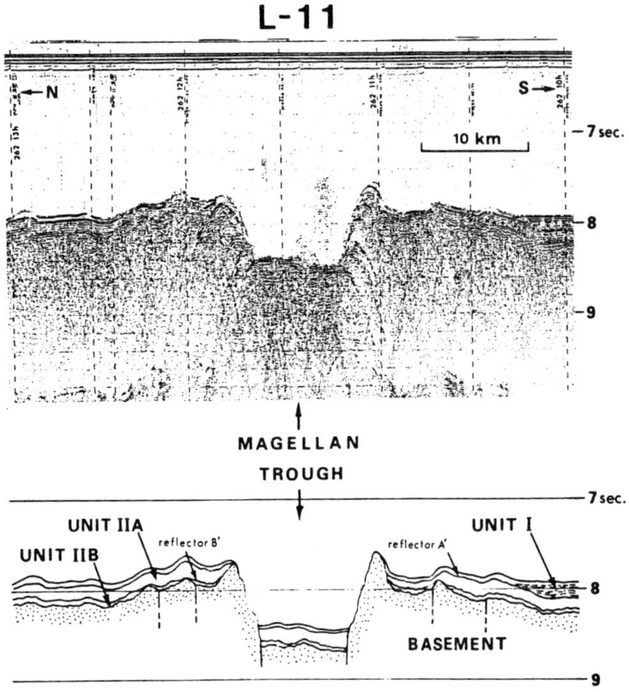

Figure 4. A typical seismic profile across the Magellan trough and its interpretation. The dotted portion shows acoustic basement. Location of the profile is shown in Figure 3.

MORPHOLOGY OF THE MAGELLAN TROUGH

Seismic reflection surveys were carried out along the ship's tracks shown in Figures 2 and 3. The sound source was an air-gun with a 120 cubic-inch firing chamber. Reflection profile tracings and reflection records across the Magellan trough are shown in Figures 5 and 6, respectively. In the following text all locations of profiles refer to Figure 3.

The Magellan trough region includes the trough itself and its flanking ridges (Figs. 3 and 4). The zone strikes approximately N65°W. The trough is structurally continuous for a distance of at least 200 km, between the profiles L-4 and L-20 (Figs. 5 and 6), and is traceable over a total distance of about 300 km (Fig. 3) between the profiles L-1 and L-21 (Figs. 5 and 6). It is never wider than 25 km and commonly is less than 15 km wide including flanking ridges. Maximum depth of the trough axis exceeds 6600 m.

East of the profile L-21 (Fig. 5) at 173°W, the trough rapidly loses its topographic expression. A narrow depression trending WNW is recognized as a segment of the trough at 173°W. This narrow depression shows a right lateral offset of about 10 km from the main axis of the trough (Fig. 3). West of the profile L-3 (Fig. 5), at 175°30', the trough loses its linear character and is divided into two segments,

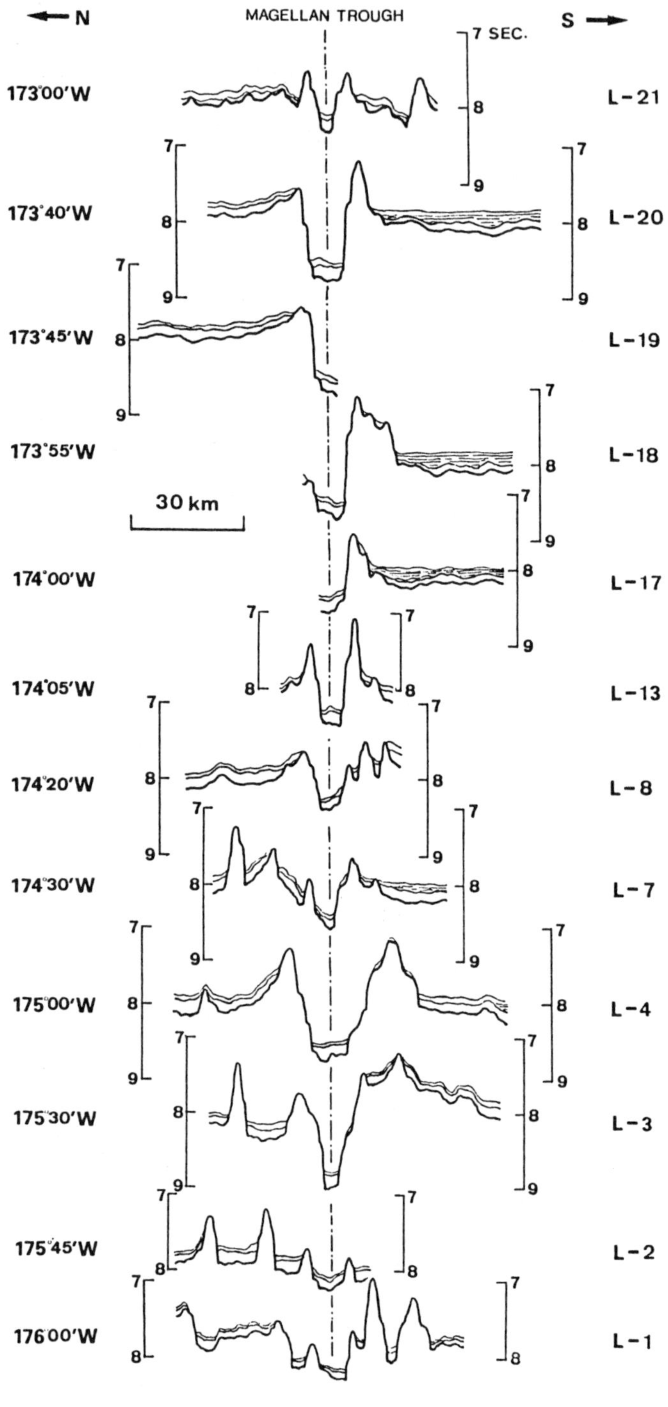

Figure 5. Interpretations of north-south seismic profiles across the Magellan trough. Upward is east. Locations are shown in Figure 3.

which show an EW alignment (Fig. 3). At about 176°W, along the profile L-1 (Fig. 5), a shallow depression oriented EW is recognized as a segment of the trough, which is topographically the western end of the trough.

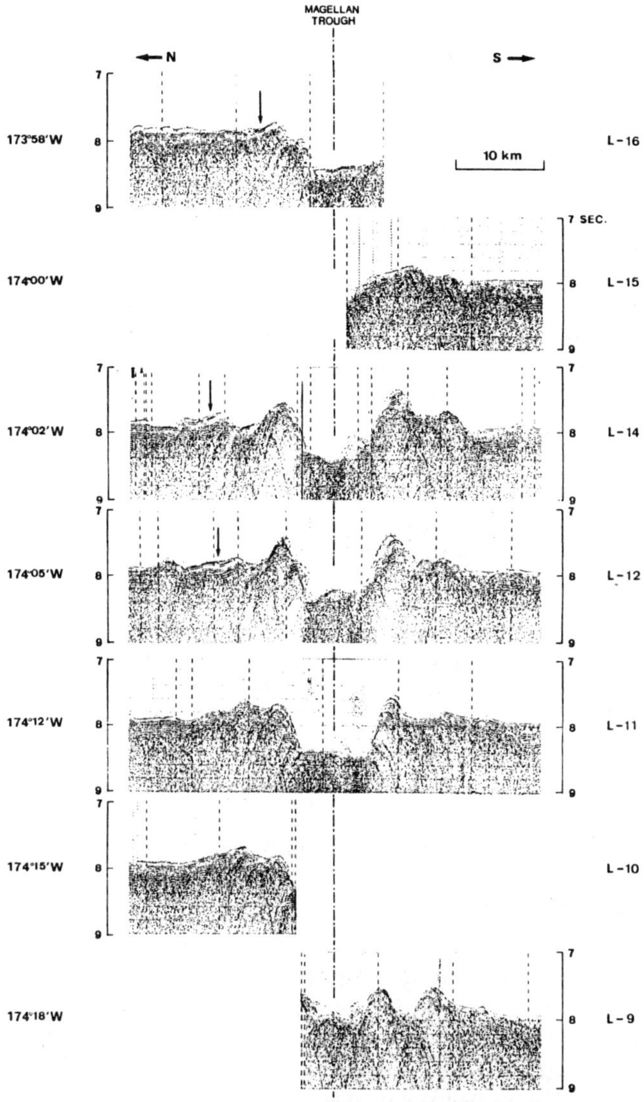

Figure 6. North-south seismic profiles across the Magellan trough. Upward is east. Arrows show the occurrences of the transparent Unit I. Locations are shown in Figure 3.

At 174°W-174°25'W, the trough axis, together with the flanking ridges, shows left lateral offsets of about 10 km (Fig. 3). Left lateral offsets of the trough are remarkable in the eastern half of the trough. In these offset areas, the trough floors are wider than in the other areas of the trough. Between 174°30'W and 174°W (the profiles L-7 to L-17; Figs. 5 and 6), the bottom of the trough is also irregular (Fig. 3). The bottom of the trough is interrupted by a small knoll between 174°10'W and 174°20'W along the profile L-9 (Fig. 6) and shows an axial elevation at 174°05'W along the profile L-12 (Fig. 6).

The flanking ridges show complicated and irregular features. Most of the flanking ridges lie at depths of 5300 to 5700 m. The relative height of the flanking ridges above the surrounding sea floor ranges from 200 m to 700 m. The southern ridge is generally higher than the northern ridge, showing the relative height from the bottom of the trough to the peak of the southern ridge as 1000 m (Figs. 5 and 6). Maximum relief occurs along the profile L-3 (Fig. 5), where peak-to-trough relief approaches 1200 m in horizontal distance of 20 km. Seismic reflection profiles show that the flanking ridges are composed generally of two or three terraces, with steep scarps facing inward toward the trough axis and with planar tops dipping gently outward (Figs. 4, 5, and 6). The slopes of the flanking ridges are more gentle in the western half of the trough area than in the eastern half. The profiles suggest that the terraces were formed by block faulting (Fig. 4). Similar block faulting is observed within acoustic basement of the north Magellan basin area, especially around the base of the north Magellan Plateau (Fig. 7).

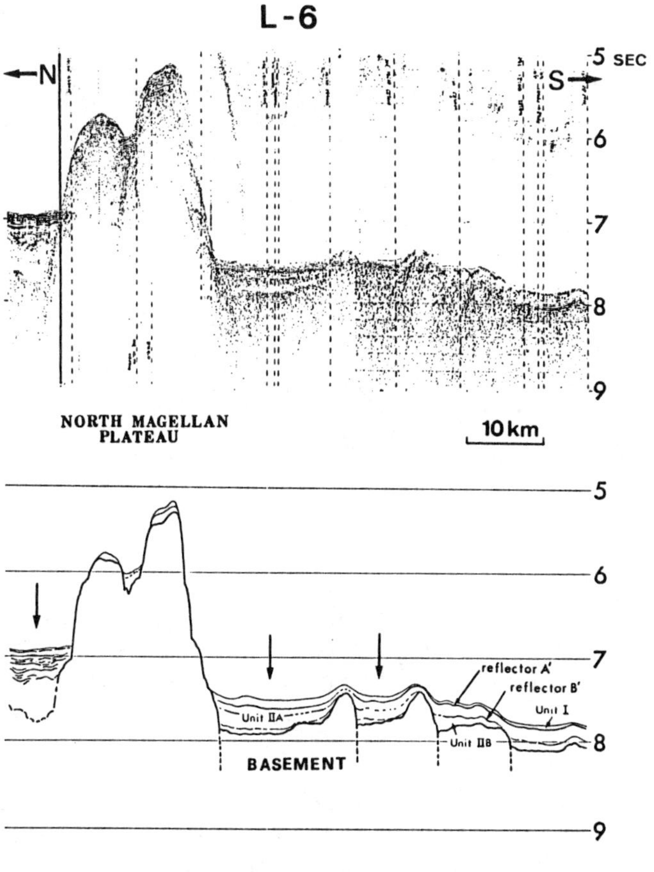

Figure 7. A seismic profile across the north Magellan plateau and its interpretation. Arrows show the occurrence of the thickened Unit II. Location is shown in Figure 3.

ACOUSTIC STRATIGRAPHY

Two prominent reflecting horizons are commonly observed above acoustic basement in the study area (Figs. 4, 7, and 8). These two horizons have been observed elsewhere in the northern Pacific area and are called reflectors A' and B' (Ewing et al., 1966). These reflectors are good tracers for identifying acoustic sequences. Acoustic sequences are divided in descending order into Unit I, Unit II, and acoustic basement. Unit II is divided into two subunits, an upper Unit IIA and a lower Unit IIB. Reflectors A' and B' represent the tops of Unit IIA and Unit IIB respectively. These acoustic units can be traced very easily throughout the study area and appear to be layered conformably.

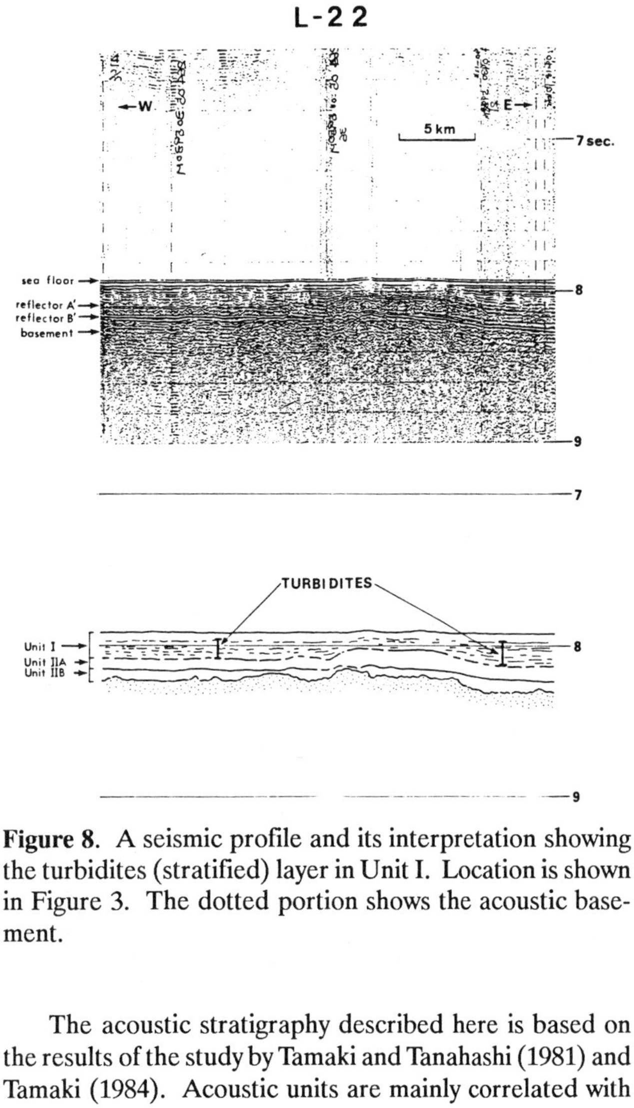

Figure 8. A seismic profile and its interpretation showing the turbidites (stratified) layer in Unit I. Location is shown in Figure 3. The dotted portion shows the acoustic basement.

The acoustic stratigraphy described here is based on the results of the study by Tamaki and Tanahashi (1981) and Tamaki (1984). Acoustic units are mainly correlated with results of DSDP Leg 17 sites 165, 166, and 170 around the study area (Winterer, Ewing, et al., 1973; locations are shown in Fig. 1). It is very difficult to estimate the lithology for each acoustic unit, especially Unit II, because of lack of drilling data in the study area.

Unit I

Unit I is generally acoustically transparent or semi-transparent, and is correlated mainly with the Quaternary to the middle Eocene brown clay or siliceous ooze. Sediment samples from most of Unit I were obtained in the GH80-5 cruise by box coring and piston coring; their geologic sequence and age combined with paleontological (Nishimura, 1984) and paleomagnetic (Joshima and Nishimura, 1984) studies yield the following sedimentary sequence (in descending order): brown, siliceous, fossil-rich clay of the recent to the late Pleistocene; dark brown pelagic clay of the late Pleistocene to the late Pliocene; very dark grayish brown zeolitic clay with siliceous ooze, siliceous clay, and calcareous ooze of the late Miocene to the late Eocene; and brown to yellowish brown, siliceous ooze with chert of the late Eocene to the early Eocene. This sedimentary sequence shows the existence of a hiatus between the late Pliocene and the late Miocene.

An isopach map of Unit I is shown in Figure 9. The thickness of the transparent to semi-transparent layer of Unit I does not exceed 0.1 sec in two-way travel time (~75 m) and is commonly observed to be less than 0.025 sec in thickness in the north Magellan basin area. The average thickness is about 0.05 sec. Unit I tends to become thinner on the topographically high area, and to become thicker in the depression. This tendency, however, is not systematic. Thinner accumulations of Unit I are sometimes observed in the depressions. In the Magellan trough, normal thicknesses of Unit II and very thin accumulations of Unit I (0.01–0.02 sec) are observed (Figs. 4 and 6). This observation indicates that the thickness of Unit I does not depend solely upon the topography, but also upon other factors, such as deep currents.

This characteristic appearance of Unit I is observed more clearly on 3.5-kHz profiles (Figs. 10 and 11). The acoustic sequence of the 3.5-kHz profile is commonly divided into two layers: an upper transparent layer and a lower opaque layer. The transparent layer on the 3.5-kHz profile is identical with the upper half of the transparent layer (Unit I) on air-gun profiles. Some 3.5-kHz profiles show the truncation (Fig. 10) and abrupt thinning (Fig. 11) of the transparent layer. These observations suggest that the transparent layer (Unit I) is subjected to submarine erosion.

Unit I in the south Magellan basin south of the Magellan trough contains an acoustically stratified layer with a thickness of 0.1–0.15 sec (Figs. 4, 5, 6 and 8). The stratified layer tends to become thinner towards the Magellan trough and is absent in the area north of the Magellan trough. The

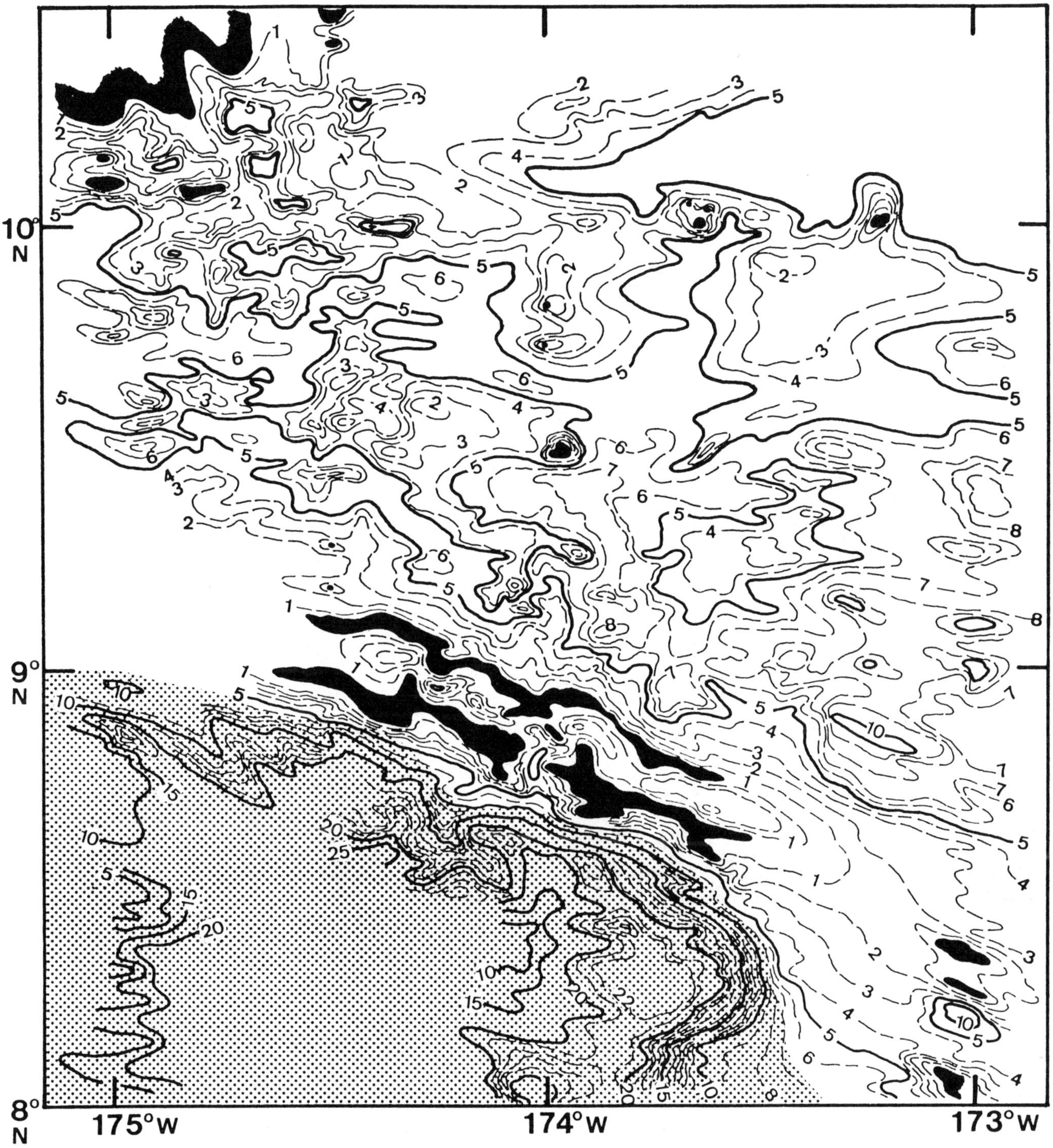

Figure 9. Isopach map of Unit I. Contour interval in seconds of two-way travel time (0.1 sec = ~75 m). The shaded area shows the distribution of the turbidites (stratified) layer in Unit I. The black area shows the area where Unit I is absent and/or acoustic basement is exposed at sea floor.

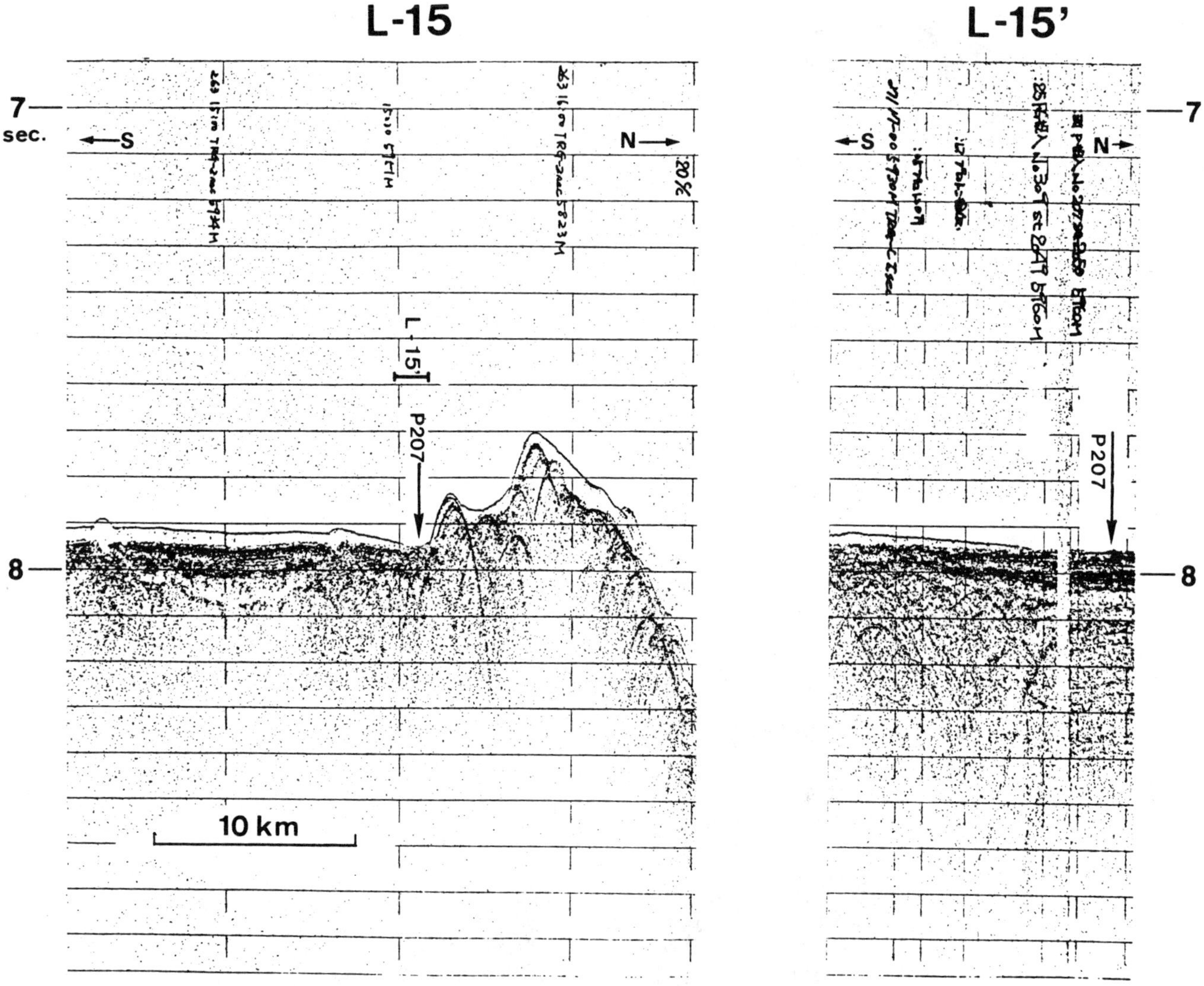

Figure 10. Acoustic character in 3.5-kHz records. Location of profiles is shown in Figure 3. Note disappearance or thinning of the transparent layer of Unit I as indicated by arrows.

southern flanking ridge of the Magellan trough, therefore, seems to have been a barrier to the deposition of this layer (Fig. 8). Similar layers are observed on the air-gun profiles near DSDP Site 165 (Winterer, Ewing, et al., 1973) and in the GH79-1 detailed survey area (location shown in Fig. 1) about 300 km northwest of the site. These similar layers (Tamaki and Tanahashi, 1981, Fig. 12) were correlated with calcareous turbidites of the middle Miocene to the middle Eocene (Winterer, Ewing, et al., 1973; Tamaki and Tanahashi, 1981). South of the Magellan trough, there is an outcrop of the stratified layer. At Station 2049 (location shown in Fig. 2), where the stratified layer outcrops (Fig. 10), the transparent layer of Unit I becomes very thin or is lacking. Piston core P207 (7.24 m long), collected during GH80-5 cruise (Station 2049), shows the existence of calcareous ooze. This ooze is mainly composed of Oligocene to early Miocene calcareous nannoplankton (Nishimura, 1984). The calcareous ooze of P207, therefore, can be correlated to the uppermost part of the stratified layer. Calcareous nannoplankton are not observed in the surface sediment samples of the basin area (Nishimura, 1984), because present calcium carbonate compensation depth of this area is shallower than 5100 m (Takayanagi et al., 1982). These results suggest that the stratified layer is composed of calcareous turbidites. Distribution of the turbidites in the study area are shown in Figure 9.

Unit II

Reflector A', the top of Unit II, is correlated with the top of the middle Eocene chert bed (Tamaki and Tanahashi, 1981; Tamaki, 1984), which suggests that the study area had

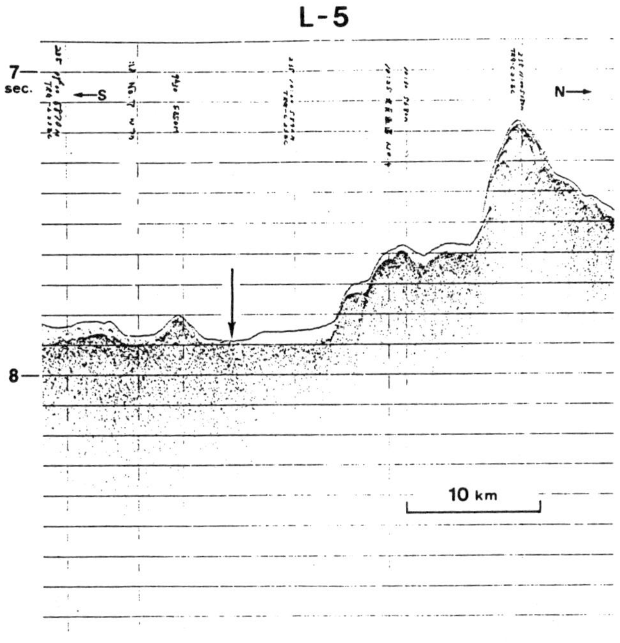

Figure 11. Acoustic character in 3.5-kHz record. Location of profile is shown in Figure 3. Arrow shows abrupt change in thickness of the transparent layer of Unit I.

been situated in the equatorial high productivity province of siliceous ooze in the pre-late Eocene time. The siliceous sediments deposited in the Eocene were probably changed to chert through diagenesis (Heezen et al., 1973). Bottom sediments above siliceous ooze of the early to late Eocene are mainly composed of zeolitic and pelagic brown clay of the late Eocene to the Recent (Nishimura, 1984), suggesting that this area passed out of the high productivity province of siliceous ooze toward the late or middle Eocene. These data suggest a northward movement of this area, concordant with that of the Pacific plate.

Unit IIA, in general, is an acoustically semi-opaque layer. The results of DSDP sites 165, 166, and 170 (Winterer, Ewing, et al., 1973) indicate that Unit IIA is likely to be composed of various sediments such as chert, marl, limestone, sandstone, and volcanogenic sand, silt, clay stone, and turbidites of the Cretaceous to the middle Eocene. Unit IIA shows two different facies in its upper and lower parts. The upper part of Unit IIA often is acoustically semi-transparent, and the lower part is acoustically stratified. Transparent Unit IIA is locally observed along the profiles L-12, L-14, and L-16 (Fig. 6). The transparent Unit IIA is distributed along the foot of the northern flanking ridge of the Magellan trough at about 174°W and 9°N, where the trough offsets (Fig. 3). The thickness of Unit IIA is generally 0.2 sec. The thickest Unit IIA, more than 0.3 sec, is observed at the foot of the north Magellan Plateau

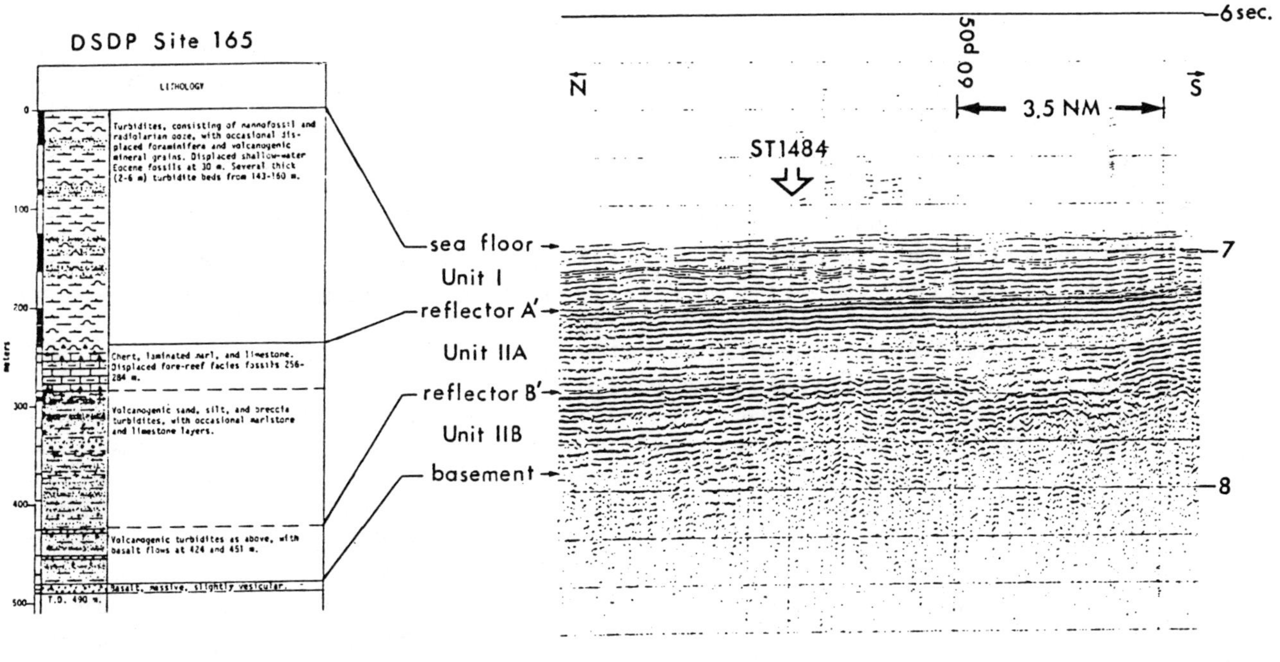

Figure 12. A typical seismic reflection record in the GH79-1 detailed survey area showing the correlation with the results of DSDP site 165 (after Tamaki and Tanahashi, 1981).

(Fig. 7). In the south Magellan basin, Unit IIA under the turbidite layer in Unit I is thinner (thickness 0.05 sec) than in the other areas, where reflector A' is sometimes lacking or is weakened, suggesting the erosion by turbidity currents (Fig. 8).

The distribution of reflector B' is more restricted than that of reflector A'; it develops in the topographically low area. Reflector B', the top of Unit IIB, is correlated with thin basalt flows above the acoustic basement, on the basis of results of DSDP site 165 (Tamaki and Tanahashi, 1981; Tamaki, 1984).

Unit IIB is acoustically semi-opaque or stratified. Results from DSDP Site 165 (Winterer, Ewing, et al., 1973) show this unit is correlated with thin basalt flows and volcanogenic turbidites (Tamaki and Tanahashi, 1981; Tamaki, 1984) which appear to have been ponded or deposited in depressions on acoustic basement. The thickness of Unit IIB is less than 0.1 sec.

Acoustic Basement

Acoustic basement, i.e., the acoustically opaque layer, is correlated with basalt (Tamaki and Tanahashi, 1981; Tamaki, 1984). The basement ages around the study area are 80 Ma, 120 Ma, and 100 Ma, based on the results of DSDP sites 165, 166 and 170 respectively (Fig. 1). From the identification of magnetic anomaly lineations, it was deduced that the basement age of this study area is M9 to M11 time in the Cretaceous (Tamaki et al., 1979; Tamaki and Larson, 1988).

A depth-to-basement map of the study area is shown in Figure 13. The topography of the acoustic basement shows EW-WNW trending structure. In the south Magellan basin, the basement tends to become deeper (6100–6200 m) away from the Magellan trough, but in the north Magellan basin, the basement tends to become shallower (6000–6100 m) away from the trough. Several EW-trending faults are observed in the basement of the north Magellan basin (Fig. 13). These faults are normal faults and do not cut the sedimentary units above acoustic basement (Figs. 4 and 7). Acoustic basement is exposed at the sea-floor in some places, e.g., along parts of the flanking ridges of the Magellan trough (Figs. 4, 5, and 6), on the north Magellan Plateau (Fig. 7), and at the top of small deep-sea knolls (Fig. 9). The EW-trending faults and the EW to WNW trending structure, which are observed within the acoustic basement, converge to the northwest.

DISCUSSION AND CONCLUSIONS

The Magellan trough strikes about N65°W and the general trend of the surrounding topography converges to the northwest. Topographic features of the sea floor (Fig. 3), morphology of the acoustic basement (Fig. 13) – which defines major features of the sea floor, and the distribution of faults (Fig. 13) are concordant with the magnetic anomaly lineation patterns (Fig. 14), the Magellan sets, which were identified by Tamaki et al. (1979) as M9 to M11 in the Early Cretaceous. These concordant features suggest that the formation of acoustic basement morphology and structure originated during crustal accretion by sea-floor spreading with an axis at the Magellan trough in the Early Cretaceous.

Depth of basement in the Magellan trough area increases from north to south. Several EW-trending faults are mainly observed in the basement of the north Magellan basin (Figs. 7 and 13). Assymmetric rotation occurred along the Magellan spreading system during M11 to M9 time (Tamaki et al., 1979) such that the northern flank

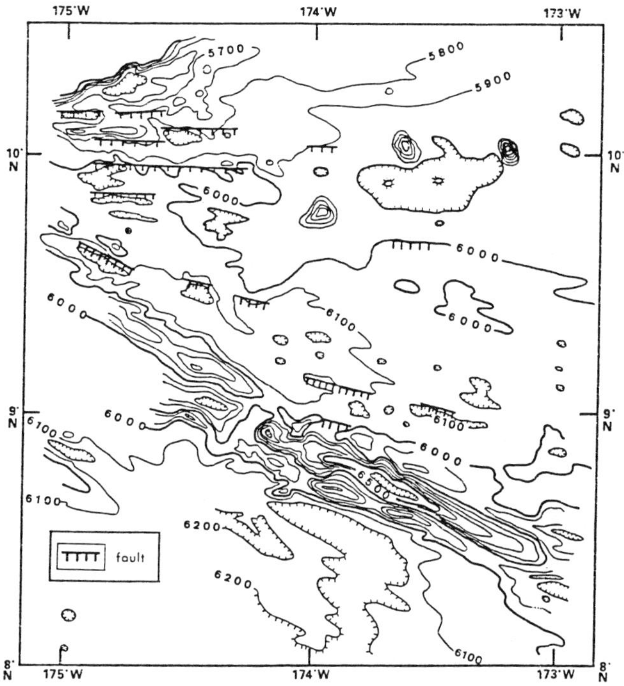

Figure 13. Depth-to-basement map and distribution of normal faults within the acoustic basement. Contour interval is 100 m.

rotated more rapidly (4.2°/m.y.) than the southern one (3.1°/m.y.). The above differences in acoustic basement morphology between the northern area and the southern area thus may be related to the difference in the spreading rates between the northern flank and the southern flank of the Magellan trough.

The Magellan trough is similar in topography and structure to some active spreading centers on the mid-oceanic ridges or some fossil spreading centers, for instance, the Escanaba trough on the Gorda Rise in the northeastern Pacific (Moore and Sharman, 1970), the Mal-

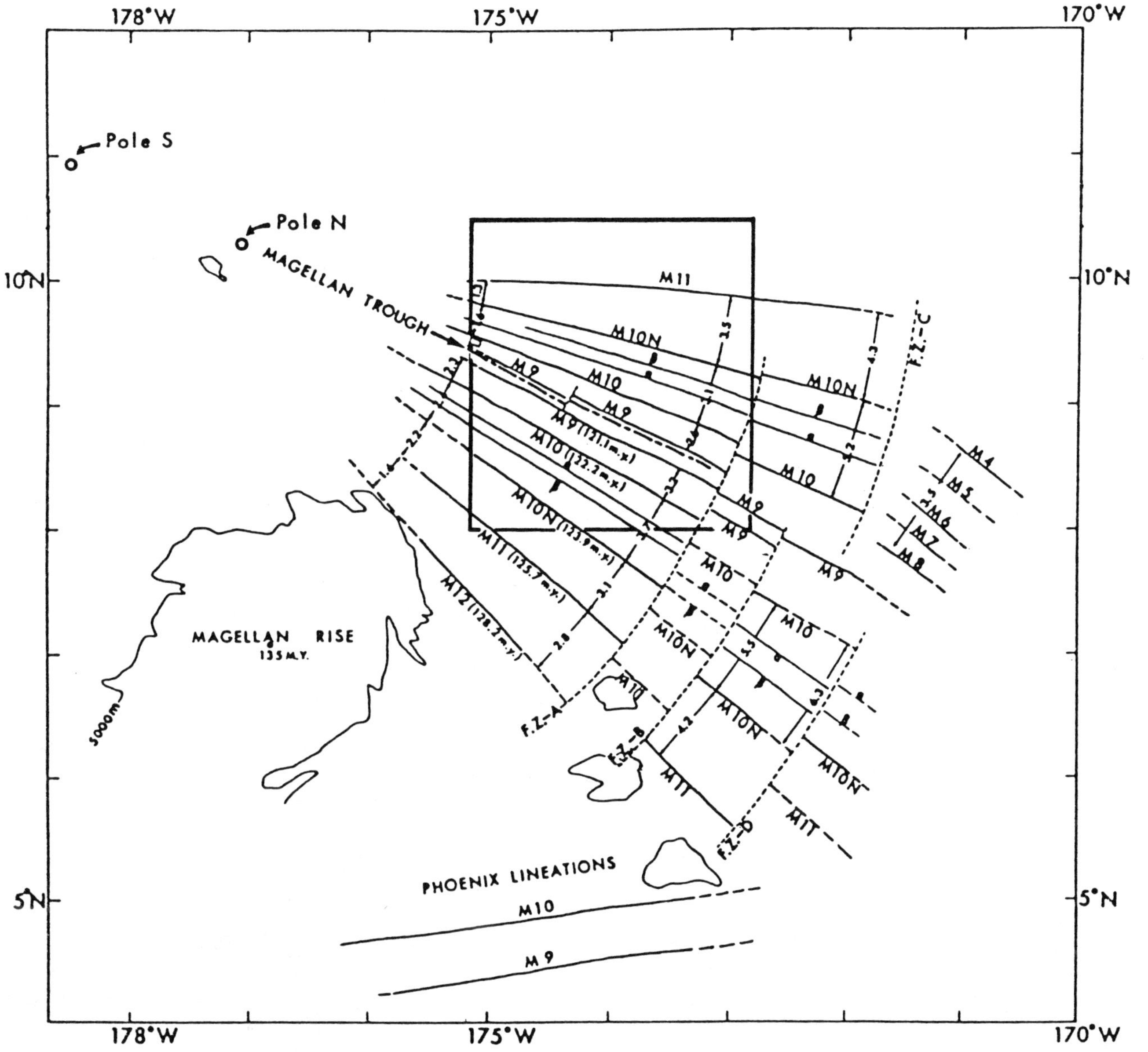

Figure 14. Magnetic lineation patterns of the Magellan trough area. The small square shows the study areas as shown in Figures 9 and 13. Tamaki, K., M. Joshima & R.L. Larson, J. of Geophysical Research, V. 84, pp. 4501-4510, 1979. © by the American Geophysical Union.

pelo Rift in the Panama Basin (Lonsdale and Klitgord, 1978), the Emperor trough in the North Pacific Basin (Rea and Dixon, 1983), and the Canton trough in the Central Pacific Basin (Rosendahl et al., 1975). These troughs and rifts have been interpreted to be slowly to moderately spreading ridge axes. Differences in axial topography between slow and fast spreading mid-ocean ridges is well known (e.g., Macdonald, 1982). The characteristic axial rift-valley and flanking crestal mountains that are formed at slow spreading rates do not exist at fast spreading ridges, where instead relatively low relief axial topography is observed. The topography of the Magellan trough to the west and the position of magnetic lineation symmetry east of the fracture zone labeled FZ-A (Fig. 14) resemble, respectively, the axial topography of the Mid-Atlantic Ridge (rift-valley) and the East Pacific Rise (subdued feature).

Magnetic anomaly lineation patterns, the Magellan sets (Fig. 14), show that half-spreading rates for the Magellan spreading system increase from west to east. The lineation patterns indicate slow half-spreading rates ranging from 1.1 to 2.4 cm/yr at the western end of the trough and intermediate rates ranging from 2.6 to 3.7 cm/yr at the eastern end, with the southern flank of the trough being faster (1.6-3.7 cm/yr) than the northern flank (1.1-3.5 cm/yr). The fastest half-spreading rates (4.2 to 5.5 cm/yr) are observed along the easternmost end of the trough, east of the fracture zone labeled FZ-A (Fig. 14), where the trough loses its characteristic bathymetric features (Fig. 3) along the profile L-21 (Fig. 5). We conclude, therefore, that the topographic variation along the Magellan trough can be accounted for largely by variations in spreading rates of the Magellan spreading system. The above morphotectonic features are good indicators of the fundamental processes of the sea-floor spreading in the Mesozoic Pacific.

An isopach map of Unit I (Fig. 9) and some seismic profiles (Figs. 4, 5, 6, and 7) show various thicknesses of the transparent layer in Unit I. Some 3.5-kHz records (Figs. 10 and 11) show the truncation and the abrupt thinning of the transparent layer in Unit I. Nemoto and Kroenke (1981) indicated the pronounced effect of bottom water current on the present-day environment of deposition in the Hess Rise, and summarized Antarctic bottom water (AABW) current patterns in the Pacific, which identify those branches of AABW current that affect this study area. The variation of Unit I may be interpreted as erosion by AABW current. In the south Magellan basin, the transparent layer of Unit I above the turbidites is thinner than that in the north Magellan basin north of the Magellan trough (Figs. 4, 6, and 8), probably due to the effect of bottom water currents.

The turbidite layer in Unit I exists only in the south Magellan basin (Fig. 9). Some seismic profiles across the trough area (Figs. 4, 5, and 6) show the layer tends to become thinner towards the trough. The distribution of the turbidites (Fig. 9) indicates that the southern flanking ridge of the trough probably acted as a barrier to the deposition of the layer. Turbidites around DSDP Site 165 originate on the higher elevations of the Central Basin rise and flow into the abyssal depths of the Central Pacific Basin, and turbidite flows appear to have aggraded continuously since Eocene time (Orwig, 1981). However, the spread of these turbidites into the study area is not observed. The turbidites of the study area, therefore, probably were derived from the Magellan Rise (Fig. 1), where deposits of calcareous sediments of the earliest Cretaceous to the Quaternary as thick as 1200 m were found to occur at DSDP Site 167 (Winterer, Ewing, et al., 1973), shown in Figure 1. The rise forms a large plateau of elliptical shape oriented in a NNE direction (Fig. 1). The relatively flat top is at 3200 m depth (Onodera et al., 1979). Some eroded features are observed at the outer edges of the rise (Murakami and Moritani, 1979). These observations suggest that the turbidity currents may have been caused by a northward-flowing bottom water current such as the recent AABW current.

The sedimentary sequence of Unit I indicates existence of a hiatus between the late Pliocene and the late Miocene. Winterer, Ewing, et al. (1973) discussed the erosion between the early Miocene and the Quaternary near DSDP Site 165 and argued that it might indicate vigorous flow of bottom water near the site. Mizuno et al. (1980) and Nishimura (1981) discussed the hiatus between the middle Miocene and the late Miocene (or the late Pliocene) from the results of piston core samples collected in the GH79-1 detailed survey area (Fig. 1). They concluded that the hiatus may have been related to repeated northward flow of the AABW associated with the development of a large ice cap on Antarctica around 12 Ma as clarified by van Andel et al. (1975). In the same way, the hiatus in Unit I of the study area may have been caused by the presence of a deep water current similar to the present day AABW current.

In the north Magellan basin area, the sea-floor topography appears to be controlled by acoustic basement morphology, i.e., sedimentary units conformably overlie the topography of the acoustic basement (Figs. 4, 6, and 7). In the south Magellan basin area, however, the sea-floor topography is diminished (Fig. 3), becoming flatter even over a rugged acoustic basement (Figs. 4, 6, and 8). In this area, even though Unit II is nearly conformable with acoustic basement, the turbidites in Unit I bury the surface topography of Unit II, eroding the upper part of Unit II in some places, destroying reflector A' (Fig. 8). In the south Magellan basin, therefore, the sea-floor morphology is more suppressed and becomes flatter than in that of the north Magellan basin (Fig. 3).

The above findings suggest that the deposition of Unit I and sea-floor topography in the Magellan basin have been affected by repeated bottom water currents similar to the present day AABW current.

ACKNOWLEDGMENTS

We are grateful to H. Ujiie and K. Kizaki of University of the Ryukyus, and N. Isezaki of Kobe University for the helpful suggestion and encouragement. We are much indebted to scientific staff of the Geological Survey of Japan and crews of R/V *Hakurei-Maru* (H. Okumura, Captain) for their assistance with the survey.

REFERENCES

Ewing, M., T. Saito, J.I. Ewing, and L.H.Burckle, 1966, Lower Cretaceous sediments from the northwest Pacific: Science, v. 152, p. 751-755.

Heezen, B.C., I.D. MacGregor, et al., 1973, The post Jurassic sedimentary sequence on the Pacific Plate; A kinematic interpretation of diachronous deposits: Initial Reports of the Deep-Sea Drilling Project, v. 20, p. 725-738.

Joshima, M., 1981, Remanent magnetization of deep-sea sediments in the GH79-1 area: Geological Survey of Japan Cruise Report, no. 15, p. 223-228.

Joshima, M. and A. Nishimura, 1984, Remanent magnetization of sediment cores in the GH80-5 survey area: Geological Survey of Japan Cruise Report, no. 20, p. 165-192.

Larson, R.K., S.M. Smith, and E.G. Chase, 1972, Magnetic lineations of Early Cretaceous age in the Western Equatorial Pacific Ocean: Earth and Planetary Science Letters, v. 15, p. 315-319.

Lonsdale, P. and K.D. Klitgord, 1978, Structure and tectonic history of the eastern Panama Basin: Geological Society of America Bulletin, 89, 981-999.

Macdonald, K.C., 1982, Mid-ocean ridges: Fine scale tectonic, volcanic, and hydrothermal processes within the plate boundary zone: Annual Review of Earth and Planetary Science, v. 10, p. 155-190.

Mammerickx, J. and S.M. Smith, 1979, North Magellan Plateau: A possible symmetric twin to the Magellan Plateau [abs.]: Eos, Transactions of the American Geophysical Union, v. 60, p. 950.

Mizuno, A. and T. Moritani (Eds.), 1977, Deep sea mineral resources investigation in the central-eastern part of Central Pacific Basin (GH76-1 cruise): Geological Survey of Japan Cruise Report, no. 8.

Mizuno, A., T. Miyazaki, A. Nishimura, K. Tamaki, and N. Tanahashi, 1980, Central Pacific manganese nodules and their relation to sedimentary history (OTC 3830), paper presented at 12th Annual Offshore Technology Conference in Houston, Tex., p. 331-340, May.

Mizuno, A. and S. Nakao (Eds.), 1982, Regional data of marine geology, geophysics,and manganese nodules: the Wake-Tahiti transect in the Central Pacific (GH80-1 cruise): Geological Survey of Japan Cruise Report, no. 18.

Moore, G.W. and G.F. Sharman, 1970, Summary of site 4: Initial Reports of the Deep-Sea Drilling Project, v. 15, p. 761-773.

Moritani, T. (Ed.), 1979, Deep sea mineral resources investigation in the central-western part of Central Pacific (GH77-1 cruise): Geological Survey of Japan Cruise Report, no. 12.

Murakami, F. and T. Moritani, 1979, Continuous seismic reflection profiling survey: Geological Survey of Japan Cruise Report, no. 12, p. 74-102.

Nakao, S. and T, Moritani (Eds.), 1984, Marine geology, geophysics, and manganese nodules in the northern vicinity of the Magellan trough (GH80-5 cruise): Geological Survey of Japan Cruise Report, no. 20.

Nemoto, K. and L.W. Kroenke, 1981, Marine geology of the Hess Rise, 1 Bathymetry, surface sediment distribution, and environment of deposition: Journal of Geophysical Research, v. 86, p. 10734-10752.

Nishimura, A., 1981, Deep-sea sediments in the GH79-1 area: their geological properties: Geological Survey of Japan Cruise Report, no. 15, p. 110-135.

Nishimura, A., 1984, Deep-sea sediments in the GH80-5 area in the northern vicinity of the Magellan trough: Geological Survey of Japan Cruise Report, no. 20, p. 67-89.

Onodera, K., F. Murakami, and T. Moritani, 1979, Submarine topography by 12 kHz PDR: Geological Survey of Japan Cruise Report, no. 12, p. 33-43.

Orwig, T., 1981, Channeled turbidites in the eastern Central Pacific Basin: Marine Geology, no. 39, p. 33-57.

Rea, D.K. and M. Dixon, 1983, Late Cretaceous and Paleogene tectonic evolution of the North Pacific Ocean: Earth and Planetary Science Letters, no. 65, p. 145-166.

Rosendahl, B.R., R. Moberly, A.J. Halunen, J.C. Rose, and L.W. Kroenke, 1975, Geological and geophysical studies of the Canton trough region: Journal of Geophysical Research, v. 80, p. 2565-2574.

Takayanagi, Y., T. Sakai, M. Oda, and S. Hasegawa, 1982, Micropaleontology of piston cores, Wake-Tahiti:. Geological Survey of Japan Cruise Report, no. 18, p. 238-263.

Tamaki, K., M. Joshima, and R.L. Larson, 1979, Remanent Early Cretaceous spreading center in the Central Pacific Basin: Journal of Geophysical Research, v. 84, p. 4501-4510.

Tamaki, K. and M. Tanahashi, 1981, Seismic reflection survey in the northeastern margin of the Central Pacific Basin: Geological Survey of Japan Cruise Report, no. 15, p. 77-99.

Tamaki, K., 1984, Seismic reflection survey in the Central Pacific Basin during GH80-5 cruise: Geological Survey of Japan Cruise Report, no. 20, p. 43.

Tamaki, K. and R.L. Larson, 1988, The Mesozoic tectonic history of the Magellan Microplate in the western Central Pacific: Journal of Geophysical Research, v. 93, p. 2857-2874.

van Andel, T.H., G.R. Heath, and T.C. Moore, 1975, Cenozoic history and paleoceanography of the Central Equatorial Pacific Ocean: Geological Society of America, Memoir 143, 134 pp.

Winterer, E.L., 1973, Introduction: Initial Reports of the Deep-Sea Drilling Project, v. 17, p. 5-16.

Winterer, E.L., J.I. Ewing, et al., 1973, Part 1: Site reports: Initial Reports of the Deep-Sea Drilling Project, v. 17, p. 17-282.

Keating, B.H., and Bolton, B.R., editors, 1992, Geology and offshore mineral resources of the central Pacific basin, Circum-Pacific Council for Energy and Mineral Resources Earth Science Series, V. 14. New York, Springer-Verlag.

SOUTHWEST PACIFIC SEAMOUNTS REVEALED BY SATELLITE ALTIMETRY

Peter J. Hill
Division of Marine Geosciences and Petroleum Geology, Bureau of Mineral Resources, Geology and Geophysics, GPO Box 378, Canberra, ACT 2601
and
Geology Department, Australian National University, GPO Box 4, Canberra, ACT 2601, Australia.

Nicolas Baudry
ORSTOM, BP A5, Noumea Cedex, New Caledonia
and
Laboratoire de Geophysique, Universite de Paris-Sud, 91405 Orsay Cedex, France.

ABSTRACT

GEOS-3/SEASAT radar altimeter images were used to detect major uncharted seamounts in the southwest Pacific. Verification of the predicted seamounts was undertaken during research cruises of HMNZS *Tui* and R/V *Moana Wave* during 1986–87. Four large seamounts were located and mapped by a variety of shipboard recording systems including 12-kHz and 3.5-kHz echo sounders, seismic, gravity, magnetics, and SeaMARC II.

Of the three seamounts located in the southern Line Islands area, the largest has a height of 4030 m and basal diameter of 40 km. The fourth seamount, in the Samoan Basin, has a height of 2825 m and basal diameter of 35 km. The summits of the seamounts are located at 1° 45.4′S 156° 10.0′ W, 1° 55.5′S 156° 17.3′ W, 3° 22.3′S 155° 41.0′W, and 12° 25.2′S 166° 27.9′W. Ferromanganese-oxide crusts on these seamounts constitute a potential mineral resource for Co, Ni and Cu. Crust up to 5 cm thick was dredged from the largest seamount.

Despite limited success by some marine expeditions in locating uncharted seamounts from satellite altimetry, our results indicate that it is a reliable and cost-effective technique for delineating major submarine topographic features, particularly in remote and poorly explored regions such as the southwest Pacific.

INTRODUCTION

The southwest Pacific is a region not well mapped by ship bathymetric surveys. Coverage is generally sparse and patchy, with large data gaps still present. In contrast, the altimeter sub-satellite tracks are fairly uniformly and systematically distributed (see Figure 1 for example). The considerable new information available from analysis of the altimeter observations facilitates detection of previously unknown bathymetric features (Lambeck and Coleman, 1982; Lazarewicz and Schwank, 1982; and Baudry et al., 1987).

Altimeter-derived geoid anomalies corresponding to possible uncharted seamounts were investigated during Tripartite II* research cruises by HMNZS *Tui* in 1986 and R/V *Moana Wave* in 1987 (Coulbourn, Hill et al., 1987). Seamounts in the central and southwest Pacific are economically significant since they host ferromanganese-oxide crusts potentially rich in Co, Ni and Cu (Halbach et al., 1982; Cronan, 1984). In this work, four uncharted seamounts were located and mapped. The first three of these seamounts, designated MWI, MW2, and MW3, are situated in the southern Line Islands area; the fourth (TI) lies in the Samoan Basin (Figure 1).

*Cooperative Australia, New Zealand, and U.S.A. marine geoscience research program coordinated by the Committee for Coordination of Joint Prospecting for Mineral Resources in South Pacific Offshore Areas (CCOP/SPAC), Suva, Fiji.

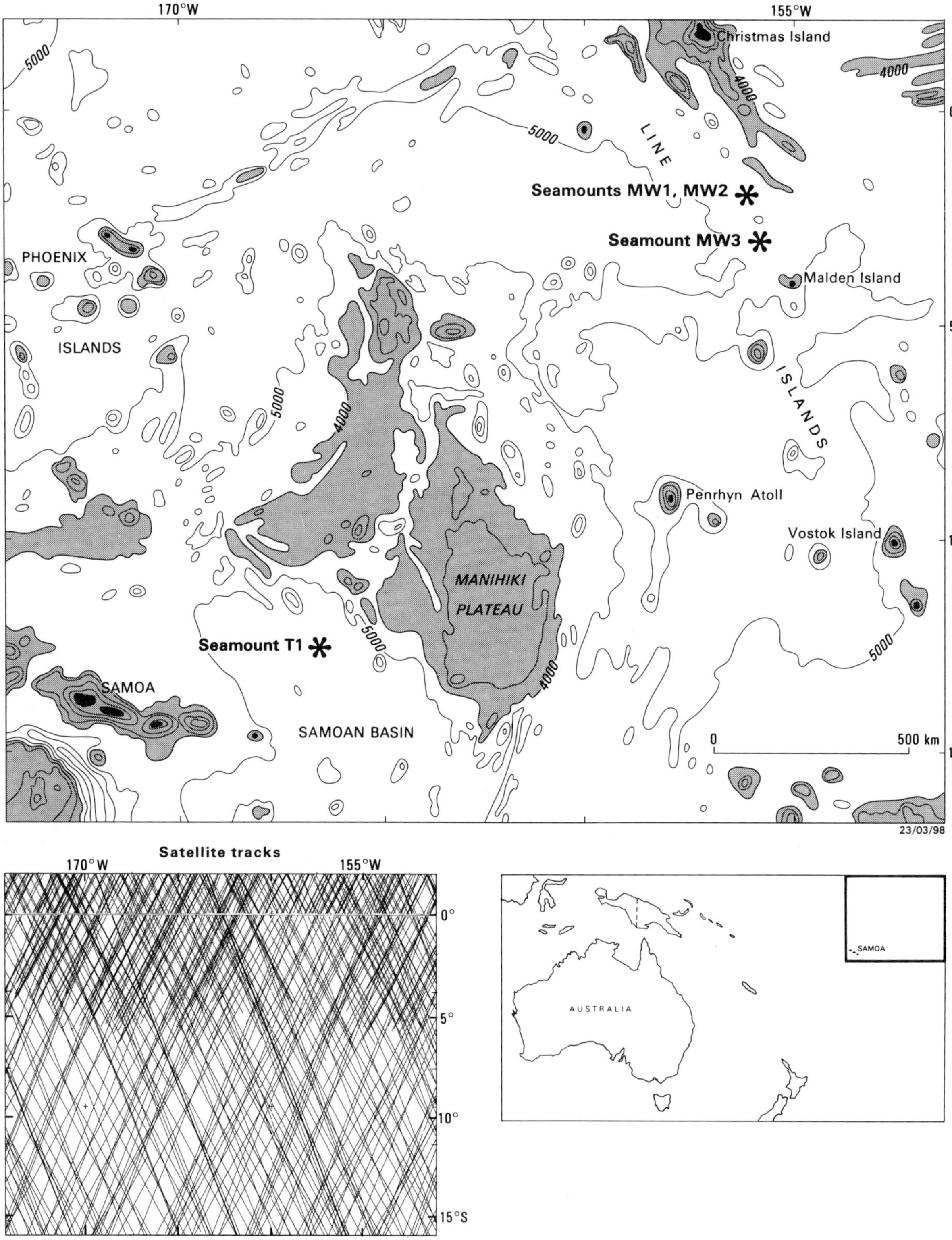

Figure 1. Location map and regional bathymetry (after Geographic Map of the Circum-Pacific Region, Circum-Pacific Council for Energy and Mineral Resources, 1978). Stippled areas indicate water depths shallower than 4000 m. The lower map depicts the same area at reduced scale and illustrates SEASAT/GEOS-3 data coverage.

Satellite radar altimeter missions GEOS-3 (Stanley, 1979) and SEASAT (Born et al., 1979) generated a vast data set of sea-surface height observations covering most of the world's oceans. GEOS-3 and SEASAT currently provide the best estimates of sea-surface heights, which are a close approximation to the marine geoid. Deviations from the geoid are produced by oceanographic effects such as currents, eddies and storms. Such perturbations generally amount to less than 0.5 m, and are typically only in the order of centimeters in the open oceans. The other contribution to local variations in the marine geoid is the gravimetric effect of lateral density variations in the crust and upper mantle. Because of the large, relatively shallow density contrast at the water/seafloor interface, seafloor topography has a major influence – hence the observed strong correlation between short-wavelength geoid anomaly and bathymetry. Seamounts, in particular, produce distinct and characteristic geoid signatures with amplitudes typically 0.3–1.0 m and diameters 25–100 km.

GEOS-3/SEASAT AND SEAMOUNT DETECTABILITY

Editing, correction, and adjustment procedures for recovery of sea-surface heights have been documented for GEOS-3 (Rapp, 1979) and SEASAT (Rapp, 1982). Our work is based on a combined GEOS-3/SEASAT data set (Liang, 1983). This data set is constructed from smoothed altimeter signal sampled at approximately 1 second intervals. This sampling rate applies to both GEOS-3 and SEASAT and represents a 7 km data spacing on the sea-surface.

The altimeter data contain an orbit determination error that can be as much as several meters, but since it is of very long wave-length, seamount signatures are not affected. For short wavelengths ($<\sim$400 km), the noise level (largely instrument noise) for the SEASAT sea-surface heights is about 10 cm and 20–50 cm for GEOS-3. Power spectra analyses indicate that the geoid signal starts to emerge from a high-frequency noise 'floor' at wavelengths of 30 km for SEASAT and 60 km for GEOS-3 (Marks and Sailor, 1986). Because of the spectral overlap of noise and geoid signature for typical seamounts (of diameter 20–60 km), little can be done in the way of further processing to substantially improve the signal/noise ratio for non-repetitive tracks.

The maximum geoid anomaly to be expected from a seamount depends on its shape, height, density, and amount of isostatic compensation of the volcanic load. This compensation mechanism governs the reconfiguration of the underlying crustal density structure. In modelling the lithosphere as a thin elastic plate overlying a weak fluid substratum, Watts and Ribe (1984) have shown that the age of the lithosphere, at the time of loading, is an important factor in the size of the geoid anomaly. As the lithosphere ages, it cools and becomes more rigid (increasing effective elastic thickness) in response to surface loading. Thus, seamounts emplaced off-ridge on the flanks show significantly higher geoid anomalies than those emplaced close to ridge crests. The difference is greatest for very large seamounts (factor of about 300%, where factor = $(\Delta G_{off} - \Delta G_{on}) / \Delta G_{on}$) but diminishes with seamount size to a factor of about 40% for more typical seamounts (30 km diameter).

Lambeck and Coleman (1982) have examined the detectability of seamounts from altimeter measurements by modelling the geoid anomaly due to conical seamounts resting on lithosphere of moderate effective rigidity. The sides of their nominal seamounts slope at 8.5° and rise from a seafloor at 4.5 km depth. These parameters are typical for the Cook-Austral region and probably apply for much of the Pacific. The results indicate that for a seamount approaching the surface (i.e., height equals 4.5 km) the maximum geoid anomaly is 2.7 m directly over the center. For a measurement noise level of 0.5 m, a perceptible geoid signal would still be obtained from sub-satellite tracks passing within 30 km of the load center. Taking the case of smaller seamounts – a 2.5 km high seamount would be detectable provided that the track passes within 10 km of its center.

Various techniques have been used to detect and locate new seamounts using satellite altimeter data. Some are 'unidimensional', using (1) matched filters to automatically detect seamount signatures (Lazarewicz and Schwank, 1982; White et al., 1983), (2) transfer functions to predict unidimensional bathymetry under each satellite track (Dixon et al., 1983), (3) computation of the deflection of the vertical (Sandwell, 1984a). These methods yield a location uncertainty ranging from 1.1 to 15 km. The cross-track uncertainty (variously quoted as 25 to more than 40 km) is thus the main factor which limits precise location of seamounts and computation of their shape and height. The accuracy of cross-track location is significantly improved for seamounts with multi-track coverage, particularly if this coverage includes non-parallel tracks (seamount MW3, discussed later, is a good example). 'Bidimensional' processing of the altimeter data, involving data interpolation between tracks (Sandwell, 1984b; Haxby et al., 1983; Haxby, 1987), reduces cross-track uncertainty. The cross-track location uncertainty associated with these methods is not precisely quantified, however.

We conducted a simple test to evaluate altimeter resolution of seamounts in our study region, using the combined GEOS-3/SEASAT data sets. Well-mapped isolated seamounts, atolls and islands with good altimeter coverage were selected, and the distance between the center of the bathymetric feature and the visually estimated center of the associated geoid anomaly high was measured. Thirty-two such measurements yielded a value of 4.0 ± 3.3 (SD) km. It seems, therefore, that a value of 15 km for the along-track uncertainty is conservative.

Seamount size cannot be determined with arbitrary precision directly from geoid or altimeter-derived gravity images (Sandwell, 1984). By modelling seamount response on two or more adjacent tracks, Baudry et al. (1987) were able to (1) improve the accuracy of the crosstrack location of seamounts to within 15 km, and (2) obtain estimates of seamount size and shape, once the emplacement environment (on-ridge or off-ridge) has been established (or assumed, in poorly known areas). Their method was applied to the Southern-Cook and Austral area, and successfully tested during two recent oceanographic cruises (Baudry and Diament, 1987; Baudry et al., 1988). Previous field investigations aimed at locating uncharted seamounts from altimeter data in the Atlantic and other parts of the Pacific had only limited success. Uncharted seamounts were detected in only about 50% of the cases tested (Keating et al., 1984).

ALTIMETER IMAGES and SELECTION OF SITES FOR INVESTIGATION

Altimeter data used during the cruises comprised, (1) black and white and colour images of SEASAT-derived gravity (provided by W. Haxby of Lamont-Doherty Geological Observatory) - both cruises, and (2) color images of along-track filtered residual sea-surface heights (Baudry, 1986) derived from the combined GEOS-3/SEASAT data sets (Liang, 1983) - R/V *Moana Wave* cruise.

The latter along-track images were produced as follows. The sea-surface height data along each track (SEASAT and GEOS-3) were band-pass filtered (35–245 km for SEASAT and 70–245 km for GEOS-3) using a Butterworth digital filter. The filtered altimeter profiles were then plotted at a scale of 1:6442192 (the same as the bathymetry map of Kroenke et al., 1983) using a color representation of the residual geoid heights, without any interpolation between adjacent profiles.

Possible uncharted seamounts were detected by visually comparing the altimeter images with available bathymetric charts, including Mammerickx et al. (1973), Kroenke et al. (1983), New Zealand Navy Hydrographic Office 1:1M plotting sheets, plots of digital data from the U.S. National Geophysical Data Center, and British Admiralty navigation charts.

NAVIGATION

Positioning aboard R/V *Moana Wave* was by satellite navigation using the U.S. Navy operated Transit and Global Positioning Systems. Though HMNZS *Tui* was equipped with a transit satellite receiver, a failure of the system meant that the initial crossing of seamount Tl was by conventional navigation. The positioning is considered reliable, however, because of a fix on Nassau Island not long before the crossing.

METHODS AND INVESTIGATION RESULTS

The three areas investigated are shown in detail in Figures 2–5. The four seamounts (MWI, MW2, MW3, and TI) are mapped with a 500-m contour interval. The GEOS-3/SEASAT sub-satellite tracks and profiles of sea-surface height derived from the Liang (1983) data set are illustrated for comparison with the bathymetry. The sea-surface heights are referred to the GRS8O ellipsoid (a = 6378137 m, f = 1/298.257222). Profiles of bathymetry and free-air gravity along R/V *Moana Wave* tracks across the seamounts are presented in Figure 5.

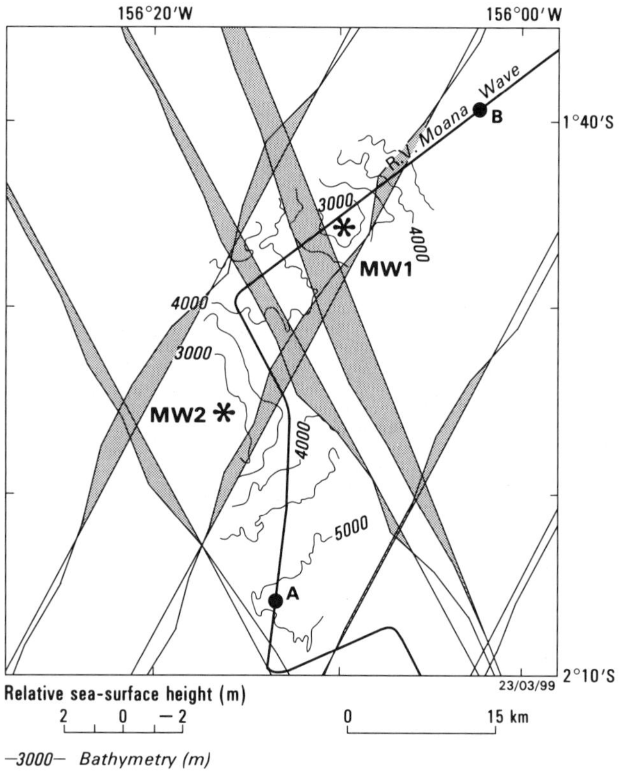

Figure 2. Area of seamounts MW1 and MW2, showing (a) research ship's track, (b) observed bathymetry (500-m contour interval), (c) GEOS-3/SEASAT sub-satellite tracks, data points, and along-track profiles of sea-surface heights. Sea-surface height increases to the left of the baselines (tracks) and positive values, relative to a base of 15.5 m, are highlighted by shading.

Table 1 provides a summary of information on the seamounts, including exact position, size, morphology, and geological/geophysical characteristics. These data are from the HMNZS *Tui* and R/V *Moana Wave* cruises.

SEAMOUNTS MWI AND MW2

The area shown in Figure 2 is well covered by a number of altimeter sub-satellite tracks. The sea-surface heights indicate a somewhat busy pattern, though there is a general geoid high of about 1.0 m over the center of the area. Using satellite altimetry, Baudry (1986) predicted a seamount at 1°54′ S 156° 12′W.

R/V *Moana Wave* crossed the area from south to north with the seafloor swath-mapping system SeaMARC II (Blackinton et al., 1983) deployed. The ship crossed the upper flank of a large seamount MW2 and then almost directly over the top of a slightly smaller, cone-shaped seamount MWI to the NNE. The geoid high thus represents a complex of at least two topographic highs. Mammerickx et al. (1973) show a relatively small high of 600 m (above surrounding seafloor) in the position of MW2 indicating that a survey may have crossed over the lower slopes of MW2 previously.

SEAMOUNT MW3

A number of criss-crossing altimeter sub-satellite tracks provide good coverage of the area in Figure 3. A well-defined geoid high of 1.3 m occurs at the center of the area clearly indicating the presence of a large seamount. A location of 3°24′ S 155°42′W was predicted by Baudry (1986).

The R/V *Moana Wave* was set on a course to intersect the center of the geoid high, with the result that the ship passed directly over the summit of the seamount on its first pass. Seamount MW3 was mapped seismically and also sampled by dredging (Table 1), with thick ferromanganese oxide crust being recovered.

SEAMOUNT Tl

As can be seen in Figure 4, seamount Tl is actually located in a 60-km gap between altimeter sub-satellite tracks. Though data along the nearest tracks to the southwest indicate a geoid anomaly of only about 0.7 m, they show a trend of increasing anomaly height towards the seamount. The possibility of a seamount was first recognized from our examination of W. Haxby's SEASAT-derived gravity images

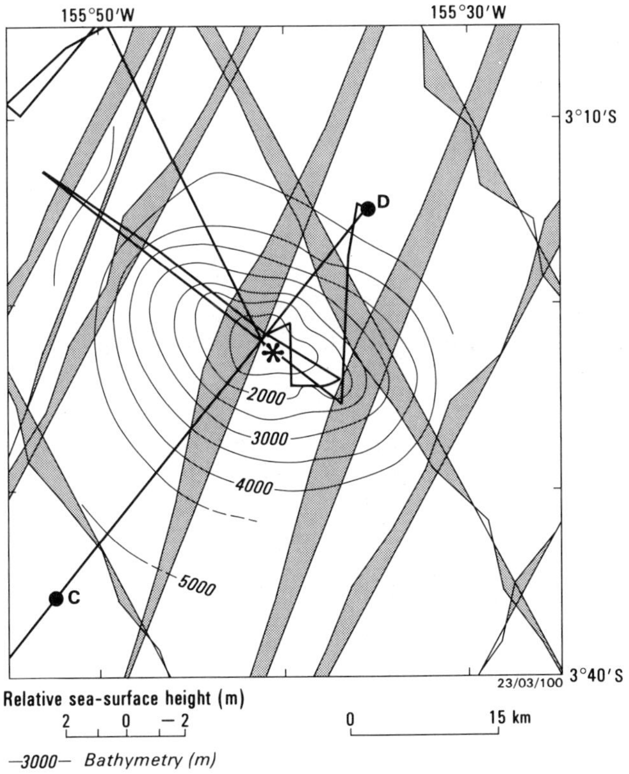

Figure 3. Area of seamount MW3. Data presentation is as in Figure 2, except that (1) some sea-surface height profiles have been omitted for clarity, and (2) the baselines represent 14.0 m.

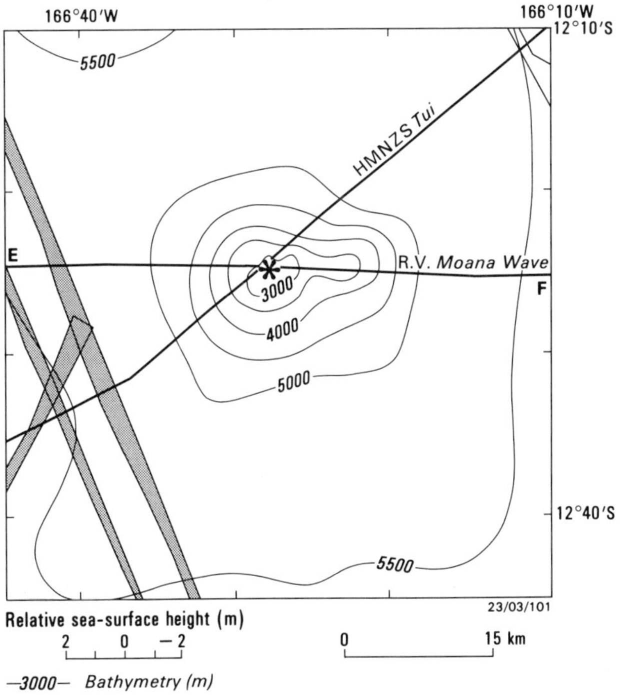

Figure 4. Area of seamount TI. Data presentation is as in Figure 2, except that the sea-surface baselines represent 17.0 m.

Table 1. Seamount statistics and characteristics.

Seamount	MW1 Southern	MW2 Southern	MW3 Southern	Tl Samoan Basin
Area	Line Islands	Line Islands	Line Islands	
Summit position	1°45.4′S 156°10.0′W	1°55.5′S 156°17.3′W (approx.)	3°22.3′S 155°41.0′W	12°25.2′S 166° 27.9′W
Depth (m)	2200	<2700	1170	2775
Depth of surrounding ocean (m)	5000	5000	5200	5600
Seamount height (m)	2800	2300	4030	2825
Max. recorded FA gravity anomaly (μm s^{-2})	710	720	1470	940
Max. recorded magnetic anomaly (nT)	220	400	700	540
Morphology	Cone-shaped with steep sides	Steep,gullied flanks	Gaussian shape with relatively flat summit area	Steep-sided with jagged peaks near the top, possible subsidiary cone on eastern flank
Approximate diameter near base (km)	25	30	40	35
Mean slope of of mid-flanks	15°	13°	14°	14°
Sediment cover	300-m thick horizontally layered sediment ponds in basement depressions adjacent to seamount	As for MWI	Capped by veneer of pelagic sediment (? foram-nanofossil ooze/chalk) 0–30 m thick	No obvious sediment cover on the seamount; 20–30 m of pelagic sediment (? red-clay) covers the basal slope and extends out onto the abyssal plain.
Dredged Samples			FeMn-oxide encrusted hyalo-clastite cobbles. Crusts to 5 cm thick.	

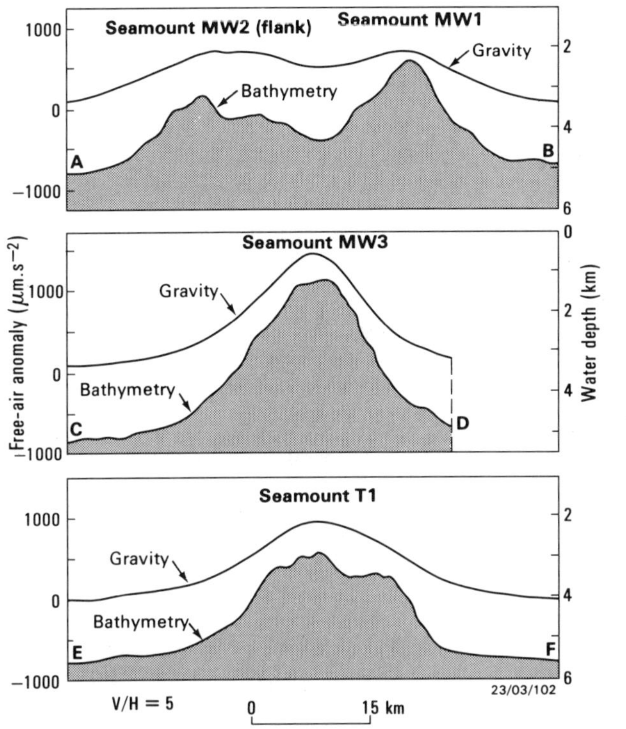

Figure 5. Ship bathymetry and free-air gravity profiles over seamounts MW1 and MW2, MW3, and T1. Track locations A–B, C–D, and E–F are shown in Figures 2, 3, and 4, respectively.

in which data gaps are filled by interpolation from adjacent tracks (Haxby et al., 1983).

Seamount Tl was initially mapped by HMNZS *Tui* with a single crossing of the summit. A subsequent crossing was made in 1987 by R/V *Moana Wave* in order to improve the bathymetric and geophysical control on the feature.

CONCLUSIONS

We have demonstrated the effectiveness of using satellite altimeter images and displays to detect and locate uncharted seafloor topographic features in the southwest Pacific. The existence of seamounts at the locations of typical geoid anomalies on different geoid / gravity data sets has been confirmed by follow-up shipboard verification surveys.

Time constraints allowed the testing of only a few major geoid anomalies during the research cruises. Examination of altimeter data covering the southwest Pacific, including previous work by others, indicates a considerable number of other geoid anomalies that could be interpreted as major uncharted seamounts. These still await confirmation by other ship surveys. This method of seamount detection has important seafloor resource implications for the South Pacific island nations, particularly in relation to deposits of metalliferous crusts.

ACKNOWLEDGMENTS

P. J. Hill thanks fellow members of the 1986 HMNZS *Tui* and 1987 R/V *Moana Wave* expeditions for their support and dedicated data acquisition efforts during the seamount searches; he is particularly grateful for the support of co-chief scientists Geoff Glasby, Maury Meylan (HMNZS *Tui* cruise) and Bill Coulbourn (R/V *Moana Wave* cruise). William Haxby of Lamont-Doherty Geological Observatory provided the spectacular SEASAT-derived gravity images which first revealed the significant potential of the satellite altimeter data and stimulated this study. Thanks also go to Richard Rapp of the Ohio State University for generously providing the processed combined GEOS-3/SEASAT digital data sets. Brian Pashley drafted the figures and Joan Brushett typed the manuscript.

REFERENCES

Baudry,N., 1986, Filtering SEASAT and GEOS-3 altimeter data in the Southwest Pacific: Office de la Recherche Scientifique et Technique Outre-Mer (Noumea), Report N 4-86, 11 pp.

Baudry, N., M. Diament, and Y. Albouy, 1987, Precise location of unsurveyed seamounts in the Austral Archipelago area using SEASAT data: Geophysical Journal of the Royal Astronomical Society, v. 89, p. 869-888.

Baudry, N. and M. Diament, 1987, Confirmation with ship data of SEASAT bathymetric predictions in the South Central Pacific, *in* B.Keating, P. Fryer, R. Batiza, and G. Boehlert, eds., Seamounts, Islands, and Atolls: Washington, DC, American Geophysical Union, Geophysical Monograph, p. 115-122.

Baudry, N., U. von Stackelburg, and J. Recy, 1988, Alignments volcaniques dans les Iles Australes: Analyse et interpretation de donnees SEASAT et Seabeam: C.R. Academie des Sciences, Paris, v. 306 (II), p. 643-648.

Blackinton, J.G., D.M. Hussong, and J. Kosalos, 1983, First results from a combination of side-scan sonar and seafloor mapping system (SeaMARC II): Offshore Technology Conference, OTC 4478, P-307-311.

Born, G.H., J.A. Dunne, and D.B. Lame, 1979, SEASAT mission overview: Science, v. 204, p. 1405-1406.

Coulbourn, W.T., P.J. Hill and shipboard party, 1987, CCOP/SOPAC Moana Wave Cruise 3 (MW87-02) to the territorial waters of Western Samoa, the Cook Islands and Kiribati - cruise report: Bureau of Mineral Resources, Australia, Record 1987/36, 42 pp.

Cronan, D.S., 1984, Criteria for the recognition of areas of potentially economic manganese nodules and encrustations in the CCOP/SOPAC region of the central and southwestern tropical Pacific: CCOP/SOPAC (Suva), South Pacific Marine Geological Notes, v. 3(1), 17 pp.

Dixon, T.M., M. Naraghi, M.K. McNutt, and S.M. Smith, 1983, Bathymetric prediction from SEASAT altimeter data: Journal of Geophysical Research, v. 88, p. 1563-1571.

Halbach, P., F.T. Manheim, and P. Otten, 1982, Co-rich ferromanganese deposits in the marginal seamount regions of the Central Pacific Basin - results of the Midpac '81: Erzmetall, v. 35(9), P- 447-453.

Haxby, W.F., 1987, Gravity field of the world's oceans. A portrayal of gridded geophysical data derived from SEASAT radar altimeter measurements of the shape of the ocean surface: US Office of Naval Research / National Geophysical Data Center, NOAA, Colorado.

Haxby, W.F., G.D. Karner, J.L, Labreque, and J.K. Weissel, 1983, Digital images of combined oceanic and continental data sets and their use in tectonic studies: EOS, Transactions, American Geophysical Union, v. 64(52), p. 995-1004.

Keating, B., N.Z. Cherkis, P.W. Fell, D. Handschmacher, R.N. Hey, A. Lazarewicz, D.F. Naar, R.K. Perry, D. Sandwell, D.C. Schwank, P. Vogt, and B. Zondek, 1984, Field tests of SEASAT bathymetric detections: Marine Geophysical Researches, v. 7, P- 69-71.

Kroenke, L.W., C. Jouannic, and P. Woodward, 1983, Bathymetry of the Southwest Pacific, 1:6442192 Chart: CCOP/SOPAC, Fiji.

Lambeck, K., and R. Coleman, 1982, A search for seamounts in the Southern Cook and Austral region: Geophysical Research Letters, v. 9(4), p. 389-392.

Lazarewicz, A.R., and D.C. Schwank, 1982, Detection of uncharted seamounts using satellite altimetry: Geophysical Research Letters, v. 9(4), p. 385-388.

Liang, C., 1983, The adjustments and combination of GEOS-3 and SEASAT altimeter data: Ohio State University, Department of Geodetic Science and Surveying, Report 346, 68 pp.

Mammerickx, J., S.M. Smith, I.L. Taylor, and T.E. Chase, 1973, Bathymetry of the South Pacific, chart 13: Scripps Institution of Oceanography.

Marks K.M. and R.V. Sailor, 1986, Comparison of GEOS-3 and SEASAT altimeter resolution capabilities: Geophysical Research Letters, v. 13, p. 697-700.

Rapp, R.H., 1979, Geos-3 data processing for the recovery of geoid undulations and gravity anomalies: Journal of Geophysical Research, v. 84, p. 3784-3792.

Rapp, R.H., 1983, The determination of geoid undulations and gravity anomalies from SEASAT altimeter data: Journal of Geophysical Research, v. 88, p. 1552-1562.

Sandwell, D.T., 1984a, Along-track deflection of the vertical from Seasat, GEBCO overlays: NOAA Technical Memorandum NOS NGS-40, 8 pp.

Sandwell, D.T., 1984b, A detailed view of the South Pacific geoid from satellite altimetry: Journal of Geophysical Research, v. 89, p. 1089-1104.

Stanley, H-R., 1979, The GEOS 3 project: Journal of Geophysical Research, v. 84, p. 3779-3783.

Tapley, B.D., G.H. Born, and M.E. Parke, 1982, The Seasat altimeter data and its accuracy assessment: Journal of Geophysical Research, v. 87, p. 3179-3188.

Watts, A.B., and N.M. Ribe, 1984, On geoid heights and flexure of the lithosphere at seamounts: Journal of Geophysical Research, v. 89, p. 11152-11170.

White, J.V., R.V. Sailor, A.R. Lazarewicz, and A.R. LeShack, 1983, Detection of seamount signatures in SEASAT altimeter data using matched filters: Journal of Geophysical Research, v. 88, P- 1541-1551.

Keating, B.H., and Bolton, B.R., editors, 1992, Geology and offshore mineral resources of the central Pacific basin, Circum-Pacific Council for Energy and Mineral Resources Earth Science Series, V. 14. New York, Springer-Verlag.

INSULAR GEOLOGY OF THE LINE ISLANDS

Barbara Keating
Hawaii Institute of Geophysics, University of Hawaii, Honolulu, HI 96822

INTRODUCTION

The Line Islands seamount chain consists of a series of ridges and seamounts which extend 4800 km across the Central Pacific Basin (Fig. 1). The seamounts form the longest intraplate submarine mountain chain in any ocean, rising from a sea floor in excess of 5000 m deep. Only twelve seamounts reach sea level to form low islands or atolls; all are presently bordered by fringing reefs. Some seamounts extend above sea level to form typical atolls, having shallow salty lagoons and a rim occasionally broken by channels to the sea, e.g., Fanning, Palmyra and Caroline. Still other seamounts reach sea level and form islands with unbroken land surfaces, e.g., Jarvis, Vostok and Flint Islands. All of the islands are low and nearly flat. Christmas Island and Johnston Island display the highest elevations with sand dunes reaching 12 m.

The Line Islands seamounts rise abruptly from great depths. Fringing reefs are present at their summits but are generally restricted to a fairly narrow width of 50-150 m, though rarely extending to 1.5 km. Usually these reef platforms are awash at low water, though around Jarvis the reef platform does dry out. The outer beaches usually rise fairly abruptly, sloping from sea level to beach crest, ranging in width from a few meters to a few tens of meters. Beach cliffs generally do not occur. Windward beaches exposed to higher surf are higher and steeper than leeward beaches, which tend to be protected from the weather and be low and

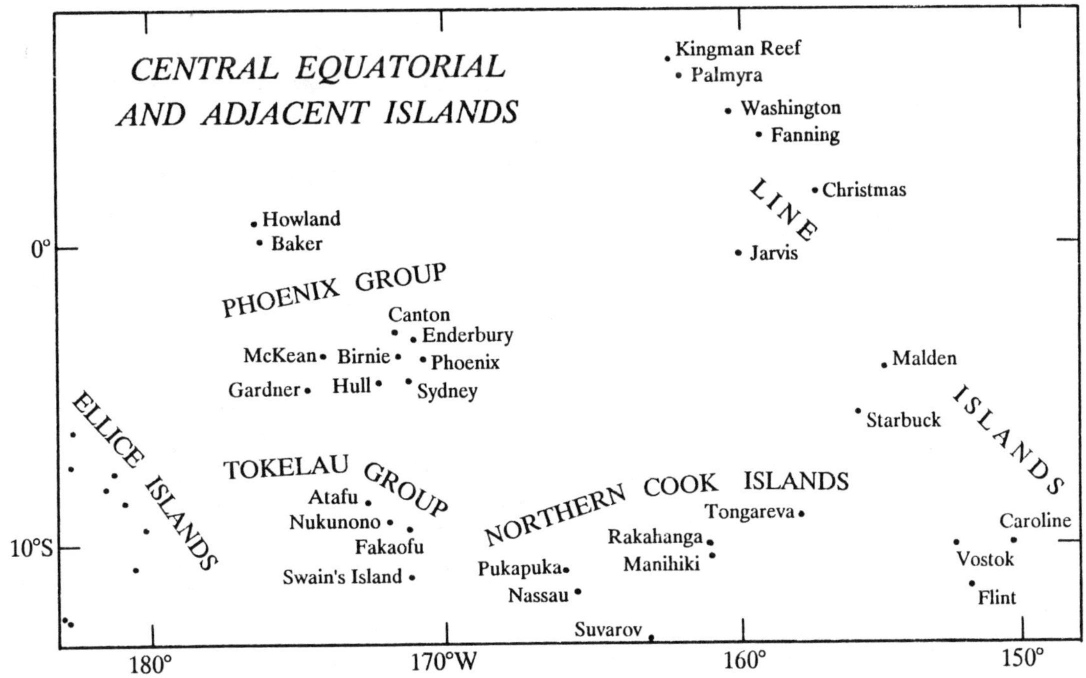

Figure 1. Map of the central equatorial Pacific islands. All of these are low islands and atolls. The figure is based upon British Admiralty charts and is redrawn from an original figure published by British Intelligence Geographic Handbook Series (1943).

gently sloping. The beaches are generally composed of short stretches of sand, alternating with shingle or rubble cover. Sandy beaches are more common on the leeward sides of islands or along lagoon shores. The sands are derived from abrasion of corals and shells and foraminiferal tests. They are wholly organic. The rocks commonly become cemented together by calcium carbonate cements. Inward of the beach crest, the land generally slopes down to a central depression. The depressions are the locations of salt-water lagoons, fresh-water lakes, peat-filled bogs or dry flat plains.

Mining of phosphate took place on many of the southern Line Islands in the late 1800's. Mining began on Jarvis Island in 1858. The mining operations on Jarvis were extensive for those days but lasted only about 20 years. Most mining by the American Guano Company ceased by 1879. An Australian firm, Grice Sumner and Company mined phosphate from Malden Island between 1860 and 1862 and continued until 1927. British firms, particularly, Messrs Holder Bros., mined phosphate from Starbuck beginning in 1870, and then extended their mining to Flint and Caroline. Vostok Island was also occupied but it appears no phosphate was mined. The deposits on these islands were exhausted by 1895. Caroline Island is reported to have yielded only about 10,000 tons or less of phosphate. Flint Island yielded 30,000 tons. Malden and Jarvis Islands yielded many times this amount. It is roughly estimated that the total production from all these islands in the south equatorial Pacific was in the range of 200 million tons. By the end of the 1800's the phosphate supply on these islands was nearly exhausted. The export from Malden Island, however, continued into the 1900's, finally ceasing in 1914 (British Intelligence, Geographical Handbook, 1943).

HISTORY OF DISCOVERY

The early history of these islands is poorly known. Stone ruins are (or were) present on some of the islands, such as Fanning, Christmas, Howland, Hull and Malden (British Intelligence, Geographical Series Handbook, 1943). The most notable stone ruins occur on Malden Island. It is likely that some Spanish ships sighted these islands. Magellan, in 1521 crossed the equator in their vicinity but no sightings were reported. The first discovery was Christmas Island reported by Captain Cook after the third Cook voyage in 1777. Just over 20 years later, Captain Fanning in the American ship *Betsy* discovered Fanning Island, Washington Island, and Kingman Reef. Broughton discovered Caroline Island in 1795. Palmyra Atoll was discovered by Sawle, in 1802. The discovery of Johnston Island is attributed to Captain James Johnston in 1807 (Stark, 1960). However, an account of the island (in several newspapers of the times) was made by Captain Pierpont of the *Sally* in 1796 as a result of the grounding of two ships, the *Sally* and the *Prince William Henry*, on the reef. Krusenstern (1811) declared that Don Jose Comisares sighted the atoll in 1786; however, this sighting is poorly documented (Amerson and Shelton, 1976). Krusenstern himself suspected land in the vicinity of Johnston Atoll, because he spotted hundreds of birds near the same location when he sailed through the area in 1804. Though a lookout was kept during the night, no land was spotted. The location of the island was well fixed by the Wilkes Expedition of 1841. The leader of a Russian expedition, Bellingshausen, discovered Vostok Island in 1820 and named it after his vessel. Starbuck, an American captain of a British ship, discovered and named Starbuck Island in 1823. Malden Island was discovered in 1825 by Captain Lord Bryon of the British ship H.M.S. *Blonde*, which searched for mutineers from the infamous British vessel, H.M.S. *Bounty*. Malden was named after Lt. Malden, who went ashore there. Jarvis Island was discovered by Captain Brown of the British vessel *Elis Francis* in 1821. Caroline Island was discovered by Clark in 1823 (Stark, 1960). The discoverer of Flint Island is unknown. It appeared on maps about 1801. Flint and other islands in the chain first had accurate positions fixed as a result of the United States Exploring Expedition, in 1839-41 (Wilkes, 1845).

ISLAND GEOLOGY

JOHNSTON ATOLL

Johnston Atoll is the best studied of any of the island and atoll groups within the Line Islands chain. It has previously been shown on maps as Cornwallis Island. The atoll is crescent shaped (approximately 17 km in width). An island, Johnston, and three small islets, Sand, Hikana and Akau are present. Only two of the islands were natural-Johnston Island and Sand Islet. The two additional islets were built using dredged rocks. Each of the islands has been extensively modified by dredging and construction activity. Originally Johnston Island was only 46 acres, by 1942 it had been enlarged to 211 acres, by 1952, the island was enlarged to 570 acres. Much of the dredging was done in the 1960's.

The highest point on Johnston Island was a sand dune ridge called Summit Peak, which reached 14.5 m above sea level (it has been altered by construction). The highest point on Sand Islet is 5 m above sea level. Both islands consist of beach rock (cemented sand and gravel) and loose sand.

Emery (1956) described the geology of Johnston Island based upon information obtained from 56 borings for building foundations, ranging to 12 m below sea level and from 6 wells drilled to depth of between 26 and 57 m below sea level. (Data from additional borings are now available, J.

Maragos, personal communication, 1987.) None of these holes reached non-calcareous rocks. Two layers of beach rock were encountered, one above sea level (which is presumed to correlate with that still exposed on Sand Island), and one at 2 m below sea level. E. H. Bryan (Field notes, 1943) notes that the coral beds on Sand Island dip 4-5 degrees to the southeast, indicating possible tilting in that direction (Amerson and Shelton, 1976). This lower layer is found at 2.4 m below sea level along the bases of the parallel reefs that extend between Johnston and Sand Island according to Amerson and Shelton (1976). Alternating loose coral, sand, or coral and sand occur beneath the beach rock, which cannot be correlated with certainty from hole to hole. Muds occur in several wells.

Guano was discovered on Johnston Island in the late 1850's. Guano, generally speaking, consists of deposits of excreta from birds (most often sea birds) and bats, as well as the bones of associated animals. In an island environment, rain and salt spray leach the deposits and remove ammonia and the soluble dibasic phosphate in aqueous solutions. Phosphatic rocks are generally found underlying guano deposits and most often are limestone or dolomite in which the carbon dioxide has been replaced by stronger phosphoric acid. The soluble dibasic phosphate, leached out of the guano, has drained down to the underlying rock and taken up sufficient lime to form the less soluble tribasic phosphate (Power, 1925).

Guano was mined on Johnston Island by the Pacific Guano Company (San Francisco) and the American Guano Company (Honolulu). Rights to mining on the island were contested since both The Pacific Guano Company and the government of the Kingdom of Hawaii claimed the island (the American Guano Company acting as its agent). The amount of guano taken off the island during 1859 and subsequent years "must have been substantial, if the number of ships and the statements of their captains can be trusted" (Amerson and Shelton, 1976). Hague (1862), however, contended that the greater part of the material contained little guano.

Several hundred pounds of pumice floated ashore in 1953, presumably from volcanic activity which took place on San Benedicto Island off Mexico in 1953 (Emery, 1956, and Richards, 1958). Pumice also floated ashore in the 1960's. No seismic activity has been reported near Johnston Island (Walker and McCreery, 1988).

Gravity studies were conducted on the island by Kroenke and Woollard (1965). Bouguer anomaly gradients on Johnston were lower by about half than in the Hawaiian Chain, suggesting that there are no volcanic necks or plugs involved in the upper structure of Johnston Island. These observations would be consistent with the presence of a thick limestone cap which deeply buries volcanic structures under Johnston Island. Volcanic fragments within a limestone matrix were dredges from a depth of 1300 m, on the southeast flank of the Johnston Island seamount (J. Hein, personal communication, 1986).

Emery (1956) charted the detailed bathymetry of the Johnston Atoll carbonate platform. A simplified version of this bathymetric map is shown in Figure 2. The live barrier reef of Johnston is restricted to the northwest, i.e., leeward margin, and extends for roughly 12 km in a northeastern direction. Emery (1956) described five distinct zones associated with the living barrier reef: a seaward slope, lithothamnion ridge, coralline algal reef flat, coral slope, and a lagoonal debris slope. Brock and others (1961; 1965) describe the reef habitats in more detail.

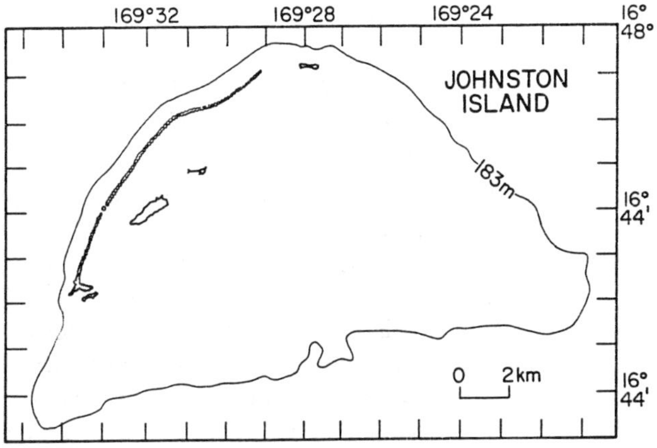

Figure 2. Map of Johnston Atoll. The figure is redrawn from an original figure published by Emery (1956).

A distinct rim (interpreted as the fossil ring reef) is revealed by echo sounding along much of the southern and western edges of the platform (Ashmore, 1973). Emery (1956) and Ashmore (1973) both provide evidence that the submerged ridge (not associated with a living reef) has a height approximately the same height as the existing barrier reef above the adjoining submarine platform. Based upon geomorphology, this ridge appears to be the drowned eastern margin of the atoll. No central lagoon is obvious in the bathymetry; instead two flat-topped reef terraces cover much of the central platform.

Ashmore's bathymetric studies (1973) describe a submarine ridge upon which Johnston Island and Sand Islet are situated. The ridge is composed predominantly of dune sand and minor beach rock units (based upon well data and outcrops on Sand Island, Emery, 1956; Maragos, personal communication, 1983). The sands forming the ridge appear to be largely deposited by wave and wind activity. Outcrops can be seen on portions of Sand Island. (Most outcrops on

Johnston Island have now been disturbed by building.) Dune features suggest a transport direction perpendicular to the current trade wind direction.

Ashmore (1973) reported two distinct submarine terrace surfaces at -8 and -18 m. These reef terraces sit on a sloping carbonate bank surface. The -8 m reef terrace overlies a deeper reef terrace. Numerous ridges and knolls are found on its surface where live coral and patch reefs are common. Ashmore (1973) reports the most striking feature of this surface to be the abundant sink holes. Many of these sink holes follow contorted, elongate paths, which are open to the -18 m level at one end. Some sink holes are approximately 0.75 km in length. The sink holes are aligned parallel to the dip of the larger carbonate terrace. The larger sink holes occur at depths of 20 m in the lower terrace. Sediment samples show that sand and fine debris cover much of the -8 m terrace, thus it is likely that these sink holes have substantial sediment fill (Ashmore, 1973). The -8 m terrace does not have an observable slope.

The -18 m reef terrace extends over an area of approximately 40 square kilometers. This terrace is morphologically very distinct from the upper terrace. It is "somewhat dished," suggesting a lagoon topography. There are few knolls, or patch reefs. Ashmore (1973) however, speculated that most minor depressions of this deeper reef terrace are likely to be filled with sands and coral debris. The on-lapping contact between the upper and lower reef terraces is distinct but very uneven. The horizontal contact extends across the carbonate bank in a north-south direction and is between 500-1000 m wide.

Sediment samples from the shallower portion of Johnston Atoll (Emery, 1956) yielded the following average composition: coralline algae 51%, coral 20%, shell fragments 2%, *Halimeda* 1%, and fine debris 16%. The coralline algae is composed of high-Mg calcite and is the dominant component of the sands on the surface of the Johnston platform. On the basis of aerial photography, Emery (1956) estimated that about half of the platform was dominated by coral growth, and the other half by sandy sediments. While coralline algae is for the most part restricted to the barrier and patch reefs, algal and coral debris contribute significantly to the sediments.

Aydemir (1985) summarized petrographic and geochemical studies of carbonate rocks collected by submersible on the slopes of Johnston Atoll. The samples were collected between 300 and 400 m, they were found to contain fossils and sedimentary structures typical of very shallow waters. Aydemir (1985) reports age dates from the rocks range from 10,590 to 11,630 ybp. Keating (1987) presented results from submersible dives on the slopes of Johnston Atoll (rocks for the Aydemir study were collected from these dives).

Johnston Atoll appears to be an example of a seamount that is undergoing a transition from an atoll to a drowned seamount (guyot). The live reef surrounding the platform is no longer present at sea level around most of the periphery. Only a limited barrier reef, approximately 12 km in length remains along the western margin of the atoll. While dominant reef growth is normally present on the windward side of islands, the dominant reef growth at Johnston atoll is reported on the leeward side (Emery, 1956). The carbonate bank slopes to the southeast. The upper surface of the carbonate platform has two reef platforms which horizontally overlie a sloping bank surface. Ashmore (1973) pointed out that the islands and islets lie on a prominent ridge which marks the axis of tilt and calculated a slope of 0.016 degrees for the carbonate platform (sink holes observed on the platform run down the dip slope). Dune structure within this ridge (Ashmore, 1973) indicates that sand transport was perpendicular to current trade wind direction, it seems highly unlikely that the dominant wind direction has shifted to that extent. Instead a tectonic or erosional process which has remobilized the sand appears likely.

Two explanations have been suggested to explain the anomalous bathymetry of Johnston. Ashmore suggested a simple erosional model in which the reefs of the eastern margin have been eroded. The second model proposes that the surface of the atoll has been tilted (Emery, 1956; Ashmore, 1973). Keating (1987) suggests the latter model is preferred, for several reasons. First, if the reef structure was eroded and killed by a combination of a lower stand of sea level and a lower water temperature during a glacial period, both margins of the atoll would have been affected. Second, the submarine ridge on the eastern margin is comparable in height to the existing barrier reef. If this ridge is the fossil reef that Emery suggested was removed by erosion, it should not be present. All of the observed major bathymetric features are consistent with a model for tilting of this atoll. The configuration of bathymetric features is inconsistent with existing environmental factors (wind, wave, and current directions) which strongly affect reef growth elsewhere.

Keating (1987) presented side-scan sonar images and bathymetry for the southern margin of Johnston Atoll. The side scan images show that the southern margin of the atoll is exceptionally steep. The margin itself is nearly linear (suggesting faulting) as opposed to the curved seamount margins commonly found on other Pacific islands and atolls. A debris apron was found on the sea floor underlying the cliffed southern flank. The topography of the debris apron is very chaotic. The adjacent basin is 700 m shallower than the depositional basin on the southwest margin of Johnston. Keating (1987) suggests structural failure of a portion of the southern margin of Johnston seamount has occurred producing a large load of locally derived detritus. Keating proposes the local loading of the sea floor along the southeastern margin of the seamount has resulted in uneven subsidence of the seamount and tilting of the carbonate atoll.

KINGMAN REEF

Kingman Reef is an atoll which for the most part is submerged (Fig. 3). A small islet is present rising only 1 m above sea level and was reported, in 1943, to extend only 75 m in length. The size of the islet varies according to season and local weather conditions. A few coconut palms were planted there in 1924. The reef was discovered in 1798 by Fanning (Stark, 1960). The island was named after the master of the American vessel who found it again in 1853. Kingman Reef was used temporarily during trial flights of the Pan-American Airways clipper flights (1937–38) as a seaplane base, but it was abandoned in favor of Canton Island.

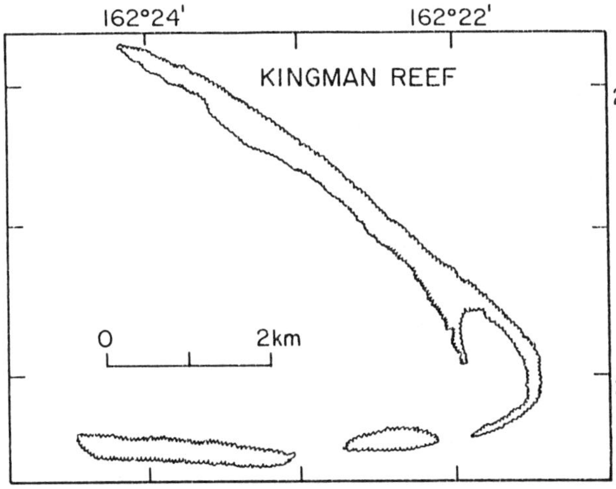

Figure 3. Map of Kingman Reef. The figure is based upon the U S. Hydrographic Office Map of Kingman Reef.

PALMYRA ISLAND

Palmyra Atoll was discovered in 1802 by Sawle. Palmyra Atoll has a horse-shoe shape (Fig. 4). There were fifty-five small islets which ringed a shallow reef platform. The four large lagoonal ponds are as deep as 40 m and one small pond is present. The western lagoon consists of reef-flat now submerged to 20 m below the sounding reef (Dawson, 1959). The more eastern lagoons are reef-flats, only slightly submerged, and showing patches exposed at low tide (Dawson, 1959). Dredging during World War II, however, has deepened the lagoons in general. The changes resulting from construction activities are outlined by Dawson (1959). The west end of the atoll is open. The islets rise to a maximum of 2 m above sea level.

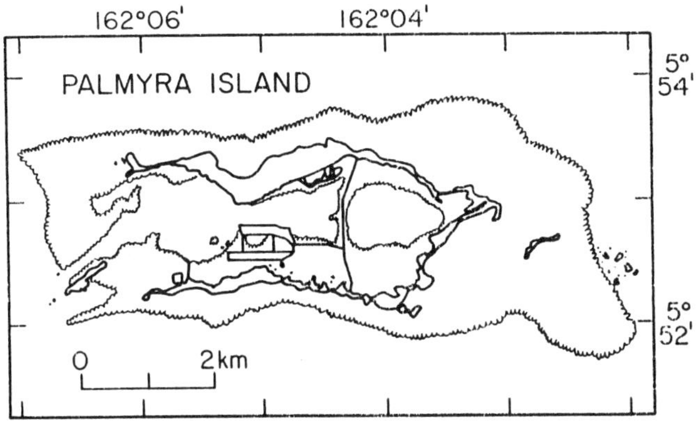

Figure 4. Map of Palmyra Atoll. The figure is based upon the U.S. Hydrographic Office Map of Palmyra Atoll.

Major modifications to the island took place during World War II. Palmyra Island was a military base for the United States and contained both land runways and seaplane ramps. Extensive dredging and bulldozing drastically altered the surficial features of Palmyra Atoll. New islets were formed near the western channel to the ocean. All of the islets were connected by causeways; many islands were enlarged. A road connected all of the islands. A ship turning basin was blasted from the western lagoon connecting the islets of the north shore with those of the south shore. Aircraft runways were built. Thus, little of the land surface on Palmyra was left unaltered. The soils exposed on the islets vary. Generally the beach crests are characterized by a light grayish brown sand with low admixtures of organic material and high water content. The soils in the most interior locations are brown in color. They consist of large amounts of organic material with high moisture content. In some areas the soil layer is very thin and overlies hard coral ground. The hard ground is reported to be calcium phosphate with a layer of yellowish white coral sand below (British Intelligence, Geographic Handbook, 1943). The vegetation, though not luxuriant, is abundant.

The lagoon, particularly the side open to the sea, contained abundant live coral prior to World War II. However the lagoons have been overwhelmed by calcareous debris carried into the lagoon since the extensive dredging and construction on the islets. Most of the coral in the lagoon is dead. Live coral exists on the seaward margins of the reef platform (K. Constantine, personal communication, 1988). The outer beaches are composed largely of sand with some coarse gravel. The inner beaches are low and are composed of fine calcareous sand and mud. No reef rock was exposed on the islets visited by Wentworth (1924). Wentworth suggested there is no indication of a former higher sea level, and that either the island had not grown to the sea level at the time of the 4-m eustatic shift of sea level, seen on most Line Islands, or that much of the island has been cut away by waves since that time. No phosphate mining has been reported here.

Dawson (1959) reports that the reef platform on Palymra has been tilted westward, with the western portion of the atoll submerged to a depth of 20 m. The submerged reef flat is bounded by a still living, but submerged reef. Dawson reports, "This submergence, however, has apparently resulted from a long period of solution on that side as growth has proceeded actively eastward." Dawson suggests the central and eastern lagoons were formed more recently than the western lagoon. And, he points out that an incipient lagoon is in the process of being surrounded by islets building in the position of Barren Island on the eastern end of the lagoon. (Barren Island was not present on original British maps of the island, but is represented on maps made in the 1940's.) Several sand banks have appeared near Barren Island between the 1940's and 1988 when visited by Keating. Dawson (1959) suggests the formation of the latter islets may be supposed to have been hastened by man's modification of the atoll.

In support of Dawson's observations, a beach ridge can be seen on the eastern coast line of Papala Island, which stands 1-2 m above sea level, that would appear consistent with a upward tilting of the eastern end of the atoll. Significant erosion has also occurred on the eastern end of Palymra. Concrete structures built on the southern side of Bird Island now stand in the surf and have been destroyed by the erosion since their construction in the 1940's. The tall trees (*Pisonia grandis*) on the south side of Sand Island also appear to be dying from the effects of salt water. Many palm trees have also fallen and now lie on the reef flat in the surf east of Holei Island.

In addition, east of "Eastern Islet" and Papala Islet, small mounds of old corals can be seen protruding through the sand flats. The mounds consist of reef rock which is deeply weathered and dark black on the surface. These appear to be erosional remnants of an older island, now for the most part removed by erosion.

WASHINGTON ISLAND

Washington Island was discovered by Fanning in 1798 (Stark, 1960). It is shown on older maps as New York Island. Washington Island is a low roughly rectangular island, reaching only 3–5 m above sea level (Fig. 5). The island is noted for its heavy seas and violent wave activity around the shores. The reef extends 50–200 m offshore. A large submerged shallow bench extends to the west of the island and appears to be the product of westward shore drift and extends nearly a km offshore. The island is surrounded by sandy beaches characterized by fine grained sand. Only one small area of beach was reported to contain gravels reaching 12 mm in diameter. This observation contrasts with other islands in the chain, which commonly display shingled beaches. The height of the beaches range from 1 to 5 meters. At the west end of the island, the beach crest occurs at 4 to 5 m, however, waves commonly break on this beach and spill over the beach crest on to the back-slope. The highest elevations are near the beach and the land slopes gently toward the interior lake. The soils of the beach crest are light grayish brown sand with only small amounts of organic material. The interior of Washington Island is covered by fresh water lakes and several bogs. The fresh water lake of 3 km width and the many bogs are probably the remnants of an ancient lagoon. The rim of the island between the beach crest and the bogs and interior lake supports a luxuriant growth of vegetation. The soils of Washington Island in the bog area are for the most part mixtures of organic material and limited sand. Where rocks are exposed they consist of coral debris. At a few spots, sandstones and conglomerates are exposed which are not eolian. No reef rock was exposed in the interior of the island. Wentworth (1924) suggested the interior sediments were deposited when the island was more completely submerged than now.

The fresh water bogs are now completely filled and can be crossed by foot. The peat is a dense inter-growth of plant matter with little or no sediment. Underlying the peat is a hard bottom of white coral sand. The deeper parts of the bogs are 25-40 cm below sea level. The surface of the fresh water lake is about 1 m above sea level. The deeper parts of the fresh water lake extends 10 m below sea level. Wentworth (1924) reports that the lake water is fresh to the taste and flows through porous rocks to the sea. Wentworth suggests Washington Island, like several other islands in the group, was first built as an atoll with reference to a higher

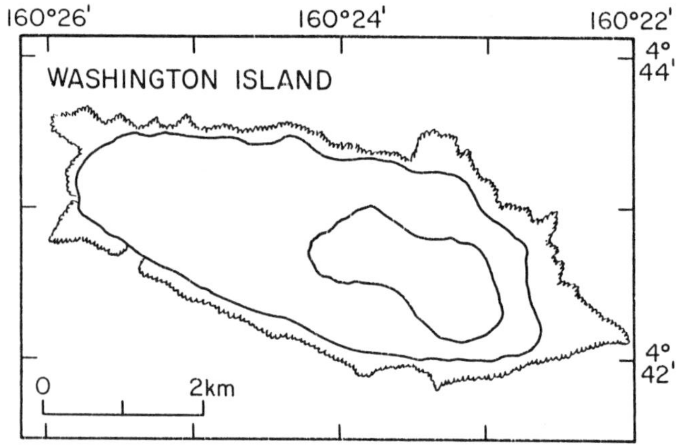

Figure 5. Map of Washington Island, based upon the U.S. Hydrographic Office Map of Washington Island.

sea level and that it became emerged through the lowering of sea level. Samples of rock from the atoll were examined for phosphate in the late 1870's but no mining is reported.

FANNING ISLAND

Fanning Island is part of a low oval atoll (see Fig. 6). The land rim is broken in three places and varies in width from a meter to approximately 1 km. A reef platform extends seaward of the shoreline, 30 to 100 meters in width. Parts of the platform are exposed at low tide. At high tide the platform is covered by approximately a meter of water. The surface of the platform is generally smooth but is cut by shallow radial channels. The outer edge of the platform is characterized by live coral and coralline algae on a coral substrate with deep under-cut channels paved with calcareous debris (Wentworth, 1924). The west and southwest shores are composed of sand, shells and coral fragments, in large part the beach is made up of discs of coral shingle.

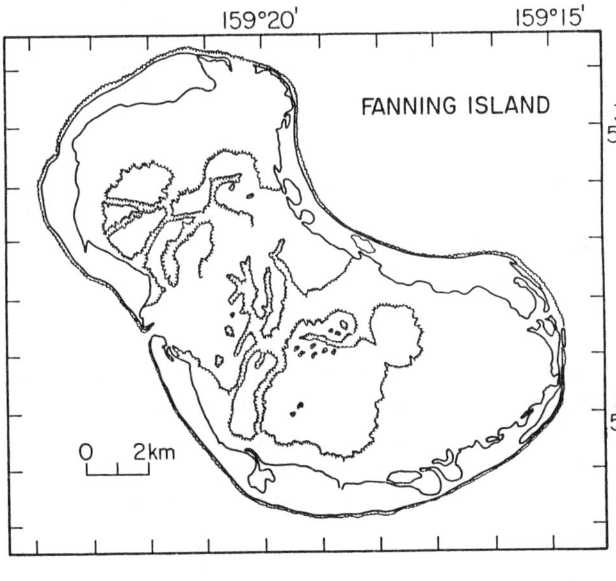

Figure 6. Map of Fanning Island. The figure is based upon the U.S. Hydrographic Office Map of Fanning Island.

These cobbles or shingles are 3 to 20 cm in diameter, and stack to depths of 12 to 40 cm. Surf washing the beaches rearranges the coral shingles and rapidly abrades the coral. Most beaches generally experience surf within 1 m of mean sea level; however waves can run up the beaches 2–3 m above mean sea level. The island reaches only 3–4 m above sea level.

Water is abundant on the island, thick vegetation is present, and trees reach heights of 30 m. Coconut palms were present on Fanning when it was discovered. During active copra-producing times, 350,000 palms were planted on the island. The surface soil within the palm groves is grayish brown sand with admixture of organic material. This sand forms a shallow surface layer which is underlain by coral fragments with only minor amounts of fine sand. On the beach crests the sand is a light, sometimes yellow-white color. Elsewhere, the surface layers consist of dark gray, etched coral and gravel which is mixed with small amounts of organic material. The weathered coral surfaces are a dark gray color. Where holes are dug or the shore line cut by wave activity, fresh white surfaces are exposed 0.5 to 1 m below the surface. Crabs are very conspicuous on Fanning Island and have heavily burrowed the land surface (Wentworth, 1924; Keating, 1979, personal observations). Bioturbation thus appears to be an active process altering sedimentation on this island.

The ground surface gradually declines from the outer beach crest toward the lagoon. The mean lagoon level is estimated to be 13 cm above mean ocean tide for an assumed ocean tide amplitude of 1 m (Wentworth, 1924; Johnson, 1929). High tide in the lagoon lags behind the outside high tide by as much as 2 hours. Salinity within the lagoon varies considerably (Guinther, 1971). The lagoon is shallow over most of its area. Holes as deep as 18 m are found in the lagoon. Roy and Smith (1970) examined sedimentation and reef growth in Fanning Lagoon. They found that extensive coral reefs intimately associated with muddy sediments are present in Fanning Lagoon both in clear and in turbid waters. Within the lagoon they found a tendency for reefs to form linear segments, up to 200 m wide, forming ponds approximately 1 km wide. The primary mode of occurrence is perpendicular to prevailing wind directions, but a secondary mode parallel to wind direction was also important. They point out, however, that this alignment could be fortuitous and the basic pattern may be related to karst topography formed during subaerial exposure of the island. The linear reefs in excess of 20 m width have a medial sand strip. The wider the reef, the wider the sand strip. Sand drifting off the medial sand strip accumulates as a sand wedge on the leeward side of the reef. These sand wedges are produced as a result of mechanical abrasion and wind transport. In Fanning lagoon, bioerosion appears very extensive, both from boring organisms and fish grazing. The suspended calcium carbonate in the lagoon appears to be the product of biological and mechanical abrasion of skeletal materials from the reefs. Wind drives surface water over the windward reefs and sediment is put into suspension. When the sands settle into the lagoons, circulation aided by the wind produces turbid waters.

The sands on reef flats are coarse- to medium-grained. Little sand is produced on the clear-water reef flats. The sand on the turbid water reefs tends to be finer and better sorted. On the lee side of distinct wedges of muddy sands extend about 60 m into the ponds before grading to a muddy bottom covered with silt-size particles.

Smith et al. (1970) examined the pattern and flux of calcium carbonate in lagoonal waters at Fanning Atoll. They estimated that during a 24-hour observation period, 10 tons of suspended calcium carbonate were transported into the lagoon. Roy and Smith (1970) estimate that the suspended calcium carbonate in the lagoon is about 65 percent aragonite and 35 percent calcite based upon x-ray diffraction analyses. Smith et al. (1970) estimated that the suspended carbonate debris plume at English Harbor represents little or no sediment loss from the lagoon. The plume debris is interpreted as material produced on the outside fringing reefs, which is washed into the lagoon on the incoming tide and subsequently expelled. They calculate that the calcium carbonate produced in the lagoon is filling the lagoon faster than sea level is rising and that the "overproducing" lagoon would eventually fill.

Coral knolls in the turbid water have about 80 percent live coral cover, according to Roy and Smith (1970). The knolls cover 37 percent of the area of the pond bottom, suggesting an average coral cover of 30 percent. The coral knolls have a low abundance of corals on the up-wind side and a high occurrence on the down-wind side, which may be controlled by the encroachment of the sand wedge in the ponds. In ponds deeper than 8 m, coral cover is further reduced by the load of fine sands settling uniformly to the bottom. The bottom of the lagoon is being covered by calcium carbonate mud, which accumulates in excess of 1 mm/yr.

Roy and Smith (1970) conclude that the reef development in turbid waters of the lagoon is of the same magnitude as that in clear waters. They found that the reefs in turbid waters tend to be structurally different than those in clear waters. Clear water reefs were primarily encrusting and massive corals, while turbid waters are characterized by ramose growth corals which tend to be more open, form more gentle slopes, and be infilled by sediments. Based upon the current patterns of sedimentation and coral growth observed by Roy and Smith (1970) predict the lagoon will eventually fill, producing a limestone body of coral biolithite surrounded and in part in-filled by micrite and calcilutite. The fine-grained rocks would be penecontemporaneous with the biolithite and the bathymetric relief would not exceed 8 m.

Roy and Smith (1970) observed that the distribution of land around the rim of Fanning is peculiar. The windward area is swampy intertidal and shallow subtidal (Guinther, 1971). The subaerial part of the atoll is on the leeward side of the lagoon, which is very unusual. The windward and leeward fringing reefs are about the same width. Roy and Smith point out the land distribution and reef widths are the reverse of that generally found on atolls (Wiens, 1962). Rapa Pass appears to have resulted from the breaching of a conglomeratic beachrock ridge. The erosional remnants of the beachrock occur some distance out on the reef flat around Rapa Pass. On the opposite side of the lagoon, an old pass has been infilled. (The cable station is situated on the site). The spits on the lagoon side of the in-filled pass have cemented shingles with opposing dips and strikes into the lagoon. The unconsolidated sediments filling the old pass are interpreted as being younger than the phosphatized conglomerates making up this part of the atoll rim. Additionally a complex set of relic tidal deltas occur at the lagoon edge west of Rapa Pass. The water path through this area is extremely complex and does not appear to have formed under the present environmental regime. A short distance from the ocean the pathway is now blocked by a boulder ridge. Keating visited the island in 1988 and suggests that Putu Tutae, southeast of English Harbor, was a pass connecting the ocean and lagoon which has subsequently been filled. The area is now a dry salt pan, which is only occasionally wetted according to local residents.

Roy and Smith (1970) suggest the atoll has been tilted. Furthermore, they suggest the tilting has produced the large areas of turbid water within the lagoon. Comparing reef growth in Fanning to other coral reef areas indicates the growth of the turbid water coral community could be on the order of 400 years.

Detrital limestones 4 m above tide level, are reported by Wentworth (1924). Wentworth suggests these are evidence that Fanning Island was probably an encircling reef, just awash at low tide, and protecting a gradual accumulation of debris on its inner margin before emergence to its present position.

Phosphate is present on Fanning Atoll. Mining of phosphate on this island took place between 1877 and 1887. It has been estimated that at least 19,500 metric tons of phosphate rocks were shipped from Fanning and that the original reserve of 30,000 to 40,000 tons is reasonable (Hutchinson, 1950). The chemical composition of the phosphate rocks was been described by Stutzer (1911), Elschner (1923), Frondel (1943) and Hutchinson (1950). The phosphate rocks present near the cable station are coarse calcarenites grading to calrrudites (Roy, 1970). The beds are generally 5 to 10 cm thick, but can range from 30 to 1 cm. They display shallow dips (less than 5 degrees). Along the lagoon the phosphatic beds outcrop in roughly one meter high cliffs. They do not outcrop on the seaward side. Fragments of red algae, corals, and shells are common in the phosphate. Tree molds are also common. The phosphate rocks are very similar in fragment composition and texture to the sands and gravels inland of the boulder ramparts near English Harbor. The sands and gravels are composed of fringing reef-flat material that was washed over the beach ridge by storm waves.

Roy (1970) reports that the wash-over gravels near the cable station are cemented by phosphate minerals to produce a phosphatic rock. Calcium carbonate cements are not found within the rock, and the clasts have not been altered. Calcite, magnesian calcite, and aragonite are still present in the rocks. Recrystallization and obliteration of the microstructure has not taken place. Many gastropod shells are still nacreous. Some solution and cementation has taken place, however, since bridges of cement are present in some voids. On some surfaces, microcrystalline box-work patterns have developed.

Roy (1970) reports most of the phosphate minerals in island deposits are in the francolite-dahllite series (Frondel, 1943; Hutchinson, 1950). The phosphate cements are a mixture of carbonate-apatites (francolite-dahllite) and whitlockite and callophane. The cements display a progressive (three-part) change in the nature of the phosphatizing fluids. Roy suggests that a change took place in the chemistry as the solutions were leached from the surficial guano. A decrease in the activity of fluoride in the leaching solution is believed to be responsible for a change from francolite to dahlite. The observed formation of late-stage whitlockite could be explained by a late major-decrease in pH in the solution. The solution would not be buffered by calcium carbonate after the clasts were sealed off by initial cementation. An alternative explanation for a change in pH would involve formation of extensive forests on the guano deposit.

Emory (1934; 1939) describes the remains of stone ruins of Polynesian style on Fanning Island. These include a dressed-stone enclosure near the cable station, ruins which were interpreted as tombs, and the quarry site for the limestone slabs used in the construction of the ancient ruins. Emory (1939) reports that human teeth, fish bones, and fish hooks where collected from the ruins near the cable station. Four basalt adzes were also collected. Emory (1939) concludes that the adzes and ruins display "very close affinities" to those found in Tonga, and suggests that western Polynesia is the ancestral land of the ancient settlers on Fanning Island.

CHRISTMAS ISLAND

Christmas Island is part of a low atoll. The atoll is shaped like an elongated claw open toward the northwest (Fig. 7). The rim of the atoll is broken at two places on the west side. It is one of the world's largest atolls. Tidal flow into interior lagoons is very restricted; thus, many ponds are shallow and have become isolated and with evaporation of surface waters have become hypersaline. The border of some lakes are encrusted with salt and the water near the margins are colored red or orange with algae (Gregory, 1924). Other ponds are spring-fed (Wentworth, 1924).

A fringing reef around Christmas Island extends 30 to 150 m seaward of the shoreline. The seaward margin of the platform consists of ridges and spurs. The spur channels can be seen at low tide but are also clearly visible under other circumstances since the white channel filling sands appear as a marked contrast to the colored coral masses of the ridges.

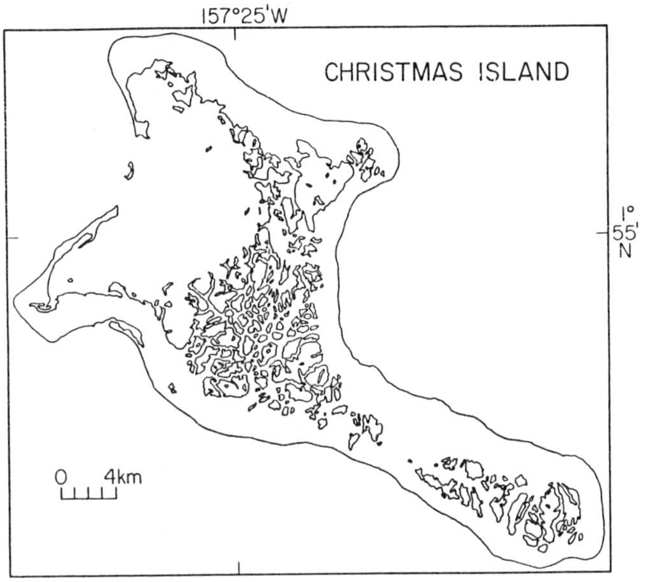

Figure 7. Map of Christmas Island. The figure is based upon a map published by Helfrich et al. (1973).

The beaches of the western side of the island are composed of sand, shell and coral debris. The beach crest is generally 3–4 m above sea level. In many places, a succession of coarse coral debris ridges extend inland from the beach. On the windward coast, a steep beach of rough angular coral debris was reported by Wentworth (1924). The beach is flanked landward by a narrow expanse of coarse debris ranging from 15 to 50 cm in diameter.

The rim of the island on the north and northeast side consists of an outer and higher plain and an inner and lower flat which is 10–25 cm above sea level. The two benches are separated by a gently sloping scarp 2–3 m high. The outer margin of the upper bench flat merges with the beach-ridge plain or is lined with low sand dunes. Coconut trees are concentrated on the slope, and in these groves small amounts of humus, soil and phosphatic material have concentrated (Wentworth, 1924).

Masses of marine reef rock stand 3–4 m above sea level at a number of places on Christmas Island. Wentworth (1924) states that none higher were seen, which indicated a probable stand of sea level greater than 4 m above present.

The lagoon makes up the larger part of the north half of the atoll. The northeast, and southeast parts of the atoll rim form a broad plain which is dotted with salt water lakes and ponds. Most are fed by seepage springs round their margins and are probably maintained at least a few cm above sea level (Wentworth, 1924). Usually several shore lines can be seen on the beaches above the existing water line. On the windward side of the island there are discontinuous sand dunes. The highest elevations on the island are associated with the sand dunes, which range from 6 to 12 m. Aeolian sediments from the windward coastal plain formed the island and lagoonal dunes which form topographic highs on the southeast part of the atoll. Aeolian as well as in-situ carbonate deposition has partially filled the main lagoon, forming a complex system of 500 isolated and interconnected ponds and lakes occupying 25 percent of the total island surface area (Valencia, 1977). Studies of several of the hypersaline lakes were completed by Valencia (1977). He found the maximum lagoon depth is 7 m and is approximately equivalent to that of many of the lakes. Continuous shoaling of the northern pass into the lagoon is indicated from its dredging history (Jenkin and Foale, 1968). Despite the high sedimentation rate, patch reef development is frequent in sheltered areas where the substrate is stable (Valencia, 1977). The water levels between adjacent lagoons differs by 30 cm to as much as 3 m. The lagoons on the south side of the atoll fluctuate with rainfall. A reducing environment was indicated (based on hydrogen sulfide odor) immediately below the surface sediments in all lakes sampled by Valencia (1977). Completely isolated lakes are accumulating precipitates (principally gypsum and halite) from evaporation, thus they continue to shoal. Relatively isolated lakes, less than 2 m deep had salinities greater than 300 $^{o}/oo$. The bottom of these lakes are rough resulting from a network of polygonal ridges of 1 m relief. Gypsum and halite crusts 5 cm thick are present on some high salinity lakes and overlay roughly 1 m thick layers of red algae interspersed with halite crystals. Other evaporitic lakes having relatively low salinity (36–41 $^{o}/oo$), have smooth bottoms composed of 1-cm thick halite cubes overlying 1 m of red algae and halite crystals. Lakes which lacked gypsum and halite deposits are interconnected by narrow channels to the main lagoon. A salinity increase is observed from the edge of the main lagoon to the interior of the interconnected lake series. Valencia (1977) associated this change in salinity with the extinction of reef and invertebrate faunas in the lakes farthest from the lagoon. The general increase in salinity in a northwest-southeast direction in the interconnected lakes is interpreted as resulting from the influence of a permanent freshwater lens in the northeast.

The shores of many of the lakes are strewn with molluscs, pelecypod and gastropod shells, which form extensive flats and beach ridges. Broad areas adjacent to the lakes and elsewhere are covered with a hard pavement 1–10 cm in thickness (Valencia, 1977). The pavement is biosparudite, dark-gray in color and when broken the inner surfaces are a light buff color. Below the pavement is moist reef rock (coral biolithite and intrasparrudite) and unconsolidated medium calcirudite (coral debris) (Valencia, 1977). The pavement is formed by the cementing of the coral sand and mud by calcareous cements deposited from evaporating water at the sediment surface. Wentworth (1924) suggests "this crust is of terrestrial origin, is being formed at the present time, and should not be confused with the massive reef rock and coral heads which were formed when the sea stood over Christmas Island." In most places where the crust is well developed there is little vegetation and that which is present grows in linear and radial arrangement along cracks in the pavement which appear to be due mainly to expansion and contraction. Soils in the upland are light grayish brown or reddish brown calcareous sand, deriving its color from admixtures of organic material. The surface layer is usually fine-grained but lower layer consists of coarse coral fragments and little fine-grained sand.

The elevation on the island generally is less than 4 m. The exception, however, is Joe's Hill, which reaches 11 m, in the coastal dune area on the windward side of the island. Where dunes are absent, coral debris and rubble are the dominant rock types. Rubble ramparts characterize the windward coast, while clastic sediments characterize the leeward side of the atoll. The village of London lies in an area of sand dune and beach terraces. Moving toward the lagoon, the topography changes to a flat plain which slopes toward the lagoon. Fresh-to-brackish water seeps are present on the north side of the atoll. Near the lagoon on the east side of the atoll, a sloping scarp 2–3 m high is present. Behind this feature are dunes and an inner flatland. On the southeast peninsula, a rubble rampart is present, upon which the road was built. The rampart is 100–150 m wide, sloping toward the lagoon. The rampart is replaced by dunes on the southeast peninsula. The dunes act as a protective barrier to the flat grasslands of the peninsula.

According to Hutchinson (1950) it is doubtful that phosphate mining ever took place on Christmas Island. Only very limited and local evidence exists for the presence of phosphate on the island. Cook (1785) wrote that the soils of the island were in some places light and others black, composed of decayed vegetation and bird droppings, mixed with sand. J. P. Payne, who visited the island in 1834 reported that the "interior of the island, and in fact most of its surface, is covered with fish bones, bird dung and decayed animal matter," according to Hutchinson (1950). Bryan (1942) remarked that the United States Guano Company worked the island for several years after 1858. However, this cannot be confirmed from shipping records. Elschner (1913) reported that the phosphate of Christmas Island was highly impure. Power (1925) states, as does Hutchinson (1950), that no exploitation of phosphate has ever taken place on the island.

Apatite was collected by Valencia (1977) and analyzed in the laboratory by W. Burnett. The apatite occurs to depths of 0.5 to 1 m as peripheral and fracture rinds in a beveled and severely eroded beach rock situated in the lagoon flat. This one occurrence of phosphate rock accentuates the sparsity of phosphate on this island. The island is situated in a zone of low precipitation but has a large population of birds. Helfrich et al. (1973) estimated, as did Hutchinson (1950), from the bird population that 200 tons/yr of guano would be produced. Drilling to 40 m on Motu Tabu by the Avian Mining Company (Australia) failed to encounter commercial deposits of apatite (Valencia, 1977; Helfrich et al., 1973). Helfrich et al. (1973) suggest the absence of guano may be due to bird behavioral patterns and more importantly to leaching by occasionally heavy rains (Helfrich, personal communication, 1988).

Other shallow exploratory holes have been drilled around the island and calcarenites are found interbedded with hard coral layers. At shallow depths coral reef deposits are present.

JARVIS ISLAND

Jarvis Island is a small roughly rectangular coral island (Fig. 8). It appears on some older maps as Bunker Island. The rim of the island is 3–7 m above sea level with steep beaches. Around most of the island, the beach is composed of calcareous sands and shingle. Old beaches of coral fragments are found inland of the present beaches. Schlanger and Tracey (1970) sampled a 6 m high shingle beach ridge which surrounds the island. The ridge overlies a cemented coral rubble 1.5 m above the present reef flat. Corals from the rubble yielded radiocarbon ages of 1800± 200 years to 2250± 400 years. Schlanger and Tracy (1970) suggest that the drying of the lagoon was related to the emergence of the atoll between 1800 and 4450 years ago. At many locations, the lower part of the beach is a well-cemented calcareous sandstone which makes a bench which extends roughly 35 m offshore except in the east side where a shoal extends offshore some distance. The seaward margin of this bench consists of ridges (20–50 cm above mean tide) and channels which cut approximately one meter below mean tide. Dipping beds of sandstone and conglomerate can be seen along parts of the seaward margin of this bench (Wentworth, 1924). A sea-level notch is cut 0.6 m above sea level on the north side of the island. The beach on the east coast of Jarvis consists of a coral shingle beach with slabs or flat or rounded coral layers at a 20–30 degree angle.

The interior of the island slopes gently to form a basin (in the lower parts reaching sea level) that is an ancient lagoon (Gregory, 1924), which has gradually filled with sand debris (Hague, 1862) and guano. Toward the end of the filling process (Hutchinson, 1950) the evaporating lagoon waters deposited gypsum, with the more soluble salts washing away into the remains of the lagoon. Concentric beach lines indicating the former positions of the lagoon margins are clearly present (particularly on the east side), as well as parallel sand ridges marking the gradually decreasing lagoon margins. The process of desiccation continued until after the thick guano deposit had formed. The island is the nesting place for sea birds. Much of the basin has been covered to a depth of 2–3 m with phosphatic earth or guano (Wentworth, 1934; Bingham, 1900). The guano deposits on Jarvis were heterogeneous according to Hague (1862). Chemical analysis of samples have been reported by Webb

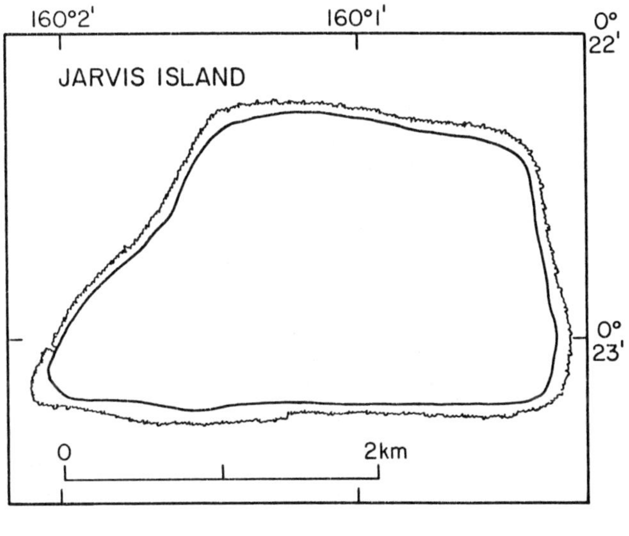

Figure 8. Map of Jarvis Island. The figure is based upon the U.S. Hydrographic Office Map of Jarvis Island.

and Sardy (1859), Liebig (1860), Hague (1862), Voelcker (1876), Heiden (1887), Christophersen (1927). Hague reports a shaft was cut vertically into the guano deposit. Underlying the deposit were 0.6 m of evaporites (which ranged from compact and crystalline to soft and amorphous). Beneath these sediments were 6 m of fine to coarse sands with an occasional stratum of coral fragments and shell as well as dolomitized algal mats interspersed. Elsewhere on the island, the guano deposits were found to rest directly on a coral formation, and in still another part of the island the phosphate deposits appear to result from mixing by water with coral muds. Many reported analysis of the Jarvis guano indicate that the samples were mixtures of phosphate crusts with underlying evaporites or phosphatic muds (Hague, 1862). It is not clear from Hague's account if the last deposit of gypsum is stratigraphically continuous with the main gypsum hard-pan under the worked guano deposit. Christensen (1927) suggested at least two periods of guano deposition occurred with a rise and fall in lagoon level, and $CaSO_4$ formation between them. Hutchinson (1950) suggests three periods of bird colonization and guano deposition occurred. The first two were separated by wind-blown (or water deposited) coral sands, and the third by the gypsum layer. Phosphatic nodules ("hummocks") were reported by Hague (1862), who argued that the features present crusts formed in a soft (nearly plastic) state. Hague also suggests that the phosphatic deposit on Jarvis "has evidently long been acted upon by sea water." During a visit to the island in 1988, the phosphatic nodules (described by Hague, 1862) were observed to be common near and on the dry lake bed.

Schlanger and Tracey (1970) took cores from the interior evaporite basin and found 0.3 to 0.6 m of gypsum crystal mush overlying 0.3 to 0.6 m of laminated algal mat, in turn overlying a couple meters of lagoonal carbonate mud containing mollusk shells, foraminifera and coral fragments. Both the lagoonal mud and the bioclastic debris in the algal mat contain slightly Ca-rich dolomite. A lagoon mud sample, approximately 50 percent dolomite, gave a ^{14}C age of 2650± 200 ybp.

Jarvis Island was mined by the Pacific Phosphate Company beginning in 1858. Much phosphate remains. In the center of the island the soil is a light to dark brown mixture of sand with a high organic content. In places, gypsum and phosphatic hard grounds are extensively developed (British Intelligence, Geographic Handbook, 1943). The mining operations have considerably altered the surface of the island. On the beach crests and landward slopes the soil consists of light grayish brown coral sand with only slight organic material. Some analysis of phosphatic crusts, guano and soils were reported by Dixon (1877). In a report to stock holders, the American Guano Company reported that 7,000,000 tons of phosphatic guano existed on the island. Hutchinson (1950) takes the view that based on the known length of exploitation, these estimates are "gross exaggeration." Hague (Hutchinson, 1950) noted reminiscences of a prospectus published in 1897 by a British Company indicated 120,000 tons of phosphate were left on Jarvis and Baker Islands together. Bryan (1942) indicates that little digging was done by the Pacific Phosphate Company, which leased the island in 1906. Hutchinson (1950) thus states that it is likely that 200,000 to 300,000 tons of material was removed from Jarvis and further states, "an estimate of this kind made in 1897 of the remaining reserve would not be unreasonable in view of the fact that some phosphatic material still exists on the island."

Hague (1862) suggests the whole island has been elevated well above sea level. Wentworth (1924) states that no evidence of reef formation was found at elevations greater than 4 m. All of the land above 4 m has been formed of detrital material. Wentworth suggests the main growth of reef which formed the atoll took place prior to the last change in sea level when the water stood about 4 m higher than at present. Other authors, for example, Gregory (1924), state the island appears to have been uplifted and the lagoonal deposits are now exposed 2–3 m above sea level. There are only two sickly looking coconut trees growing on the island, a few scrubs and little grass.

MALDEN ISLAND

Malden Island is a triangular shaped coral island (Fig. 9). The island rises 5–15 m above sea level. The rocks forming the interior of the island are for the most part in situ corals. Patches of guano overlie the carbonates and vary greatly in thickness and composition. Near the south side of the island is a fissure extending in an east-west direction, which is filled with debris. The water level in the open part of the fissure rises and falls with sea level. The brackish-water ponds are completely enclosed. Small isolated ponds are present where guano mining took place. These hypersaline ponds contain algal mats around their margins. Large deposits of calcium sulfate and calcium carbonate are developed, especially on the west side of the lagoon farthest from the recharge area (Dixon, 1877; British Intelligence Geographic Handbook, 1943). The beach crest and several ridges (a total of nine were present be-

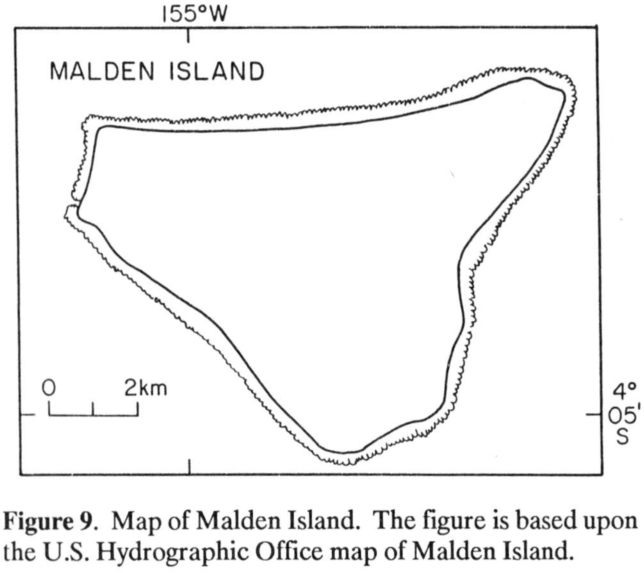

Figure 9. Map of Malden Island. The figure is based upon the U.S. Hydrographic Office map of Malden Island.

tween 1866–69) form a rim around the lagoon. These ridges consist of coral fragments and are dark gray to black in color. The lowest part of the rim is about 5 m above sea level, the highest about 15 m. The beach rises sharply from the waters edge on the north shore and is generally about 60 m wide. It grades from coarse coral sand through fine gravel to coarse gravel and coral pebbles, 2–3 cm in diameter and is a brilliant white color. The upper beach consists of still coarser coral debris interlocking with shingle set at steep slopes, often at an angle of 35 degrees. Water-lain pumice was abundant on the beaches in the late 1860's (Dixon, 1877). A narrow fringing reef, cut by narrow channels, surrounds the island and extends approximately 20 to 200 m offshore. Only two palm trees were present on the beach crest in 1988. Limited shrubs were seen as well as grass which covered approximately 50 per cent of the interior.

Keating visited the island in 1988. At that time, the mud flats of the island were dry and formed large areas of dry salt pan surfaces, which could be easily traversed. On the southern end of the island, the irregular indurated limestone surface was exposed. The limestones displayed extensive karst morphology. Limestone pedestals, bridges and pools were found. During this visit, the water filling these karstic ponds was salty but pleasant for swimming; no fish or mollusca were present in the water. One fissure was seen in the three parallel karst ponds but did not extend over 15 m in length. Some of these ponds display limited algal growth and in isolated segments appear to be stagnant.

Because phosphate deposits are irregular and much of the phosphate has been removed, the underlying carbonate reef and lagoonal features are much better displayed than on any of the other Line Islands. The exposures of patch reefs and sand channels are clearly evident. The geologic exposures in the interior of Malden Island are indisputable evidence of the emergence of an atoll (presumably caused by the lowering of sea level). The karstic features of these limestones clearly show that the rocks have experienced substantial subaerial karstic erosion and dissolution.

Phosphate was mined from this island beginning between 1860 and 1862 and continued until 1914. Considerable alteration of the island surface resulted from the mining. Dixon (1878 a,b) reported some analysis of wind-blown phosphatic material. The guano deposits occur irregularly within the encircling ridge on a nearly flat surface about one meter above sea level (Hutchinson, 1950). Underlying the wind-blown sands the coral rock surface is irregular. The phosphate fills pockets in this irregular surface. The sediments consist of soft yellowish brown to dark chocolate grains mixed with white specks of calcium carbonate. The bottoms of the pockets were lined with indurated phosphate. Analysis of the indurated phosphate crusts were made by Voelsker and reported by Hutchinson (1950).

The United States Guano Company claimed the phosphate deposits on Malden Island; however, British phosphate diggers already were operating on the island. According to Bryan (1942) mining had begun by 1856. In 1842, a "Report to Stockholders" estimated there were 5,000,000 tons of reserve on the island. Hutchinson (1950) used figures reported by Gussefeld in Heiden (1887). He suggests that Malden Island yielded 8,000 to 10,000 tons of phosphatic material per year. If the mean yield for 70 years is half that, on the order of 350,000 tons of phosphatic material was exported. Hutchinson (1950) states, "The British Phosphate Commissioners state that in 1943 there were an estimated 100,000 tons left of material containing 40 to 50 percent phosphate. With the development of the phosphate industry on Ocean and Nauru Islands, Malden Island was abandoned prior to exhaustion of the phosphate deposits."

Fresh water is scarce on Malden Island. Laborers on the island used condensers for drinking water. In the rainy season, depressions formed by former excavations of guano fill to form fresh-water ponds around the lagoon margin. The anchorage off this island is precarious. During mining operations vessels were moored off buoys on the western end of the island. Only small boats were landed on the sandy beach since strong currents at the site make landing hazardous.

Limestone slabs were used to build stone enclosures, platforms, and paths on Malden Island resembling maraes or shrines elsewhere in Polynesia (Fig. 10). Graves from the days of active mining are also present. Emory (1934) suggests the ancient remains are probably associated with an early habitation of between one hundred and two hundred people.

Figure 10. Drawing of one of the ancient Polynesian stone structures on Malden Island, based upon a photograph by K.P. Emory (1934). The figure originally appeared in the British Intelligence Geographic Handbook Series (1943).

STARBUCK ISLAND

Starbuck is a low atoll (Fig. 11). The islets, of white sands, rise 5 m above sea level with low weeds, making them difficult to see from a distance. The atoll rim consists of low ridges of coral blocks and fine powdered coral, enclosing guano beds in the central part of the atoll. Behind the beach crest the center of the island is depressed but there is no major lagoon; instead there are several small salt lagoons near the eastern end (Arundel, 1890) that communicate with the ocean by a tunnel (S.C. Ball, cited by Hutchinson, 1950). Schlanger (1989; personal communication) recalls that in 1968, these "lagoons" were crusted over by large salt crystals. Hague (1862) also reports large quantities of salt have crystallized around these. The steep beaches are surrounded by fringing reefs, which extend 900 m offshore. There is no safe anchorage. Landings are made through a small channel blasted into the reef near the western end. Heavy surf breaking against the reefs makes landing hazardous.

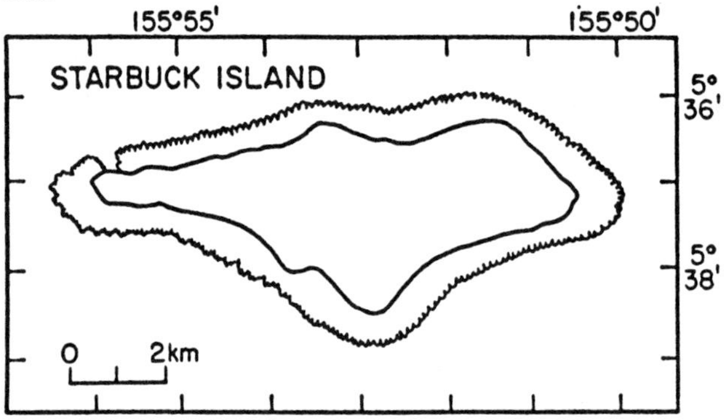

Figure 11. Map of Starbuck Island. The figure is based upon the U.S. Hydrographic Office Map of Starbuck Island.

Voelcker (1876) reports analysis of guano. The guano was found resting on, or was mixed with a crust. Phosphate was mined from this island beginning shortly after 1866 and continued until 1874 (Hutchinson, 1950). In 1872 one hundred laborers worked on the island. Hutchinson (1950) suggests that the original deposit contained tens of thousands of tons.

VOSTOK ISLAND

Vostok Island is a low triangular-shaped island (Fig. 12). The greatest elevation on the island is 5 m above sea level. The island can easily be seen from offshore due to the large number of high trees. The soil on the island is rich in organic matter and moisture. The island has no central lagoon. It is surrounded by a fringing reef 100 m wide, which extends 500 m further as submarine platforms to the north, west and east, reflecting the triangular shape of the island.

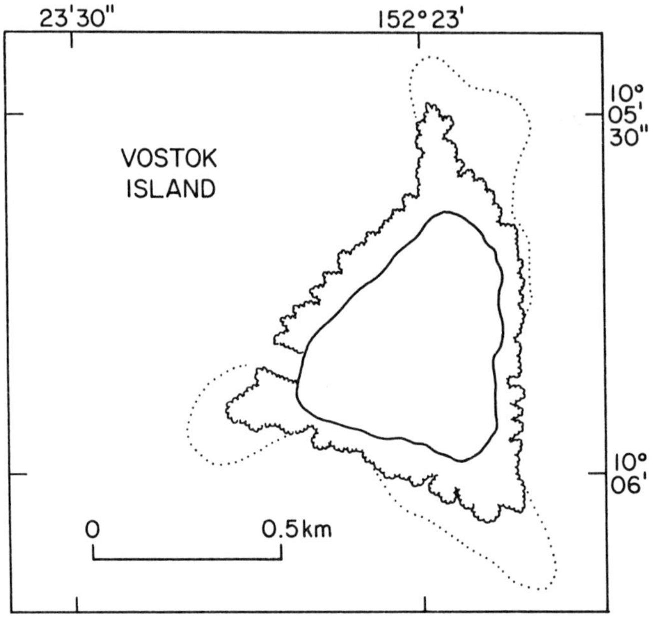

Figure 12. Map of Vostok Island. The figure is based upon the U.S. Hydrographic Map of Vostok Island.

The beaches consist of gravels and the surf is heavy. Phosphate was examined on the island around 1870 by Arundel but no phosphate appears to have been extracted.

CAROLINE ISLAND

Caroline Island consists of a crescent-shaped atoll with about 24 islets (Fig. 13). South Island rises 4–6 m above sea level. It has been shown as Clark Island on some older maps. There is no anchorage and no ships pass into the lagoon. Vegetation is heavy, including palm and pandanus trees. Phosphate was mined from this island between 1873 and 1895. About 10,000 tons were reported to be shipped from the island.

Temple platforms and graves containing stone adze blades have been reported from the northern islets. These structures are similar to Polynesian structures elsewhere in the southern Pacific and indicate early habitation of these islands.

FLINT ISLAND

Flint Island is a low coral island, which is shaped like a kite (Fig. 14). The maximum height of land is 7 m. The interior of the island was mined for phosphate. The excavations have filled with brackish water to form two and some times three small lakes. The island is surrounded by a narrow fringing reef with a reef platform extending 2 km northward from the island. The surf break is heavy on the

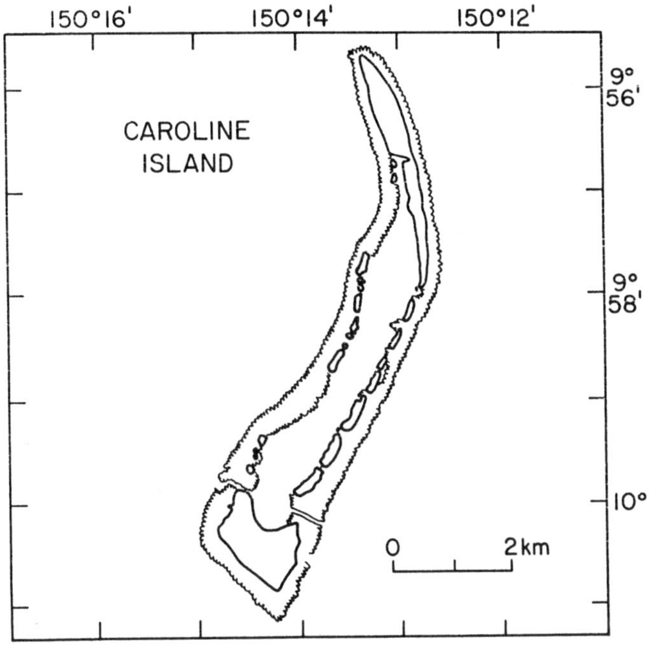

Figure 13. Map of Caroline Island. The figure is based upon the U.S. Hydrographic Office Map of Caroline Island.

beaches. Anchorage is not safe and landing hazardous. Phosphate was mined from this island between 1872 and 1890. A total of 30,000 tons was reported to have been shipped from the island (British Intelligence Geographic Handbook, 1943), with an average of 45-60 percent phos-

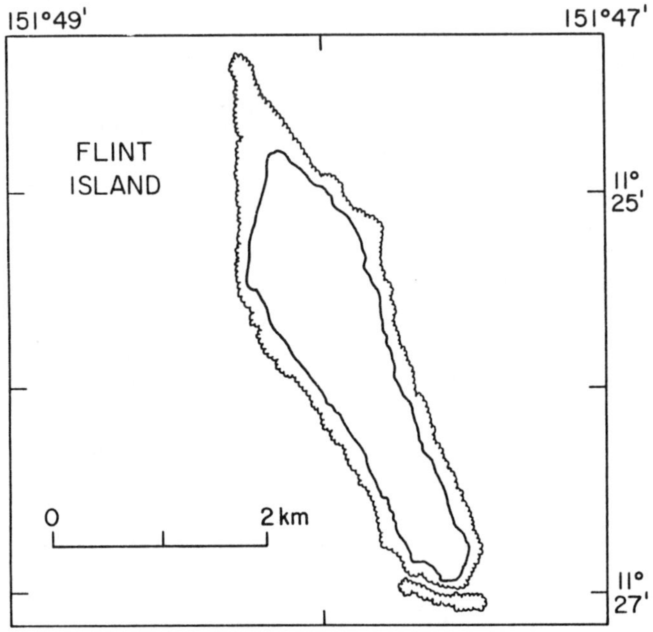

Figure 14. Map of Flint Island. The figure is based upon the U.S. Hydrographic Office Map of Flint Island.

phate of lime. The island was heavily forested prior to mining and subsequent coconut plantation development.

DISCUSSION

The earliest geologic studies of the Line Islands (Hague, 1862; Wentworth, 1931) pointed out that the islands are emergent. According to these authors, in-situ reef materials are exposed, up to 4 m above present sea level on many of the islands. The youngest radiometric dating of dredged basalts from seamounts in the Line Islands chain are of Eocene age (Schlanger et al., 1984). Thus, the emergence of the islands is probably not due to tectonic uplift involving recent volcanism or nearby plate boundaries. The Line Islands have become emergent as the result of a eustatic drop in sea level. Since emergence, various environmental, climatic, and geologic processes have acted on the islands, producing a range of physiographic forms.

ISLAND EMERGENCE

Hague (1862) observed that Jarvis, Malden, and Starbuck Islands in the southern Line Islands were emergent islands similar to Baker, Howland, Enderbury, and McKean Islands (in the Phoenix Island Group). Wentworth (1931) described the geology of several islands in the Line Islands group and provides further evidence for emergence. Wentworth (1931) states, "it is believed that Washington Island, like several others of the group, was first built as an atoll with reference to a higher sea level and that it became emerged through the lowering of sea level." Jarvis Island, in particular shows clear examples of in-situ beach rock 4 m above sea level.

Wentworth suggests all of the Line Islands with the exception of Palmyra Island are emergent, and most stand roughly 4 m above the present sea level. Palmyra Island is an exception. Wentworth suggested that either Palmyra island had not existed as an island at the time of higher sea level or because it was imperfectly developed, much of the island has since been cut away by waves, leaving no record of emergence.

More recent studies in the region suggest a longer, more complex record of eustatic sea level oscillations in these islands. Valencia (1977) suggests that "as sea level rose to a maximum 2 m level within the last 5000 years (Fairbridge, 1952; 1961), Christmas Island was almost entirely submerged." During the brief still stand of sea level, Valencia (1977) suggests the reefs of Christmas Island were planed off. Power (1925) studied Nauru and Ocean Islands, which lie west of the Line Islands. Power suggested both islands "have risen and sunk, and that in so doing have affected the formation of the phosphate deposits... These islands have not only risen, but have also sank since guano was deposited on them, and this more than once....As the island sank, fresh coral growths would form round it, and so protect....The island rose a second time, and conditions again became favorable for birds to visit...The island sank again; some of the phosphate sandstone became broken up and rolled about, forming boulders. Such boulders are found on the top of the islands, showing that the islands have risen yet again. This rising and sinking may have been repeated still more often."

This description of the complex emergent history of Ocean and Nauru Islands results from petrographic studies of samples collected early in this century prior to extensive mining on these islands. Most of the phosphatic deposits in the Line Islands were removed by mining in the late 1800's. The island surfaces have been physically disrupted prior to visits of geologists. Because of this disruption and removal of surface deposits, it is doubtful that details of the emergence of the Line Island chain will ever be resolved. (Jarvis Island, however, may be an exception. When visited in 1988, relatively abundant phosphate deposits remain and some are undisturbed.) Based upon the presence of abundant in-situ reef limestones, it is clear however that most of the Line Islands seamounts are emergent and sea level now lies at least 4 m below the highest reef and beach deposits.

MORPHOLOGIC ARGUMENTS

Early studies of island and atoll morphology reflect the heated discussions surrounding Darwin's coral reef subsidence theory. Daly (1910) expanded Darwin's hypothesis and suggested that much of the reduction of coral atolls was carried out mechanically by wave abrasion. It was not until the 1940's that Japanese workers recognized that karstic dissolution may explain the morphology of atolls. It was even later, when Tayama (1952) translated the works of Yabe (1942) and Asano (1942) into English that the model of incremental accretion of thin limestone sequences on intermittently emerged and karst eroded older reefs was made widely known. Hoffmeister and Ladd (1945) suggested that the morphology of atolls was not due to the original growth form (i.e., inherited) but instead was due to solution subsequent to exposure. Hoffmeister and Ladd simulated many atoll forms by placing a slab of limestone under a shower of dilute acid. Kuenen (1947) proposed intertidal and subaerial solution as the main processes shaping the morphology of atolls. Jennings (1971, 1972b) indicates that higher temperatures and rainfall, rapid plant growth and decay and intense microbial activity in high PCO_2 soil produce high solution activity. The young reef limestones which are highly porous, poorly consolidated and mechanically weak thus are quickly acted upon by karst processes.

In the case of the Line Islands, karstic features are present in two forms. On Malden Island, karstic sink holes or dolines are present on the south side of the island. There, limestone pedestals and bridges stand at roughly 1 m above water level among the sink holes, which are roughly 10 m deep. On Johnston Island, sink holes are present on the carbonate bank (Ashmore, 1973). In addition, numerous caves were observed from a submersible by Keating (1987) and the submersible was taken into one large cave containing stalagmites, stalactites, and shelfstones.

PHYSIOGRAPHIC DEVELOPMENT

Wentworth (1931) visited several islands within the Line Islands chain. Speaking physiographically, he stated "Fanning is distinctly younger than Christmas, Jarvis or Washington Island." He observed that the islands displayed surface features that appear to represent various stages of lagoon in-filling.

Wiens (1962) presented a model for the physiographic development of atolls during times of changing sea level (Fig. 15). Wiens suggests the lowest reef level presents the appearance of an atoll at the end of the last low stand of sea level. He suggests that the emergent reef once stood 100 to 300 m above the Wisconsin sea level (Wiens, 1959; 1962) and has since been mechanically eroded and chemically dissolved so that only portions of the reef periphery (A) and small topographic highs in the lagoon (D) remained above sea level, with the lagoon floor (C) at sea level. At various stages of sea level change, wave cut benches were formed. Subsequent to the main erosional period, fringing reefs developed with algal ridge growth and a live reef-front, characterized by windward spur and groove development and wave-cut benches.

Utilizing the initial profile of Wiens (1959), a sequence of physiographic stages observed in the Line Islands are outlined in Figure 16. During the post-Wisconsin eustatic rise of sea level, upward growth occurred along the entire fringing reef flat as well as outward growth of the reef margins. Rectilinear patterns of patch reef developed on the immersed lagoon floor, both parallel and transverse to the primary wind direction. Linear patch reef development was possibly due to more rapid reef growth in divergences of Langmuir cells or to a rectilinear base of sand dunes developed during the Wisconsin low stand (Valencia, 1977). As sea level rose, sedimentation within the lagoon began filling the basins.

When the level of the sea rose to 2 m above present during the last 5000 years (Fairbridge, 1952, 1961), most of the Line Islands would have been almost entirely submerged (Figure 16A). Lagoon patch reef and fringing reef development on most islands kept pace with the rising sea level. During the brief still-stand of sea level and the sub-

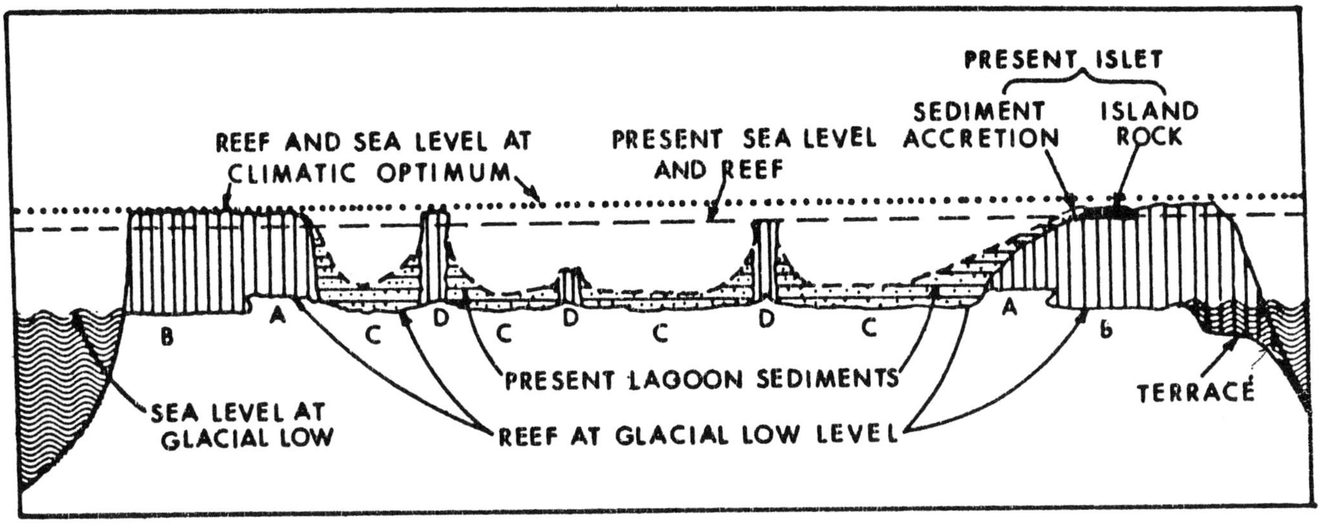

Figure 15. Model for atoll evolution suggested by Wiens (1962) and discussed in text. "A" represents undercut "raised" reef remnants. "B" represents the old reef flat. "C" represents the old lagoon bottoms. "D" represents coral patches, knolls and pinnacles. The figure illustrates that many equatorial Pacific islands are now emergent and once formed reefs at a stand of the sea higher than present sea level. As the sea level dropped, the islands have become emergent. Infilling of the lagoon has resulted from their most recent emergent condition. During glacial lows (sea level was less than that at present), notches and terraces were cut into the carbonate banks.

sequent eustatic fall of sea level, the reefs were planed off and supplied sediments to the emerging lakes and lagoons (Valencia, 1977). Beach rock formation took place on exposed reef tops at this time. Aeolian sediments, which were derived by wind and wave erosion of the emerged windward coastal plains and seaward beaches, formed the coastal inland and lagoon dunes (Figure 16B). These aeolian sediments helped fill the lagoons along with autochtonous biogenic sediments.

As sea level dropped, wave and subaerial erosion planed the reef flat. Karst caves and sink holes were formed. The eroded sediments progressively filled passes within and between lagoons, and together with growth of linear patch reefs, isolated the lagoons. The lagoons, once isolated from tidal flow, formed hypersaline ponds (Figure 16C). The increased salinity in these ponds decreased reef and invertebrate faunas occurred. As a result, shells and reef fragments accumulated as berms on the leeward shores of the ponds. The hypersaline ponds continued to shoal with the precipitation of evaporites, principally gypsum and halite (Figure 16D). With continued aeolian deposition lagoons filled. Most islands then accumulated significant deposits of guano and phosphatized sediments (Figure 16E). On those islands where rainfall was sufficient, fresh water lakes and bogs developed in depressions. The fresh water features have proved in the last century to be transient features, very dependent upon local rainfall, except on Washington Island. Since the volume of the fresh water lake on Washington is several times the volume of water falling on the island during a year, and the process of reducing salinity is a slow one, it is likely that the freshening of the water on Washington Island has taken several hundred years (Wentworth, 1931).

Christopherson (1927) and Dixon (1878a) suggest that once the equatorial islands become emergent within a dry belt and are relatively barren of vegetation they remain dry. These authors note that precipitation is often prevented on dry and barren islands, due to the heating of the air by the highly reflective coral rock and sands. The heated air rises in a column above the island. Often, when conditions are favorable for precipitation over the surrounding ocean, rain falls on the ocean but fails to fall on the island as the relative humidity is decreased by the warm air.

PHYSIOGRAPHIC PROGRESSION

The progression of physiographic stages can be identified in the Line Islands. The physiographic stages range from submerged reef banks to typical atolls with ring reefs and finally to high islands with in-filled lagoons. This sequence is shown in Figure 16. Johnston Island and Kingman Reef are good examples of nearly submerged reef platforms (Figure 16A). Palmyra Island has a partially in-filled lagoon (Figure 16B). Fanning Atoll has an atoll

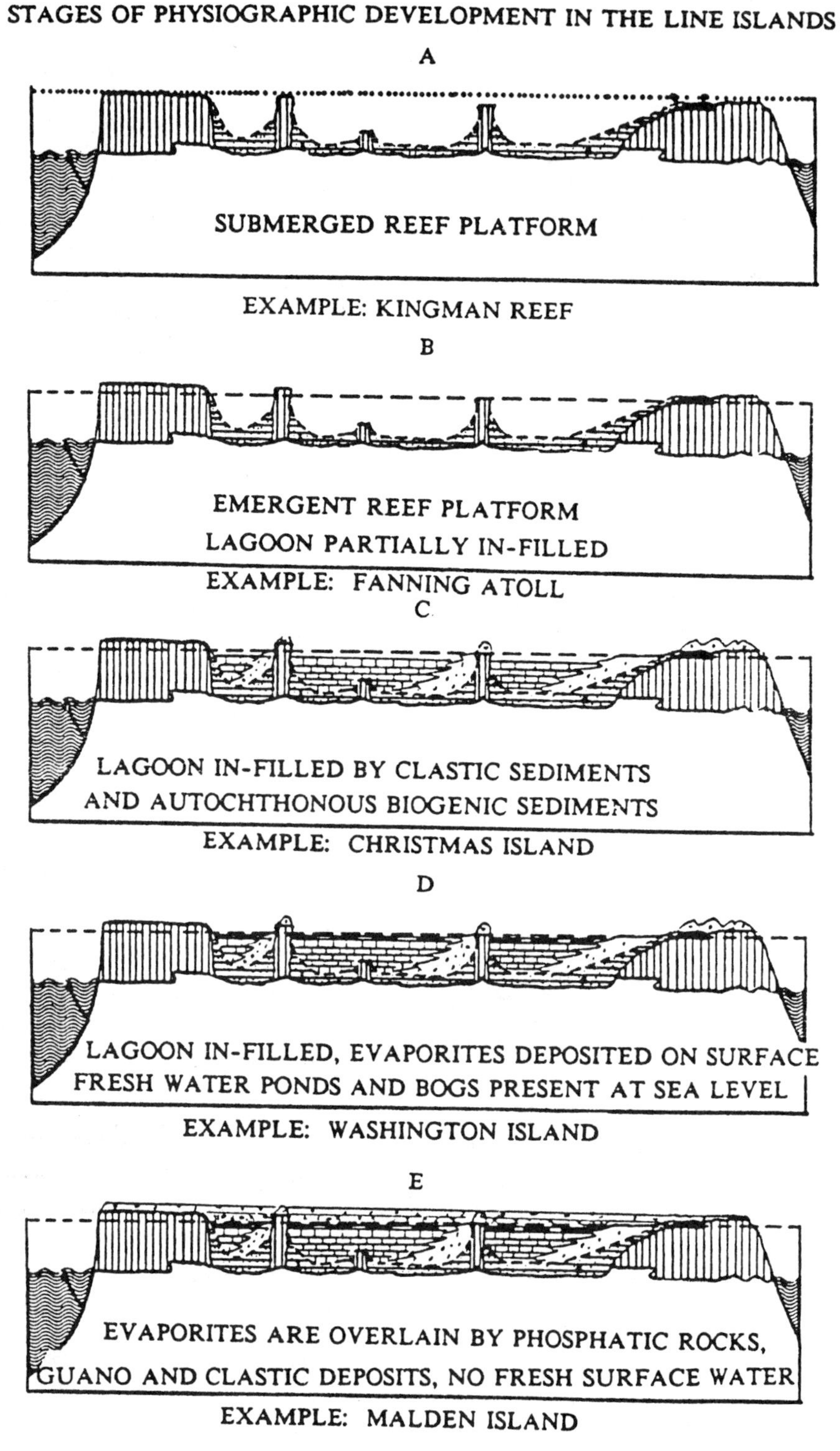

Figure 16. The cross sections in this figure illustrate the stages of physiographic development displayed by islands and atolls in the Line Islands chain. The profiles begin with an idealized atoll cross-section (redrawn from Wiens, 1962) at a time when the majority of the reef platform was submerged. The atoll subsequently became emergent as a result of eustatic sea level change. The lagoon was progressively infilled by aeolian sands and in-situ biogenic sedimentation. When the lagoon was sufficiently infilled evaporites were deposited close to sea level within the central topographic depression. Eventually the central depression was infilled by aeolian deposits, guano, and phosphorites.

with a well developed lagoon (Figure 16C). Christmas Island is an atoll with a partially infilled lagoon and abundant hypersaline lakes and ponds (Figure 16D). Washington Island has no lagoon but has fresh water lakes and bogs instead. Jarvis Island and Malden Island have totally infilled lagoons, abundant phosphatic hard grounds, and poorly or unconsolidated phosphatic deposits which are underlain by evaporites (Figure 16E).

Climatic conditions, namely rainfall and aeolian processes, appear to control the development of these islands. The annual rainfall accumulation is shown in Figures 17 and 18 (Taylor, 1973). The rainfall on Washington Island ranges from 70 to 180 inches per year, and is the highest rainfall observed in the Line Island chain with a maximum of 210 inches in a year. Palmyra Island generally ranges from 80 to 170 inches per year with a maximum of 200 inches per year. Fanning Island rainfall accumulations have ranged from 40 to 120 inches per year with a maximum of 180 inches a year. The maximum annual accumulation at Christmas Island ranges from 20 to 60 inches per year with a maximum of 100 inches. The rainfall ranges from 5 to 50 inches a year with the maximum accumulation at Malden of 90 inches. The rainfall on Jarvis Island ranged from 20 to 40 inches per year with a maximum accumulation at Jarvis Island of 40 inches in a year.

Within these islands, the high islands are those exposed to the minimum rainfall. The low islands are those islands exposed to high rainfall. The physiographic progression observed in the Line Islands chain appears to reflect: (1) the degree of weathering and dissolution experienced at these sites, (2) the high rates of carbonate deposition filling the lagoons, (3) a strong contribution from aeolian processes in the dry equatorial islands in redistributing detrital material from beaches and beach ridges inland and building the interior of the island, and (4) strong biologic contributions from nesting sea birds.

The physiographic development of the more northern islands is dominated by the destructive influences of solution subsequent to emergence. These islands are situated within the tropical convergence zone, receive high rainfall, and experience the greatest karst weathering. Near equatorial islands have less rainfall, rapid deposition of calcium carbonate lagoon sediments, and important aeolian deposits. On these islands the constructive influences of rapid sedimentation appear to outweigh the destructive effects of rapid weathering and dissolution. The equatorial islands are high islands, situated in low rainfall zones. Clearly, the destructive karst weathering occurs very slowly. These destructive processes are overwhelmed by the constructive processes of aeolian redistribution of beach derived sediments and the voluminous deposition of biogenic material (guano).

Other unseen but well-demonstrated geological processes have also influenced the surficial features of an island. Keating (1987) suggested that Johnston Atoll in the northern Line Islands is tilted upward to the northwest. Side-scan sonar surveys of the southern flank of the atoll indicate that much of the southern margin of the seamount appears to be a fault scarp and that 700 m of debris covers the sea floor below the scarp. Keating (1987) suggests that mass wasting of the southern flank of the seamount has locally loaded the sea floor adjacent to the seamount producing the observed tilt. Keating further suggests that the local structural failure and differential subsidence is one mechanism by which reefs drown and guyots form. Roy and Smith (1970) suggest Fanning Atoll, also in the northern Line Islands, is tilted upward to the west. Valencia (1977) suggests Christmas Atoll may be experiencing a progressive northwest tilt. Dawson (1959) and Helfrich (personal communication, 1988) suggest Palmyra Island is tilted westward. Recent studies of the sea floor around the Hawaiian Islands by Normark and Searle (personal communication, 1988) using side-scan sonar imaging indicate that very large scale mass-wasting features, similar to those seen at Johnston Island, are common. The small scale tilting of atolls is likely to have drastic consequences on the flora and fauna of these atolls and significantly impacts the physiography of islands. The gradual transition from atoll to guyot appears to be underway at Johnston Atoll and Kingman Reef and perhaps at Christmas, Fanning, and Palmyra Islands.

PHOSPHATE DEPOSITS AND RAINFALL

Hutchinson (1950) reviews the available information on the distribution and abundance of phosphates in the atolls of the Pacific Ocean. Hutchinson concludes that phosphates accumulate preferentially on islands which are situated in climatic dry belts. Since the Line Islands extend from roughly 20°N to 15°S the islands are situated in several climatic belts. An equatorial wet belt runs across the Pacific just north of the Equator. North and south of the wet belt are regions of lower precipitation (Taylor, 1973). In these drier belts the rainfall increases irregularly from east to west.

Johnston Island lies in a relatively dry belt. The northern Line Islands fall within a wet belt. And, the southern Line Islands (and most of the Phoenix Islands) lie within a relatively dry belt. Hutchinson (1950) points out, "the principal commercially significant phosphate deposits occurred, in the relatively dry belt, while ... certain adjacent atolls without phosphate deposits lie in another wet belt." Hutchinson points out that the three northern Line Islands, Palmyra, Washington, and Fanning are well covered with groves of vegetation, and Washington Island has a fresh water lake and bogs. Christmas Island is considerably drier

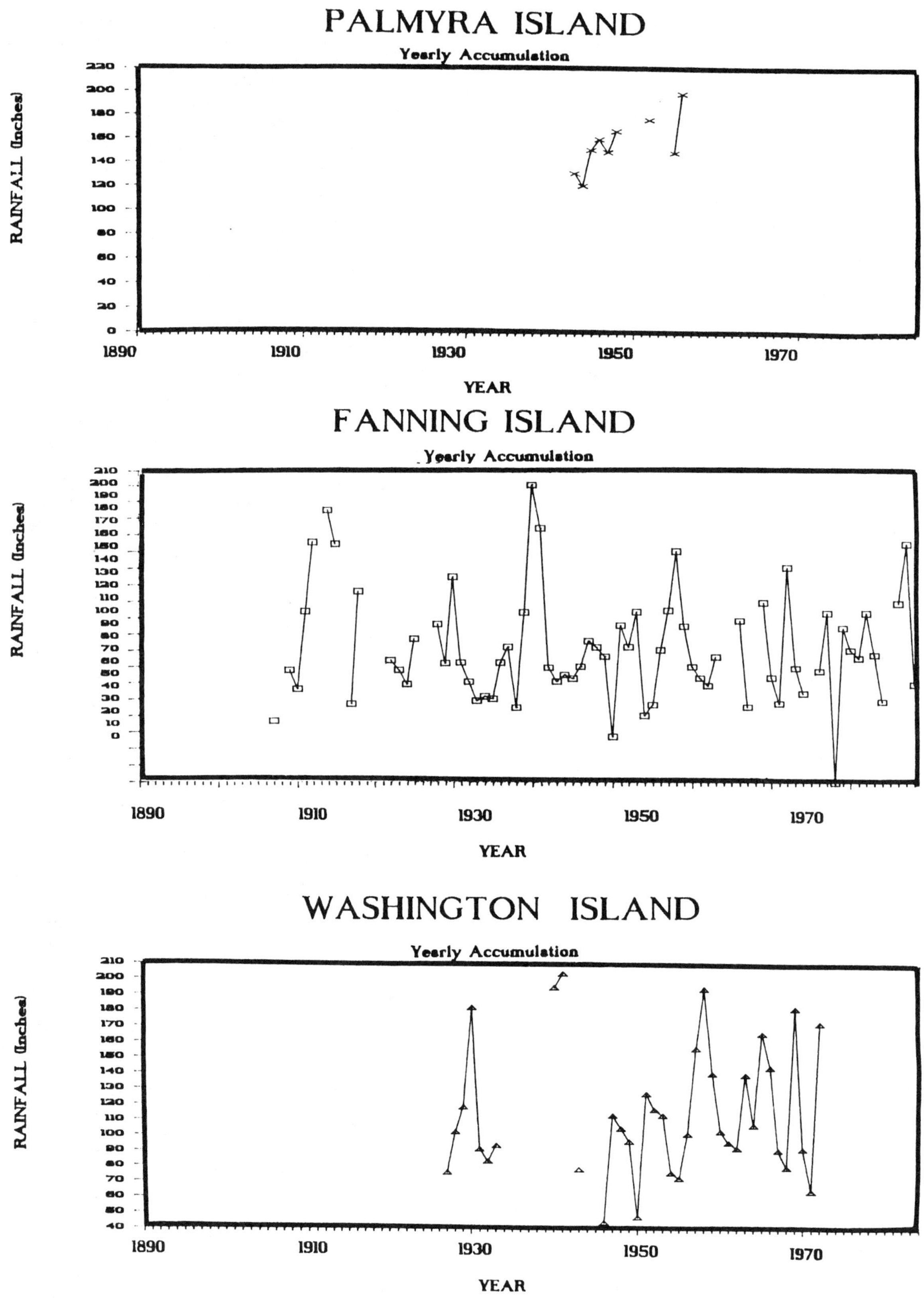

Figure 17. Plot of annual rainfall for Palmyra, Fanning, and Washington islands of the Line Island chain from data compiled by Taylor (1973).

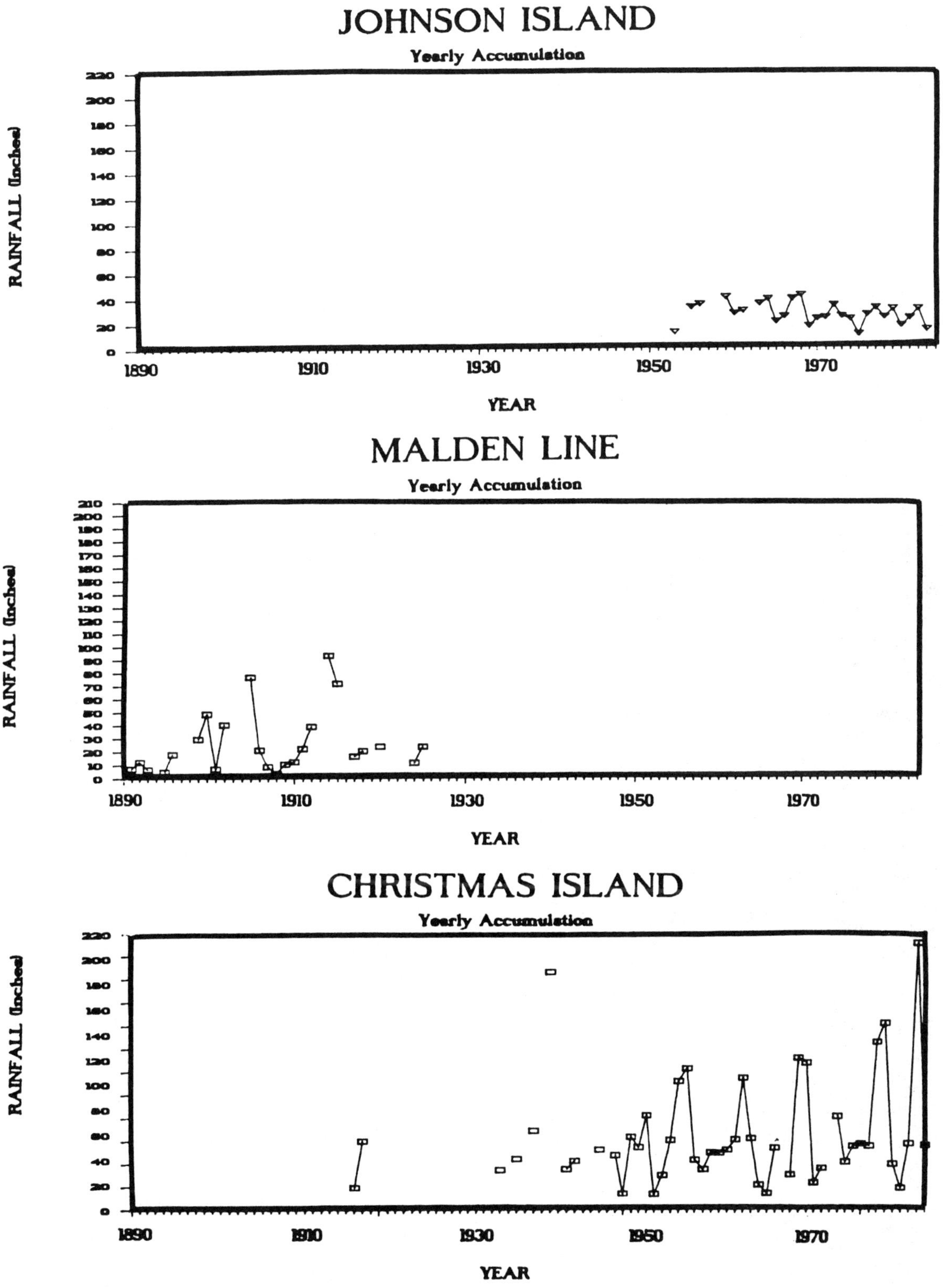

Figure 18. Plot of the annual rainfall (in inches) from Johnston, Malden, and Christmas islands. The data was compiled by Taylor (1973) and supplemented by the New Zealand Meteorological Service Data (C. Coutts, personal communication, 1989).

with vegetation limited for the most part (prior to introduction of coconuts) to scrubs (Christophersen, 1927). Jarvis, Starbuck, Malden, Howland, and Baker are situated in a very dry belt (Hutchinson, 1950). Their vegetation consists of limited herbs and grasses (Christopherson, 1927). Jarvis Island appears the driest of the group, based upon vegetation.

Numerous ornithology studies have been conducted in the Line Islands and continue to present. Johnston and Jarvis islands are bird sanctuaries. The drier barren islands within the Line Islands group form natural breeding grounds for ground-nesting birds, the most important source of guano. Hague (1862) mentions breeding colonies of millions of terns covering acres of ground on Baker Island. Dixon (1878b) speaks of large bird colonies on Malden Island. Beck (1921) mentions large colonies on Christmas Island. Keating observed vast colonies on Palmyra and Jarvis islands in 1988. Where vegetation is much more abundant and continuous, for example on Washington and Palmyra Islands, bird colonies are much more restricted, to littoral areas (or inland plains). Hutchinson (1950) states, "under these conditions, it is very improbable that any permanent accumulation of guano would take place, as the rainfall is heavy and the slope of the beach would lead to loss of material into the sea." Hutchinson (1950) and Hague (1862) suggest that not only is freshly deposited guano more easily eroded on the wetter islands, but the forest vegetation may establish itself and make the island unsuitable for large colonies of ground breeding sea birds. Hutchinson states, "where large numbers of arboreal species occupy an island, phosphatic soils are probably always developed, but such evidence as is available, notably from Rose Atoll, indicates that neither workable guano nor significant phosphatization of consolidated coral rock occurs on such islands."

SUMMARY

The equatorial Line Islands are a prime example of a series of emergent islands. The islands span the equator in a northwest-southeast direction, covering roughly 35 degrees of latitude. Each of the islands is affected by varying oceanographic, climatic and geologic conditions that result in physiographic development ranging from submerged banks to classic atoll, with open circulation of lagoon waters, to islands, that are dry with lagoons completely infilled. These islands were major sources of phosphates used as fertilizers throughout North America, Europe, Australia, and Hawaii during the late 1800's and early 1900's. Eustatic sea level oscillations have played a very important role in shaping these islands and providing environments in which phosphates could accumulate. Recent climatic studies suggest that global warming is taking place and sea levels are currently rising (Moberly and Mackenzie, 1985). Wiens (1962) suggests that over the next 5,000 to 6,000 years "it is possible that periods of rising sea levels may inundate most present land on atolls and possibly destroy most present reef islets." Looking at a longer time frame, however, Wiens suggests renewed emergence of these atolls and the formation of "high limestone islands." It appears very likely that the repeated emergent/submergent history outlined by Power (1925) will continue to shape these islands for the foreseeable future.

ACKNOWLEDGMENTS

Support for this project was provided by the United States Agency for International Development. I thank the reviewers for their helpful suggestions, particularly Philip Helfrich. I also thank Roger Lukas and Steve Chiswell for the opportunity to take part in the MW88-10-05 cruise to the Line Islands and the opportunity to visit Palmyra, Fanning, Jarvis and Malden islands. I thank the captain and crew on the R/V *Moana Wave* for support and assistance in landing on these islands. I thank the government of Kiribati for permission to land and investigate these islands and I thank the Fullard-Leo family for permission to go ashore on Palmyra Island. Jarvis Island is a wildlife refuge protected by the U.S. Wildlife Service.

REFERENCES

Amerson, A.B., Jr. and P.C. Shelton, 1976, The natural history of Johnston Atoll, Central Pacific Ocean: Atoll Research Bulletin, v. 192, p. 1-382.

Arundel, J.T., 1890, The Phoenix groups and other islands of the Pacific: New Zealand Herald, Auckland, July 5 and 12, 1890.

Ashmore, S.A., 1973, The geomorphology at Johnston Atoll: Technical Report TR-237, Naval Oceanographic Office, Washington, DC, 23 pp.

Atlas of the South Pacific, 1986: Government Printing Office, Wellington, New Zealand, 2nd edition.

Aydemir, V., 1985, A Geological Study of Reef Margins of Johnston Atoll, Central Pacific Ocean: M.S. Thesis, University of Tulsa, Tulsa, Oklahoma, 114 pp.

Beck, R.H., 1921, Visiting the nests of seabirds by automobile: National History, v. 21, p. 399-401.

Brigham, W.T., 1900, An index to the islands of the Pacific Oceans: Bishop Museum Press, Honolulu, HI, 300 pp.

British Intelligence Geographical Handbook, Pacific Islands, Volume II, Eastern Pacific, B.R. 519B, Geographical 1943 Handbook Series: Great Britain Naval Intelligence Division, p. 453-503.

Bryan, E.H., 1942, American Polynesia and the Hawaiian Chain: Tongg Publishing Co., Honolulu, HI, 254 pp.

Christophersen, E., 1927, Vegetation of Pacific equatorial islands: Bulletin, B.P. Bishop Museum, B44, 79 pp.

Cook, Capt. James, 1785, A voyage to the Pacific Ocean... 1776-1780: London, Printed by H. Hughs for G. Nicol and T. Cadell, 3 vols.

Dawson, E.Y., 1959, Changes in Palmyra Atoll and its vegetation through the activities of man, 1913–1958: Pacific Naturalist, v. 1, p. 3-5.

Dixon, W.A., 1878a, Notes on the meteorology and natural history of a guano island: Journal of the Royal Society of New South Wales, 1877, v. 11, p. 165-175.

Dixon, W.A., 1878b, The guano and other phospatic deposits occuring on Malden Island: Journal of the Royal Society of New South Wales, 1877, v. 11, p. 176-181.

Elschner, C., 1922, Killoide Phosphate: Kolloid-Zeitschrift, v. 31, p. 94-96.

Elschner, C., 1923, Beitrage zur Kenntnis der Koralleninseln des Stillen Ozeans: Zeitschrift fuer Praktische Geologie, v. 31, p. 69-73.

Elschner, C., 1933, Corallogene Phosphat-Inseln. Austral-Oceaniens und ihre Produkte: Lubeck, Max Schmidt, 120 pp.

Emory, K.P., 1934, Archaeology of Pacific equatorial islands: Bulletin, B.P. Bishop Museum, 123, 43 pp.

Emory, K.P., 1939, Additional notes on the archaelogy of Fanning Island: Bernice P. Bishop Museum, Occasional Paper No. 15, p. 179-189.

Emery, K.O., 1956, Marine geology of Johnston Island and its surrounding shallows, central Pacific ocean: Geological Society of America Bulletin, v. 67, p. 1505-1519.

Fairbridge, R.W., 1952, Multiple stands of the sea in post-glacial times: Proceedings of the 7th Pacific Science Congress, 3.

Fairbridge, R.W., 1961. Eustatic changes in sea level: Physics and Chemistry of the Earth, v. 4, p. 99-185.

Frondec, G., 1943, Mineralogy of the calcium phosphates in insular phosphate rocks: American Mineralogist, v. 28, p. 215-232.

Gregory, H.E., 1925, Report of the Director for 1924, Bernice P. Bishop Museum, Bulletin 21, p. 20-36.

Guinther, E.B., 1971, Ecologic observations on an estuarine environment at Fanning Atoll: Pacific Science, v. 25, p. 249-259.

Hague, J.D., 1862, On phosphatic guano islands of the Pacific: American Journal of Science, Ser. 2, v. 34, p. 224-243.

Heiden, E., 1887, Lehrbuch der Dungerlehre, 2nd Ed.: Hannover, Philipp Cohen, V. 2.

Helfrich, P., in collaboration with J. Ball, A. Berger, P. Bienfang, S. A. Cattell, M. Foster, G. Fredholm, B. Gallagher, E. Guinther, G. Kransnick, M. Rakowicz, and M. Valencia, 1973, The feasibility of brine shrimp production on Christmas Island: Sea Grant Tech. Rept. TR-73-02, University of Hawaii, 173 pp.

Hutchinson, G. E., 1950, The biochemistry of vertebrate excretion: Bulletin American Museum Natural History, v. 96, 554 pp.

Jenkin, R.N. and M.A. Foale, 1968, An investigation of the coconut growing potential of Christmas Island, Volume 1. The Environment and the Plantations: Land Resources Division Directorate of Overseas Surveys Land Resources Study 4, v. 1, p. 1-34.

Jennings, J.N., 1971, Karst: Australian National University, Canberra, 252 pp.

Jennings, J.N., 1972, The character of tropical humid karst: Zeits. F. Geomorph, N.F., v. 16, p. 336-241.

Keating, B.H., 1987, Structural failure and drowning of Johnston Atoll, Central Pacific Basin, *in* Keating, B., Fryer, P., Batiza, R. and Boehlert, G., eds., Seamounts, Islands, and Atolls: Washington, DC, American Geophysical Union, Geophysical Monograph 45, p. 49-59.

Kroenke, L.W. and Woollard, G.P., 1965, Gravity investigations on the Leeward Islands of the Hawaiian Ridge and Johnston Island: Pacific Science v. 19, p. 361-366.

Krusenstern, I.F., 1811, Reise um die Welt: Hande und Spener, Berlin.

Kuenen, P.H., 1947, Two problems of marine geology: Atolls and canyons: Verhandelingen de Koninklijke Nederlandse Akademie van Wetenschappen, Afdeling Natuurkunde, v. 43, 69 pp.

Liebig, J. von, 1860, Ueber Baker, Jarvis und Howland-Guano: Zeitschrifte fuer Landwwirtschaftliches Versuchs- und. Undersuchungswesen in Baiern, Sept 1860 in Chem. Centralbl., 1861, new ser, v. 6, p. 264-269.

Meinecke, K.E., 1876, Die Inseln des Stillen Oceans, eine geographische Monographie: Leipz, Paul Frohberg, v. 1, 1875, 382 pp; vol. 2, 1876, 487 pp.

Moberly, R. and Mackenzie, F.T., Climate change and Hawaii: Significance and recommendations: Hawaii Institute of Geophysics Report HIG-85-1, 33 pp.

Power, F.D., 1925, Phosphatic deposits of the Pacific: Economic Geology, v. 20, p. 266-281.

Richards, A.F., 1958, Transpacific distribution of floating pumice from Isla San Benedicto, Mexico: Deep-Sea Research, v. 5, p. 29-35.

Roy, K.J., 1970, Note on phosphate rock at Fanning Island: Hawaii Institute of Geophysics Report 70-23, p. 193-199.

Roy, K.J. and Smith, S.V., 1970. Sedimentation and coral reef development in turbid water: Fanning Lagoon: Hawaii Institute of Geophysics Report HIG-70-23, p. 61-76.

Schlanger, S.O. and Tracey, J.I., Jr., 1970, Dolomitization related to recent emergence of Jarvis Island, Southern Line Islands (Pacific Ocean) [abstr.]: Geological Society of America Abstract Volume, Annual Meeting, Milwaukee.

Smith, S.V., Roy K.J., Schiesser, H.G., Shepard, G.L. and Chave, K.E., 1970, Flux of suspended calcium carbonate ($CaCO_3$), Fanning Island Lagoon: Hawaii Institute of Geophysics Report 70-23, p. 41-59.

Stark, A., 1960, The discovery of the Pacific Islands: Oxford at the Clarendon Press, 259 pp.

Taylor, R.C., 1973, An Atlas of Pacific Islands Rainfall: Hawaii Institute of Geophysics Technical Report 73-9, Honolulu, 50 p.

Valencia, M. J., 1977, Christmas Island (Pacific Ocean) reconnaissance geologic observations: Atoll Research Bulletin v. 197, 17 pp.

Voelcker, R.R.S., 1876, On phosphatic guanos: Journal of the Royal Agricultural Society, Ser 2., v. 12, p. 440-459.

Walker, D.A. and McCreery, C.S., 1988, Deep ocean seismology, seismicity of the Northwestern Pacific Basin interior: Eos, Transactions of the American Geophysical Union, v. 69, p. 737-742.

Webb, W.H., and Sardy, J.B., 1859, Guano as a fertilizer: New York, Slote and Jones, 22 pp. (Trade pamphlet)

Wentworth, C.K., 1931, Geology of the Pacific equatorial islands: Occasional Papers of Bernice P. Bishop Museum, Honolulu, IX, 15, 25 pp.

Wiens, H.J., 1962. Atoll environment and ecology: Yale University, 532 pp.

Keating, B.H., and Bolton, B.R., editors, 1992, Geology and offshore mineral resources of the central Pacific basin, Circum-Pacific Council for Energy and Mineral Resources Earth Science Series, V. 14. New York, Springer-Verlag.

COASTAL GEOLOGY OF UPOLU, WESTERN SAMOA

Bruce M. Richmond
U.S. Geological Survey, 345 Middlefield Road, MS 999, Menlo Park, California 94025

ABSTRACT

Upolu is a volcanic island surrounded by fringing and barrier reefs. Maximum altitude of the rugged interior is about 1150 m. The volcanic rocks range in age from late Pliocene to Holocene and include basaltic lavas, scoria cones, pyroclastics, and tuffs. The coast of Upolu can be subdivided into the following three basic types based on reef characteristics, depositional features, and hinterland geology: wide fringing reef transitional to a shallow barrier reef; cliffed coast with little or no reef development; and fringing reef and narrow coastal strip consisting mostly of beaches, barrier spits, and coastal swamps associated with streams. Engineering structures such as seawalls and groins are common along many areas of the coastline, indicating a widespread coastal erosion problem. Causes for the erosion appear to be (a) possible relative rise of sea level enhanced by island subsidence, (b) reclaimed areas and engineering structures extending beyond the natural shoreline in areas prone to wave attack interrupting littoral transport of sediment, and (c) beach mining of sands and gravels for construction purposes. The well-developed fluvial drainage systems and the wide fringing and barrier reefs are conducive to the formation of numerous sites for potential nearshore construction materials. The potential for economically viable placer mineral resources appears to be limited because good source rocks are not present and the nearshore depositional systems are of relatively small size.

INTRODUCTION

Coastal Mapping Program

At the request of the Government of Western Samoa, a coastal mapping program was initiated by CCOP/SOPAC (Committee for Co-ordination of Joint Prospecting for Mineral Resources in South Pacific Offshore Areas) on Upolu Island, Western Samoa. This mapping was to provide a framework within which to identify broad geologic patterns. The first stage of the program involved interpretation of aerial photographs of coastal deposits, followed by field verification at selected locations. A reconnaissance survey was made of the shoreline, reef flats, and coastal plain of Upolu to investigate nearshore mineral resources (primarily construction materials), potential geohazards, and coastal processes. A technical report and ten 1:20,000 scale maps (Richmond, both in press) of the Upolu coastal zone were prepared. Previous nearshore studies on Upolu by CCOP/SOPAC included surveys at the following locations: Aleipata (Richmond, 1985; construction materials), Manono (Richmond and Roy, 1989; bathymetry and sediments), Mulinu'u Peninsula (Carter, 1987; engineering applications), Solosolo Beach (Eade, 1979; construction materials), Mulifanua and Salelologa (Gauss, 1982; bathymetric surveys), and Apia and northwest Upolu (Rubin, 1984, and Gauss, 1981; bathymetry and construction materials).

Location and Island Geology

Western Samoa is the larger and more westerly segment of the Samoan Archipelago (Fig. 1). It consists of two main islands, Savai'i and Upolu, two smaller islands, Apolima and Manono, and several uninhabited islets. The islands lie between 171°20' and 172°50' W longitude and 14°10' and 13°20'S. Total land area is approximately 2,850 km^2, Upolu being just over 1,100 km^2. Upolu is roughly

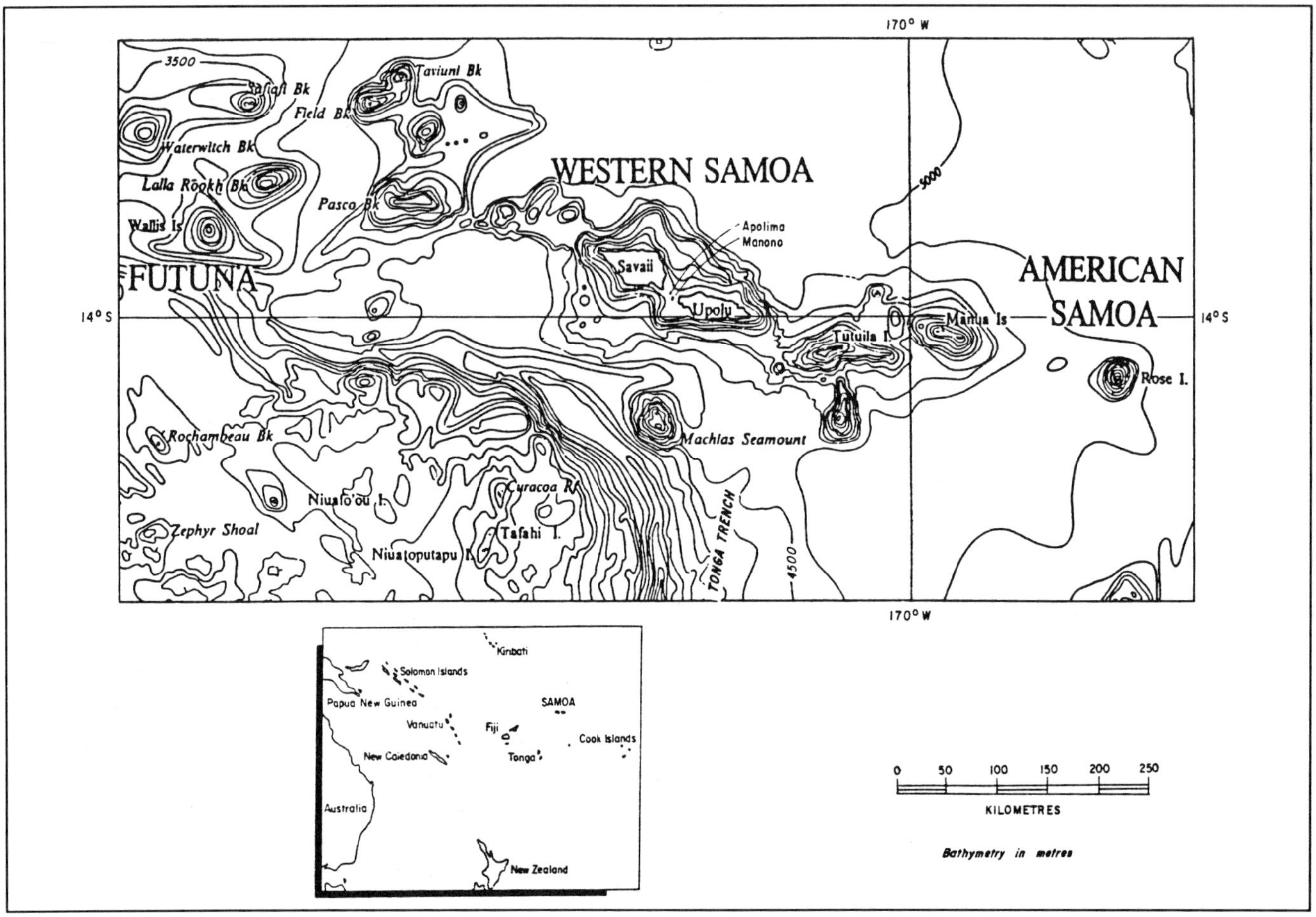

Figure 1. Location and bathymetry of Western Samoa (adapted from CCOP/SOPAC Chart 1: Bathymetry of the Southwest Pacific)

elliptical in shape. It is 75 km long and 25 km wide, and has a central ridge of volcanic cones as much as 1,100 m high. Samoa has about 155,000 inhabitants, 35,000 residing in the capital city of Apia. About 75% of the people live in villages on or near the coast.

The Samoan islands are part of a 1,200 km long volcanic chain produced by bending of the Pacific Plate and rupturing associated with subduction at the Tonga Volcanic Arc (Hawkins, 1976). Most of the volcanic rocks are composed of alkalic olivine basalts and their differentiates. Upolu evolved through a stage of shield building and caldera collapse followed by post-caldera eruptive phases, and then eruption of highly differentiated lavas (Hawkins and Natland, 1975). The petrology of Samoan rocks has been discussed by Kear and Wood (1959), Hedge et al. (1972), Hawkins and Natland (1975), and Ishii (1984).

The age of the volcanic rocks on Upolu ranges from late Pliocene(?) to Holocene (Kear and Wood, 1959). Table I lists the names of the volcanic formations in Upolu, their ages, and lithologies. The ages are poorly known and are based mainly on field relationships between units, the relative amount of erosion and weathering, and the degree of adjacent reef development.

Of particular importance to the present study is the relationship between the volcanic formations and the coastal features. For example, barrier reefs occur adjacent to the Mulifanua Volcanics, and cliffed coasts are typical of the Puapua and Fagaloa volcanics (Kear and Wood 1959). Well-developed fluvial systems are usually restricted to the Salani and Fagaloa volcanics. The distribution of the major volcanic formations are shown in Figure 2.

Climate

Western Samoa enjoys a mild tropical climate (Streten and Zillman, 1984). The mean annual temperature for Apia is 26.2°C with little variation from the mean (~23– 30°C). Average annual rainfall is about 2,900 mm/yr with October through March being the wet season. Bright sunshine averages 2,575 hours/yr and the most frequent wind direction is from the east (tradewind direction).

Table 1. Volcanic Formations of Upolu (after Kear and Wood, 1959).

Formation	Age	Lithology
Puapua Volcanics	Middle to late Holocene	Basaltic lavas (pahoehoe) and well-preserved cinder cones with scoria and scoriaceous basalt.
Lefaga Volcanics	Early Holocene	Feldspathic porphyritic basalts (picrite basalts and dolerites) and thick irregular scoria beds (a'a) with many bombs and lapilli.
Mulifanua Volcanics	Last Glacial (Late Pleistocene)	Vitreous, porphyritic, and non-porphyritic basalts (olivine basalts, dolerites, and analcite basalts) interbedded with a'a. The upper flow surfaces typically consists of pahoehoe.
Salani Volcanics	Last Glacial and older (Late Pleistocene?)	Picrite basalts, and olivine cloerite and basalts that typically grade upward from porphyritic basalt through vesicular basalt to rubbly a'a.
Vini Tuff	Late Pleistocene(?)	Hard, thick-bedded, calcareous tuff (vitric).
Fagaloa Volcanics	Late Pliocene or Early Pleistocene	Steep weathered slopes of intercalated a'a and pahoehoe flow rocks, rubbly scoria, ash beds, vitric tuffs, and contemporaneous basaltic dykes.

The frequency and magnitude of major storms is of importance to the development of the coastal zone. Although Western Samoa lies to the north and east of the areas that experience regular tropical cyclones, major storms have been reported (Kerr, 1976). The most recent severe cyclone occurred in January 1986. Maximum sustained winds of 60 kts were reported, with gusts to 82 kts at Mulinu'u Peninsula (Carter, 1987). A storm of this magnitude has a recurrence interval of about 40 years according to Carter (1987). He also calculated that a significant storm can be expected to strike Samoa about once every seven years*. The storms typically travel in a southward direction and can cause damage along north-facing and south-facing coasts.

Methods Used In The Survey

This survey was conducted by examining published maps and aerial photographs, utilizing student maps prepared during the 1985 CCOP/SOPAC-USGS Coastal Mapping Workshop held on Upolu, reviewing videotape images of the shoreline, collecting samples, and making field observations.

Vertical aerial photographs taken in 1981 (including the negatives) of Western Samoa are held at the Lands and Survey Office, Apia. They were used in stereo pairs to identify morphological and surficial features of the coast and to provide details not available elsewhere. Earlier photographic surveys completed in 1954 and 1970 were used in the production of the published topographic and orthophoto maps.

As part of the 1985 CCOP/SOPAC - USGS Coastal Mapping Workshop held in Apia, Upolu, an aerial videotape image of the Upolu coastline was produced. The airplane altitude was approximately 500 ft and the hour-long tape focusses primarily on the shoreline. It was used to provide information on sediments, man-made features, coastal stability, and rock outcrops of the shoreline.

Field checking and sample collection were limited to three days during the initial visit (March, 1987) and five days in April, 1989. During the initial sampling survey, 37 surface samples were collected from beaches, rivers, and reef flats. Field checking consisted of visual observations at all sample locations and intervening coastal areas that are accessible from the main coastal road.

*Since the time of writing, a severe cyclone struck the northeast coast of Upolu, resulting in severe erosion and development of an extensive hurricane bank.

DEPOSITIONAL ENVIRONMENTS AND DISTRIBUTION OF COASTAL FEATURES

Coastal Classification

Depositional environments that formed during the Holocene – the period of time when sea level reached its approximate present position, about the last 4,000 to 6,000 years – are important geologic features to recognize. They are the sites of accumulation of construction materials and other mineral resources. Mapping of depositional environments can also provide much useful information regarding coastal processes in the area. A number of interrelated coastal depositional environments are present on the island of Upolu including: a variety of reef environments, beaches, stream mouths and deltas, swamps, and mangrove communities.

The coast of Upolu can be broadly subdivided into three basic types, based on reef characteristics, depositional features, and inshore geology. The distribution is shown below in Figure 2.

Wide fringing reef, transitional to a shallow barrier reef. Coastal deposits are generally poorly developed and consist of mixed carbonate/terrigenous sediments that form a narrow coastal fringe. The hinterland consists of the gently seaward sloping Mulifanua and Salani Volcanics.

Cliffed coast with little or no reef development and few depositional features. The cliffs are composed of either the Holocene Puapua Volcanics (lava flows) or the older (Pliocene and Pleistocene?) Fagaloa Volcanics.

Either: (a) **fringing reefs and narrow coastal strip** consisting mostly of storm-derived carbonate-sand beach ridge(s) with smaller amounts of terriginous material, or (b) **beaches, barrier spits, and coastal swamps associated with streams.** These coasts are fronted by a slightly wider fringing reef; prominent gaps occur where freshwater output is high. The inshore rocks consist predominantly of the Salani Volcanics but also may include the Lefaga and Fagaloa Formations

Nearly half of the coastlines are of this last category; cliffed coasts are the least common type.

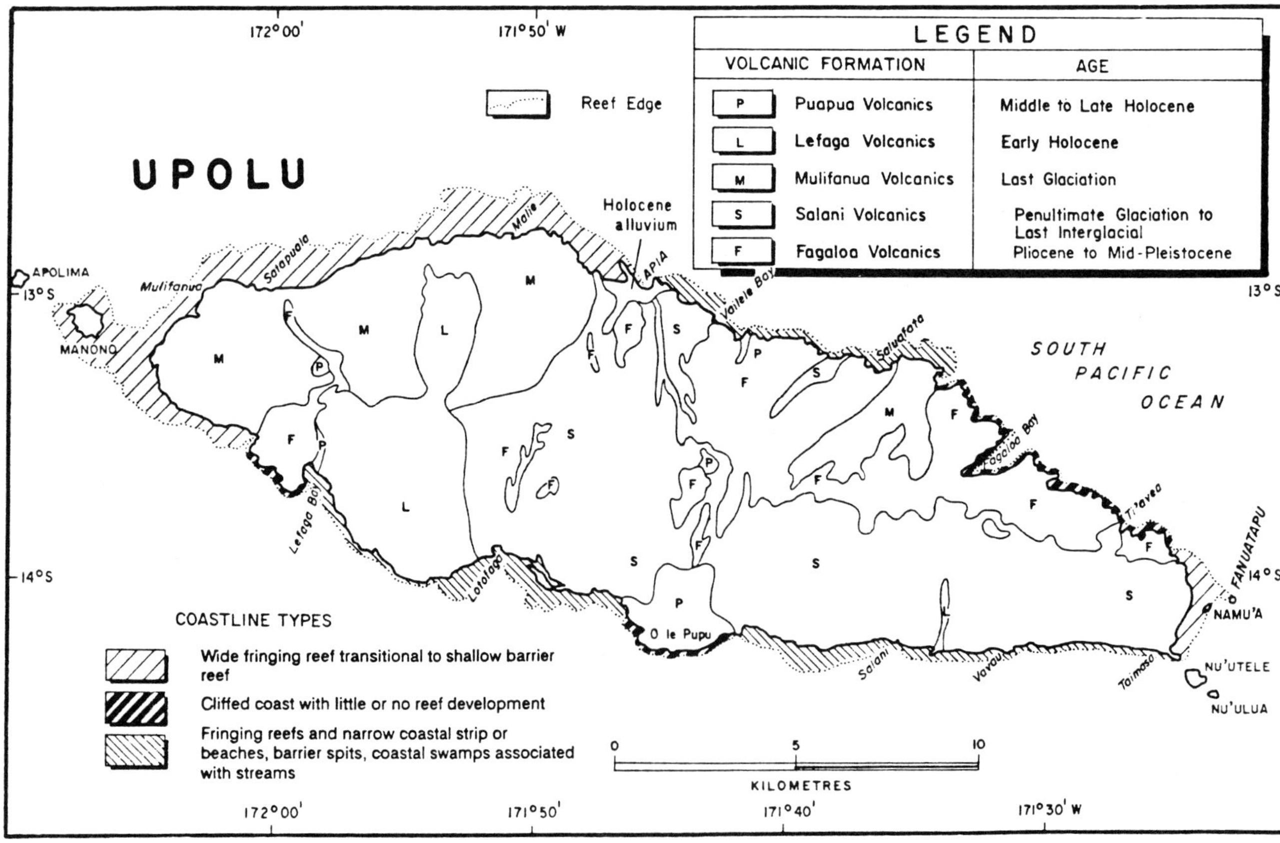

Figure 2. Volcanic formations and coastline types for Upolu. Geology adapted from Kear and Wood (1959).

Reefs

Reefs encircle nearly the entire coast of Upolu. They include fringing reefs—the most common type—barrier reefs, and patch reefs. The barrier reefs of Upolu are typically shallow, generally less than 5 m, except for deeper blue holes. These barrier reefs appear to be transitional between true fringing reefs and the barrier types—i.e., transitional between a deep fringing reef and a shallow barrier reef. Because they have already been described as barrier reefs (for example, Kear and Wood, 1959), that terminology will be followed here. Reef components that were mapped include: reef flats, reef crests, submarine terraces, blue holes, and shallow lagoons.

Reef Flat

Reef flat is a general term for near-planar reef surfaces; vertical relief is on the order of 1 m. Reef flats are commonly graded to a level near mean low spring-tide level. Subtle channels or moats may be present. On Upolu, reef flats are widespread and are composed of a variety of substrates such as coral, algae, carbonate and terriginous sand and gravel, cemented reef pavement, and a variety of marine plants. Typically there is a seaward to landward gradation of the following substrates (Fig. 3):

(a) A coral zone containing numerous compact coral communities. Interspersed with the coral may be thin patches of coarse sand and gravel. Encrusting coralline algae can cover extensive areas. Large storm-derived reef blocks can also be present. Drilling in this zone along the south coast near Mulivai (Matsushima and others, 1984) revealed small sections of in-situ coral and calcareous algae separated by larger zones of gravelly sand.

(b) A sediment-rich zone that grades landward from gravel/rubble to sand. These unconsolidated sediments can be more than 3 m thick and are predominantly derived from the adjacent reef front, reef crest, and reef flat calcareous communities. In more exposed locations, the upper surface is usually rippled.

(c) A marine-flora zone characterized by broad expanses of seagrass, and green and brown algae overlying a sandy substrate. The content of silt and organic matter in the sediment is higher than in bare sand areas because the plants retard water currents and stabilize bottom sediments. A prolific burrowing infauna is supported, and these areas are important habitats for commercially important juvenile faunas. As a general rule, the marine flora zone occupies inner reef flat areas and is slightly raised above adjacent sandy areas.

(d) Cemented reef material forms a widespread crust or pavement on the reef flats. It is common along the inner reef flat; but its presence farther offshore is not well documented. The overall areal distribution of the pavement is poorly known, in part because it is not readily identifiable on aerial photographs. The stratigraphy at the dredging site for the Mulifanua Ferry indicates that a pavement-like crust about 0.9 m thick has formed in less than 3,000 years (Jennings, 1974).

The foregoing surficial facies descriptions are greatly simplified—in practice, reef flat areas are very complex structures responding to a variety of physical, chemical, and biological controls. A detailed account is beyond the scope of the present study; however, two recent studies of Upolu reefs were made by Morton and others (1988), and Andrews and Holthus (1988; Aleipata District).

Reef Crest

The reef crest is the highest part of the reef and is composed of either compact coral colonies, encrusting coralline algae, or a rubble pavement. Where an algal rim is present, it is usually characterized by red corraline algae (*Porolithon*). In shallow pools, coral colonies such as *Acropora*, encrusting *Montipora*, and *Porites* can be present. Sediments are coarse and sparse, occupying the interstices between corals and algae. Wave energy is high, and the boundary with the reef flat is often marked by a debris slope of reef-derived material transported landward. The relationships between the reef crest and the adjacent reef flat and reef front/submarine terrace is shown in the echosounder profile of Figure 4.

Submarine Terrace

Wherever the fringing and barrier reefs are well developed, they are usually fronted by a submarine terrace. The terrace can be identified on aerial photographs, appearing as a spur-and-groove zone of coral growth often with a highly irregular seaward margin. Depth information is sparse but the terrace is most prominent at water depths between 5 and 25 m. The terrace is separated from the reef crest by a very steep reef front. Formation of the terrace may be related to former sea level positions; however without precise depth and morphology information, the uniformity of the terrace and the cause of its formation are uncertain.

Based on diver observations, most of the terrace is covered by wave-resistant compact corals and corraline algae. Sediment is mostly algal-encrusted coral gravel except in the sand-floored channels that traverse the terrace. The seaward edge drops abruptly, and additional terraces may be present downslope. For example, along the south coast, in the vicinity of the Salani River, Carter (1988) measured a terrace between 30 and 60 m depth to a distance of nearly 1,000 m from the reef. This deeper terrace is faintly visible on aerial photographs of the area. The approximate submarine terrace distribution for northwest Upolu is shown in Figure 5.

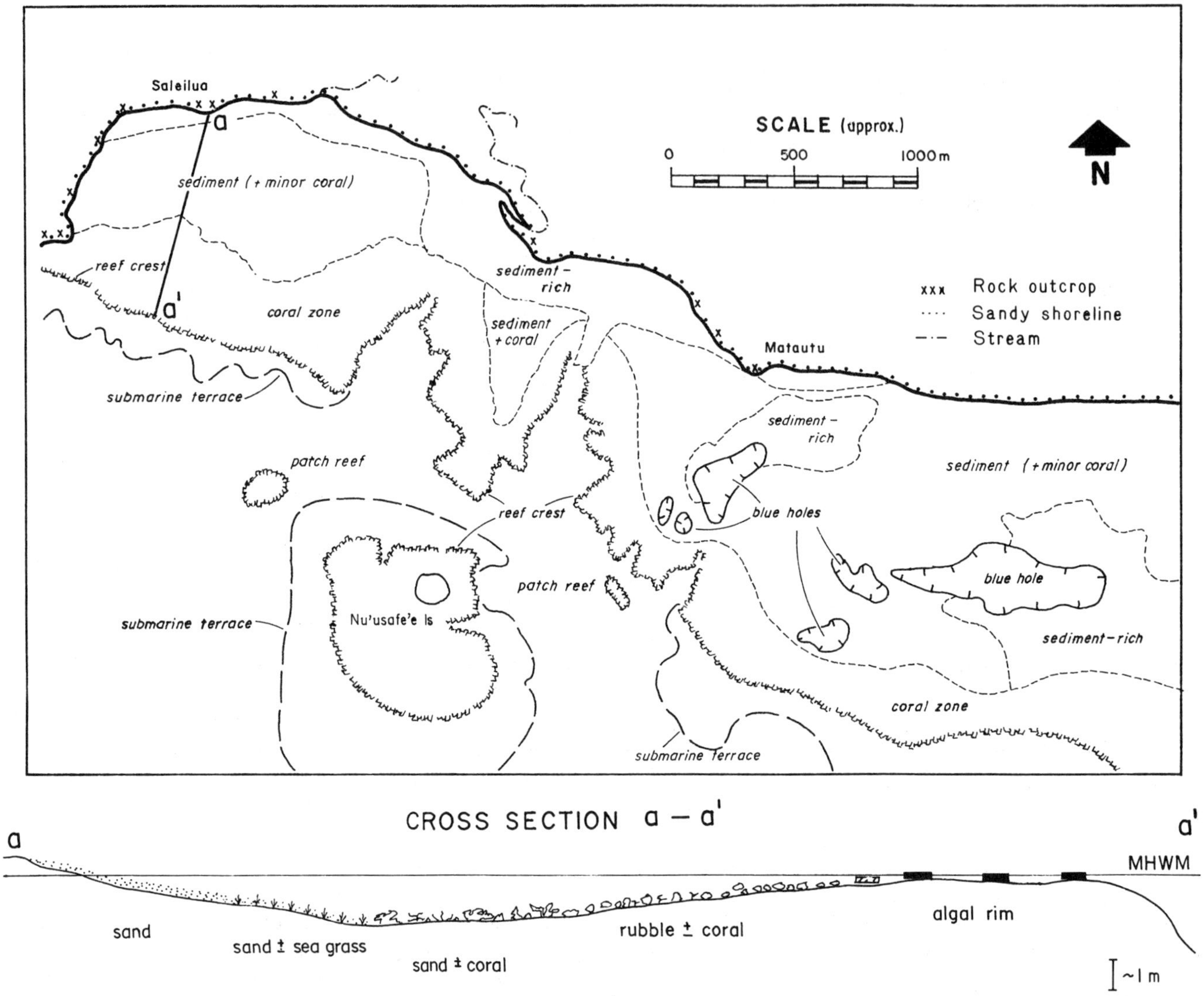

Figure 3. Sketch and schematic cross section (vertical relief about 2 m) showing the fringing reef and associated features near Saleilua-Matauta, Falealili, south Upolu. The deep reef reentrants are related to onland drainage systems.

Andrews and Holthus (1988) used diver observations and aerial photographs to subdivide the Aleipata reef front and submarine terrace into five distinct morphologies: (1) Spur-and-groove with drop off. A dissected reef terrace supporting well-developed coral growth with a steep drop off which grades into a sand/rubble debris slope at a depth of about 25 m. (2) Spur-and-groove with a broad dissected lower slope. Similar to (1) but with a wider spur-and-groove zone and a gently sloping terrace to about 25 m depth. (3) Scoured reef platform. A wide submerged platform extending from depths of 10 to as much as 30 m and consisting of a scoured surface indicating strong current activity. (4) Spur-and-groove with short slope. A narrow shallow terrace fronted by a steep slope. Reef-rubble flats occur at depths of about 20 to 25 m. (5) Reef channels (passages). The nearshore ends are characterized by a debris slope of reef talus that extends offshore to a sand floor at a depth of about 20 m.

Blue Holes

Blue holes are sinkholes created in limestone reefs by freshwater during lower sea level stands (Backshall and others, 1979). Blue holes are fairly common in the western half of Upolu. They are best developed within the barrier reef between Upolu and Manono but they also occur along the north and south coasts. They vary from a few tens of

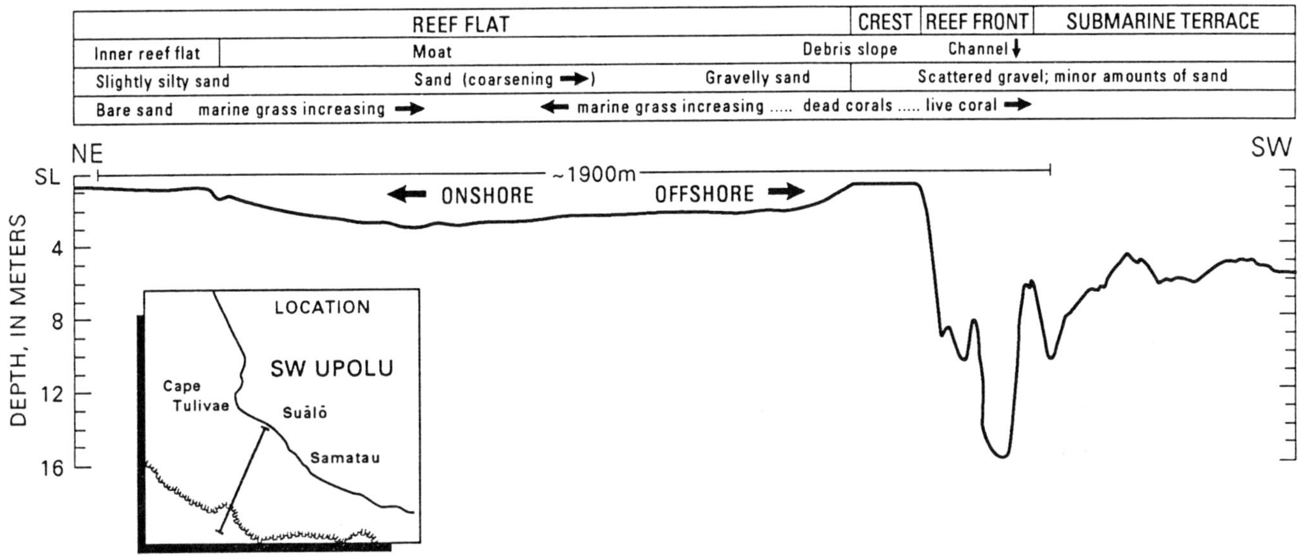

Figure 4. Precision echosounder profile showing sedimentary environments from shore to submarine terrace at Sualo, southwest Upolu. Profile taken at high tide; same as line 6 in Richmond and Roy (1989).

meters to hundreds of meters in width. Depths as much as 14 m have been measured (Fig. 6a).

Blue hole sedimentation patterns follow a typical sequence (Fig. 6b): (a) A rim of coarse coral debris, in-situ dead and living coral colonies, and sand. (b) A steep (~33°) debris slope of sand and gravel; sediment fines downslope. (c) A silty-mud-filled basin. Variations in this sequence include finer rim sediments on the landward side, or the debris slope can be replaced by near-vertical limestone walls covered with corals.

Shallow Lagoons

Lagoons, situated behind the barrier reef, are as much as several kilometers or more in width. Although depth information is limited the lagoons probably average between 2 and 5 m in depth. Bottom type varies from smooth-floored sediment or marine-floral blanketed substrates to very irregular areas composed of coral pinnacles interspersed with sediments. Morton and others (1988) described a general sequence of coral variation from higher to lower energy conditions: *Pocillopora* (+ *Millepora*)... *Acropora*...*Porites* + *Pavona*. Andrews and Holthus (1988) subdivided the shallow lagoon at Aleipata into eight different habitats based on substrate type, dominant seagrass, dominant coral, and geomorphology. In general, the lagoons are characterized by the deposition of large volumes of reef-derived sediment (for example at Aleipata and opposite Faleolo International Airport) which makes them potential construction material repositories.

Beaches

A beach is an unconsolidated inter- to supratidal deposit of sediment formed at the shoreline: typically it consists of sand or, more rarely, gravel. A beach acts as a buffer between the sea and the land, and constantly changes in profile and alignment as wave conditions vary. During storms, beaches are prone to erosion and during fair weather they tend to build outward. The immediate source of the beach sediment is the adjacent reef flat, which in turn derives its sediment from the reef and nearby streams.

Natural beaches are common on Upolu and constitute a large portion of the south, east, and west coasts. Appendix I lists the beach samples collected during the present survey; the sample locations are shown on Figure 7. Sand is the dominant textural component of Upolu beaches. Composition varies from those composed of nearly pure carbonate sand to sand composed primarily of volcanic material and mixtures of both. Volcanic-rich beach deposits are generally restricted to areas in close proximity to river or stream mouths. Carbonate sediments are produced in reef environments around the entire perimeter of the island; however, the finest examples of carbonate sand beaches occur along the high-energy south coast.

Beach height above mean sea level is related to the incident wave energy and the width of the adjacent reef flat. The highest beaches occur along the southeast coast in the Lepa district. Here, the beach ridge crest is several meters above mean sea level—a condition which prompted Kear

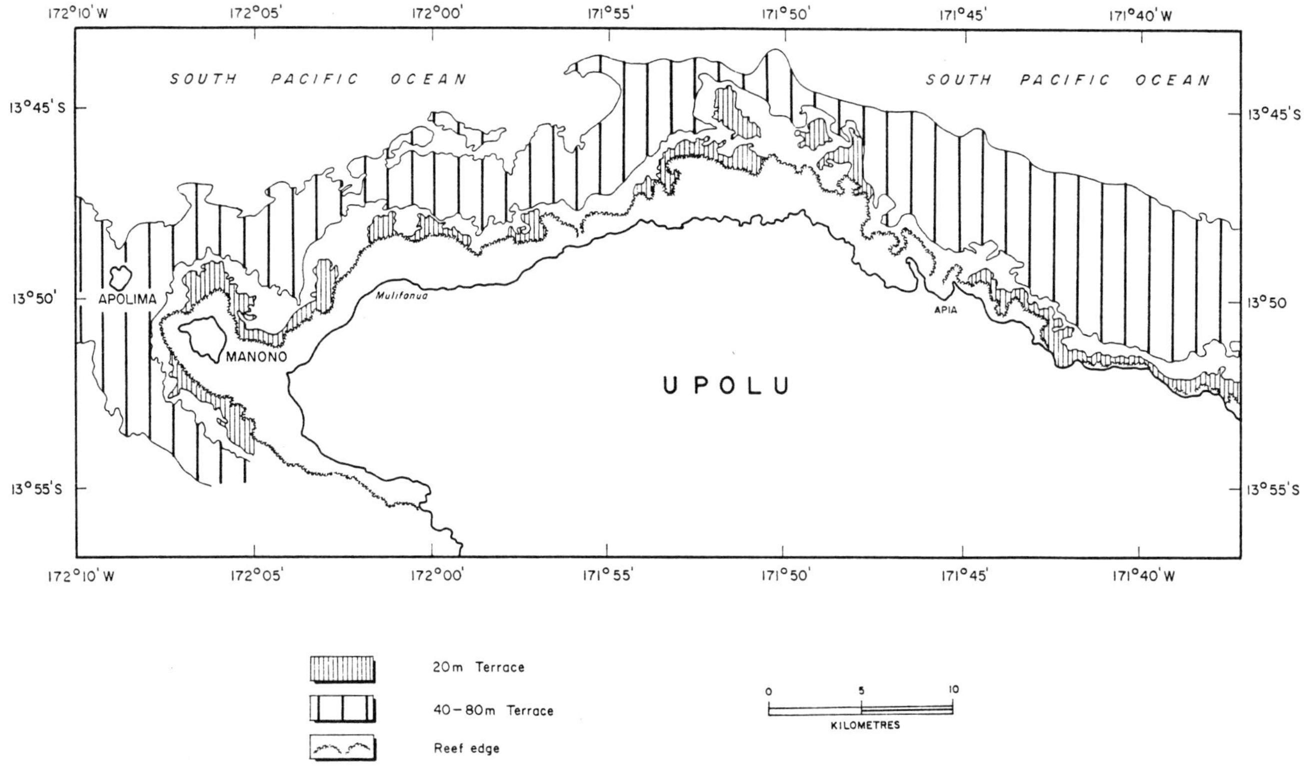

Figure 5. Distribution of submarine terraces for northwest Upolu, based on the map "Bathymetry of Apolima Straight and Apia Coast" by Kear and Wood (1959).

and Wood (1959) to hypothesize a higher sea level stand. However, the ridges are breached occasionally during storms, indicating that they could have formed under present sea level conditions (see section on sea level).

The beaches usually occur either as (semi-)continuous stretches of coast or as discrete pocket beaches bounded by rock headlands (Fig. 8a). Pocket beaches are the most common type on Upolu. The longest continuous beach on Upolu is the ~4 km beach fronting the Vaie'e Peninsula barrier spit in the Safata District (Fig. 7).

If coastal progradation proceeds regularly, beach ridges are typically formed. However, extensive beach ridge plains are not common features of Upolu. This could be due in part to human activities (such as house building) disturbing the ground surface and making beach ridge recognition difficult. Another possibility is that the narrow and generally steep coastal plain is not conducive to the development of beach ridges . One of the wider coastal plains is situated along the east coast in the Aleipata district. Although no prominent beach ridges are visible, the ground surface in undisturbed areas is hummocky, and carbonate sand underlying 0.8 m of alluvium was found nearly 300 m inland (Fig. 9). The absence of beach ridges at this site may be due to human activity, aeolian modification, or erosion of the summits and infilling of the swales. Also it is possible that ridges never formed.

A common feature of tropical coasts composed of carbonate sediment is beachrock—calcium carbonate cemented beach sediment. Although its formation is not entirely understood, beachrock cementation proceeds both intertidally and at depth within the sediment column. Therefore, surficial beachrock exposures indicate erosion has taken place. Beachrock exposures on Upolu are most prominent along the south coast (Fig. 8b) but they also occur elsewhere.

Stream Mouths

Streams have two pronounced effects on the coastline. First, they provide a mechanism for delivering large amounts of sediment to the coast, and second, the freshwater flow and increased turbidity inhibits coral growth. Where streams enter the sea current flow usually decelerates resulting in deposition of sediment. On Upolu, stream-mouth spits are the most common fluvial depositional feature; more rarely small deltas are formed. The deltas, where present, are usually small, wave-dominated, and typically submerged. Occasionally flood-deltas are developed on the reef flats.

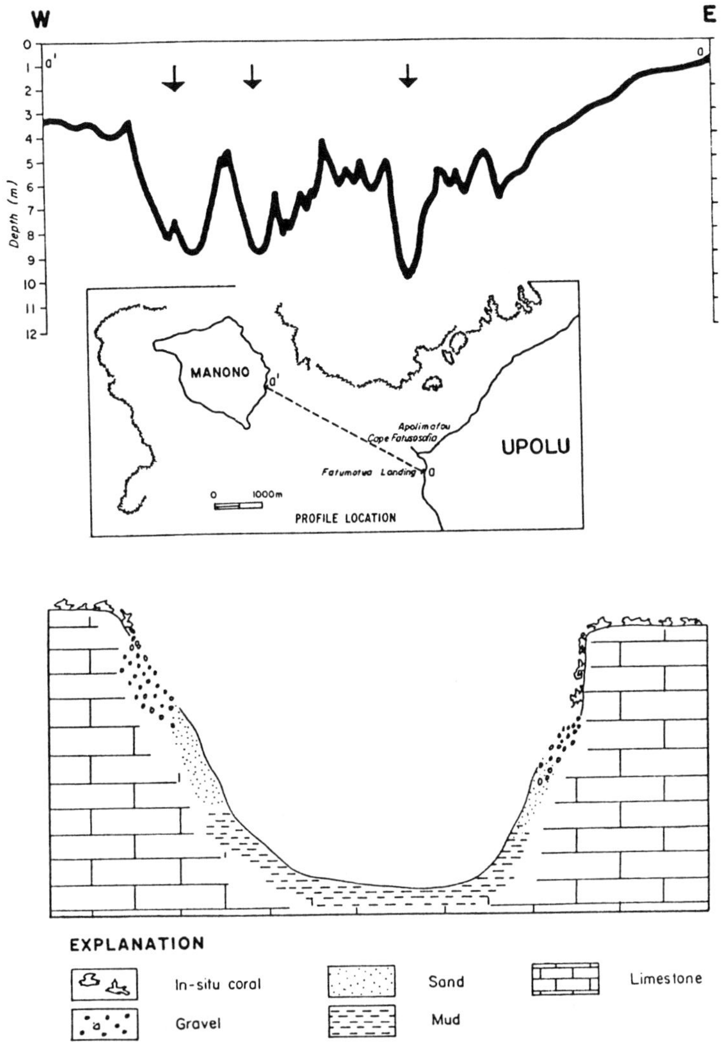

Figure 6. (a) Echosounder profile beween Upolu and Manono islands showing several blue holes (arrows) developed within the barrier reef. Average lagoon depth is about 4 or 5 m. Profile a–a′ from Richmond and Roy (1989). (b) Schematic sketch of a blue hole illustrating facies variations.

Gaps in the reef may be present opposite the larger rivers. Large deltas are not formed, primarily because wave energy is high along the coast, and the shelf is narrow .

We can divide the stream mouths on Upolu into three different types:

(1) The stream discharges directly to the open sea through a prominent gap in the reef. If a shallow offshore terrace or platform exists, a submerged delta may form. Figure 10, showing the mouth of the Vaisigano River, is an example of this type of delta. A 500-m-wide break in the reef was presumably caused by freshwater outflow. The overall surficial sediment pattern fines seaward out to mid-harbour. Here an influx of reef-derived material produces a coarsening trend. A "soupy" mud covers much of the bottom over the delta front (Gauss, 1981). Where the river first enters the sea, an extensive gravel body, as much as 2 m or more in thickness, has developed. The gravel body is progressively covered by finer material but is still detectable at depth in the sediment column more than 125 m from the mouth. Terrigenous sands and gravels at the mouth are periodically excavated for construction purposes; however, they are replaced during floods.

This reef-gap type of stream mouth is prevalent along the northeast coast. Examples of this type of stream mouth include: Vaivase Stream (Fagali'i, Apia); Letogo Stream, Lauli'i Stream, Maepu Stream, Namo River, Falefa River (Luatuanu'u); and several small streams in the Lepa and Fagaloa Districts (Fig. 7).

(2) Barrier-impounded stream mouths are characterized by barrier spits, meandering channels near the mouth, and well-developed swamps or mangrove communities. There may be an adjacent gap in the reef, and submerged deltas may be present. This type of stream mouth differs from the reef-gap type in that it has more extensive barrier/swamp deposits and a less prominent break in the reef, although transitional types exist.

The best examples, the Mulinu'u and Vaie'e barrier spits, do not presently front large rivers. They are composed mostly of terrigenous sand and may represent older Holocene deposits; therefore they may be partially relict features. The generalized sediment distribution in the vicinity of the Mulinu'u barrier spit is shown in Figure 11. Other examples of barrier-impounded river/stream mouths include (Fig. 7): Mataloa Stream, Vaifaliuga Stream, Vailoa Stream, and Mulifagatoloa River (Falealili); Vaiania Stream (Safata); and Vaisala Stream (Lepa).

3) Ephemeral streams deposit directly onto the reef flat, and during times of high discharge, a flood-delta may form rapidly and deposit large amounts of mixed sediment (Richmond, 1988). Reworking of the flood deposit by waves and currents often leaves behind a lag of coarse debris (Fig. 12). Reef re-entrants commonly occur nearby, but usually a reef flat fronts the mouth. This type of stream mouth is typical of the ephemeral streams along the south coast.

Sediment discharge

The amount of sediment delivered to the coast by streams is unknown. Unpublished streamflow records from the Hydrology Section of the Apia Observatory were ex-

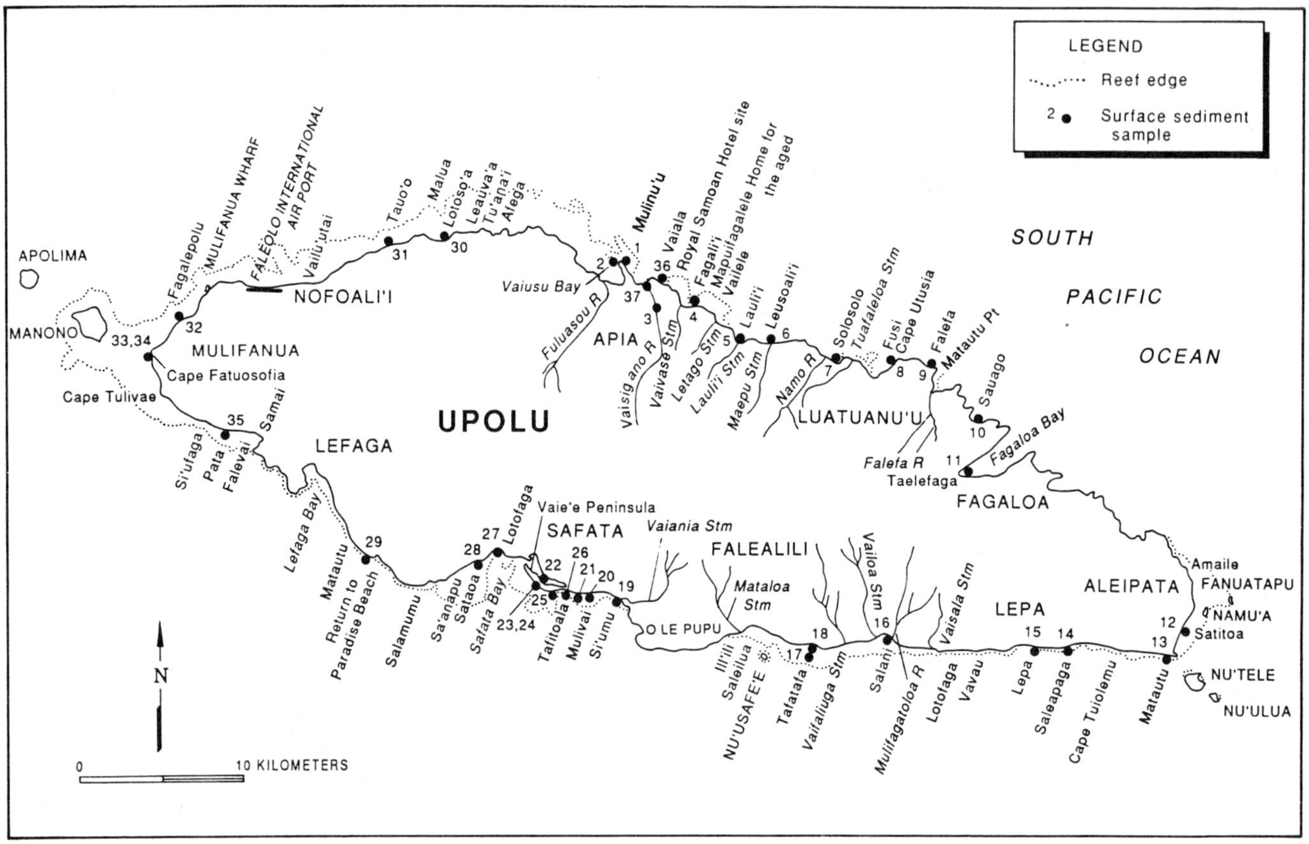

Figure 7. Place names and location of Upolu surface sediment samples collected during this study.

amined to obtain peak discharges for a number of streams on Upolu. Although incomplete, the records provided an indication of the potential discharge during floods. Using relationships between discharge and sediment load established in the United States (Nordin and Beverage, 1965) approximate values of bed-load transport were determined (Table 2). These values give an order-of-magnitude estimate of the amount of material transported to the coast during

Figure 8. (left) Pocket beach of carbonate sand bounded by outcrops of Lefaga basalts, southeast of Matautu, Lefaga (Return to Paradise Beach). View to northwest. (right) Carbonate sand beach with prominent beach rock expsoures (center and center right). The beach rock marks a former shoreline position that is slightly oblique to the present shoreline indicating that some realignment has taken place. Photograph taken near Salepaga, Lepa. View to east.

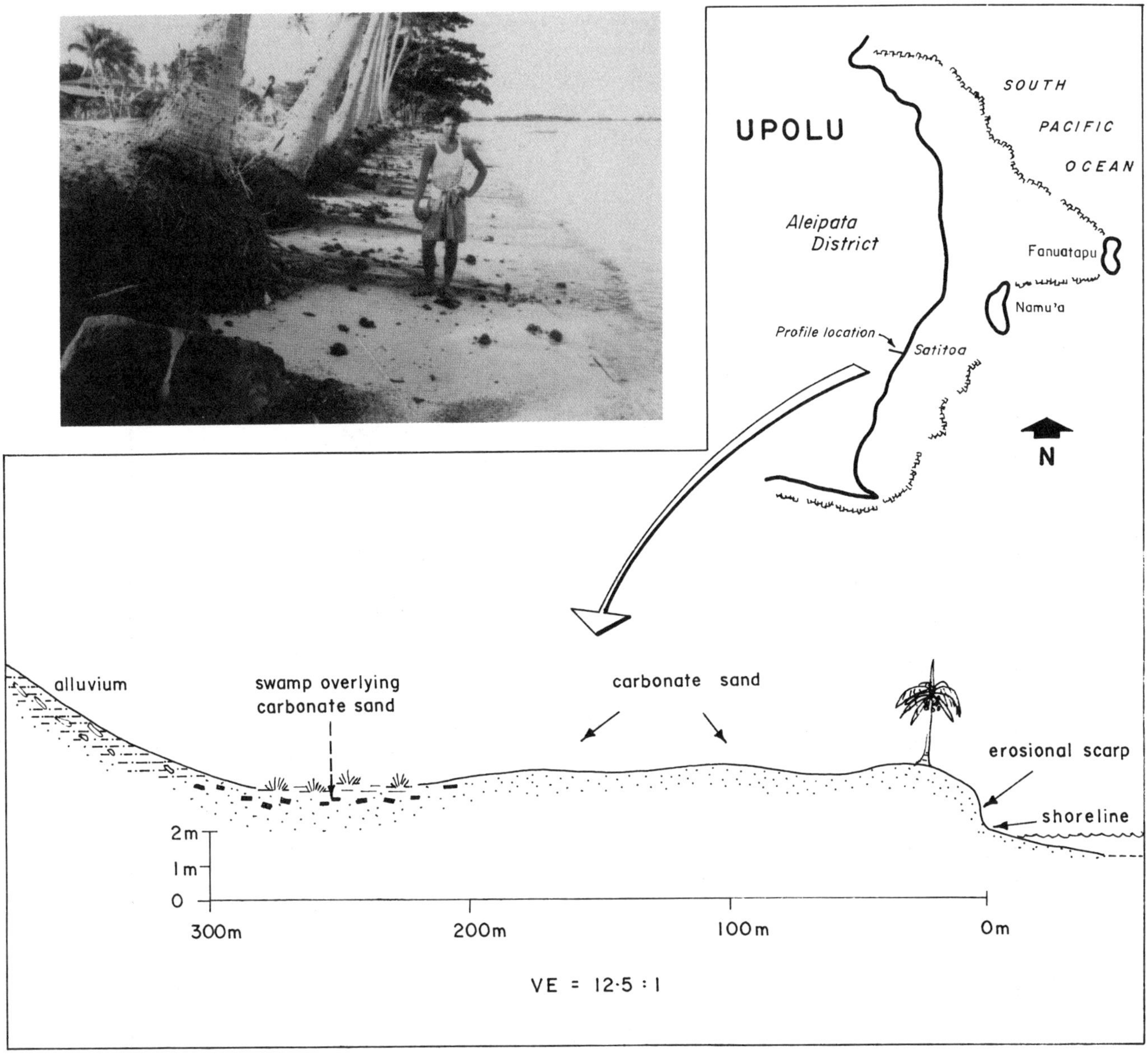

Figure 9. Beach and coastal plain near Satitota, Aleipata . (a) Carbonate sand beach is backed by an erosional scarp along the base of the coconut palms. View to north. (b) Sketch map of traverse across the coastal plain. The surface topography is hummocky; individual crests could only be traced for short distances.

floods, which are a major contributor of sediment to the coastal zone. Some of this material forms depositional coastal features, but most of it is probably transported offshore into deeper water.

Rubin (1984) estimated the rate of harbour infilling at Apia (~ 1,500 m^3/yr), and the rate the Vaisigano River could supply sediment during a flood (~10,000 m^3/day). These calculations show that the Vaisigano could easily supply enough material in one day of flooding to account for a yearly rate of harbour infilling. The excess material is presumably transported offshore.

Swamps and Mangrove Communities

Low-lying coastal swamps and mangrove communities are common features of the drowned valleys and barrier-impounded stream mouths of Upolu. The largest mangrove swamps are situated behind the large barrier spits at Mulinu'u and Vaie'e. Swamps also are present in many low-lying areas and stream mouths backing beach ridges of Tafagamanu Sand. They form important habitats for many commercial fish and other species of marine fauna (Bell, 1985). Historically, mangrove swamps were more extensive,

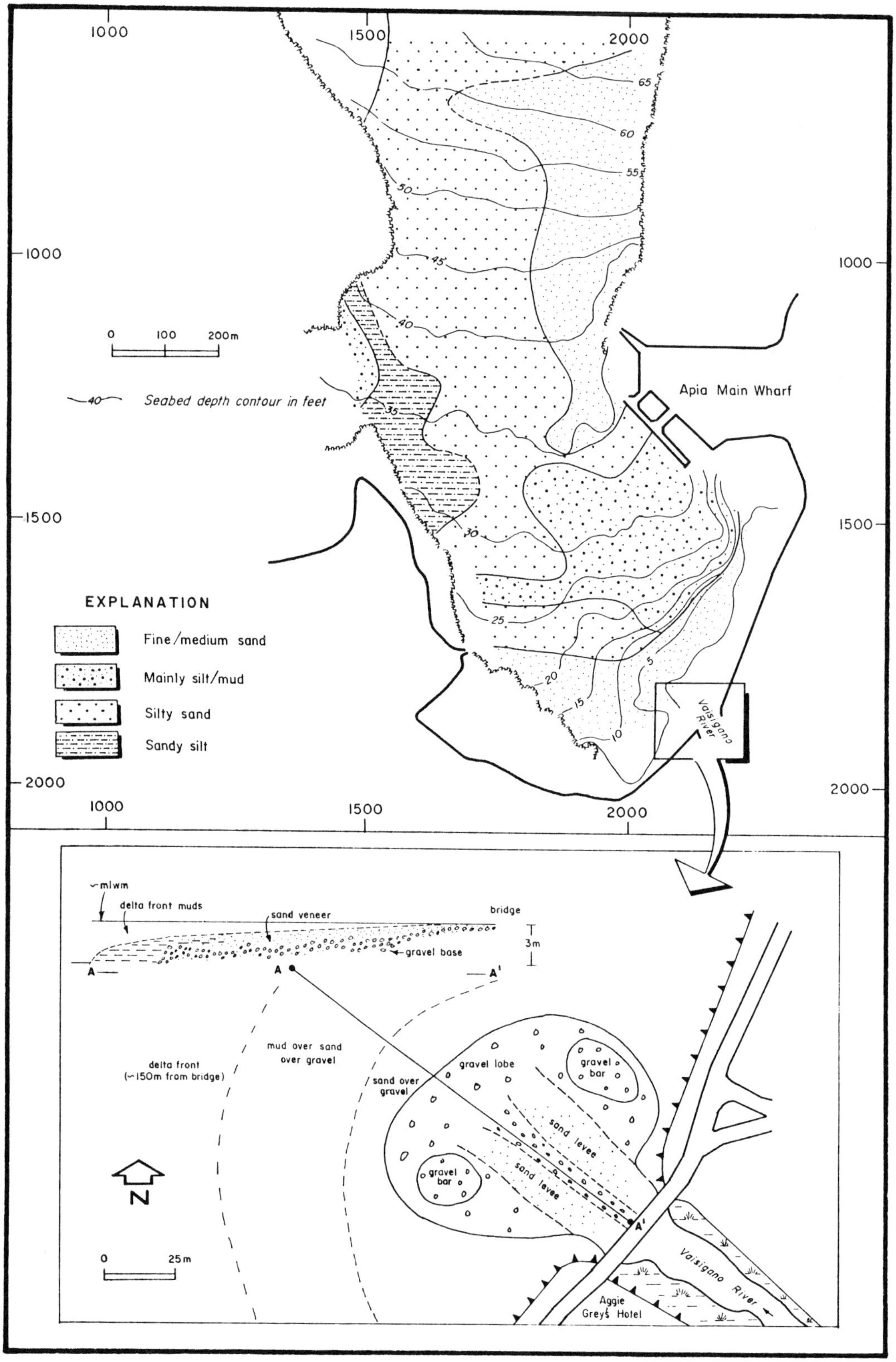

Figure 10. Delta development at the mouth of the Vaisigano River, Apia. (a) Offshore bathymetry and surface sediment distribution (adapted from Gauss, 1981). The harbor was dredged before the survey was made; therefore, the contours may not be representative of a natural situation. (b) Sketch map and cross section of the intertidal to shallow subtidal delta at the river mouth (based on data collected in September 1988, Harper and Richmond, in preparation).

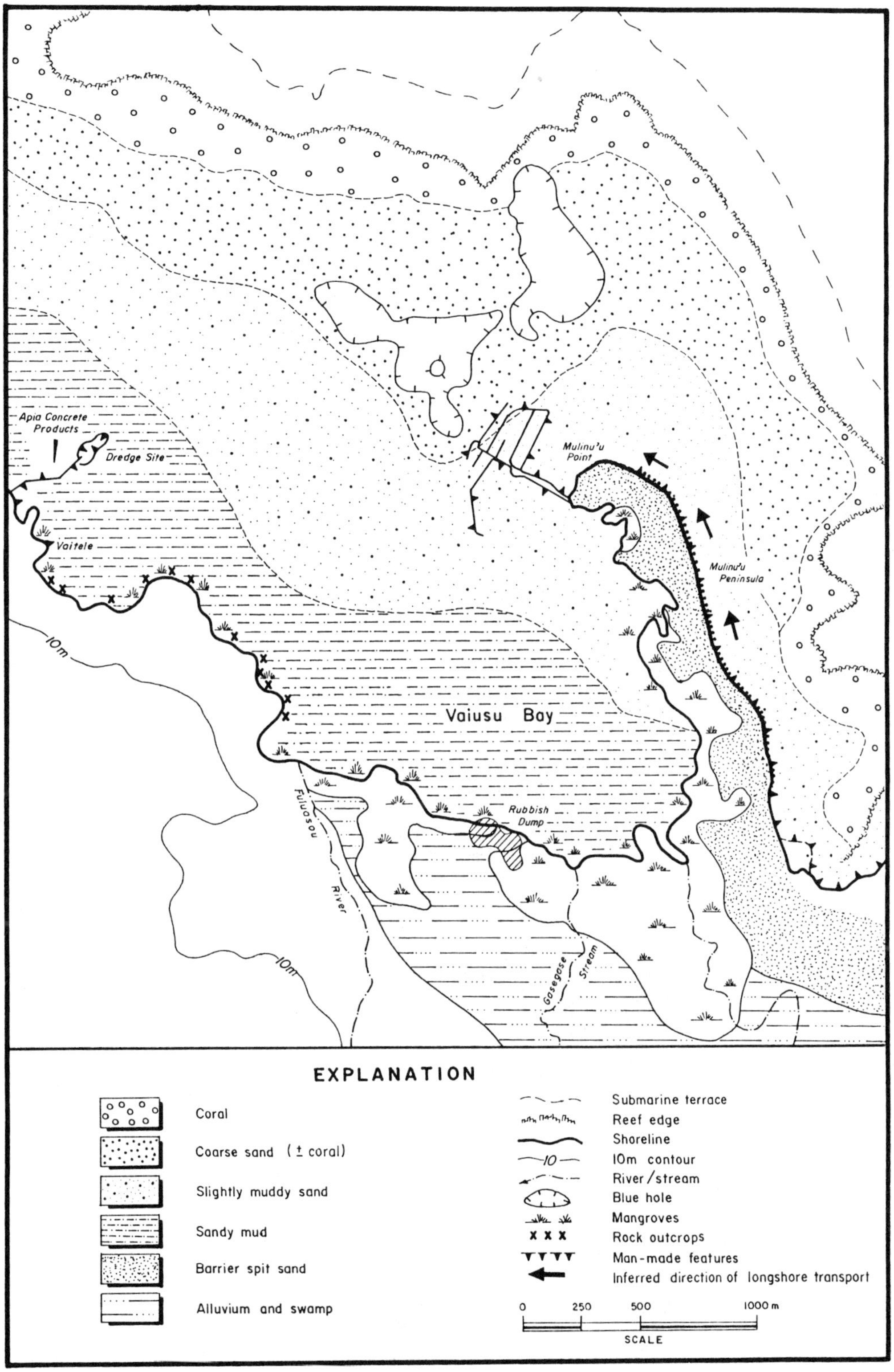

Figure 11. Sediment facies of barrier-impounded river/stream mouths at Vaiusu Bay, Apia. The Fuluasou River and Gasegase Stream mouths are protected from the open sea by the Mulinu'u barrier spit. The spit is composed of terrigenous and carbonate sands. Extensive mangrove communities are developed within the bay. Sediments coarsen and increase in carbonate content seaward. The blue holes are probably related to paleo-drainage systems. The material from the dredge site is a mixture of terrigenous (~30-50%) and reef sand and gravel and a small amount of fine materials.

Figure 12. (top) Flood delta composed of terrigenous sediment (volcanics) at the mouth of an unnamed stream, Vava'u, Lepa. Presumably the initial flood delta was much larger and the overlying sands have been winnowed away, leaving behind a gravel base, view to west. (bottom) Volcanic gravel flood delta base overlain by carbonate sands. Mouth of Sinoi Stream, Lepa, view to south.

often fronting protected sand beaches, but were removed for firewood – a practice that may have contributed to coastal erosion (Wright, 1963). The presence of mangroves in front of sand beaches suggests a recent change in coastal hydrodynamics. Mangroves are usually associated with lower energy environments and beaches with higher energy environments. Mangrove swamps are also common targets for land reclamation activities.

Wright (1963) recognized several types of coastal swamp communities: (1) Tidal forests that are dominated by mangroves (*Rhizopora mangle*, and *Bruguira conjugata*) and occur in sheltered bays, estuaries, and lagoons where fine sediment is accumulating. (2) Lowland rush and reed swamps that typically form a fibrous peat deposit. Ferns may grow in the drier areas. The ground water varies from brackish to almost fresh. Several large swamps occur to the south of Mulifanua. (3) Swamp forests in low-lying coastal areas where Pandanus are usually conspicuous.

Figure 13 illustrates the idealized vertical distribution of facies within a mangrove swamp at Lefaga Bay on the south coast. It is based on three core sites within the swamp and is modified from Sugimura and others (1988). A vertical succession from barrier spit (medium sand with abundant shell and coral fragments) to estuary (silty-sand, silt, and shell fragments) to swamp (silty peat with some shells, coral, and lapilli) indicates subsidence and seaward progradation of the facies. This sequence is probably typical of other swamp deposits on Upolu.

COASTAL PROCESSES AND GEOHAZARDS

Sediment Transport

Because of a lack of data for the direction, strength, and variability of nearshore currents around Upolu, a discussion of coastal processes must be based on transport patterns inferred from the morphology (Fig. 14) and a knowledge of the general climatic circulation for Samoa. Overall, the nearshore waves and currents are driven by the predominant easterly winds. Consequently, a simple model of longshore transport infers east-to-west movement of nearshore sediment. The orientation of some of the major depositional features, such as spits, confirms this direction of transport. The relative contributions to shoreline deposition by longshore versus cross-shore (on/offshore) transport are unknown although both processes undoubtedly contribute. Elongated barrier spits are typical longshore-generated features, whereas pocket beaches probably indicate a significant amount of cross-shore sediment movement. Figure 15 summarizes the inferred directions of longshore transport for Upolu.

Factors which complicate a simple model of east-to-west circulation and transport include (1) local bathymetry and exposure; (2) passage of storm fronts; and (3) extended shifts in weather patterns, such as periods of strong westerly winds associated with ENSO events (El Niño–Southern Oscillation).

Bathymetry and Exposure

Incident waves, and the currents they generate, are one of the primary driving forces for sediment transport in the nearshore. Wave approach at an angle to the coast imparts a longshore component of transport. Wave refraction around reef reentrants and passages can cause local variations in the angle of wave approach. Reef passages are

TABLE 2. Peak discharge and predicted sediment loads for some Upolu rivers and streams.

Location	Year	Peak discharge (m^3/s) *	Predicted sediment load (m^3/day) +
Salani R., Sopo'aga	1978	164	34,300
Salani R., Vaipu	1974	47	4,600
Vaisigano R., Alaoa W.	1974	90	13,100
Vaisigano R., Alaoa E.	1976	186	45,700
Vaisigano R., Tiapapata	1974	55	6,900
Mulivai S., Ti'avea	1978	41	4,000
Falefa R., Sauniatu	1976	144	25,700
Falese'ela S., Falese'ela	1975	15	1,500

*Peak discharge values from unpublished records of the Hydrologic Section, Apia Observatory. The values are determined from several different measuring techniques and do not represent a comprehensive analysis of all available records. It is not known at what frequency these discharges are likely to occur.

+Predicted sediment load by comparing peak discarge values with published sediment bed-load data for the Rio Grande River, New Mexico (Nordin and Beverage, 1965). Assumptions used include a particle density of 2.65 g/cm^3 and a porosity of 40%. Not corrected for temperature differences. The variation between erosion potential between Western Samoa and the temperate climate site are unknown; however, data from Fiji suggests the lowest erosivity rates there equal some of the highest rates in the United States (Nelson, 1987).

generally situated in broad lows; water passing over adjacent reef crests is driven by gravity toward these lows, which then serve as conduits for seaward return flows during most of the tidal cycle (Fig. 16a). Consequently, predominant current directions on the western side of a passage may be toward the east.

Shielding of the coastline by prominent headlands may allow only waves/currents from certain directions to be effective (Fig. 16b). For example, the Holocene basalt flows of 'O le Pupu' on the south coast have created an impediment to westward transport—spit orientations near Si'umu suggest that an easterly drift predominates. Farther to the west, the principal transport direction returns to the west.

Storms

Storms, and in particular tropical cyclones, can cause significant changes in nearshore circulation patterns in very short periods of time. They can be very erosive, but alternatively they are perhaps the main mechanism for the shoreward transport of reef and fore-reef material. Major storm deposits (hurricane banks) occur on the reef flat near Si'ufaga, Mulifanua. Other minor deposits are scattered on reefs around the island. Storms can have their greatest impact by (a) increasing sediment input to the coastal zone through flooding and shoreward transport of reef material; (b) creating strong wave/current activity in normally protected areas; and (c) eroding the shoreline and flushing sediment from reef passages.

ENSO (El Niño - Southern Oscillation) Events

ENSO events typically create strong westerly conditions for as much as several months during which time important changes in nearshore circulation can take place. Areas protected from easterly storms but exposed to the west may undergo significant changes. For example, the west coast near Mulifanua would become the windward side of the island and areas that are normally sheltered can be exposed to higher incoming wave energy.

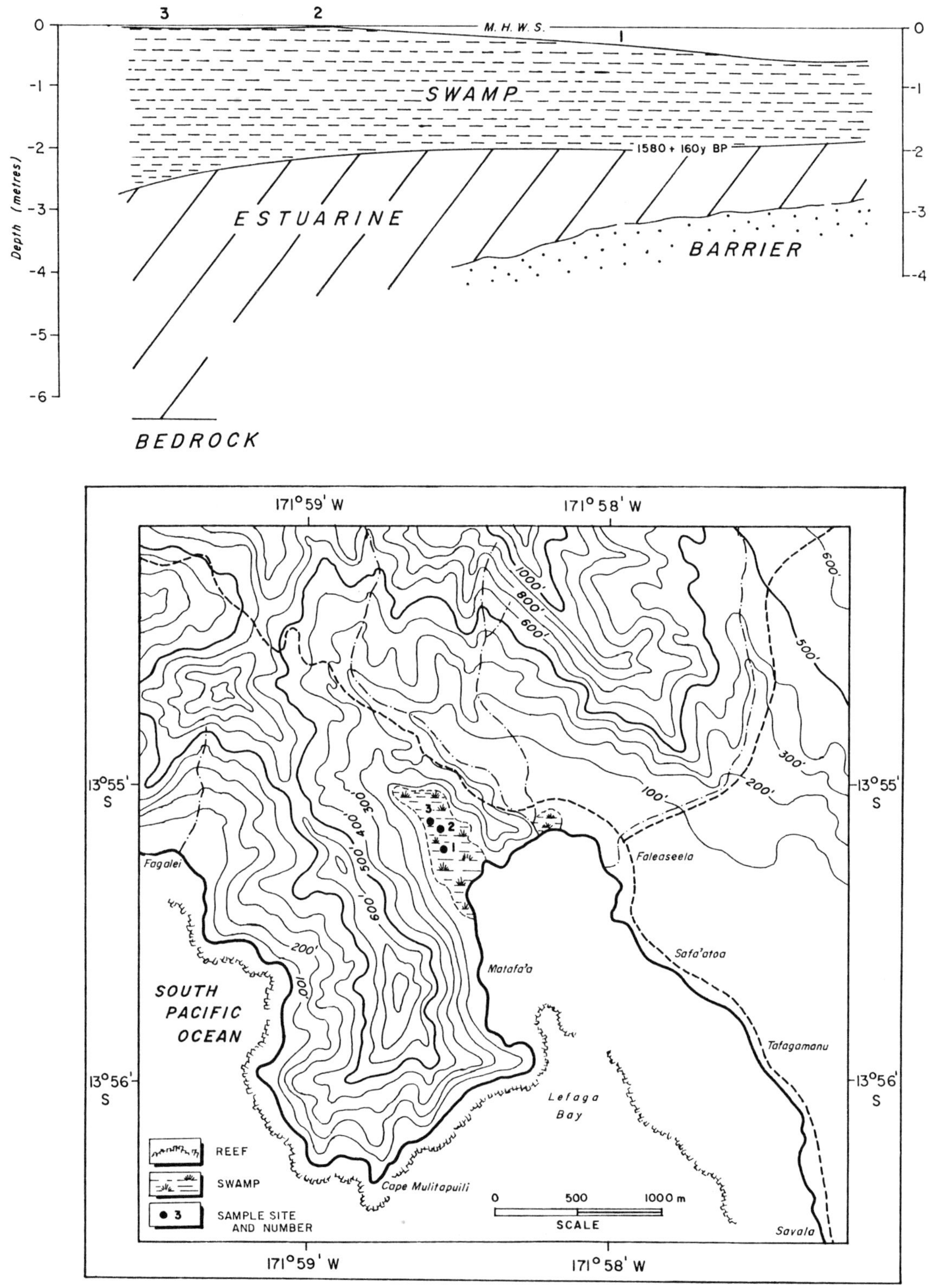

Figure 13. (top) Simplified cross section through the mangrove swamp at the head of Lefaga Bay. The swamp deposits consist of a black-brown silty peat with transported shells, coral, and lapilli. The underlying estuarine unit is a blue-brown to grey silty sand with mollusc and coral fragments. The boundary between the swamp and estuarine units was dated by radiocarbon analysis at 1.58 ka. The basal sand cored at site 1 is probably a buried barrier spit. Modified from Sugimura et al. (1988). (bottom) Location of swamp and core sites; numbers refer to core locations.

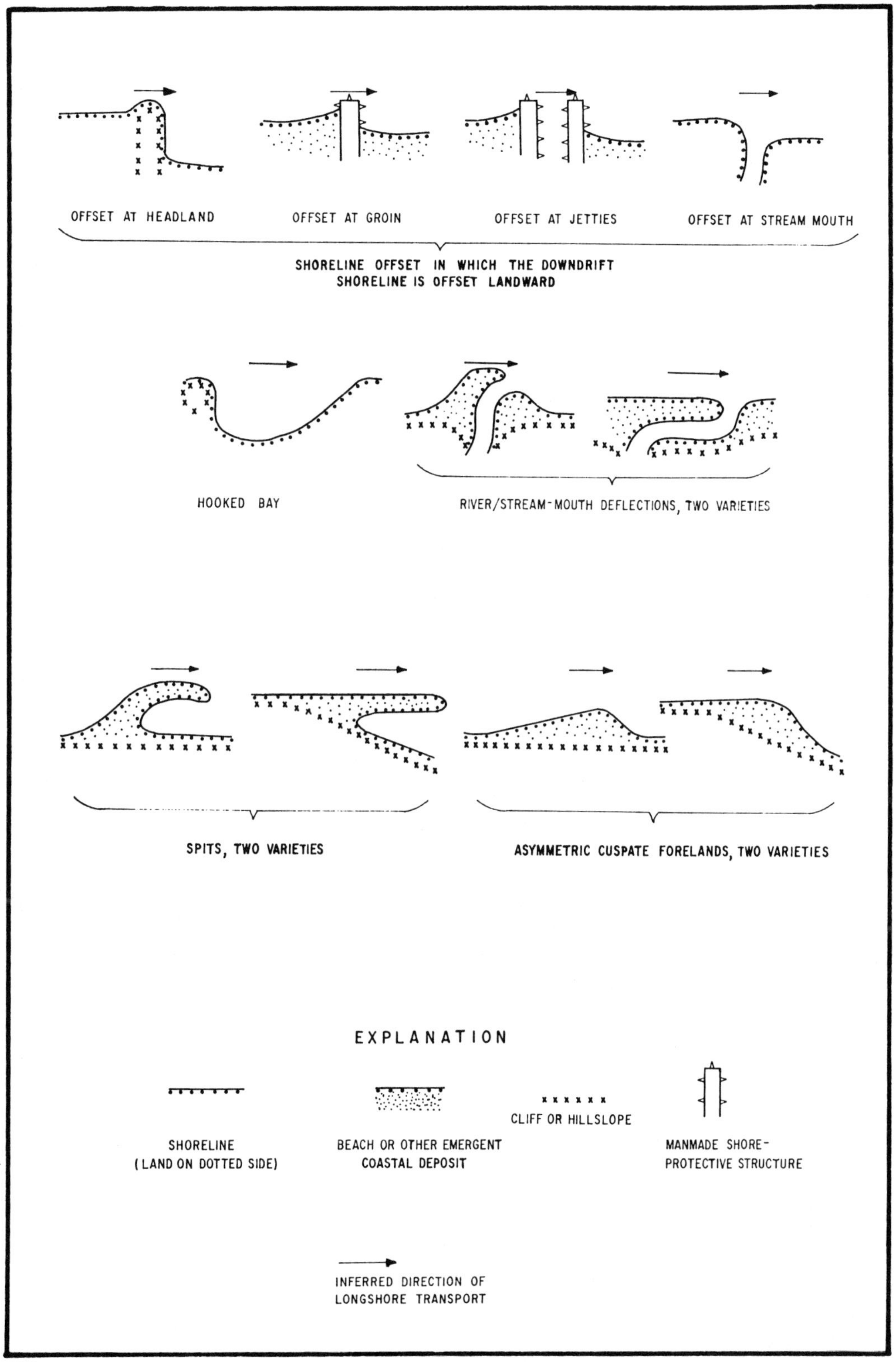

Figure 14. Diagram illustrating coastal morphology features used to infer direction of local longshore transport (modified from Hunter et al., 1979).

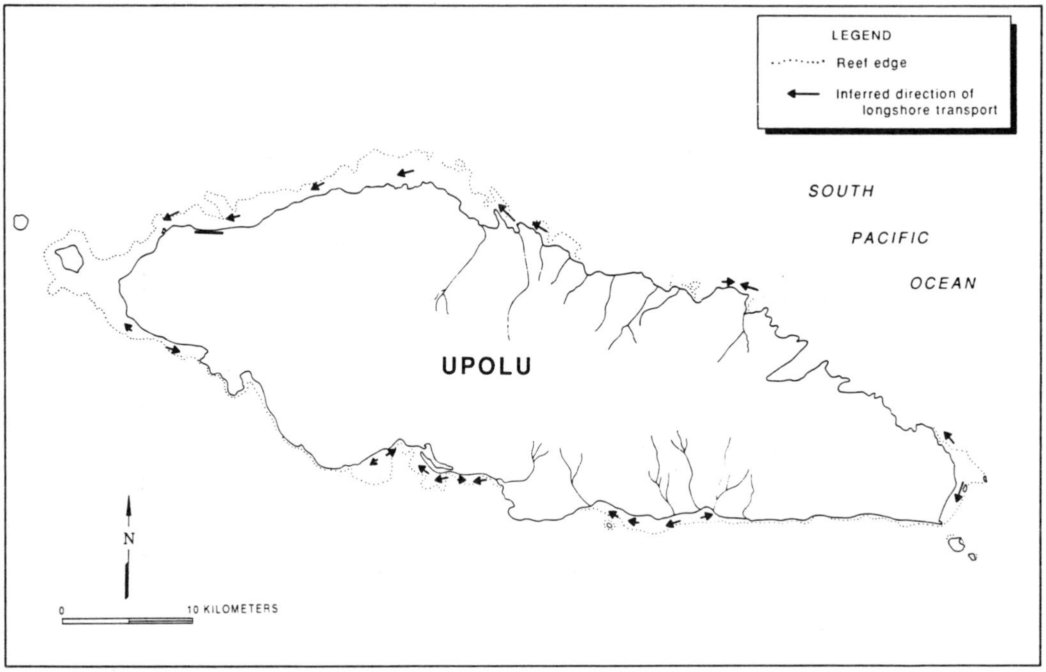

Figure 15. Summary of inferred net longshore transport directions for Upolu based on the orientation of coastal morphology features.

Holocene Sea Level

Published information regarding the local Holocene sea level history for Upolu is somewhat conflicting. Kear and Wood (1959) relate the Tafagamanu Sand to a "5-foot" stand of sea level during the Holocene. A coral sample from the base of a 1.5-m sand bank at Tafangamanu yielded an age of 1.18 ka (Grant-Taylor and Rafter, 1962). These deposits, however, undergo periodic overtopping at the present time and, therefore, do not require a higher sea level for their formation. Sugimura and others (1988) inferred a high stand because of the presence of "emerged" beachrock along the south coast. They measured beachrock up to 0.95 m above mean sea level; however, all beachrock observed is within intertidal limits and could form at this position in relation to the present mean sea level. "Emerged" reef remnants on the reef flat in Fagali'i Bay on the north coast (Rodda, 1988; Stearns, 1944) were briefly examined during the present study. The outcrops border a main reef passage, and appear to consist of reef-flat pavement that has been undercut and tilted. They are not in-situ remnants of a reef formed at a higher sea level.

Lapita pottery shards, which are decorated with a distinctive punctate pattern, were recovered from submerged "beach" deposits during dredging at the Mulifanua Ferry site (Green and Richards, 1975; Jennings, 1974). Typical stratigraphy of the site consists of a 0.8 m cemented hardground overlying about 5 m of coral sand, staghorn coral, and basalt pebbles that, in turn, overlie basalt boulders (~3 m) and basalt basement. The pottery occurs about 2.5 to 3 m below

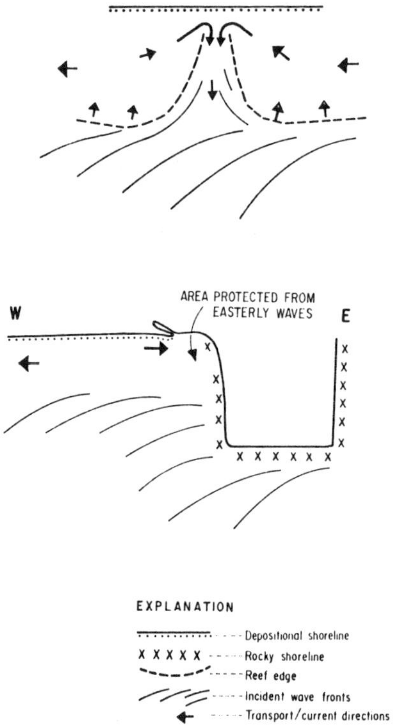

Figure 16. Conceptual models of wave/current flow around reef passages and headlands. (a) Flow in the passages is typically ebb dominant; it is mostly gravity driven by wave set-up at the reef crest. Flow is toward the passage from both sides regardless of the direction of wave approach. (b) Transport reversal in the lee of a large headland. The headland protects a section of coast from easterly conditions, allowing westerly flows a relatively greater impact.

the present mean sea level. Shells, from the level where the shards were found, have radiocarbon ages of around 2890 yr B.P. (Green, 1979). Lapita pottery throughout the southwest Pacific typically is associated with beach ridges a meter or two above sea level. If the lapita sherds are from an in-situ beach ridge deposit, then about 3 m of relative subsidence is implied over the last 3,000 years. This subsidence could be gradual, or more likely, is relatively rapid and possibly related to co-seismic activity. Bloom (1980) presented a Holocene sea level curve for Upolu based on radiocarbon ages from basal mangrove peat deposits along the south coast (Lefaga?). He considered Upolu to be undergoing relative subsidence at a rate of about 1m/1000 yrs in later Holocene time.

Published radiocarbon age dates are presented in Table 3 and age/elevation relationships are summarized in Figure 17. In the present study we found no unequivocal evidence indicating a recognizable higher-than-present Holocene sea level stand. The archeological evidence, extensive shoreline erosion, and age/elevation relationships of dated samples suggest that Upolu is undergoing relative subsidence. Stearns (1944) recognized subsidence on Upolu and nearby Tutuila in American Samoa based on barrier reef development. The subsidence is presumably tectonically controlled and perhaps related to crustal loading by recent volcanism.

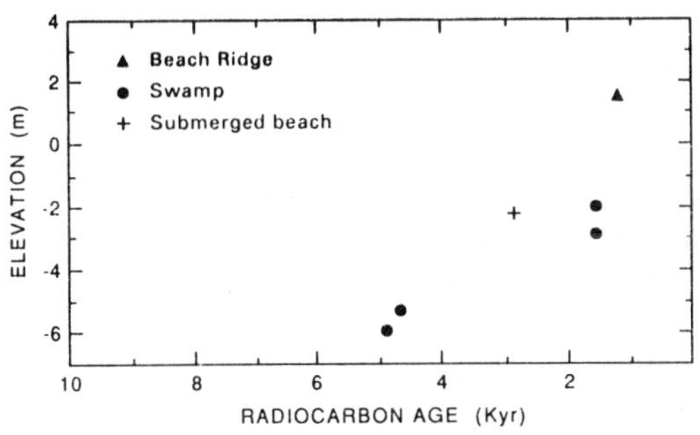

Figure 17. Radiocarbon age versus elevation plot for published age dates from Upolu. Circles from swamp deposits, triangle from modern beach ridge, and cross from buried beach ridge deposits. Details of the age dates are shown in Table 3.

Coastal Stability

Bakx (1987) noted that Western Samoa is presently experiencing widespread, locally severe, coastal erosion. He identified coastal sites requiring shore protection, such as rock revetments and revegetation. The cause of the shoreline erosion was cited as "direct wave attack". Results from the present survey support the existence of widespread coastal instability and the need for shore protection. However, the cause of the erosion is probably due to a number of interrelated factors. If the assessment regarding relative subsidence of Upolu is correct, then the relative sea level rise associated with island sinking could be responsible for the increased erosion. The reefs theoretically can keep pace with the change in sea level by vertical growth. The shoreline, however, needs to readjust to the new level. During a relative rise in sea level this readjustment involves shoreline retreat coincident with nearshore deposition. Such retreat could be the case in Upolu since "direct wave attack" does not explain why coastal depositional landforms would start to erode unless there has been a change in the coastal regime to promote greater wave attack. A relative rise in sea level would accomplish this.

Other factors responsible for shoreline erosion include:

(a) Sand mining, particularly along beaches. This practice is widespread in Samoa, but the amounts taken are probably small when compared to the overall sediment budget. However, in areas where there already is a net loss of sediment, mining will accelerate the problem.

(b) Poorly designed sea walls and revetments. Coastal engineering structures with vertical walls tend to scour at the base, increasing erosion. Rock groins deprive downdrift beaches of sediment and, in effect, displace the erosion problem farther alongshore.

(c) Destruction of vegetation and habitats. Removal of coastal plants exposes the substrate to increased erosion susceptibility by storm waves, precipitation, and wind. Wright noted (1963, p. 33) "A certain amount of coastal erosion may have resulted from the destruction of the mangrove screen although the main cause of shoreline regression has undoubtedly been the constant removal of sand and shingle to pave the village houses." Modification of habitats, such as coastal swamps, may have unforeseen consequences on local sediment budgets and transport pathways. In general, the severest erosion problems occur along the coasts fronted by wide fringing and barrier reefs. These problems could be the result of deeper lagoonal/reef-flat water levels which allow more wave energy to reach the shoreline. Fringing reefs and beaches are also experiencing shoreline retreat, whereas cliffed coasts appear to be relatively stable because they consist of resistant volcanic rock.

NEARSHORE MINERAL RESOURCES

Placer Minerals

The non-living nearshore and coastal mineral resources of Upolu fall into two categories – placer minerals and construction materials (there is no indication of phosphorites). Placer deposits are concentrations of economically valuable,

TABLE 3. Published radiocarbon age dates for Upolu.

Locality	Material	Elevation (m)	Age (yrs bp)	Reference
S. Upolu (Lefaga?)	peaty mud	-3.6 to -4.3	3060 ± 95*	Bloom, 1980
" "	" "	-5.2 to -5.5	4655 ± 95	"
" "	" "	-5.6 to -6.1	4845 ± 95	"
" "	" "	-2.7 to -3.0	1595 ± 85	"
Lefaga Bay	coral	-1.9 to -2.2	1580± 160	Matsumoto and Togashi (1984)
Tafagamanu	coral	+1.5	1180± 55	Grant-Taylor and Rafter, 1962
Mulifanua	shell	-2.25	2890 ± 80	Green, 1979

(*) Reported as depth below the present equivalent environment (mangrove swamps = slightly below high-tide level).

weathering-resistant minerals having a density greater than that of quartz (2.65 g/cm^3). Exploring for placer minerals was not a major component of the present study, and there does not appear to be any existing detailed information pertaining to placers in Samoa. However, based on the geology of Upolu, certain broad assumptions can be made regarding potential placer deposits. The alkalic basalts and associated volcanics can be expected to produce titanomagnetite-rich weathering products. For example, whole-rock chemical analyses (Hawkins and Natland, 1975; Ishii, 1984) show a relative enrichment in titanium.

Sediment samples collected from some of the beaches and stream mouths are high in heavy mineral content (Table 4). Separations of grains less than 1 mm in diameter yielded concentrations of up to 92.8% heavy minerals (>2.86 g/cm^3), most of which were opaque. The larger terrigenous-rich sand deposits are found at the barrier spits at Vaie'e and Falefa.

Eade (1979) estimated there was approximately 12,000 m^3 of very fine dark sand at Solosolo beach. Stach (1969), in discussing the same beach, indicated that no detrital heavy mineral concentrations of economic value are likely. For placer deposits to be economically viable, high concentrations and very large deposits are usually required.

Associated with volcanic activity are potential epithermal and/or hot-spring deposits. There are unconfirmed reports of a gold mine that operated in the nineteenth century. Over 200 stream bed samples were collected as part of a recent gold survey (F. Malele, Apia Observatory, written communication). Gold was detected in about 10% of the samples, however, in all cases but one, the gold content was near the detection limit (approx. 0.01 ppm). The one anomalous sample (~ 0.9 ppm) occurred in the Fuluasou River.

The absence of surficial ultramafic rocks on Upolu severely limits the potential for placers containing nickel, platinum, and chromium. There have been no reports of placer deposits on Upolu containing these minerals.

Construction Materials

Sand and gravel deposits for construction and landfill represent a major nearshore mineral resource on Upolu. Most of the earth materials used in the greater Apia area are presently mined from the nearshore in the Mulinu'u Peninsula–Vaiusu Bay area. Conservative estimates for the amount of material excavated at Mulinu'u are about 28,000 m^3/yr, a total of about 420,000 m^3 having been removed in the 15 years of mining (Maatafa and others, 1988; estimate based on conversations with the dredging operators). The material mined is a poorly sorted, silty, gravelly sand composed of a mixture of back-reef and terrigenous deposits. Apia Concrete Products, on the western shore of Vaiusu Bay, removes nearly 8,000 m^3/yr of material (or approximately 64,000 m^3 in their 8 years of operation). The deposits are similar to those at the Mulinu'u operation

TABLE 4. Percentage of heavy minerals for selected samples *

Sample no.	Location	% > 2.86 g/cm^3
5	Lauli'i	74.6
6	Leusoali'i	40.6
7	Namo River	82.5
9	Matautu Pt.	73.8
15	Lepa	75.9
19	Si'umu	92.8
21	Mulivai	18.2
27	Sataoa	57.9
28	Sataoa	47.7
37	Vaisigano River	80.7

* Written communication, F. Wong, U.S. Geological Survey

except that they are overlain by as much as 1 m of fine sediment. Dredged areas are commonly partially infilled after storms by westward-flowing currents (oral communication, Anthony Brighouse, Apia Concrete Products).

In many areas nearshore deposits have been utilized for the dual purpose of land reclamation and channel construction, for example: Mulifanua Ferry site; Royal Samoan Hotel; near the 'Home of Old People', Vailele, Luatuanu'u; and the wharf and access road near Satitoa, Aleipata. In all of these locations back-reef deposits have been used as a fill for land reclamation whereas the dredge sites are left as deep-water access areas.

Both private and commercial sand mining of beach, beach ridge, and barrier spit deposits is widespread, but there are no permanent quarries established at present. Areas that have been periodically mined include: the spit at the mouth of Tuafaleloa Stream near Solosolo, Luatuanu'u; beach and beach ridge deposits near Salamumu, Lefaga; and flood delta deposits at the mouth of the Vaisigano River, Apia. Private mining using shovels and trucks takes place at nearly every beach that is accessible by road.

Promising nearshore areas for further earth material resources are scattered around the island. Previous CCOP/SOPAC surveys indicate that some potential sites include the area opposite Faleolo airport (Rubin, 1984), the barrier reef between Mulifanua and the offshore island of Manono (Richmond and Roy, 1989), the shallow lagoon at Aleipata (Richmond, 1985; Lewis and others, 1989), and the Vaie'e Peninsula.

The environmental consequences of nearshore mining have not been fully assessed in Western Samoa. Potential problems created by nearshore mining include accelerated coastal erosion due to an imbalance in the local sediment budget, and a change in the nearshore profile allowing greater wave energy to reach the shoreline; and increased turbidity damaging the local ecosystem and fisheries. Although there is some shoreline erosion on the Mulinu'u Peninsula, there is no strong evidence indicating that either coastal erosion or the destruction of fisheries results from the dredging operations. The operation is sited in an area where periodic flooding creates natural turbid conditions, therefore coral and other species inhabiting the area are probably tolerant of such conditions. However, a well-designed monitoring program is needed to determine the effects of the dredging.

SUMMARY

1. *Coastal Types.* Three general types of coast occur on Upolu: Wide fringing and barrier reefs backed by gently sloping volcanic rocks; cliffed coast with little or no reef development; and fringing reef and narrow coastal plain typically consisting of a storm ridge or stream mouth deposits.

2. *Reefs.* Fringing, barrier, and patch reefs surround most of Upolu. Reef features that were mapped include submarine terraces, reef crests, reef flats, shallow lagoons, and blue holes. Reef surfaces can be categorized into three types: coral/algal/pavement, sediment covered, and a marine flora zone.

3. *Beaches.* Beaches are widespread and consist of sands and gravels composed of carbonate or terrigenous material, or a mixture of the two. Terrigenous composition is related to the proximity of a fluvial source. Beach height is directly related to the exposure and width of the adjacent reef.

4. *Stream and river mouths.* Barrier spits, swamps, submerged deltas, and flood deltas occur at stream mouths. Where discharge is particularly high or persistent, gaps in the adjacent reef may be present.

5. *Swamps and mangroves.* Swamps and mangroves are widespread in low-lying protected coastal areas. They are sites of fine sediment deposition and a valuable fisheries resource.

6. *Sediment transport.* Longshore and cross-shore transport of sediment is primarily controlled by the dominant east to west wind/wave circulation. Variations from this pattern are created by storms, interaction of morphology and coastal currents, and ENSO events.

7. *Sea level changes.* Available evidence suggests that Upolu is undergoing relative subsidence, perhaps at a rate as high as about 1 m/1,000 yrs. There is no unequivocal evidence for a sustained Holocene sea level stand that is higher than the present stand.

8. *Coastal stability.* Coastal erosion appears to be widespread on Upolu, but it is only severe in limited areas. Widespread erosion of the shoreline may be related to the relative subsidence of the island enhancing wave attack, improperly designed coastal engineering structures, and sand and gravel mining.

9. *Placer minerals.* Potentially economic placer mineral deposits are limited to those derived from basaltic rocks: magnetite (iron) and ilmenite and rutile (titanium). There are no ultramafic rocks on Upolu and no confirmed gold deposits.

10. *Construction materials.* The nearshore region is the site of extensive sand and gravel deposits, which are composed of both carbonate and terrigenous materials. Extensive reserves occur in the Vaiusu Bay area and several promising prospects are located throughout the island, notably at Mulifanua/Manono barrier reef, opposite Faleolo airport, Aleipata, and near Vaie'e Peninsula.

Acknowledgments

The field work and subsequent Technical Report and map preparation (Richmond, in press a&b) were supported by CCOP/SOPAC as part of work program CCSP/WS.11: "Mapping of coastal and nearshore areas in Western Samoa". The staff of the Apia Observatory and the Department of Lands and Survey, Apia, are gratefully acknowledged for their assistance in obtaining aerial photographs and maps and for logistic support in the field. In particular, I would like to thank the following people for their contributions: Aleni Fepuleai (base map digitization), Fa'atoia Malele and Tana Sua (driver, guide, and assistance with sediment sampling), and the Superintendent of the Apia Observatory, Ausetalia Titimaea (logistic support). Lui Bell of the Fisheries Department provided assistance in the field and in discussions on mangrove communities. Dr. Abby Sallenger of the U.S. Geological Survey produced the videotape images of the Upolu shoreline. O.M.W. Richmond assisted with sediment sampling and preliminary drafting of field maps. Peter Roy of the New South Wales Geologic Survey was instrumental in initiating the mapping program and provided numerous helpful discussions in the early stages of the project. John Harper, James Kamsoo, Sekoue Motuiwala, Phil Woodward, Nika Naibitakele, and Ed Saphore of CCOP/SOPAC assisted in various stages of the project; their help is greatly appreciated. [CCOP/SOPAC has changed its name to SOPAC–South Pacific Applied Geoscience Commission.]

REFERENCES

Andrews, G.J., and P.F. Holthus, 1988, Marine environment survey: Proposed Aleipata Islands National Park, Western Samoa: unpublished SPREP report, 61 pp.

Backshall, D.G., J. Barnett, P.J., Davies, D.C. Duncan, N. Harvey, D. Hopley, P. Isdale, J.N. Jennings, and R. Moss, 1979, Drowned dolines - the blue holes of the Pompey Reefs, Great Barrier Reef: BMR Journal of Australian Geology and Geophysics, v. 4, p. 99-109.

Bakx, M.A., 1987, Consultancy on shore protection works, Western Samoa: United Nations Development Programme/International Labour Organisation, Geneva, Switzerland. 101 pp.

Bell, L.A.J., 1985, Case study: Coastal Zone Management in Western Samoa: Third South Pacific National Parks and Reserves Conference Report, v. 2, p. 57-72.

Bloom, A.L., 1980, Late Quaternary sea level change on South Pacific coasts: A study in tectonic diversity, *in* N-A. Morner (ed.), Earth Rheology, Isostasy, and Eustasy: New York, Wiley Interscience, p. 505-516.

Carter, R. 1988, Site location for waverider buoy in Western Samoa: CCOP/SOPAC Preliminary Report 3, 9 pp.

Carter, R., 1987, Design of the seawall for Mulinu'u Point, Apia, Western Samoa: CCOP/SOPAC Technical Report 78, 18 pp.

Eade, J.V., 1979, Solosolo beach report: CCOP/SOPAC Cruise Report 27, 5 pp.

Gauss, G.A., 1981, Apia Harbour survey, Samoa, 19 January - 6 February 1981, 20-31 March 1981, Cruise WS 81(1): CCOP/SOPAC Report 55, 10 pp.

Gauss, G.A., 1982, Mulifanua and Salelologa Harbour surveys, Samoa, 7-31 March 1982: CCOP/SOPAC Cruise Report 66, 12 pp.

Grant-Taylor, T.L., and T.A. Rafter, 1962, New Zealand radiocarbon age measurements - 5: New Zealand Journal of Geology and Geophysics, v. 5, p. 331-359.

Green, R.C., and H.G. Richards, 1975, Lapita pottery and lower sea level in Western Samoa: Pacific Science, v. 29, p. 309-315.

Harper, J.R., and B.M. Richmond, in prep., Investigation for nearshore sand and gravel deposits, Western Samoa: CCOP/SOPAC Technical Report.

Hawkins, J.W., 1976, Tectonic setting and petrology of Samoa-Tonga-Fiji region: CCOP/SOPAC Technical Bulletin No. 2, p. 141-152.

Hawkins, J.W. and J.H. Natland, 1975, Nephelenites and basanites of the Samoan linear volcanic chain: Their possible tectonic significance: Earth and Planetary Science letters, v. 24, p. 427-439.

Hedge, C.E., Z.E. Peterman, W.R. and Dickinson, 1972, Petrogenesis of lavas from Western Samoa: Geological Society of America Bulletin, v. 83, p. 2709-2714.

Hunter, R.E., A.H. Sallenger, and W.R. Dupre, 1979, Maps showing directions of longshore sediment transport along the Alaskan Bering Sea Coast: U.S. Geological Survey Miscellaneous Field Studies Map, MF-1049 (5 sheets).

Ishii, T., 1984, Volcanology and petrology, *in* A. Sugimura (ed.), Sea-level changes and tectonics in the middle Pacific. Report of the HIPAC Project in 1981, 1982, and 1983: Department of Earth Sciences, Kobe University, Nada, Kobe, Japan, p. 63-83.

Jennings, J.D., 1974, The ferry berth site, Mulifanua District, Upolu, *in* R.C. Green and J.M. Davidson (eds.), Archaeology of Western Samoa 2: Auckland Institute and Museum Bulletin, v. 7, p. 176-178.

Kear, D., and B.L. Wood, 1959, The geology and hydrology of Western Samoa: New Zealand Geological Survey Bulletin 63, 92 pp.

Kerr, I.S., 1976, Tropical storms and hurricanes in the Southwest Pacific: New Zealand Meteorological Service Miscellaneous Publication 148, 114 pp.

Lewis, K.B., P.J. Hill, W. deL. Main, and J.S. Mitchell, 1989, Lagoonal sand and gravel deposits of Eastern Upolu and Eastern Savai'i, Western Samoa: Consultant report to CCOP/SOPAC, 101 pp.

Maatafa, M., B.M. Richmond, and J.R. Harper, 1988, Construction material field survey, Upolu, Western Samoa - 1988: Paper presented at the CCOP/SOPAC-ICOD Nearshore Minerals Workshop, Savusavu, Fiji, 29 Sept - 12 Oct. 1988.

Matsumoto, E., and S. Togashi, 1984, Radiocarbon dating, *in* A. Sugimura, (ed.), Sea-level changes and tectonics in the middle Pacific. Report of the HIPAC Project in 1981, 1982, and 1983:

Department of Earth Sciences, Kobe University, Nada, Kobe, Japan. p. 103-112.

Matsushima, Y., A. Sugimura, K. Berryman, T. Ishii, Y. Maeda, E. Matsumoto, and N. Yonekura, N., 1984, Research report of B-party: Holocene sea level changes in Fiji and Western Samoa, *in* A. Sugimura, (ed.), Sea-level changes and tectonics in the middle Pacific. Report of the HIPAC Project in 1981, 1982, and 1983: Department of Earth Sciences, Kobe University, Nada, Kobe, Japan, p. 137-185.

Morton, J., M. Richards, S. Mildner, and L. Bell, 1988, The shore ecology of Upolu, Western Samoa: Unpublished draft, Study Booklet, 168 pp.

Nelson, D.V., 1987, Watershed Management Study: Land conservation in the Rewa and Ba watersheds: UNDP/FAO Field Document 1 (FIJ/86/001).

Nordin, C.F., and J.P. Beverage, 1965, Sediment transport in the Rio Grande, New Mexico: U.S. Geological Survey Professional Paper 462-F, 35 pp.

Richmond, B.M., 1985, Reconnaissance survey for construction and landfill materials at Aleipata, Western Samoa: CCOP/SOPAC Cruise Report 110, 14 pp.

Richmond, B.M., 1988, Ephemeral delta development at the mouth of Aptos Creek, northern Monterey Bay, *in* S.D. Ellen and G.F. Wieczorek (eds.), Landslides, Floods, and Marine Effects of the Storm of January 3-5, 1982, in the San Francisco Bay Region, California: U.S. Geological Survey Professional Paper 1434, p. 265-282.

Richmond, B.M., in press (a). Coastal morphology, shoreline stability, and nearshore mineral resources of Upolu, Western Samoa. CCOP/SOPAC Technical Report, 45 pp.

Richmond, B.M., in press (b). Coastal Morphology of Western Samoa. Scale 1:20,000. SOPAC Coastal Map Series (WS/1 - West Upolu; WS/2 - North upolu; WS/3 - East Upolu; WS/4 - South Upolu).

Richmond, B.M., and P.S. Roy, 1989, Sedimentologic and bathymetric studies of offshore Manono and Apolima Islands, Western Samoa:. CCOP/SOPAC Technical Report 64, 12 pp.

Rodda, P., 1988, Visit to western Samoa with the HIPAC team, *in* N. Yonekura (ed.), Sea-level changes and tectonics in the middle Pacific. Report of the HIPAC Project in 1986 and 1987: Department of Geography, University of Tokyo, Hongo, Tokyo, Japan, p. 85-90.

Rubin, D.M., 1984, Landfill materials and harbour surveys at Apia Harbour, Mulinu'u Point, Faleolo Airport, and Asau Harbour, Western Samoa, 15 May - 4 June 1984: CCOP/SOPAC Cruise Report 98, 8 pp.

Stach, L.W., 1969, The mineral potential of Western Samoa, Preliminary Report: Mineral Resources Development Section, ECAFE, Bangkok. 6 pp.

Stretan, N.A., and J.W. Zillman, 1984, Climate of the South Pacific Ocean, *in* H. Van Loon (ed.), Climates of the Oceans: Elsevier, Amsterdam, p. 263-427.

Stearns, H.T., 1944, Geology of the Samoan Islands: Geological Society of America Bulletin, v. 55, p. 1279-1332.

Sugimura, A., Y. Maeda, Y. Matsushima, and P. Rodda, 1988, Further report on sea level investigation in Western Samoa, *in* N. Yonekura (ed.), Sea-level changes and tectonics in the middle Pacific. Report of the HIPAC Project in 1986 and 1987: Department of Geography, University of Tokyo, Hongo, Tokyo, Japan. p. 77-84.

Wright, A.C.S., 1963, Soils and land use of Western Samoa: New Zealand Soil Bureau Bulletin 22, 191 pp.

APPENDIX 1. Location, depositional environment, sample descriptions, and remarks for Upolu coastal samples, 1987.

No.	Location	Environment	Texture*	Color#	Remarks	Exposure$
1	Mulinu'u	Beach	MWS/CS-VCS	Black + orange	Eroded shore	ME
2	W Mulinu'u	Lagoon fill	PS/FS-VCS + G	Grey	Dredge spoils	MP
3	Vaisigano R, Apia	River bank	MPS/MS-VCS	Brown		
4	Fagali'i	Beach	WS/FS	Brown	Eroded shore	ME
5	Lauli'i	Beach	PS/FS-VCS	Brown	Eroded shore	ME
6	Leusoali'i	Beach	MWS/MS	Black + cream	Gravel on beach	ME
7	Namo R	Beach/river mouth	MPS/MS-VCS	Brown	Near Solosolo, gravel	E
8	SW Cape Utusi'a	Beach	MWS/MS-VCS	Cream/orange	Former sand mining site	MP
9	Mata'utu Pt.	Beach	MPS/MS-VCS	Black	Eroded shore	MP
10	Sauago	Pocket Beach	WS/MS-CS	Orange	Some basalt boulders	E
11	Fagaloa	Beach/river mouth	MWS/CS-VCS	Cream + black	Reworked river deposits	P
12	Satitoa	Beach	VWS/MS	Cream	Some erosion	ME
13	Mata'utu, Aleipata	Beach	WS/MS	Cream	Exposed beachrock	ME
14	Saleapaga	Beach	WS/MS-CS	Cream	Exposed beachrock	ME
15	Lepa	Mountain stream	VPS/FS-VCS	Black	Dry stream bed	
16	Salani	Beach/River mouth	WS/MS	Cream + black	Submerged flood delta	ME
17	Tafatafa	Reef flat (2mWD)	MS/MS-VCS	Cream + black	Sandy reef flat	ME
18	Tafatafa	Beach	WS/MS	Cream + black	Some beachrock	ME
19	Si'umu	Beach/stream mouth	MWS/MS-CS	Brown	Muddy reef flat	MP
20	Mulivai	Beach	MPS/MS-VCS	Cream + black	Artificial beach	ME
21	Mulivai (W)	Beach/flood delta	PS/MS-VCS	Brown	E facing spit	ME
22	Safata	Lagoonside, spit beach	MPS/FS-CS	Cream	Mangrove area	P
23	Safata	Beach	MS/MS-CS	Cream + black	Exposed beachrock	ME
24	Safata	Beach scarp	WS/FS-MS	Lt. gray	Windblown (?)sand	ME
25	Safata	Beach	PS/CS-VCS + g	Cream + black	Exposed beachrock	ME
26	Tafitoala	Beach	MWS/VCS	Cream + black	Exposed beachrock	ME
27	Sataoa	Beach/stream mouth	PS-WS/MS-VCS	Brown	Small flood delta	ME
28	Sataoa	Beach	MS/MS-VCS	Brown + cream	Distal end of spit	ME
29	Matauta, Lefaga	Reef flat	WS/MS	Cream	Pocket beaches	E
30	Lotoso'a, Apia	Beach	WS/FS-MS	Cream + black	Thin veneer, erosion	ME
31	Tauo'o	Beach	MWS/FS-MS	Cream + black	Thin veneer, erosion	ME
32	Fagalepolu	Beach	WS/FS	Cream	Thin veneer, erosion	ME
33	Cape Fatuosofia	Reef flat	WS/VFS	Gray/cream	Thin veneer	ME
34	Cape Fatuosofia	Pocket beach	WS/FS	Cream	Thin veneer	ME
35	Pata	Spit beach	MS/MS-CS	Cream + orange	Erosion	ME
36	Vaiala, Apia	Beach	WS/MS	Cream + black	Seawall	ME
37	Vaisiqano R	Beach/river mouth	WS/MS	Brown	Seawall	ME

Abbreviations used:

*Texture

Sorting terms:

WS = well sorted
MWS = moderately well sorted
MS = moderately sorted
PS = poorly sorted
VPS = very poorly sorted

Grain size terms:

Z = silt
VFS = very fine sand
FS = fine sand
MS = medium sand
CS = coarse sand
VCS = very coarse sand
G = gravel

examples of usage

MWS/M-VCS = moderately well-sorted medium to very coarse sand

Color - Color of the damp sediment.

In general, the cream colour is indicative of pure, clean, carbonate sands. The orange colour is attributed to the tests of a benthic foraminifera. The grey, brown and black shades are caused by either the presence of organic matter or terriginous sediment, primarily volcanics.

$ Exposure - Relative exposure to waves and currents at the sample site

E = Exposed; generally an open coastline with no fringing reef.
ME = Moderately exposed; a high energy coastline fronted by a fringing reef.
MP = Moderately protected; a low energy coastline or an area with a very wide fringing reef.
P = Protected; shorelines with low wave and current activity except during major storms.

Keating, B.H., and Bolton, B.R., editors, 1992, Geology and offshore mineral resources of the central Pacific basin, Circum-Pacific Council for Energy and Mineral Resources Earth Science Series, V. 14. New York, Springer-Verlag.

THE GEOLOGY OF THE SAMOAN ISLANDS

Barbara H. Keating
Hawaii Institute of Geophysics, University of Hawaii, Honolulu, HI 96822

ABSTRACT

The Samoan chain consists of high volcanic islands, atolls, and submerged reef banks near the southwest margin of the Pacific plate. The chain is unusual, particularly when compared with the Hawaiian chain, because the islands are volcanically active on both the eastern and western ends of the chain, the islands are larger westward, the easternmost edifice is an atoll not an active seamount, and the chain consists dominantly of alkali rather than tholeiitic lavas. While geological studies of the Samoan group are limited, the existing results are consistent with a hot spot origin similar to Hawaii, complicated by continued reactivation of volcanism on Savaii. The continuing volcanism on Savaii is believed to be the result of deformation of the margin due to lithospheric dilation, as the plate bends where it approaches the Tonga Trench subduction zone. The dominance of alkalic volcanism in this island chain has recently been associated with a geochemical heterogeneity in the underlying mantle.

INTRODUCTION

The Samoan islands are a chain of high volcanic islands in the southwest Pacific Ocean. The chain consists of three high volcanic islands (Tutuila, American Samoa; Upolu and Savaii, Western Samoa) and numerous islets (Fig. 1). Reef banks and seamounts continue westward from the chain, constituting the Northern Melanesian Borderland (Brocher, 1985). The first geologic studies of the islands were conducted from sailing vessels exploring the south Pacific. Several early geologic studies reported volcanic eruptions (1902-1911). Early geologic mapping took place in the 1940's and 1950's but little more was done until the 1980's. The islands of the chain, other than Rose Atoll, are young (less than a few million years in age). Previous workers have compared these islands with the Hawaiian islands because the geologic relations, ages, and orientation of the islands and submerged seamounts are similar (Fig. 2). The purpose of this chapter is to summarize the current state of our understanding of the origin and evolution of the Samoan island chain, and bring together the geologic information from diverse publications into a single summary.

Political Division of Samoa

The Samoan Islands have been divided politically since the mid-1800's. The islands of the Manua Group and the island of Tutuila are territories of the United States of America. Scenic Pago Pago harbor in Tutuila was used by

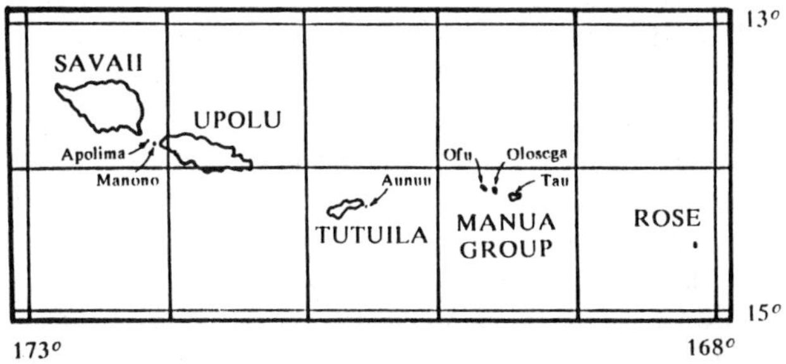

Figure 1. Map of the main islands in the Samoan island chain. Rose Atoll is a low coral atoll with two islets. The rest of the island chain consists of high volcanic islands. The islands of Samoa are larger to the west in the chain. The trend is the opposite of that observed in the Hawaiian chain.

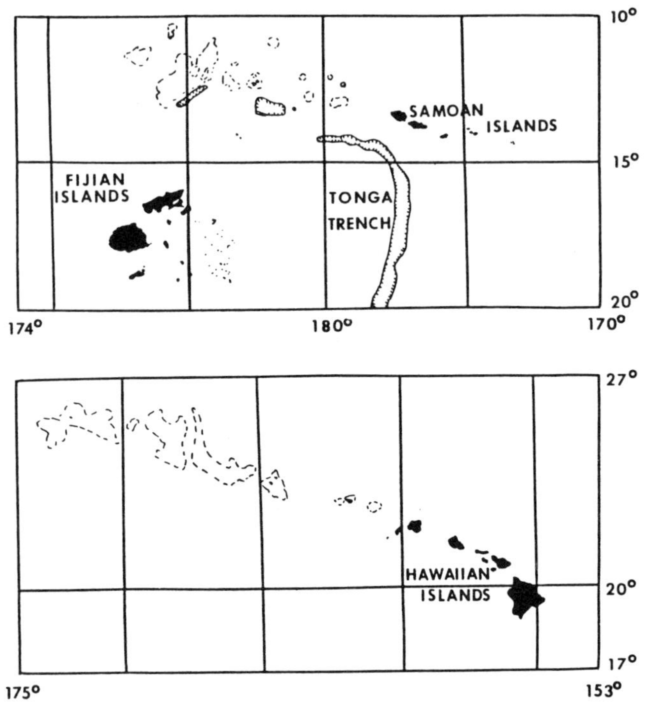

Figure 2. A comparison of the trend of islands and seamounts of the Samoan group (top) with the Hawaiian group (bottom). The seamount chains are parallel on the Pacific plate.

American ships as a coal fueling station; the harbor is now the home of the Pacific tuna fleet. In the 1800's, Western Samoa was held under a tripartite protectorate comprised of Germany, England, and the U.S. In 1900, England withdrew from the agreement and the American and German interests divided the islands into American Samoa and Western Samoa. The League of Nations granted New Zealand a mandate for Western Samoa in 1914 and German holdings were expropriated. Western Samoa is now an independent nation but still maintains strong political and economic ties with New Zealand. American Samoa has remained politically linked to the U. S.

The Samoan island group has been politically divided for many years, and as a result the early geological studies and mapping of the islands were carried out by two groups with different approaches to defining geologic units. The geologic studies in Western Samoa were done by hydrologists and the basic mapping units were based upon geomorphologic relationships observed from aerial photographs with limited field mapping. In American Samoa, the early geologic studies were carried out by a volcanologist and the mapping units were based upon field relationships of volcanic units. Because the definitions and mapping techniques used are so different, the descriptions of the geologic units of these islands cannot readily be combined. Therefore, the geology of the islands of Western Samoa are described, followed by the description of the geology of American Samoa. Because the political division of the island group directly affects the geologic studies of the group, a brief summary of the historical background of the islands has been included in this text. If the reader is not interested in this background, the reader should proceed directly to the section entitled Geologic Exploration.

BACKGROUND

Volcanic Heritage

The word Samoa comes from the words *sa ia moa* in the native Samoan language. The legendary name is derived from the volcanic origin of the land itself. According to Samoan legend, "the rocks cried to the earth, and the earth became pregnant. Salevao, the god of rocks, observed motion in the *moa,* or center of the earth. The child was born and named sa ia Moa, from the place where it was seen moving. Salevao said he would become loose stones, and that everything which grew would be *sa ia Moa* or sacred to Moa. Hence the rocks and earth were called sa ia Moa or as it is abbreviated, SAMOA," according to Turner (1979).

Discovery and Early Exploration

Jacob Roggeveen is considered to be the modern discoverer of the Samoa islands in 1722. The islands were inhabited by natives at the time of the first encounter with Europeans. Roggeveen called them the "Baumann Islands" after a captain in his squadron of ships. In 1768, L. Bougainville visited the islands, naming them the "Isles des Navigateurs" after observing the frequent use of canoes by the natives.

Captain James Cook heard of these islands while in Tonga in 1773 and recorded their names but did not visit. In 1789, a visit by La Perouse to Tutuila proved eventful. His second-in-command, Captain de Langle, died along with many crew members in a scene very similar to that in which Captain James Cook lost his life in the Hawaiian Islands. After the death of his crew members, La Perouse gave the island the name "Massacre Island" – now Tutuila. The bay on the north shore of Tutuila is still known as Massacre Bay (near Aasu). The historic event is of interest since the geologic features there played an important part in the events which transpired at the bay.

On December 12, Captain de Langle and 61 men in two longboats and two pinnaces went to a village in a cove near their anchorage to collect water. Instead of a spacious and convenient cove, the men found a coral-filled cove with a twisting, narrow entry channel. The captain had observed the bay at high tide and was not aware of the tidal range within these islands. The water collecting went well, but within an hour almost fifteen hundred natives had assembled, leading to great confusion. The confusion was compounded when de Langle began giving gifts to natives he thought were chiefs. Conflict resulted and de Langle ordered his men back to the boats. The men were stoned, de Langle responded by firing his musket in the air, and this provoked the natives to a general attack. The captain was knocked down, fell overboard, and was clubbed to death. Many sailors abandoned some of the longboats and scrambled to the pinnaces. Others steered the boats through the narrow passage for open water.

Unfortunately, the boats ran aground at the narrowest part of the channel. The reef terrace formed by a higher sea level stand allowed their pursuers to close in again. The volleys from the muskets scared the attackers away. The boats cleared the reef and returned to the frigates. La Perouse considered returning, but decided against it when he saw the coral reef platform which contributed to the loss of lives.

The Samoan islands were visited by Edwards in 1791, as part of the search by H.M.S. *Pandora* for participants in the mutiny on H.M.S. *Bounty*. A book by the captain and physician on this voyage (Edwards and Hamilton, 1915) describes many visits to South Pacific islands. Von Kotzebue visited the islands in 1824.

REGIONAL SETTING

The Samoan islands consist of a series of high volcanic islands, atolls and submerged reef banks, and seamounts which form a linear chain in the southwest Pacific Ocean. The chain trends in a southeast-northwest direction (Fig. 2) beginning near the international date line and extending westward, roughly 100 km north of the termination of the Tonga Trench, into a region known as the Northern Melanesian Borderland (Fig. 3, from Brocher, 1985). The Northern Melanesian Borderland is a complex region within the southwest Pacific where island arcs like the Tongan and Lau Islands, mid-plate features like Peggy Ridge, and seamounts of hot spot origin like the Samoan group, occur near the convergent margin of the Pacific plate and the Fiji Plateau.

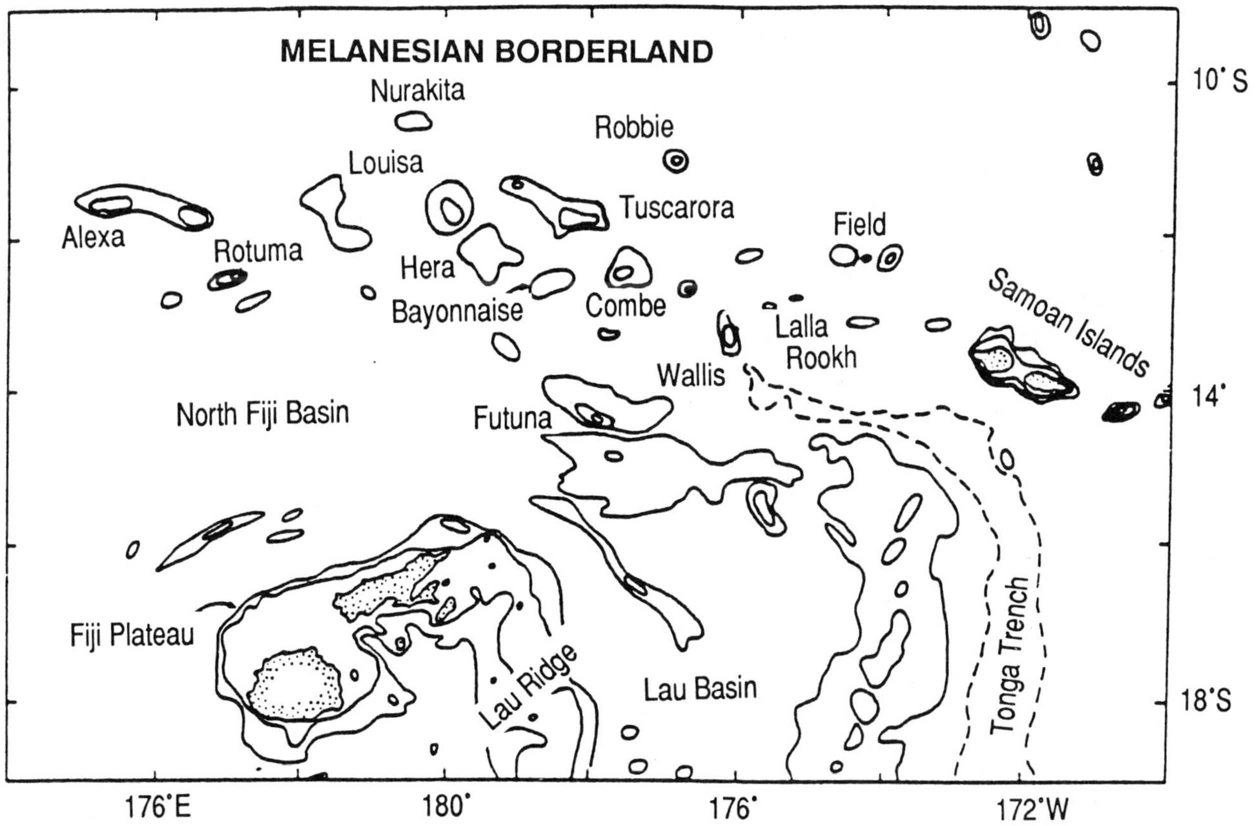

Figure 3. Map of the Northern Melanesian Borderland (from Brocher, 1985) illustrating the major bathymetric features in the region. The Samoan and Fiji Islands are shown with a stippled pattern. The island contours and 1,000 and 2,000-m contours are shown.

The easternmost member of the island chain is Rose Atoll. Rose Atoll is a low carbonate bank. Other than the alignment of Rose with the remainder of the chain, no evidence exists to directly link the atoll to the Samoan group. Thus, previous workers have suggested it may be an old seamount of unrelated origin. West of Rose Atoll are the high volcanic islands that form an island chain similar in nature to the Hawaiian or Society island groups, with the exception of the island of Savaii. Savaii, on the western end of the chain, is still volcanically active. Farther to the west are the submerged seamounts and reef banks described by Brocher (1985). Age dating of rocks dredged from these banks suggests these banks are a continuation of the Samoan seamount group.

GEOLOGIC EXPLORATION

Previous Studies

The first geological explorations of Samoa were reported by Dana (1849) as part of the U.S. Exploring Expedition under Lt. Charles Wilkes. Dana spent a very limited time on Samoa; his visit to the islands was restricted to the time required by the expedition to generate a hydrographic map of the island. Despite his short visit, the observations of Dana show an obvious insight into the geology of these islands. As a result of Dana's visit to these islands and others, he published numerous manuscripts, including "Geology of the Pacific Area" (1846) in the Wilkes expedition volumes, "Corals and Coral Islands" (1872) and "Characteristics of Volcanoes" (1890).

Dana noted the youthful appearance of the western district of Samoa (Savaii), which contrasts with the greater age of the central portion of the chain (Tutuila). He felt that these islands formed from two volcanic fissures, one of age equivalent to Tahiti (in the Society Islands) and Kauai (in the Hawaiian islands), and a second contemporaneous with the present reefs. Chemical and petrographic studies of the lava flows have been reported by Mohle (1902), Kaiser (1904), Klautsch (1907), Jensen (1908), Daly (1924), Bartrum (1927), Macdonald (1944, 1968), Stice (1968), Hubbard (1971), Hedge et al. (1972), Hawkins and Natland (1975), and Natland (1975, 1980).

Descriptions of volcanic eruptions include those of Angenheister (1909), Anderson (1910), Freidlander (1910), Grevel (1911), von Bulow (1906), Friederici (1910), Reinecke (1905, 1906), Sapper (1906, 1909, 1911a, b, 1912, 1915), Schmittman (1911), Wegener (1902, 1903a, b), and Bryan (1941). Geologic studies of the islands include those by Friedlander (1910), Park (1914), Thomson (1921), Stearns (1944), Kear and Wood (1959), Stice and McCoy (1968), Hawkins and Natland (1975) and Natland (1975, 1980). Isotopic and geochemical studies have been conducted by White and Hoffman(1982), Newman et al. (1984), Rison and Craig, 1982, Matsuda et al. (1984), Wright and White (1987) and Wright (1987). Dating studies include those of Richard (1962), Matsuda and others (1984), Natland and Turner (1985), Duncan (1985), and McDougall (1985). Geophysical studies include a paleomagnetic study by Tarling (1962; 1965) and Keating (1985a), gravity studied by Machesky (1965) and Robertson (1987) and an earthquake report by Needham et al. (1982). A sedimentological and bathymetric survey of the Samoan archipelagic apron, discussing deformation of the apron in the Tonga Trench, was published by Lonsdale (1975). Nearshore sedimentologic studies were reported by Daly (1924) and Dingleret al. (1986). Offshore pelagic sediment distribution is summarized in the geologic map of the Circum-Pacific Region, Southwest Quadrant (Palfreyman, et al., 1988).

Many of the early geological reports of Samoa are written in German and are not readily available. Fortunately, Thomson (1921) published a thorough review in English of most of these texts. Thomson summarized the petrology as known in 1921 (Table 1).

Table 1. Volcanic Rock Types on Samoa (From Thomson, 1921)

Savaii:	Olivine basalt, olivine-enstatite basalt, olivine tachylite, nepheline basanite, phonolite.
Apolima:	Nepheline basalt.
Upolu:	Limburgite, olivine basalt, olivine basalt- porphyrite, trachydolerite, nepheline basanite.
Tutuila:	Limburgite, olivine basalt, andesitic basalt, spilite, nepheline basanite, trachydolerite, nepheline basanite, trachyte, alkali trachyte, phonolitic trachyte.
Aunuu:	Trachydolerite.
Ofu:	Olivine basalt.
Olosega:	Olivine basalt.
Tau:	Olivine basalt.

GENERAL DESCRIPTION OF THE ISLANDS

Western Samoa

Kear and Wood (1959) reported on the geology and hydrology of Western Samoa. They recognize a general structure on Upolu of deeply eroded and dissected volcanic terrains (assigned a Pliocene or early Pleistocene age), largely buried by late Pleistocene and Recent lavas. They assign the name Fagaloa volcanics to the oldest of these rocks, largely on the basis of their weathered appearance, consisting of a'a and pahoehoe flows along with associated dykes, tuffs and cone deposits. They point out that these rocks characteristically form steep-sided high mountains with erosional slopes of 25–50°. Of considerable interest is their observation that the original dips on the exposed surfaces suggest that these volcanics were extruded from vents that are oriented approximately parallel to those of the younger lavas.

Rocks from each of the younger mapped units uncomformably overlie and fill valleys eroded into Fagaloa volcanics. Most of these units are olivine-rich basalts which strongly resemble one another petrologically. The features used to differentiate between the rock units include (Kear and Wood, 1959):

- Salani Volcanics - deep soil and weathering; evidence of pre-Mulifanua canyon cutting.
- Mulifanua Volcanics - the existence of wide barrier reefs existing offshore of these outcrops; only shallow stream channels.
- Lefaga Volcanics - lack of dissection; only narrow fringing reefs present offshore of these outcrops.
- PuaPua Volcanics - thin soils; lavas flow offshore and form rocky (or "ironbound") coasts; ubiquitous aa and pahoehoe structures form broad domes.
- Aopo Volcanics - cones erupted in last 200 years, fresh porphyritic pahoehoe flows and a'a flows common only around cones which fill older valleys and spill out over coasts to fill lagoons and cover the barrier reef.

Kear and Wood (1959) note that the volcanic cones are plentiful on Upolu and Savaii and that the degree of weathering, dissection, and decay of the cones varies, reflecting their Salani to Aopo ages. The largest cones occur between 600 and 900 m above sea level on Upolu, and up to 1800 m on Savaii. The cones vary widely in form, some having minor amounts of ejecta and giving rise to major lava floods, while others form simple single cones of cinders and scoria. The younger cones are generally black with veneers of glassy cinder. As the rocks weather, they tend to turn reddish in color, and the finer-grained materials are altered to clays, forming a reddish clay soil littered with blocks of scoria.

Savaii

The island of Savaii is by far the largest island in the Samoan chain. Using the definitions and mapping unit of Kear and Wood (1959), the oldest rocks observed on the island are situated on either side of the Vaipouli River on the north shore of Savaii (Fig. 4). This unit displays considerable relief and deep weathering. Paleomagnetic studies of Savaii (Keating, 1985a) indicate that these rocks are characterized by normal polarity magnetization. Based upon the magnetic reversal time scale, we believe these rocks are likely to correspond to the Gauss Normal Chron and therefore are 2.5 million years old or older. Radiometric dating of these rocks has been undertaken by Ian McDougall. The results of these studies when published will much more accurately define the ages of these rocks.

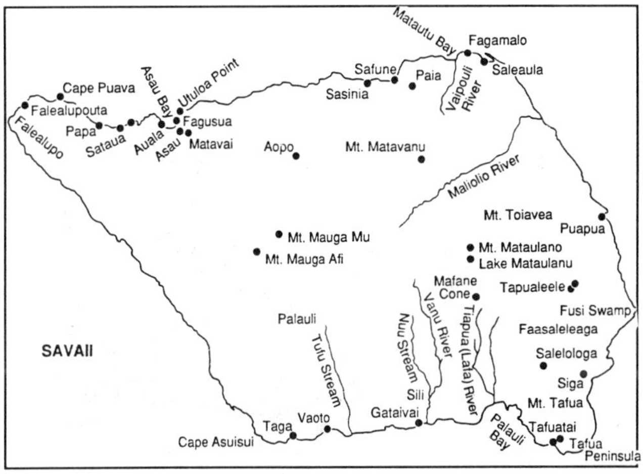

Figure 4. Site map for the island of Savaii showing the location of sites referred to in the text. Several site names have changed since their publication in geological texts early in the century. The new names are used in this map and where possible both the older and newer names have been included in text.

An erosional unconformity separates this old unit from overlying rocks; however, all of the overlying rocks sampled on Savaii are also normally magnetized. The younger rocks (Salani, Mulifanua, PuaPua, and Aopo formations) are believed to correspond to the Brunhes Normal Chron and be less than 700,000 years old (Keating, 1985a). No rocks of reversed polarity were found on this island in Keating's study. Tarling (1966) published a map of sampling sites,

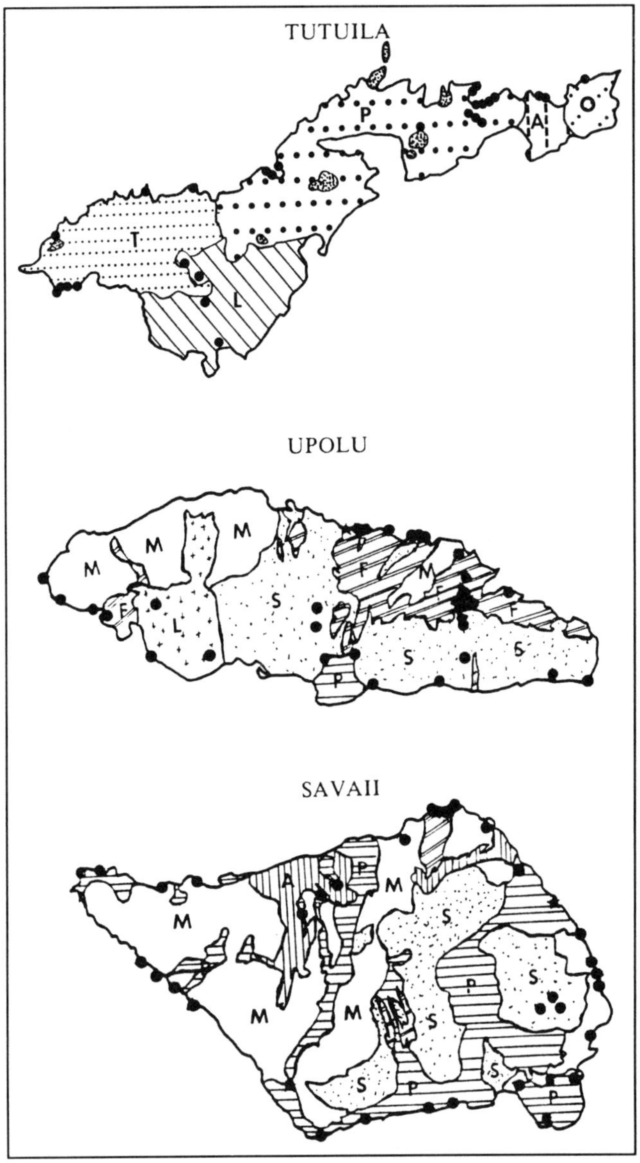

Figure 5. Geologic maps of the islands of Upolu and Savaii based upon the mapping of Kear and Wood (1959). The formation names are abbreviated in the figure (F = Fagaloa, S = Salani, M = Mulifanua, L = Lefaga, P = PuaPua, A = Aopo). Also shown is the geologic map of Tutuila based upn mapping by Stearns (1944). The formation names are abbreviated in the figure (T = Taputapu, L = Leone, P = Pago, A = Alofau, and O = Olomoana).

where oriented hand samples were collected. Reversed polarity was found on the western coast within a unit mapped as Mulifanua. His summary table however indicates that all of the igneous formations on Savaii are normally magnetized. Tarling (1966) suggested all the rocks on Savaii are less than 1 m.y.

The Salani volcanics are exposed in a wide swath extending north-south in central Savaii. These volcanics are moderately weathered, and a thick soil cover is present. Much of the Salani volcanic unit is drained by the Maliolio River, Lata River, Tiapua or Lafa River, and Faleata River. The large area in eastern Savaii centered around Tapueleele is mapped as Salani volcanics (Fig. 5). However, no large drainage systems like those elsewhere in Salani volcanics are present, which is suggestive that these volcanics are not the same age. Evidence from paleomagnetic studies of outcrops in this area confirm this observation (Keating, 1985a), and indicate that this may be a new (presently unnamed) volcanic unit.

Outcrops of the Mulifanua formation are extensive on Savaii. Most of the western half of Savaii is characterized by this rock unit, as well as large coastal areas on the north shore near Fagamalo and on the east coast from PuaPua to Saleleoga. According to Kear and Wood (1959), "the Mulifanua may be distinguished from the Salani volcanics largely on their lesser erosion and weathering. The lack of deep water courses and the angularity of surface boulders are the most important criteria." In western Savaii, the Mulifanua volcanics "rest on a weathered basalt that is considered to be Salani."

The outcrops of Mulifanua and Salani units appear concentrated on the axis of a spoke-like rift system developed on the eastern end of the island (Fig. 6). The PuaPua and Aopo volcanics, however, appear to originate from a later rift that is longitudinal to the island.

The PuaPua volcanics are distributed almost radially around the island of Savaii (Figs. 5 and 7). The PuaPua flows are extensive. The evidence from the historic erup-

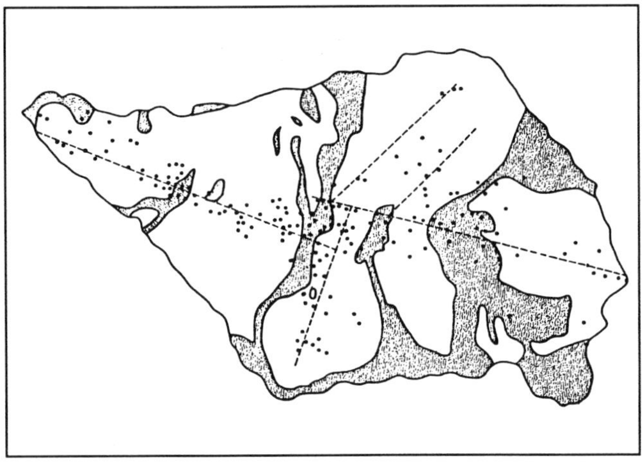

Figure 6. Map of Savaii showing the location of historic lava flows, cones, and the inferred rift zones. Based upon geologic map of Kear and Wood (1959). The dots represent cones and the dashed lines are inferred rift zones.

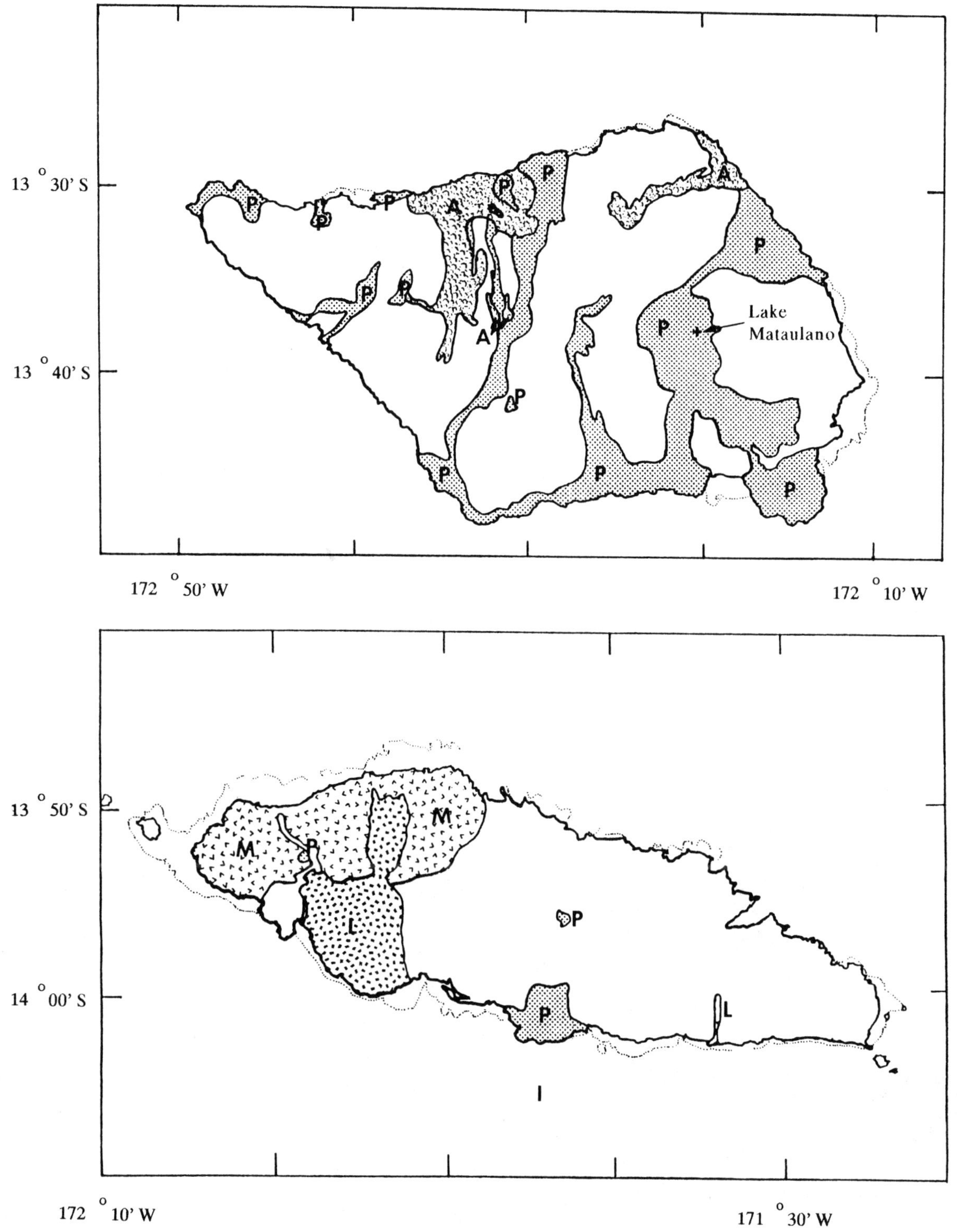

Figure 7. Maps of Savaii and Upolu showing the location of the young lava flows. The dotted lines indicate the location of coral reefs off shore, based upon mapping by Kear and Woods (1959) using aerial photographs. The formation names are abbreviated in the figure (F= Fagaloa, S= Salani, M= Mulifanua, L=Lefaga, P= PuaPua, A= Aopo).

tions indicates that these extensive lava fields can be generated in a matter of only a few years. The paleomagnetic evidence suggests that the PuaPua volcanic unit could be divided into two chronostratigraphic units. Flows from the Siga (Figs. 4 and 5), the Sili area (south coast), the Sasinia area (north coast), and Falealupo area (western tip of the island) are likely to be grouped into a new mapping unit (Keating, 1985a).

The Aopo volcanics (Fig. 7) are historic lava flows. The paleomagnetic studies of Keating (1985a) show one flow near Aopo, mapped as Aopo volcanics, giving directions similar to the PuaPua volcanic unit (senso stricto). Since the area, however, had been bulldozed and the flow material removed and used for road mantle, it seems likely the overlying Aopo flow material was removed and the PuaPua unit sampled rather than Aopo unit.

Asau Bay

Asau Bay is located on the northwestern coast of Savaii. The eastern margin of the bay (Utuloa Point) is bounded by PuaPua volcanics that are flat-lying pahoehoe lavas which filled the pre-existing lagoon and buried the reef. The remainder of the bay is formed by Mulifanua lavas. These lavas are highly vesicular nearly flat-lying basaltic lava flows. Near the water the lava flows are very fresh in appearance, lacking substantial weathering.

A well-developed coral reef exists offshore from Matavai (or Utuloa Point, depending on the map used) to Fagasua. There another PuaPua flow buries much of the lagoon (Fig. 8), while much of the area from Fagasa to Sataua is mapped as Mulifanua volcanics. The lack of a barrier reef westward suggests it is a younger unit, perhaps PuaPua. A barrier reef is reestablished near Papa.

Figure 9 illustrates a typical cross section of the reef structure buried by lava flows in the Asau Bay area. Figure 10 illustrates the structure likely at Matavai or Utuloa Point, where the most recent flows now bury the reef. If this area of Savaii has subsided substantially due to crustal loading, a complex structure similar to that in Figure 10 would develop. Menard (1986) reported similar areas of the island of Hawaii have subsided at a rate of 2 to 5 mm/yr. Local water well drillers have reported multiple repetitions of the basic reef and lava sequence occurring at Asau Bay. The multiple repetitions of reef material and lava flows have been observed in Hawaii in core samples from geothermal holes at depths 1615-1767 m on the southeast rift zone of Kilauea volcano.

Thomson (1921) pointed out that the geomorphology of Savaii is similar to that of Mauna Loa. The original observations by Thomson, however, result from arduous field work by Friedlander and R. Williams. Williams made a crossing of the island in 1907 from Matautu (north coast) to Turu (south coast) (assumed to be Tufu stream on modern maps) crossing the ridge which forms the "backbone" of Savaii east of its highest point. Friedlander traversed the ridge from west to east. Both Williams and Thomson used the altitude they gained along this ridge in order to view the cones and craters which mark the ridge. In the western part of the ridge (*tuasivi*, in Samoa) where the vegetation is poor, well-preserved volcanic cones can be observed. The central part of the ridge is covered by dense forest and contains cinder cones separated by old lava fields. To the east, numerous cones are present which can hardly be seen in the bush until reaching their base.

In one of these eastern cones, a small lake, Mataulano, is found. The crater rises 40 m above lake level. The altered ash of the cone makes the lake bottom impermeable. This cone would make an excellent site for paleomagnetic secular variation studies (using pol-

Figure 8. Aerial photograph of the southern coast of Savaii. In this photograph the lava flows cover the reef. Subsidence is occurring and new reef is growing on top of the lava flow. The cross section of the geologic structure woud be similar to that shown in Figure 10.

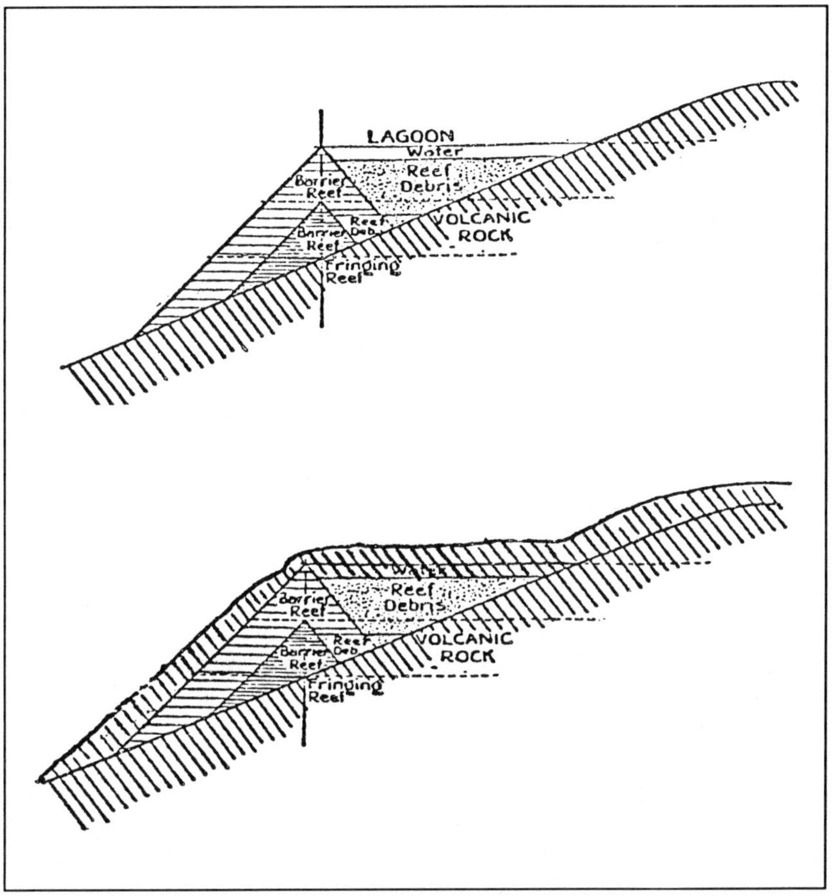

Figure 9. Cross section of a coral reef (top) showing the structure of a typical barrier reef on the margin of a Pacific Island, after that of Holmes (1945). An idealized cross section of a coral reef off Savaii or Upolu is shown in the bottom figure, where the lagoon has been filled by lava flows.

len for dating purposes). Mafane Cone south of Toiavea also contains a swampy flat bottom (shown as a lake in road maps) which would be suitable for similar work. The cone is densely forested.

On the south coast near Taga, between Nu'u and Tufu Stream are fresh-looking lava fields which appear to come from a neighboring small cone. These flows are likely to be slightly older than the written record, since some old Samoan songs are said to refer to these eruptions (Thomson, 1921).

On the peninsula of Tafua, there are small well-preserved cones. The outer slopes consist of well-stratified tuff. Thomson (1921) reports that at the bottom of these cones, a meter above sea level, is the entrance of a lava tunnel. These cones which dot the *tuasivi,* (prominent axial ridge) mark the fissure zone which is the area of concentrated

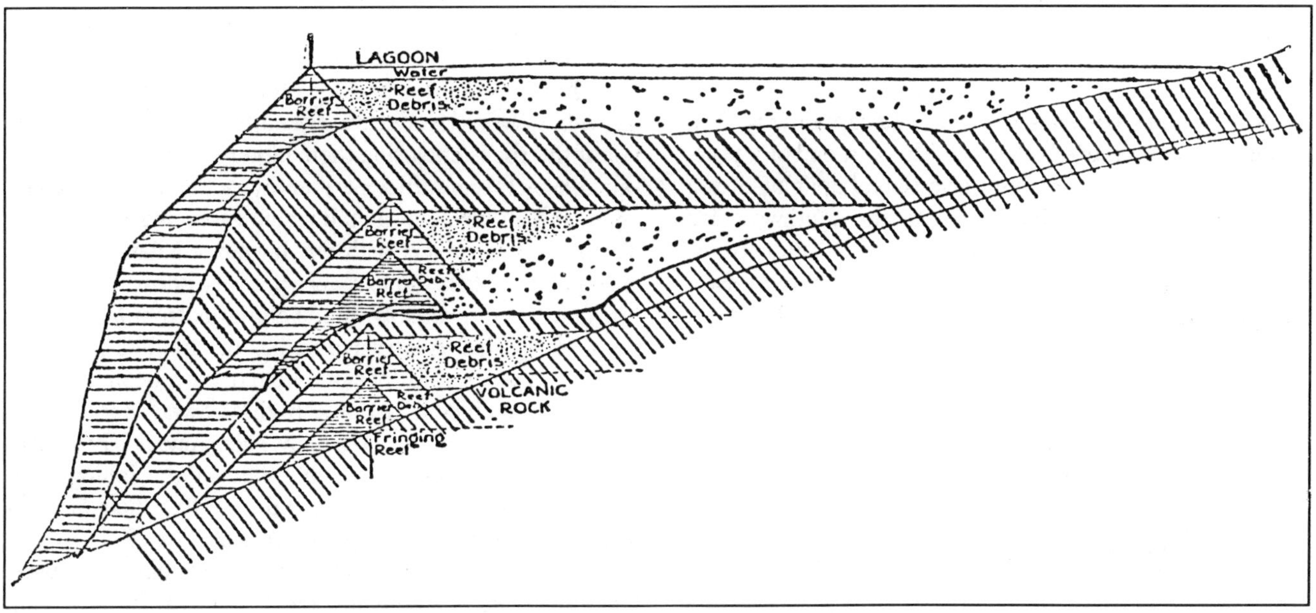

Figure 10. Cross section of the structure off Asau Bay, in northwest Savaii. There, several episodes of reef growth and subsequent burial by lava flows have occurred, making the geologic structures very complex. The sequence includes alternating layers of lava, coral, beach sands, and aquagene tuffs. This cross section was drawn on the basis of conversations with water well drillers who have drilled numerous wells on Savaii (personal communications, 1985).

volcanic intrusion and extrusion (Fig. 6). Dana, during the Wilkes Exploring Expedition in 1907, hiked to the top of Tafua Cone near Tafuatai and found a dry interior.

Apolima

Apolima is a small islet about halfway between Savaii and Upolu. An excellent view of the islet is gained by taking the ocean ferry from Upolu to Savaii. A single crater wall opens to the north. The bottom of the crater is only a few meters above sea level. The inner slopes of the crater wall slope at 30-40° while those of the outer wall are about 60°. The tuff is a compact brown palagonite tuff containing boulders of massive and porous lavas, fragments of coral, and marine mollusks (Thomson, 1921, derived from an account by Friedlander, 1910).

Thomson (1921, p. 58) notes that "owing to the impervious nature of the tuff, the island has continuously running springs and a small stream draining to the north across the bottom of the crater. It is this fact which renders the island habitable. Weber describes the tuff as palagonite, the glass lapilli being penetrated by pores filled with zeolites, calcite, and some sideromelane. There are inclusions of large olivines and microlites of augite, but no iron-ores, and lapilli of basaltic nature. A dense brownish stone, evidently one of the massive boulders mentioned by Friedlander, is described as a nepheline basalt, with large phenocrysts of rounded and somewhat serpentinized olivine in a groundmass consisting of brown titaniferous augite, some biotite and iron ores, a few needles of apatite, and much nepheline."

Manono

Manono lies between Apolima and Upolu (Fig. 11), within the barrier reef which lies off the west end of Upolu. Nulopa, to the west, contains a lava tunnel opening at sea level, but no crater. The slopes to the southwest are dominantly coral sands with occasional outcrops of vesicular basalt. The rocks at the summit are more massive but highly weathered. The springs on this islet are brackish.

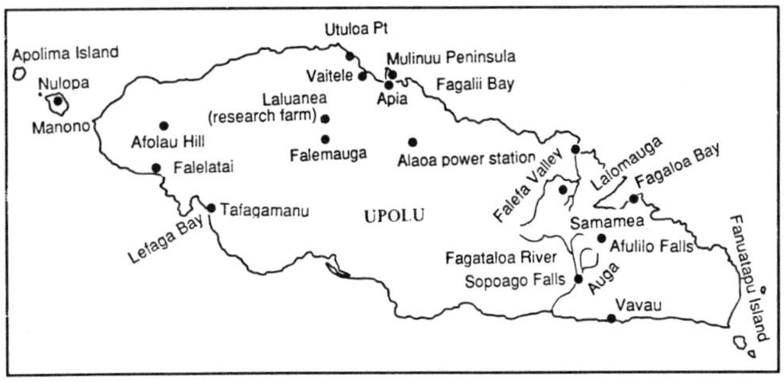

Figure 11. Site location map for localities in Upolu referenced in the text.

Upolu

Like Savaii, the island of Upolu has a prominent *tuasivi* or axial ridge, which appears to be the major axis of fissure eruptions. Cones are distributed abundantly along this central axis of the island. In the northwestern part of the island are a series of well-formed eruption cones, and neighboring lava flows that appear fresh. Afolau Hill, however, is composed of a finer-grained, light colored, less vesicular rock.

The south shore has numerous steep promontories of volcanic rock. Along the eastern shore similar outcrops exist. On the southeast coast there are several bays with steep-walled valleys and beautiful high waterfalls coming from high valleys. Thomson reports that a dense basaltic dyke cuts through a coarser, lighter rock of andesitic appearance (trachyte?) at Samamea.

Thomson (1921) also provides an excellent description of lava tubes on Savaii. He states that "the formation of lava-tunnels in the Matavanu eruption has already been mentioned. Such tunnels are a characteristic of basaltic eruptions of the Kilauea type, and have an almost circular cross-section while the flow is active, but when the supply of lava ceases that in the tunnel drains out, leaving only a little in the bottom, so that a cooled tunnel has an almost circular section except for a flat bottom. In any tunnel, the height and breadth are remarkably constant, except when branching takes place. Stalactites (composed of fused lava formed by the burning of the inflammable gases discharged from the lava flowing below) may often be observed hanging from the roof, and young tunnels also have stalactites on the walls, and coatings of soluble salts, chiefly sulphates."

Thomson (1921) suggested that the hollow spaces of lava tubes, "would lend themselves admirably to ore-deposition if the rocks enclosing them either contained the more valuable metals in sufficient quantity, or the area later came under the influence of metallogenic gases or solutions. In discussing this subject with Dr. Jaggar of Kilauea, he surmised that the lava tunnels were not very permanent structures and were probably filled with lavas from later eruptions, giving rise to intrusive bodies of pipe-like form. In Samoa, however, this does not seem to have been the case, and in the extinct lava flows, both of Savaii and Upolu, lava-tunnels are of fairly frequent occurrence. They have, however, been considerably modified by the falling in of rock from the roof and sides, and have lost their original nearly circular and very regular cross-section."

Two large caves in Samoa are known to have been used for refuge, one at Falemauga on Upolu and one near Paia on Savaii. The one on Upolu has not been inhabited in recorded history; however, the floors of the cave are stacked with rocks that create sloping terraces similar to the terraces of contemporary typical Samoan homes, called *fales*. In addition, charcoal, marine shells and adzes have been found in this cave. Thomson suggests that up to 100 individuals could have slept there.

Von Bulow (1906) describes the Savaii cave as "the cave of the non-fighting tribe." At the time of warring between the people of Safune and the rest of the island, about the end of the eighteenth century, the tunnel was supposed to have been used as a refuge with food stored there.

At Safune, Thomson described a pool formed by the downbreak of a blow-hole a short distance from the sea. There he describes a pool of water which is affected by the tides. The tunnel lies roughly a meter below the surface and opens at the seaward side, giving rise to a bathing pool. At low tide, the pool has fresh water. Similar pools are seen around Savaii and Upolu. It is interesting to note that similar pools are present on the Puna and Kona coasts of the island of Hawaii.

Another interesting cave is reported in old lava at Tapualeele on the east side of Savaii. Here, in a tunnel which has fallen in over some distance, a stream bed has formed that, when followed, ends abruptly in the tunnel. The mouth of the stream is occupied by a deep pool. Evidently you must swim the deep pool to proceed further in the tunnel, and therefore Thomson did not explore it further. He noted, however, in another tunnel nearby, that the floor of the tunnel was covered with a combination of guano (from local birds) and the chitinous remains of insects. Elsewhere, Kear and Wood (1959) reported other tunnels free of guano and containing only insect remains. The bird population in Savaii is still quite a varied one, which stands in marked contrast to most Pacific islands. This may well reflect a much less dense population of animals and humans on Savaii.

Lava Slopes

The lava surface slopes on Upolu are generally moderate, between 1° and 15°, and slope seaward. These slopes flatten as they reach sea level, where the lava flows have spread out onto the lagoon and reefs. Daly (1924) observed that the lava slopes on Tutuila ranged from 0° to 20°, with an average of about 10°. He suggests that the lava flows were extruded from fissures nearly parallel to the axes of the island. Most flows dip away from these central fissures and sea cliffs cut their down-dip edges. Isolated central vents can be associated with two tuff cones at Steps Point, Aunuu Island, the eastern offshore vent, offshore vents near Coconut Point, and off Taputapu islet on the western coast of Tutuila.

Streams and Erosional Features

Kear and Wood (1959) used stream erosion to measure the age of rocks in Western Samoa. They summarize with the following observations:

- Fagaloa Volcanics - youthful graded valleys
- Salani - amphitheater-headed canyons; deep, poorly-graded valleys
- Mulifanua and Lefaga - dry water courses
- PuaPua - few, short, weakly eroded gulleys, which give rise to small springs where the lava thinly covers older rocks (e.g., Siga Springs in southeast Savaii)
- The streams on Fagaloa rocks appear perennial. The stream patterns are dendritic. The initial radial drainage patterns appear to continue in younger rocks since the older volcanics and younger units seem to have similar drainage divides.
- Many valleys appear to be drowned, e.g., Fagaloa Bay, Falefa Valley and several valleys on the north coast of Upolu.
- The perennial streams on Salani rocks produce three landforms: widely spaced, deeply entrenched gulches with several falls (e.g., Fagaloa or Salani river); complex multiple gully systems; deep amphitheater-headed canyons.

Corals Reefs in Western Samoa

Stearns (1944) reported that living reefs around Upolu form fringing and barrier reefs. The fringing reefs range from narrow shelves to shelf flats 2 km wide. The fringing reefs consist of coral sands, shells, and detritus that in large part are bound together by algae growing on the surface. The fringing reefs tend to be exposed at low tide.

The barrier reefs are intermittent along stretches of the coast (Figure 7). In some places, they have been buried by lava, in other places rivers discharge into the sea and the turbidity is too high for corals to survive. In other locations, the flanks of the volcano have probably subsided as a result of fault slumps which drown the reef. A lagoon separates the barrier reef from the fringing reef. Stearns (1944) argues that "living barrier reef is not found along coasts composed of Pliocene rocks. Scarcely any fringing reef exists either... Barrier reefs are absent from the latest lavas also." Stearns suggests that Upolu was submerged so rapidly that reefs only grew to the surface where the Pleistocene lavas were gently sloping. The rapid submergence was probably the combined result of a rapid sea level rise and island subsidence.

Stearns reports remnants of emerged reef at 1.5 m above the sea on the reef flat at Fagalii Bay. Wave-cut flats at 1.8-2.4 m above sea level are also reported. Matsushima and others (1984) collected sediments from shallow cores in Lefaga Bay in an attempt to date the sediments. The samples analyzed were not autochthonous coral. The age determinations suggest they were Holocene deposits. Sugimura and others (1988) reported on additional studies of sea level change based on outcrops and drilling in Western Samoa. Several wave cut notches in lava flows and occurrences of emerged beach rock were documented.

Kear and Wood (1959) have suggested that a correlation may exist between the age of the volcanic rocks and the distance offshore to the barrier reef. They suggest the following relationships:

1. Fagaloa- no living barrier reef or scarce fringing reef,
2. Salani and Mulifanua- wide reef,
3. Lefago and younger- little or no reef.

Off the coasts of Salani and Mulifanua outcrops, the reefs tend to be barrier types where the lava slopes are gentle and fringing reefs form on steeper slopes. The extensive reefs around Savaii are in the process of being buried. In the area betweeen Palauli and Faasaleleaga, the lava flows have now built out beyond the old barrier reef.

Thomson (1921) concluded that many areas of the Upolu coast lines had been depressed or downfaulted below the levels of coral growth causing the steep coasts and absence of bordering reefs. Thompson suggested downfaulting was the best explanation of the steep-walled harbors like Fagaloa Bay, particularly since valleys are absent around the bay, and no crater-like features are observed.

Sediments

The oldest sedimentary rocks in Western Samoa are the alluvium at Lalomauga on Upolu. Kear and Wood (1959) report that a tributary of the Falefa Stream was blocked by Mulifanua lavas, and was subsequently filled with alluvium. The alluvium is preserved as terraces up to 9 m above stream level. Kear and Wood infer the alluvium is in part contemporaneous with the Pua Pua volcanics.

Sand Beaches

Few beaches in the Western Samoa islands have abundant sand. Thomson (1921) suggested that tsunamis (tidal waves) may remove the sand. (This suggestion is not consistent with observations of tsunamis which struck Hawaii in 1946 or 1960). Two factors obviously affect the formation of dunes: supply, and wind to transport the sands. On most islands there are sufficient breezes to move sand. Thus it is more likely a problem of there being too little sand. Sand off the coasts of most islands is a product of erosion of reef material. In Samoa, the barrier reefs may provide a sufficient barrier to the sea that little wave energy reaches shoreward of the reef; thus little mechanical erosion of reef material occurs inshore and little sand is produced. Alternately, sand may be produced continually and lost down the reef slope at the breaks in the reef. Beach sands are also limited within the Hawaiian Islands, where these processes are well documented.

Coral sands, soil and boulders form low cliffs, commonly in bays, about 1.5 m above sea level. Kear and Wood report a radiometric date of 2,300 years from a sample of this sand. Present day beaches of coral sand are locally cemented with calcite. The cemented beds, referred to as beach rock, dip seaward at about the same angle as the beaches.

Stearns (1944) reported that soundings off Upolu indicate a submarine shelf exists 46-55 m along the shore and 73-91 m, at a distance of 1.5-5 km offshore, which is similar in size and width to the drowned barrier reef on Tutuila. Stearns also correlates the submarine slope break to the shore line at the time of the "great erosional period" at least 182 m below the present shore line. Stearns concludes the drowned barrier reef rests on a thick section of marine and land deposits on an older platform.

Emerged Shorelines

Traces of a Recent +1.5 m stand of sea level are present in various places in the Western Samoan islands. These emerged shorelines form a strip of coral sand upon which most coastal villages occur and form spits (e.g., Mulinuu peninsula). In fact, the contact between these Tafagamanu Sands and the PuaPua volcanics is well exposed at the north end of the beach near PuaPua village (see Kear and Wood, Fig. 20). This +1.5 m stand of sea level is commonly observed in the Pacific and is the result of a postglacial sea level rise accompanied by a warmer climate.

Traces of a +4.5 m stand of sea level are also found in Samoa. The associated deposits of Nuutele Sand occur at Gataivai in southeast Savaii. At this site, the PuaPua volcanics have a prominent bench cut into them at 4.5 m above sea level (Kear and Wood, Fig. 21). An overhanging cliff is found at the bench. At the foot of the low, irregular cliff is a small berm of fine, lightly cemented gravel, containing worn coral. Also exposed at the site is an ash bed within the basalt at about 4 m above sea level containing fragments of coral. This +4.5 m sea level is believed to be post-glacial because of its occurrence on reef covered PuaPua volcanics; also, because of the correlation with similar benches around the Pacific.

Kear and Wood (1959) stated that there is possible evidence of a +10 m sea level in the Vini Tuff which could have corresponded with the last interglacial. They also report benches near Fagaloa Bay southeast of Falelatai between 39 m and 60 m, which could be traces of a Tyrrhenian (last interglacial) or Milazzian level (penultimate interglacial).

On the east coast of Savaii a wide barrier reef 22 km long, as well as a fringing reef, is found. On the north shore, off Safune, a narrower reef occurs. Also, a narrow barrier reef occurs off Auala in northwest Savaii. The ends of these reefs are buried by younger volcanics. Thus, the remainder of the coasts are rock-bound. The extensive reefs around Savaii are in the process of being buried. In the area between Palauli and Faasaleleaga, the lava flows have now built out beyond the old barrier reef.

Soils

Hamilton and Grange (1938) conducted the studies of the soils of Western Samoa. The studies along with those of Seelye and others (1938) show that the soils are relatively shallow, heterogeneous with frequent stones and boulders, and are rich from iron oxides derived from the basaltic lavas.

During the 1950's, a soil survey was conducted on both Upolu and Savaii (Wright, 1963). A local classification system was subsequently devised based on the soil survey of Hawaii by Cline (1955); the system has been developed and reported by Morrison and others (1986). Schroth (1970) reports a number of soil analyses. A study of soils on the Laluanea research farm of the University of the South Pacific has been reported by Morrison and others (1986). Kear and Wood (1959) suggest that a good correlation exists between the depth of soil and the age of pahoehoe flows. On Savaii most of the historic flows have scarcely formed soil. The deepest soils occur on the Fagaloa volcanics in Western Samoa.

Alluvium

Alluvium from the mouths of streams was analyzed by Bramlette (1926). Most of the material is rock fragments, fine basalt, trachyte, volcanic glass, and grains of magnetite, olivine, and feldspar. Bramlette reports that all of these are remarkably fresh and "show no evidence of weathering." A small percentage of calcareous or siliceous organisms is also included; the material is poorly sorted.

Offshore Manganese Deposits

The land area of the Samoan islands is roughly 3,000 square kilometers while the offshore area within the Exclusive Economic Zone or EEZ (the area within 200 nautical miles of the islands) is about 150,000 square kilometers (Exon, 1982). In order to evaluate potential manganese (Mn) nodule deposits, sampling programs have been conducted in the basins surrounding the Samoan islands. Manganese nodules were found to be rare, and to be of low ecomonic grades. Exon (1982) ascribes the rarity to the high sedimentation rates caused by volcanic input of debris, which is detrimental to Mn nodule formation. Exon also concludes that the low grades of nickel and copper observed reflect the low production of plankton generally associated with deposits. Because the seafloor is well above the calcite compensation depth, calcareous plankton are not dissolved and do not release metals to the seafloor where Mn concentration takes place. Manheim and Lane-Bostwick (1989) point out the importance of the chemical character of the water mass. They have found that cobalt distribution within Mn crusts reflect regional highs and lows in the distribution of hydrothermal fluid discharge in the ocean. Commonly, at moderate depths (1-2 km) Mn crusts form on the upper slopes of sea mountains. In general, on relatively young seamounts (a few m. y.) these crust formations are very thin and do not constitute economic resources. Studies in the Hawaiian chain (McMurtry et al., 1986) show thin Mn crusts of little economic value compared to the thick Mn crusts present on Mesozoic seamounts within the Central Pacific.

Two seamounts of possible Mesozoic age are present within the EEZ of Samoa, Machias Seamount and Rose Atoll. Machias Seamount is situated southwest of the Samoa Islands on the flexure arch of the Pacific plate near the Tonga trench. Early work on the seamount included dredges of the shallow summit (Hawkins and Natland, 1975). Cobbles, gravel, pahoehoe fragments, and coral were found in the dredges, indicating the crest of the seamount was formerly at sea level. A K/Ar age on a phonolite yielded an age of 940,000± 20,000.

Machias Seamount was surveyed using side-scan sonar as part of a Mn crust resource assessment. The images produced by the sonar shows that the seamount has been dissected by faults (Coulbourn and others, 1989). The faults (Fig. 12) are parallel to the Tonga Trench and the flank of the seamount has dropped downward into the trench. The abundant faults yield a very complex bathymetry that would make mining on this seamount extremely difficult if a resource were present. Combining the evidence for a young age for the seamount (Hawkins and Natland, 1975) with the complexity of the bathymetry, the prospects for economic deposits of Mn crusts on Machias Seamount do not appear large at the present time. The age of Rose Atoll is uncertain. Without additional information, a resource assessment of the Mn crust is not possible.

Historic Volcanism in Western Samoa

The island of Savaii has experienced wide-spread volcanism within historic time. The flows associated with historic and geologically recent volcanism are shown in Figure 7. Thomson (1921) reviewed the volcanism and much of his description of the activity is included here.

Mauga Afi (about 1760)

Thomson (1921) writes,

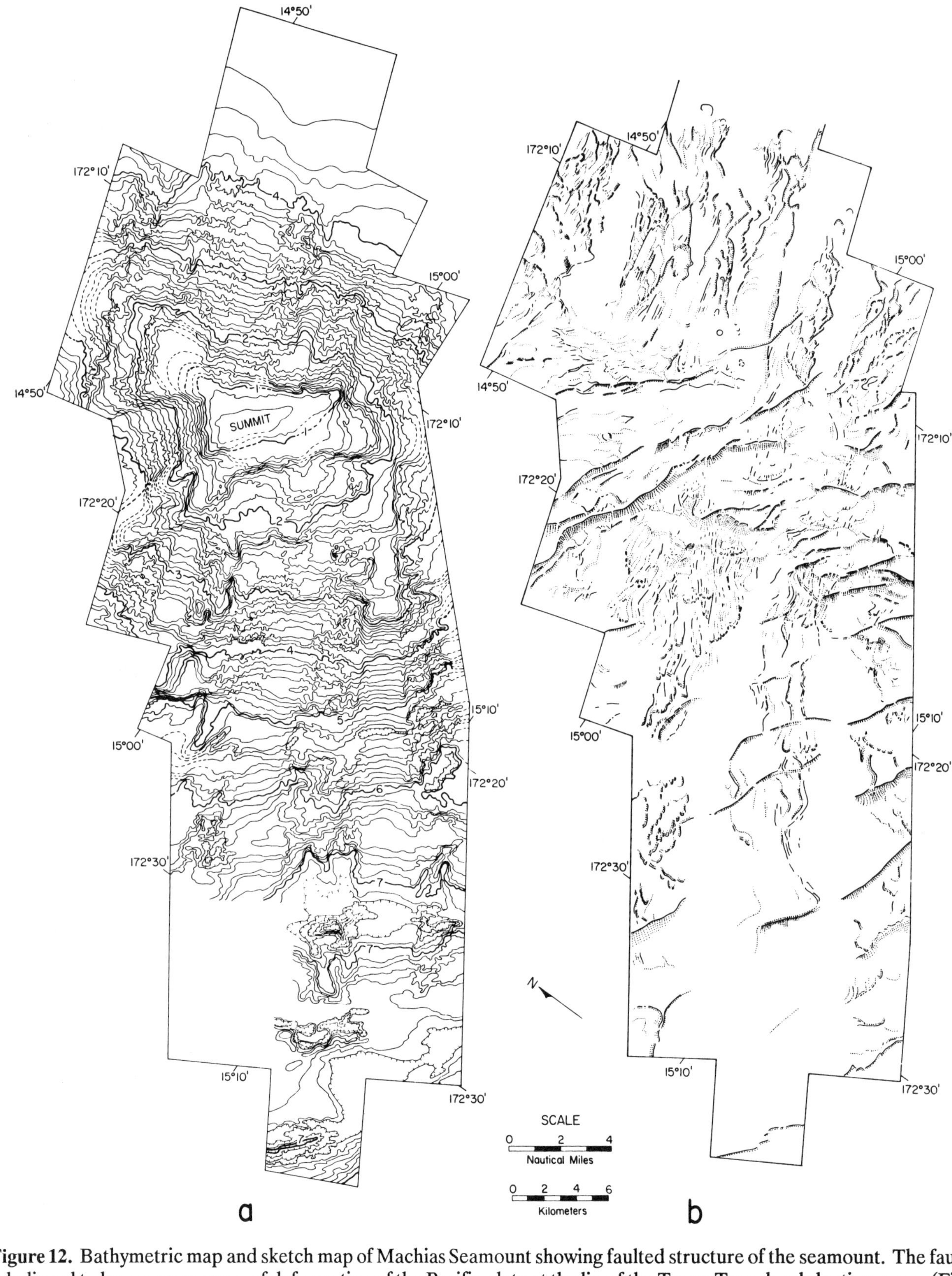

Figure 12. Bathymetric map and sketch map of Machias Seamount showing faulted structure of the seamount. The faulting is believed to be a consequence of deformation of the Pacific plate at the lip of the Tonga Trench subduction zone. (Figure from Coulbourn et al., 1989).

"According to von Bulow and Tempest Anderson, the Samoans preserve a tradition of eruptions "about one hundred and fifty years ago" (A.D. 1760) that gave rise to a rugged and very extensive lava field, called O le Mu, between the villages of Asau and Aopo, on the north side of Savaii. This field is said to be more extensive than that recently created by the Matavanu eruptions, but is shown on the German Admiralty Chart with a length of eight miles and a half, and a breadth of only one mile and a half. It is still comparatively unaltered, preserving the "wrinkled, knobbed, ropy, and tapestry-like folds, and the general characteristics of the 'pahoehoe' type of lava flow" (Jensen), and is free of bush, which is rather surprising in view of the growth that has taken place already on the Matavanu lavas. Jensen believes the flows came from Mauga Afi, a crater 5,249 ft (1,600 m) high, on the western slopes of the main ridge of the island, and ascribes to the same source as a flow on the southern side of the ridge.

"Aopo was surrounded and partially destroyed by these eruptions, and other villages were totally destroyed. The present village of Aopa occupies what is known in Hawaii as a kipuka, an area which a lava stream has flowed around and left as an island.

"Friedlander describes the cone of Mauga Afi as a fairly steep slag cone, 100 m high, with an elliptical crater, elongated east and west, 70 m deep. The western margin of the crater is broken, and here the lava flowed out in a westerly direction, turning to the north after a short distance."

Mauga Mu of Aopo (1902)

Thomson (1921) states,

"On October 30, 1902, eruptions commenced at a spot about three km northeast of Mauga Afi and some six and a half km southwest of the village of Aopo. They were preceded and accompanied by violent earthquake shocks, and for three weeks great detonations took place and flames were reported by Pere Mennel. The volcanic activity ceased after a few months. No previous crater was known at this spot, but according to Friedlander (1910) two hills were formed with an east-west extension, the larger with three well-formed oblong craters, and the smaller in the shape of a horse-shoe. From both cones, lava streams issued and flowed 1 to 2 km in the direction of Aopo. Lava also welled out from a fissure on the side of an old crater to the south, and partially filled it. Wegener (1903) states that a first crater, formed on October 30, furnished only lavas, but a second, about one km to the west, was formed on November 1 and was explosive for a short time.

Jensen describes the material of the 1902 eruption as a'a, consisting of fragments of all sizes, from cinders the size of peas to blocks many meters in diameter piled in wild confusion, and states that the lava is vesicular and scoriaceous. Friedlander (1910) describes the latter as black and metallic in lustre, and very light and porous. Weber, who studied Friedlander's material, describes the lavas as a light porous slaggy form, referred to as feldspar basalt consisting of phenocrysts of olivine, augite, and plagioclase in a brown glass matrix with magnetite. Mauga Mu is shown on the German Admiralty Chart as "Parasit 1902 referring to its position as parasitic on Mauga Afi."

Matavanu (1905-1911)

Thomson (1921) states that,

"Just as Mauga Mu is parasitic on Mauga Afi, so Matavanu may be described as parasitic on Pule, an old crater occupied by a lake and lying a few km to the north of the main mountain tops of Savaii and a little over 2,000 ft [600 m] high. Before the eruption, the place which is now the crater of Matavanu was a sort of elevated plain surrounded by mountains, about 11 km south of Matautu. The eruptions began on August 4 1905, and at first were of an explosive nature, but no severe earthquakes were experienced. From September 2 to 4, molten lava poured out and the flow advanced 3 km. The lava flowed at first to the northwest, filling up the upper ends of some valleys draining to Matautu Harbor. Later it flowed both to the west and the northeast, in the latter direction following a tortuous valley draining to the sea several kms east of Matautu; and the lava itself reached the sea in December, 1905, at Foapaipai, filling up the lagoon between the coral reef and the coast and turning westward along the reef. Early in 1906 there was a great increase of activity; the lava destroyed the villages of Salago and Saleaula to the west, and it also flowed east and overwhelmed the villages of Taputapu and Maleola. The distance from the crater, following the winding and turnings, was about 13 miles [21 km], and the seafront covered was nearly 9 miles [15 km]."

The villages destroyed by volcanism were not rebuilt; their names do not occur on modern maps of Savaii and they are not shown on the locality map (Fig. 4).

According to Anderson (cited by Thomson, 1921),

"the large, fresh lava streams soon got crusted over on the surface with solidified lava, and the liquid lava continued to flow underneath. Even at the crater it seldom flowed over the lip, but generally entered holes and tunnels in the sides and flowed underground. The lava field thus became honeycombed with channels of liquid or pasty lava, which occasionally came to the surface and flooded it with fresh sheets of lava; at other times the surface frequently floated up and was raised by the intrusion of fresh lava underneath, so that what had previously been the course of the valley now became the highest part of the field. Mr. Williams thinks that the lava must be in places 400 ft (120 m) thick."

Thomson (1921) states,

"where the coast was bordered by a coral reef, the lava quickly filled in the lagoon and thus extended the coast line. For a stretch where it was previously "ironbound" i.e., formed of old lava not protected by a coral reef, as at Asuisui, the lava flowed directly into deep water, and did not materially alter the outline of the coast. The flow into the sea continued, with only a day's intermission, from 1905 until 1910 or 1911. An immense amount of lava disappeared in this manner, estimated by Friedlander as many cubic kilometers, and four times the volume of that forming the visible lava fields (Sapper, 1911)."

"When the discharges into the sea were most active, Anderson related to Thomson (1921) that explosions were almost continuous, and the whole was obscured by clouds of steam from which fragments of red-hot lava and showers of black sand were seen to fall. This black sand formed beds capping the lava. Angenheister, in 1909, remarks that the explosions took pace every five to ten minutes. When the lava was flowing in lesser quantity, Anderson notes that explosions were much less noticeable, and the lava extended itself into buds and lobes, and cooled in the form of a 'pillow lava'."

"The crater of Matavanu has been described in various stages by Jensen (1907), Angenheister (1909), Anderson (1910, 1911, 1912), Friedlander (1910), and Grevel (1911). The crater is a "broad slag-covered cone, with a steep-walled crater about 250 ft (75 m) deep, narrowly elliptical in form, and elongated about a quarter of a mile (0.4 km) from south southeast to north northwest. The northern end of the crater is partially fallen in, and the depression is extended in this direction by two fallen-in tunnels for another 100 yards (90 m) or so. Farther to the northeast, there is another disconnected downbreak, with a tunnel showing at the bottom in the northern end," according to Thomson (1921). Angenheister, noted in 1909, that the lava passed out through two tunnels to the north and one to the south, though there was no flow to the surface in the latter direction.

"The eastern side of the crater is composed of red ash below, grey ash above, capped by a 5 ft (1.5 m) layer of lava, which is covered on the outer surface by spatter slag. The western side shows lava flows with red ash between and above them, and the same 5 ft (1.5 m) layer of lava on the top. At the northern end, lavas come in wedge-shaped fashion, lying unconformably on the ash beds. Sulphur fumaroles are still active, both near the bottom at the north end and round the top of the talus slopes at the south end."

The sulphur crystals are of the unstable monoclinic variety. In the 1980's, sulfur from the cone was being gathered by natives for medical purposes.

"The lava field from the base of the crater to the sea has an average slope estimated by Anderson at about 6°. The lava is mostly of the typical pahoehoe type, smooth over a wide area, though very irregular in detail, with typical small cracked domes with corrugated and ropy surfaces. Near the crater, areas of rough, broken lava simulating a'a are not uncommon, but they are made of broken pahoehoe, and I saw no typical a'a. Anderson's explanations of the origin of the a'a type of lava are diametrically opposed to those of Jaggar, and the reason lies probably in the fact that he mistook areas of broken pahoehoe for a'a. For a more detailed description of the lava surface, the descriptions of Jensen (1907) and Anderson (1910, 1911, 1912) should be consulted". Furthermore, Thomson (1921) reports, "it is worthy of note that the growth of vegetation on the lavas of 1905 and 1906 greatly exceeds that on the 1894 or even the 1860 flows of Kilauea than on the Matavanu flow presumably because of the moister climate."

"The lavas of Matavanu have been examined microscopically by Jensen (1908) and Weber (1909), but no analyses have been made. They are olivine-rich feldspar basalts with titaniferous augite, and generally with a considerable proportion of glass. The chilled surfaces of the pahoehoe flows are typical tachylites. Jensen states that the Matavanu lavas are richer in iron ores than any of the earlier flows," according to Thomson (1921).

GEOLOGIC MAPPING UNITS OF WESTERN SAMOA

Kear and Wood (1959) produced the original geologic maps of Western Samoa. Their division of geologic mapping units was based upon geomorphology. They identified six geologic units and assigned them ages based upon relative weathering of Holocene to Pliocene.

Fagaloa Volcanics

The stratigraphically lowest, thus oldest, unit mapped is the Fagaloa volcanics. This unit is deeply weathered. The mapped units include "two somewhat different landforms." Kear and Wood (1959; p. 36) report that, "one landform, possibly the older, has no part of the original cone form remaining and forms steep high mountains with slopes up to 50°, rising as inliers above the gentle slopes of the later lavas. It includes dykes, which form bold vertical cliffs and steep narrow ridges, often several hundred feet high. The other landform, though steep locally, includes ridges that descend steeply seawards and could be the surviving parts of the original cone surfaces. Some of the gently sloping ridges have narrow flat tops on which the soil is strongly leached, acidic, and now supports little more than wiry grass. The steep offshore slope has prevented the formation of coral reefs over much of the Fagaloa coastline, and where a reef is present, it is fringing."

Salani Volcanics

The Fagaloa volcanics are overlain by Salani volcanics. Valleys and gorges cut deeply into this volcanic unit. This morphological characteristic was used to map the unit aerially. Often the Salani valleys are filled by younger lava flows.

Salani volcanics consist of low angle lava flows that most frequently can be traced to the cone or crater from which they originated. The cones themselves are usually weathered to the point that they are breached or contain sufficient decomposed material to seal the bottom of the crater and form crater lakes or swamps. The surfaces of flows are often deeply weathered, and soil is often more than 10 cm thick. On the distal flows, over 30 cm of soil has formed. Relict boulders are often observed to show onion skin weathering. The gorges formed by the rivers on Savaii and Upolu cut deeply into Salani rocks. Often these gorges are filled by younger lava flows. Valleys are common in rocks of the Salani formation and are a distinctive mapping feature. In many places, these valleys contain permanent rivers. Elsewhere, the valleys can be dry much of the year.

Typically, a barrier reef exists offshore of Salani outcrops. However, along the eastern part of the south coast of Upolu, cliffs are found at the coast and the reef is close to shore. Near Vavau on Upolu, blowholes and sinkholes are present, probably resulting from the collapse of blowholes.

The Salani rocks are fine grained gray to black porphyritic basalt which grade upward to vesicular basalt with a rubbly a'a surface. The thickness of flows generally average 1-2 m. The more weathered outcrops and presence of large green olivine phenocrysts were used in the field mapping to distinguish these rocks from the lithologically similar but younger Mulifanua rocks (Kear and Wood, 1959).

Rocks of the Salani volcanics include picrite basalts, olivine dolerite, and basalts showing late-stage deuteric alteration and zeolitization. Olivine basalts constitute more than 50% of the Salani volcanics sampled (Brothers, in Kear and Wood, 1959).

The Salani volcanics are differentiated from the Fagaloa volcanics by their less-weathered appearance, less mature landform (partial cones and craters), the absence of dykes, less common occurrence of large olivine nodules and phenocrysts, absence of andesites and trachytes, and the presence of a well developed reef offshore. These lavas fill valleys formed in Fagaloa rocks, and top Fagaloa ridges and terrain. A well-developed soil horizon separates the Fagaloa and Salani volcanics at the Alaoa power station on Upolu. This deep soil horizon and the deep gorges cut within the Fagaloa are thought to reflect a substantial erosional interval. Barrier reefs are generally present offshore of Salani volcanics.

Mulifanua Volcanics

Areas mapped as the Mulifanua unit are characterized by dry shallow valleys and gullies, where flowing water is rare. The weathering in this unit is "moderate." The flows descend from "well-formed" cones, and are "almost uneroded." Boulders on the surface of the flows are angular and onion-skin weathering has not developed. In distal portions of the flows, weathering can reach 30 cm. These flows sometimes occur in valleys cut into pre-existing rocks.

Lefaga Volcanics

The Lefaga volcanics are the next stratigraphically higher unit. According to Kear and Wood (1959), "The surface expression is similar to that of Mulifanua rocks, except that a'a at the ground surface is more common, the lavas appear to have flowed out into the lagoonal area, and the reef is relatively close inshore. Onion-skin weathering occurs locally, due possibly to the high feldspar content."

PuaPua Volcanics

The PuaPua volcanics consist of young basalt flows and perfectly preserved cinder cones. The rocks have very little weathering and the soil is thin. The ropy pahoehoe flows cover pre-existing flows, fill gorges, and flow over pre-existing erosion scarps. The flows have low dips and form relatively even surfaces. Extensive swamps have developed in low depressions. Locally, where slopes are steep, the flows consist of a'a flows. Reefs are not developed off shore of PuaPua basalts. Instead, the basalts fill lagoonal areas and cover pre-existing reef.

Aopu Volcanics

The final unit defined and mapped by Kear and Woods is the Aopo volcanics. These are "ropy, vesicular, porphyritic (feldspar and olivine) basalts, with but little a'a. Where the a'a flows occur, it is blocky, loose and appears to have been transported on the surface of the flows and to have accumulated along the margins." These are the historical lavas.

Hydrology- Western Samoa

Kear and Wood (1959) attempted to establish a framework for understanding the ground-water gradients in Western Samoa. They attempted to map similar volcanic units, on the basis of the degree of erosion, in order to use the known depth of the water table (established by drilling) in a limited number of sites, and extrapolate away to areas where the water table was unknown. They suggested very low ground water gradients "should be suspected in the younger volcanics that cover the majority of the Territory." They suggested that wells be drilled to sea level to obtain water, because "apart from areas of the oldest rocks (Fagaloa or older Salani), where the streams are virtually permanent, the water table will be close to the surface only near the coast, or close to "buried hills" of older rocks".

The Fagaloa volcanics are the oldest rocks exposed in Western Samoa and are severely weathered. Kear and Wood (1959) state that the ground water "would generally be shallower, and available in smaller quantities than from younger less weathered formations . . ." Where Fagaloa volcanics have been buried by younger lava flows, the water trapped above these older volcanics would provide a shallow water source.

The Salani volcanics are typified by large rivers which carry water to the sea. Significant losses of water occur along stream courses due to seepage. Inland springs occur where these rocks are covered by thin lavas of younger rocks.

The Mulifanua rocks absorb water readily and rarely is surface water present. Kear and Wood (1959) note that the scoriaceous nature of parts of the flows provides abundant water. Wells however must be drilled to sea level to obtain water.

Water obtained from the Lefaga, PuaPua and Aopu volcanics is characterized by high iron content. In general, the areas covered by these rocks have the poorest water supplies in Samoa except where the units are only a thin cover which overlies older rocks. On the island of Apolima, a surface stream emits water from the Vini Tuffs. The stream never dries. On other islands, where the Vini Tuff is present, a non-brackish water supply is uncertain (Kear and Wood, 1959).

Geological History of Western Samoa

Kear and Wood (1959) outlined a possible geologic history for the islands of Western Samoa. They postulate two massive volcanoes built on a fissure or fissures trending 11° . The two volcanic piles merged undersea, but emerged above sea level as two islands, Savaii and Upolu. The volcanoes would have first appeared as broad shield volcanoes, similar to Mauna Loa, Hawaii. The shoreline at that time has now subsided to 450 fathoms (823 m). Kear and Wood (1959) referred to the early volcanoes as Fagola volcanics. When volcanism ceased the volcanoes may have reached 1.8 km above sea level, during the late Pliocene or early to middle Pleistocene.

During the early to middle Pleistocene these volcanoes were heavily eroded (Fig. 13), taking on an appearance similar to Oahu in the Hawaiian Islands, with steep rugged terrain and sharp peaks. Debris from the islands and reef growth produced offshore platforms, which have subsided and are 182 m below present sea level.

Kear and Wood suggests that renewed volcanism coincided with the penultimate glaciation (Fig. 14). The Vini Tuff is believed to coincide with the last interglacial period. Littoral cones and tuff cones were built along the coasts and Salani lavas flowed down valley walls, across reefs, and out onto the submarine slopes. Kear and Wood (1959) suggest the Salani volcanics reached 1 km above sea level in Upolu and over 1.2 km in Savaii. The Salani lavas covered most of the pre-existing landscape with the exception of eastern Upolu. Kear and Wood suggest the Salani volcanism in-

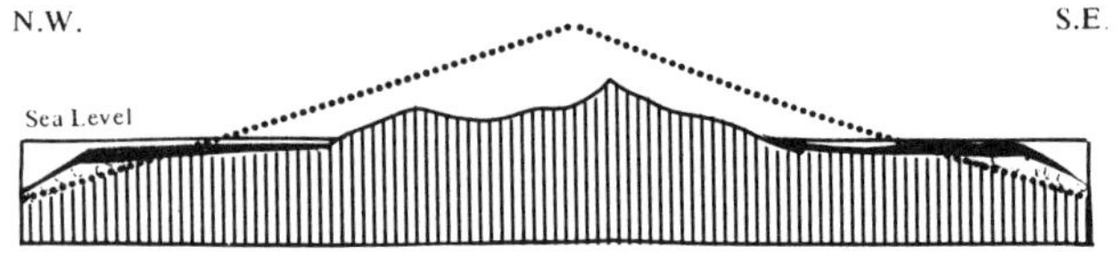

Figure 13. Profile of the island of Tutuila (American Samoa) illustrating the mature erosional surface, fringing reef, drowned reef platform and barrier reef, as shown in illustration by Mayor (1920). Using the submarine profile of the islands of Western Samoa, Kear and Wood (1959) estimate the volcanoes reached 1.8 km above sea level in the late Pliocene to middle Pleistocene.

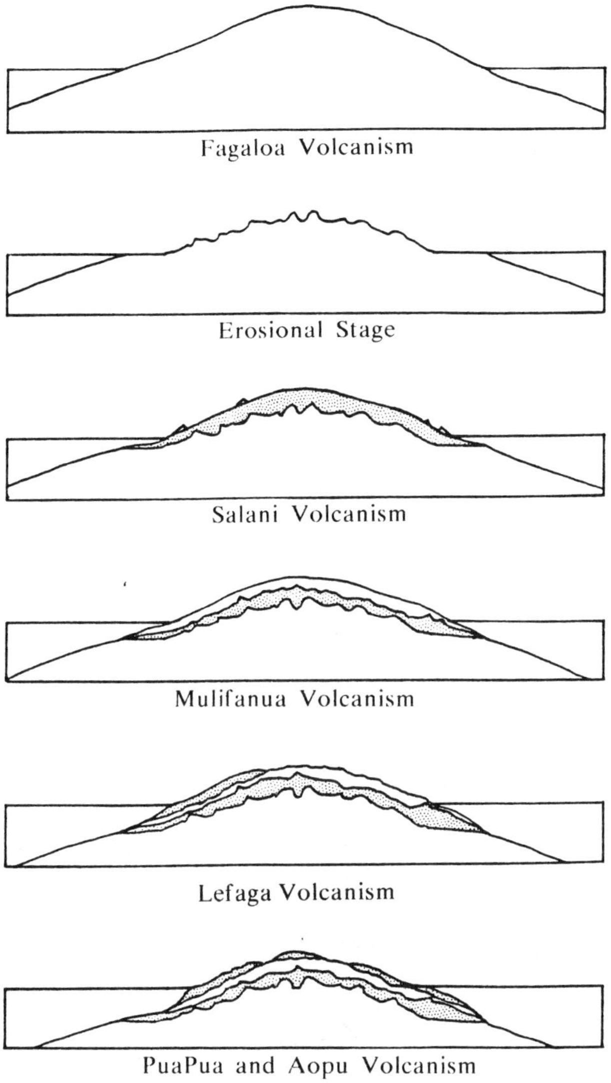

Figure 14. Profile of the development of Upolu and Savaii. Kear and Wood (1959) propose that the two volcanoes merge below sea level, but form two separate islands. During the shield building stage the volcanoes formed broad shields similar to the Hawaiian Mauna Loa volcano, during the late Pliocene or early to middle Pleistocene. During the middle Pleistocene the volcanoes were heavily eroded. Volcanism was renewed, with valley filling lavas reaching 1- 1.2 km above sea level. Erosion again took place and amphitheaters developed on Savaii and Upolu. Extensive volcanism occurred during the last glaciation, particularly on western Upolu. During post-glacial times, barrier reefs grew, and volcanism covered some of these reefs. Historic volcanism continues to present.

creased the width of the islands to roughly the present outline. New coral reefs grew along the coastlines, and have since subsided 16 m.

Erosion of Salani volcanics left gullies in the landscape. Where concentration of ground waters in Fagaloa volcanics reached the ground surface and were perched, streams eroded more rapidly into the Salani volcanics. Warm climate and increased rainfall in the last interglacial allowed amphitheatre development in canyons of south Upolu and Savaii.

During the last glaciation, extensive Mulifanua lavas were erupted. The Mulifanua lavas covered much of western Upolu. On the Savaii uplands, nearly 150 m of lavas accumulated. They also built a submarine fan on the southwest slope of Savaii. Littoral and submarine cones in Apolima Strait occurred where lava entered the sea. Kear and Wood suggest a large part of the northern part of Savaii collapsed, as well as parts of the southeast and southwest flanks.

During the post glacial rise in sea level, barrier reefs grew (Fig. 15). Limited volcanism occurred, referred to as Lefaga volcanics. When sea level reached the present level, volcanism again occurred in southeast Savaii and buried parts of the wide barrier reef. A few small flows also filled valleys in Upolu.

In historic time, volcanic activity occurred in Savaii, which constitute some of the largest historic lava flows in the world. These lavas buried older cones and cascaded into the sea.

GEOLOGIC EXPLORATION OF AMERICAN SAMOA

General description of American Samoan islands

Tutuila

Tutuila is a long, narrow island with narrow irregular ridges forming the backbone of the island. Along its southwestern shore, a low unweathered volcanic peninsula has been added by relatively recent volcanic activity. The volcanic peaks of the older portion of the island are deeply eroded. In many parts of the islands, closely spaced streams erode the volcanic slopes, creating steep slopes with thick accumulations of laterized soils, separated by inter-valley ridges. Near the coast, the valleys widen and small coastal flats are developed. Drowned valleys form several embayments such as Pago Pago, Fagaituo, Afono, Vatia, Fagasa, and Nua Seetaga Bays (Fig. 16).

Volcanic craters are absent over much of the island, since substantial weathering of the main island has left the island in a very "mature" geomorphologic stage. Volcanic intrusions such as Mt. Pioa, however, are clearly exposed by

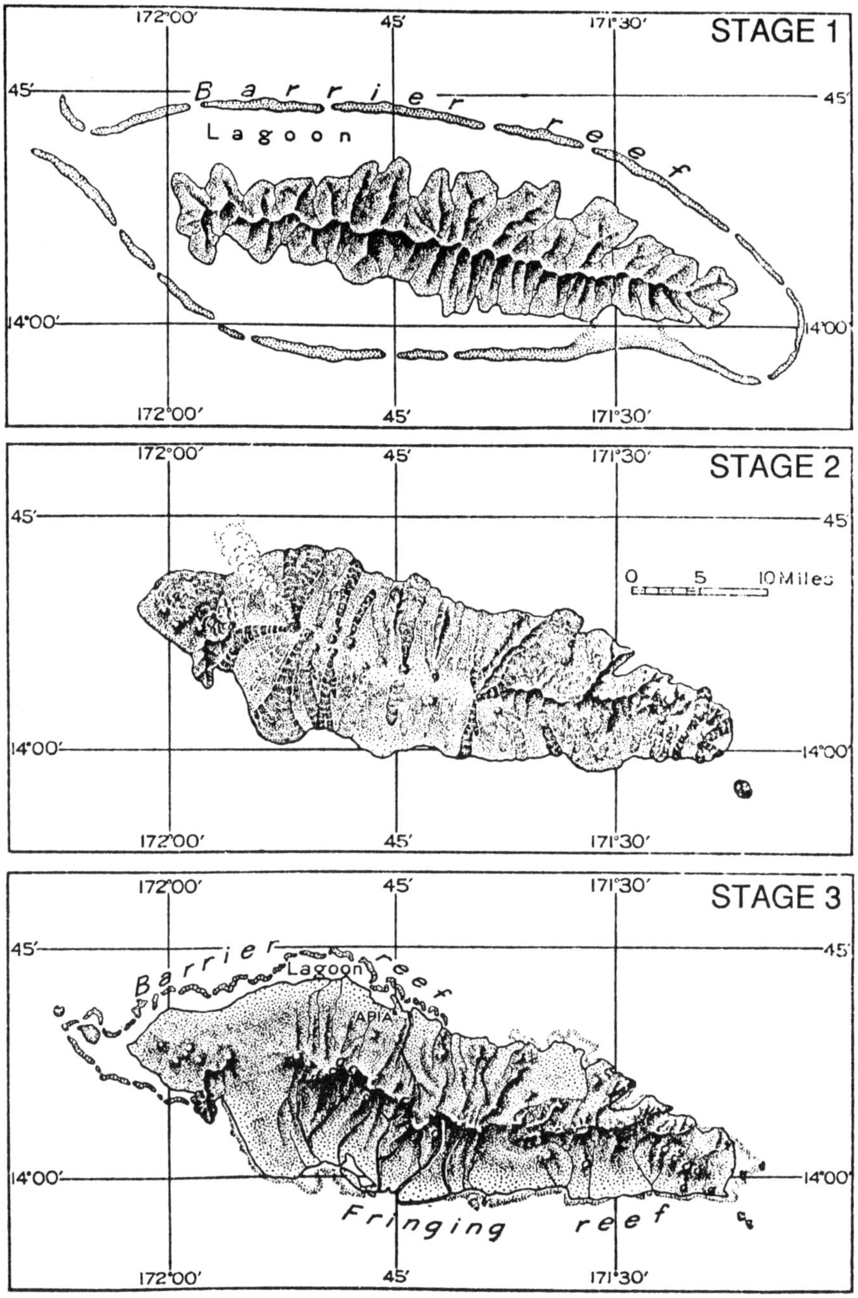

Figure 15. Outline of the island of Upolu, through time, as proposed by Stearns (1944). Stage 1 illustrates the proposed outlines of the island after extensive erosion of the shield building lavas. Stage 2 illustrates the outline of the island after the barrier reefs were drowned and volcanism was renewed. Stage 3 illustrates the present configuration of the island and the partial development of fringing and barrier reefs around the margin of the island.

the weathering, and thus form dominant landforms. The Leone Peninsula is a relatively low plain on the southwest margin of Tutuila that is generally undissected by streams. There are prominent craters, Oloava Crater, Fogamaa Crater, and Fagatele Crater. Oloava Crater near the summit of Aoloaufou is slightly eroded. Nearly 200 streams are present on Tutuila (Davis, 1963). Many of the smaller streams are "ephemeral," i.e., they carry water only after precipitation and are dry during most of the year. The flow in perennial streams comes from springs and seeps which discharge ground water. Many only supply a trickle of water during dry seasons. The largest streams have average flows of a few million gallons (tens of thousands of cubic meters) of water per day in wet weather, which is decreased to a few hundred thousand gallons (a few hundred cubic meters) per day in dry weather. Annual rainfall is approximated at 61 m/yr (Davis,1963).

Aunuu

Aunuu is a small island situated off the southeastern coast of Tutuila. The island was formed by submarine volcanic activity. A tuff cone forms the eastern half of the islet. It is breached on the eastern margin, forming Maamaa Cove. The weathering of Aunuu tuffs has produced an impermeable layer in the bottom of Aunuu Cone where a lake and marsh have formed. The Taufusitele marsh occupies a portion of the coastal flat west of the cone. Seaward and west of this marsh, the coastal flat is covered by an appreciable volume of calcareous sand and gravel (Davis, 1963).

Ofu and Olosega

Ofu and Olosega islands (Fig. 17) are remnants of a large volcano eroded to such a state that only two small steep islands remain, separated by a 200-m wide sea channel (Stearns, 1944: Stice and McCoy, 1968). Stearns (1944) suggests a double

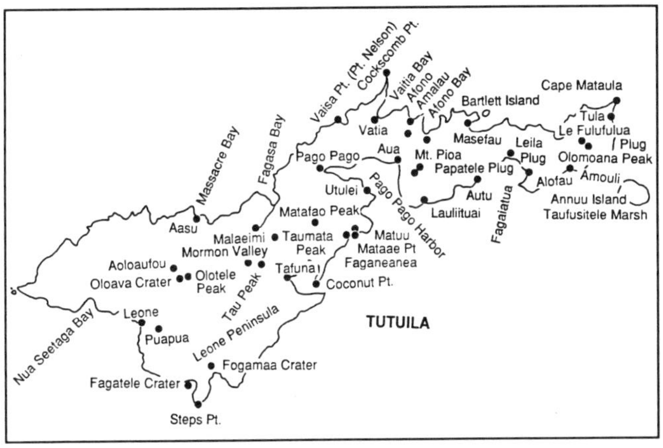

Figure 16. Location map for sites on the island of Tutuila discussed in the text.

caldera was formed; lava flows and dikes are exposed. Pyroclastic beds are thick and numerous and deeply weathered to laterized fragmental volcanics. Seeps and a small spring discharge water along the dikes on the western end of Olosega. Seeps and small springs flow from lavas into beach deposits locally along the shores. Coastal flats are generally covered by sands and gravel. The east side of Olosega and the west side of Ofu are cut by cliffs 100-200 m high. A cliff 300 m high also forms the north coast of Ofu.

Nuutele Islet, which is situated off the western shore of Ofu, is the remnant of a Recent tuff cone formed by a shallow submarine eruption. Three tunnels 70-100 meters in length have been cut through the islet by the tidal surge (Daly, 1924). Stearns (1944) and Stice and McCoy (1968) state that pre-caldera volcanics, a dike complex, and post-caldera volcanics are all present and of probable Pliocene to early Pleistocene age.

Tau Island is a remnant of a volcanic shield (Stearns, 1944; Stice and McCoy, 1968). It is very young in appearance and numerous volcanic cones are observed. The island consists mainly of thin lava flows and, on the northwest corner, tuff deposits form several small tuff cones. Seeps and springs occur in cliffs on the top of dense lava flows or from lava lying on tuff beds. Seeps also discharge fresh water at the shore line.

Rose Atoll

Rose Atoll is geographically part of the Samoan island chain. Mayor (1924) describes the atoll as a narrow ring of limestone which is nearly awash at low tide. The central lagoon is about 3 km wide and has a maximum depth of 15 m. The floor of the lagoon is sandy and generally free of living corals (Couthouy, 1842). A channel through the atoll rim is present on the northeast side. West of this main entrance to the lagoon, the live reef grows in "long nearly parallel flat-topped, overarching ridges all parallel with the line of the wave fronts of the breakers as they surge over the reef.. . . " Lithothamnium is the dominant coral. Porolithon is the dominant algae.

Mayor (1924) reports two small islets on the atoll rim, Sand Islet and Rose Islet. A map made in 1839 shows Rose Islet the width of the atoll rim and forested. At the time of Mayor's visit in 1920 the islet covered only the inner half of the reef rim and only the southern half of the islet was covered by trees. In 1839 the island was described as rising 10 m above sea level. In 1920, the islet stood only 3 m above high tide with the tallest trees (*Pisonia grandis*) 24 m high. The ground under the trees was covered by chocolate-colored humus which permeated the underlying lime coquina 2-3 m. Limited sand beaches were present around the atoll, 0.3–1.5 m in width at low tide. Cliffs of coquina 1-2 m high front the sea. Inland of these low cliffs was a rocky ledge 3–4 m above sea level. The grove of trees was confined to the region of coquina rock and did not extend over areas of loose calcareous breccia.

Sand Island is an accumulation of fragments of lithothamnium, shells and coral, which reach a height of 1.5 m above sea level in 1920. Mayor suggests the islet is submerged at times of storm induced high seas.

The upper surface of the atoll rim is a hard flat surface with little sand. Mayor (1924) states, " . . . in most places it is awash at low tide, although in others it projects as a hard, smooth ledge about a foot [0.3 m] above low tide of the neap tides."

The atoll rim is cluttered with hundreds of large limestone blocks, generally 2 m high. Most of these blocks are probably storm-driven fragments of the reef, ripped off the outer rim of the barrier reef at times of severe storms or hurricanes. Mushroom-shaped pedestals "still remain attached to the floor of the atoll rim . . ." Mayor argues, "the appearance of these boulders supports the view that the atoll-rim was once about 2–2.5 m higher, in respect to sea level, than at present, and has been cut down to present sea-level in recent times."

Mayor (1924) reports that after a diligent search he was unable to find any volcanic rocks on the atoll. Balazs (personal communications, 1984) also reports that the only volcanic rocks he has observed on the island were cooking stones, brought to the atoll by Samoans when fishing and hunting turtles on the atoll. Couthouy (1842) however reported a number of volcanic boulders scattered on the sandy bottom of the lagoon. One weighing about 20 lbs was found in about 1 m of water, and appeared similar to those of the Samoan and Fijian islands. Wilkes (1852) reported boulders of vesicular lava were seen on the coral reef and were 20 to 200 pounds in weight, found among blocks of coral conglomerate. Mayor (1924) suggests Couthouy and

MANUA GROUP

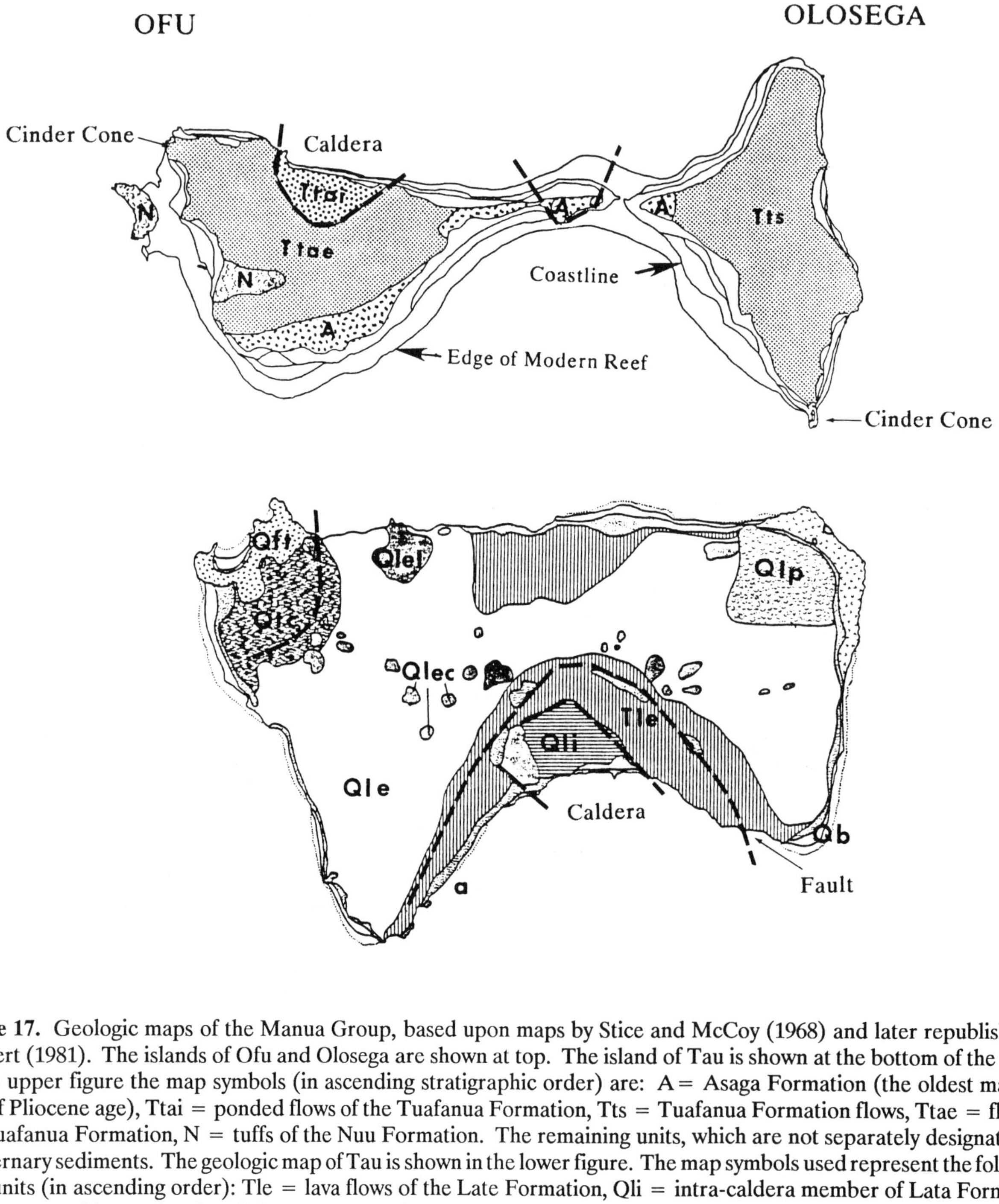

Figure 17. Geologic maps of the Manua Group, based upon maps by Stice and McCoy (1968) and later republished by Wingert (1981). The islands of Ofu and Olosega are shown at top. The island of Tau is shown at the bottom of the figure. In the upper figure the map symbols (in ascending stratigraphic order) are: A = Asaga Formation (the oldest mapping unit of Pliocene age), Ttai = ponded flows of the Tuafanua Formation, Tts = Tuafanua Formation flows, Ttae = flows of the Tuafanua Formation, N = tuffs of the Nuu Formation. The remaining units, which are not separately designated are Quaternary sediments. The geologic map of Tau is shown in the lower figure. The map symbols used represent the following map units (in ascending order): Tle = lava flows of the Late Formation, Qli = intra-caldera member of Lata Formation, Qle = post-caldera volcanics of the Lata Formation, Qtc = ash, tuff, and olivine basalt of the Tunoa Formation, Qlc = ponded lavas of Luatele Formation, Qlp = flows of Luatele Formation, Qft = Faleasao Tuff Formation, Qa = alluvium, Qb = modern beach sediments. The remaining units which are not separately designated are Quarternary sediments.

Wilkes "mistook some dark-colored scoriaceous-looking, weather-worn limestone boulders for lava."

In 1939, Schultz collected samples of lava from Rose Atoll (Schultz, 1940) and sent them to the U. S. National Museum for analysis by Gilbert Corwin. Schultz related to Sachet (1954) that a dozen or more volcanic blocks the size of a man's head were observed scattered on the reef. The rocks collected by Schultz are described as "compact olivine basalt." Schultz concluded they had been thrown up on the atoll surface from the shoulders of the atoll much like the other reef blocks.

Pickering (1876) also reported several blocks of vesicular lava and stated that "in all instances [they were] resting on the coral-shelf, not imbedded."

The reports of basaltic boulders are very important. It is assumed that Rose Atoll is an isolated old seamount, unrelated to Samoan volcanism. If it is an old seamount, a very thick carbonate cap would be expected. Johnston Atoll for comparison, appears to be capped by 1600 m of sediments (Keating, 1985b). No basaltic fragments would be expected on the shallow slopes of an ancient atoll, and the observed basalts are likely to be transported lava rock fragments from the high Samoan islands by fishermen as suggested by Balazs (personal communications, 1984). By contrast, the numerous reports of blocks of lava on the atoll are suggestive that lava outcrops are present on the upper slopes of Rose Atoll, which would be consistent with a young age for the seamount that could be linked to volcanism of the Samoan chain.

Lipman and Taylor (1924) analyzed the soil within the Pisonia grove on Rose Atoll. From the descriptions of Mayor (1924) and Lipman and Taylor (1924), it appears that an upper layer of humus is present. Underlying this is an intermediate layer of loose, very porous material that is dark brown in color, which extends to a couple meters depth. This is underlaid by a unit referred to as bedrock that is "a compact, fine-textured, almost pure calcium carbonate, which shows virtually no vital structure. It is pure white, fairly soft," according to Lipman and Taylor (1924). Mayor describes the same material as coquina. According to Sachet (1954), Lipman and Shelley regard the intermediate layer as being an intermediate product in the decomposition of the bedrock, to form, with the addition of much humus, the surface layer of "fine textured, mellow, organic soil." The analyses of soil by Lipman and Taylor (1924) indicate that there are increasingly high percentages (from bedrock to soil) of Al, P, S, Na and K, compared to decreasing percentages of Ca and Mg, and little change in Si. They explain that aluminum silicate in the original rock undergoes decomposition through reaction with ammonia, formed from the decomposition of organic material or bird excreta, followed by removal of the ammonium silicate by leaching, while accumulating alumina in the soil. This would prevent silica accumulation. The authors explain the increased sodium, potassium and sulfur as resulting from the great absorptive capacity of the soil, differential leaching, and contribution from spray. Sachet suggests a 36-fold reduction in the weight of soil decomposed is required to yield the observed amount of alumina. Stone (1951) rejected this idea, since basalt fragments are reported to be present on the reef. Sachet (1954) states that analyses may reflect contamination by decomposed pumice or basalt.

Studies of soils from Pisonia forests in the Marshall Islands led Stone to suggest that the bird excreta was acidified by humus as it washed down through the soil, dissolving calcium phosphate. As the residual reaches the sands and rock below, the aqueous solution becomes alkaline and insoluble, precipitating out and cementing the loose material together. The acid solution dissolves calcium carbonate and replaces it with calcium phosphates, producing a hardpan, immediately below the humus layer. Hutchinson (1950) describes similar hardpan development on the atolls within the Line Island chain. These hardpans, however, do not have high aluminum contents.

The high aluminum content and lack of hardpan is interpreted by Hutchinson (1950) as being significant. Hutchinson says, "It is plausible to suppose that the profile on Rose Atoll represents phosphatization due to concentration of phosphate in the parent rock, presumably with enrichment from boobies nesting in the trees, while the Palmyra profile represents a Pisonia stand growing on the site of a pre-existing guano deposit." Hutchinson points out that, in general, phosphates are better developed on dry equatorial islands than on wet ones. It is difficult to estimate the rainfall on Rose, since it is an uninhabited island. Records are available for Puka-Puka and Aitutaki, which probably reflect the conditions on Rose. Sachet reports 2 m of rain on Aitutaki as the yearly average (Seelye, 1943). On Puka-Puka the yearly totals for 1930-1942 range from 2.2 to 3.9 m. Sachet (1954) suggests, "Rose Atoll probably has a similar rainfall, although the number of rainy days may be great." These rainfall levels are significantly higher than those for the dry equatorial Line Islands (Keating, this volume). The substantial rainfall could wash guano into the lagoon, in the manner suggested for Christmas Island by P. Helfrich (personal communications, 1988) removing the surface phosphatic accumulations.

Hutchinson suggests, "the fertility of many wet atolls doubtless depends on such phosphatization, followed by bacterial nitrogen-fixation, as seems to be the case on Rose Islet (Lipman and Taylor, 1924). Nitrite and nitrate producing bacteria are present in the soils. Their nitrifying activity is more or less proportional to the amount of organic matter in the soil, according to Lipman and Taylor (1924).

Volcanic Cones

Volcanic cones are relatively rare in American Samoa as opposed to Western Samoa. On Tutuila they are concentrated on the Leone Peninsula and the flank of the Taputapu volcano (Fig. 18) adjacent to the peninsula. Older cones on Tutuila have been removed by long-term erosion. No crater lakes are present on Tutuila. One crater lake, Red Lake, is found on Aunuu Island. Nuutele and Nuusilaelae Islands west of Ofu are tuff cones. Cinder cones are present at Tauga Point on northwest tip of Ofu, and Lemaga Point on the southeast tip of Olosega. Pliocene breccia cone material makes up much of the Asaga Formation (Fig. 17). In addition, several large and small craters are present on the island of Tau (Stice and McCoy, 1968).

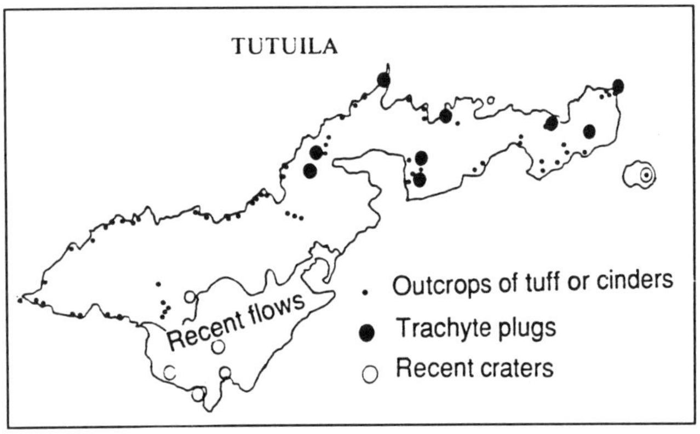

Figure 18. Map of the Recent craters and flows, outcrops of tuff and cinders and locations of trachyte plug on Tutuila, American Samoa. The map is based upon the geologic map of Tutuila by Stearns (1944) republished by Wingert (1981).

Coral Reefs in American Samoa

The coral reefs around Tutuila show considerable variation in the distance to the shoreline and degree of development (Fig. 19). Mayor (1924) studied the structure and ecology of reefs around Tutuila. He suggested a drowned barrier reef at 58 to 72 m exists around Tutuila, forming a shelf.

Daly (1924) suggests that the western part of the Tutuila shelf sank 9-18 m less than the eastern part, and that the northern part is perhaps 9 m less than the southern part, with the old shelf being tilted toward the southeast.

Sand Beaches of American Samoa

Sandy beaches are present in limited areas along the coastline of Tutuila. The abundance of sand seems to be directly related to the smaller number of concrete buildings present on that island. The sands occur as loose sand and lithified beach rock. The sands vary from fine to medium and coarse grain. The sand commonly consists of fragments of coral, shells, and calcareous algae without siliceous material. A few chitinous grains are commonly found. The sand generally has approximately equal amounts of high- and low-Mg calcite. The lithified sands commonly occur as lens-shaped bodies (now referred to as beach rock) which terminate in loose sands along the strike of the beach (Daly, 1924). The lithification is believed to result from cementation by calcite-rich fresh water percolating through the beach deposits which dip gently seaward (Daly, 1924).

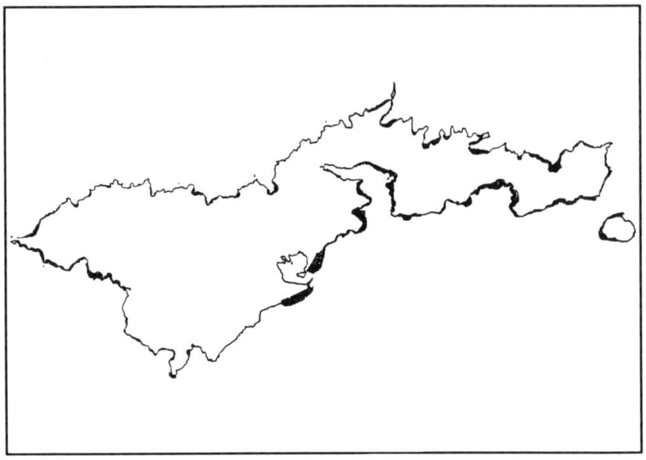

Figure 19. Map of the island of Tutuila, showing the locations of the modern reef edge as mapped by Stearns (1944) and Wingert (1981).

Streams and Erosional Features of American Samoa

If we examine the topographic features on Tutuila in the same manner as Kear and Wood (1959) did for Western Samoa, the volcanic cores exposed in American Samoa display several levels of erosional degradation. The relationships can be characterized with the following observations:

- Pago volcanics, caldera unit - amphitheater-headed canyons, deep, poor graded valleys.
- Pago volcanics, extra-caldera units - deeply incised canyons with highly laterized canyon fill, entrenched gulleys.
- Taputapu volcanics - moderately incised canyons, isolated canyon ridges, deep laterization.
- Alofau volcanics - limited canyon development.
- Olomoana volcanics - few, short, shallowly eroded gulleys.

- Aunuu volcanics - little sign of erosion other than alteration of tuffs to clay in the base of the volcanic cone.
- Leone volcanics - only weakly developed drainage patterns present, few signs of erosion and soil development.

Sea Cliffs and Caves

In American Samoa, sea cliffs are very common. Where the protective coral reefs are absent, the sea cuts against the rocks and erodes the headlands. Continuous sea cliffs ring the Leone Peninsula. The islands of Ofu and Olosega are strongly cliffed on all sides. The sea cliffs reach heights of 100 m or more (Daly, 1924; Stearns, 1944; Stice and McCoy, 1968). Many corrections were made to the topographic maps by Stice and McCoy (personal communications, 1990).

In American Samoa, sea-cut caves and benches are present on Tutuila at 6 m and 3 m above present high tide level. The 6 m stand of sea level produced notable erosion of lavas on all sides of Tutuila, producing cliffs of 10 to 100 m in height. Caves at Mataae Point, near Faganeanea, and Fagaitua have floor heights 5 m above high tide level.

Emerged rock benches are currently being cut on the eastern shore of Aunuu Island, Ofu, Tau, and Tutuila. These are 1-2 m below high tide level and correspond to similar height benches on Rose Atoll, 140 km east (Daly, 1924; Stearns, 1944).

Off Tutuila in American Samoa, the fringing reef and barrier reef merge to form a very narrow reef structure only 10 1/2 m across. The reef occurs intermittently along the southern coast of Tutuila, and is very narrow and in some places absent, around Ofu, Olosega, and Tau.

Economic Deposits

Construction Material

Stearns (1941) reported on potential sources of construction material on Tutuila (American Samoa). Stearns suggested suitable material for crushed rock is exposed in road cuts of the Pago caldera complex along the coast where the ridges plunge into the sea. The rock is dense and suitable after crushing for concrete aggregate and road metal.

Cinder for road surfacing is available at Futiga cone. Cinders for light-weight concrete are available at Mapusaga cone. Stearns suggested trachyte on the island had enough silica content to be of use in making cement, when crushed, provided additional iron was used and fused with coral sand. He stated that similar rock had been used for cement on Maui, Hawaii, during World War I. He stated, "it is economically feasible on Tutuila only if shipping space becomes scarce and fuel is obtainable at a reasonable price.

HISTORIC VOLCANISM IN AMERICAN SAMOA

Manua

Thomson (1921) reports about Manua in 1866,

> "a submarine eruption took place between the islands of Olosega and Tau, but few details have been recorded. Coleman Phillips states that on September 12, 1866, dense masses of smoke arose from the sea and continued until the middle of November. The outbreak was preceded by repeated shocks of earthquake. Friedlander ascertained from an old inhabitant of the group that dense clouds of steam and water with slag and pumice were ejected, and at night flames were plainly visible. Stewart's *Handbook of the Pacific Islands* gives the date as 1867, obviously in error, and mentions that the submarine volcano vomited forth rocks and mud to the height of 2,000 ft (610 m), killing the fish and discoloring the sea for miles (kms) around. The German Admiralty Chart, according to Friedlander [1910], mentions a submarine volcano some 46 m below the water surface."

GEOLOGIC MAPPING OF AMERICAN SAMOA

In Western Samoa hydrologists used physiographic features to map the geology. The strategy used was to map units using physiographic development in order to establish units of roughly equal age. In American Samoa Tutuila was mapped by a volcanologist, Harold Stearns, who coincidentally was a specialist in groundwater resources. Stearns carefully observed the geology of the American Samoan islands, identified the intrusive centers, calderas, and extra-caldera lava flows. The relationships he observed allowed him to map volcanic centers. Often onlapping relationships allowed Stearns to establish relative ages for volcanism on an island. It was only after radiometric dating was attempted decades later that some absolute age relationships have been established.

Stearns (1944) concluded that Tutuila is built from five volcanic centers (Fig. 20). The volcanic centers are aligned on two or possibly three rifts, trending northeast-southwest. The majority of the lavas are basic lavas, capped by limited trachyte, alluvium, coral reefs, and beach sands. Stearns suggests that the bulk of the volcanoes appear to be Pliocene, based on the weathered and eroded conditions. Fresh tuff cones and lava flows on the southwest side of the island, and on Aunuu Island are assigned Recent ages.

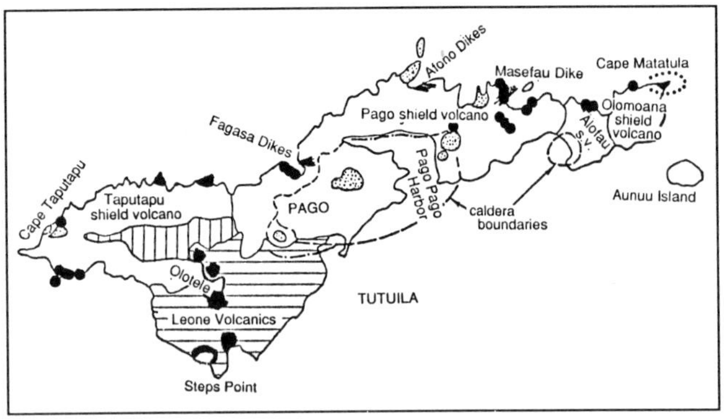

Figure 20. Geologic map of the volcanoes forming the island of Tutuila, based upon the field mapping by Stearns (1944). (Figure from proposal by Natland et al., 1984)

GEOLOGIC MAPPING UNITS

Masefau Dike Complex

On the north coast of Tutuila near Masefau village is a complex of hundreds of basaltic dikes ranging from a few centimeters to 2 meters in width. The dike rocks are vesicular and platy; some are amygdaloidal. The dike complex strikes N 70° E and dips slightly southward. These rocks are interpreted to be the oldest rocks exposed on the surface and were part of the deep structure of a volcanic rift zone.

The north side of the dike complex is exposed at Afono Bay, in a cliff on the north side of a promontory opposite Bartlett Island (now called Nuusetoga Island). Stearns (1944) suggests that the cliff may be an eroded fault scarp or a sea cliff. Resting against the face of the cliff is talus breccia overlain by 3-23 m of vitric basaltic pumice and cinders on which aa basalt cascaded over the cliff.

Stearns suggests the north side of the dike complex may be the eroded caldera wall of a volcano older than the Pago volcano, whose vent lies north along the present coast; alternately, it may be an eroded early rift zone horst block of the Pago volcano, later buried by flows. Based upon analogies with dike complexes elsewhere, Stearns suggests that the complex was apparently formed not more than 610 m nor less than 92 m below the surface of a rift zone and that 300 m or more of lavas which formerly overlaid the dike complex have been removed by erosion.

The Masefau dike complex cuts thin basaltic flows dipping at moderate angles (from 10 to 20° NW). These lavas are in large part shattered and brecciated by faults. Stearns suggests that much of the rock in the rift zone shatters slowly, during repeated intrusion of the dikes, and that the shattered brecciated material settled downward into the underlying magma chamber as breccia.

Olomoana Volcano

The Olomoana volcanic rocks are areally restricted olivine basalts. Stearns interpreted the outcrops as volcanism on the northeast rift of the Alofau Volcano during the closing phase of Pliocene activity on the island. The Alofau volcanic flows overlap the west slope of the Olomoana volcanic cone. Thus it is suggested that volcanoes may have been essentially contemporaneous.

The lavas of the Olomoana volcanics center cover 2.4 square kilometers on the east end of Tutuila, surrounding Olomoana peak. Several intrusive volcanic plugs, crater fills, and cinder cones are exposed along the eastern coast. Cape Matatula is the largest intrusive plug. Flows are interbedded with palagonitized vitric tuffs interbedded. The Lefulufulua trachyte plug cuts these rocks. A large cinder cone can be seen partly exhumed in the stream bed draining to Tula.

Alofau Volcano

The Alofau volcanics center consist of thin bedded aa and pahoehoe flows, dikes, breccias, and tuffs exposed in a shield-shaped dome. The dome covers about 2.4 square kilometers on the east side of Fagaitua Bay. The volcano is built over a rift zone trending northeast-southwest. The lava flows are thinly bedded primitive olivine basalts, dipping 10 to 20° away from Alofau village. A dike complex is exposed on the road southeast of Fagaitua village. Another 130 individual dikes are exposed in a promontory on the north side of Alofau. Large numbers of dikes are also seen exposed south of Alofau village. The rim of the caldera decreases in height rapidly to the southwest away from Alofau and extends partially submerged offshore.

Pago Pago Volcanic Series

The Pago volcanic series includes extra-caldera lavas, dikes, plugs, cinder cones, vitric and lithic tuffs, and breccias as well as intra-caldera lavas, dikes, plugs, cones, tuffs and breccias. The Pago volcano caldera was situated at the center of Tutuila Island. Stearns (1944) estimates that the caldera extended 9.6 km in length and 4.8 km in width.

Intra-Caldera Volcanic Units

Within the Pago caldera, lava flows formed thick ponded lava sequences to the north, east, and west. Pahoehoe flows are scarce. The lavas are generally aphanitic and olivine poor. Dikes are observed, ranging from 1 to 15 m wide.

The tuff and breccia member is the explosive debris that accumulated during the excavation of the inter-caldera lavas. The deposits contain lithic, crystal lithic, and lithic-vitric debris. The beds range in texture from fine-grained, laminated, silty tuffs through lapilli tuffs to coarse blocky beds of breccia 6 m thick with blocks of ejecta 1.5 m wide. Thick deposits of fire-fountain debris are interbedded with lithic beds of cataclysmic explosions. Pisolitic beds are present. A few beds of carbonaceous plant remains are occasionally found, which could be used to determine radiocarbon ages, the ages are less than roughly 40,000 years. Beds containing pumice fragments are also preserved. Other beds contain both basaltic pumice and dense fragments of trachyte, indicating that the trachyte eruptions were followed by basaltic eruptions.

The beds are nearly horizontal near the source. Where they have steep dips (reaching 32°) they interfinger and merge with talus breccia, suggesting they accumulated concurrently with the talus at the foot of the caldera wall.

An unconformity exists between the Pago tuffs and the pre-caldera lavas. The unconformity is well exposed in a small stream channel at the head of Pago Pago valley, where the caldera rim is seen. Thin bedded olivine basalts are uncomformably overlain by beds of talus breccia and tuffs. Two dikes cut the lavas but do not cut the tuffs. Faulting is present but the lack of deformation of the tuffs suggests that the faulting motion pre-dates the accumulation of tuffs. The stream's channel, which is a series of cascades, cuts down through the exhumed caldera wall.

Another unconformity can be seen on the north side of Taumata Peak. There, the massive columnar jointed ponded lava forms a peak resting on talus breccias, thin bedded lithic tuffs, and basaltic pumice and cinder beds of the Pago Tuff Member. The intra-caldera lavas on the northeast side of the peak rest on red baked vitric tuff. The tuffs of the Pago volcanic series east of Taumata Peak are underlain and overlain by massive ponded lavas.

A dike complex crops out in a valley on the northeast side of Tau Peak. The dikes are terminated by massive lavas, suggesting an unconformity or caldera wall between the caldera-filling lavas and the dike complex. A small spur of dense rock is present on the road just west of Aua. Stearns suggests it is a remnant of caldera lavas.

The Pago Caldera

Stearns suggests the Pago caldera was 4.8 km wide and 9.6 km long. The geological evidence indicates the caldera formed by collapse following the building of a cone. A water development tunnel near Pago Pago Bay cuts a major fault. Along the fault a "gouge and splinter" zone is observed 30 cm wide. There, 6.3 m of friction breccia is seen which contain faceted blocks up to 60 cm across and smaller fragments which have been ground to balls by friction.

The lavas inside and out of the caldera differ. Stearns suggests at least 480 m of collapse took place. Projections of the original height of the mountain, using the dip of extra-caldera lavas, indicate the mountain would rise 1200 m above present sea level, if the cone had the same slope at the summit. The southern rim of the volcano was much lower than the northern rim. Stearns suggests the caldera may have been horse-shoe shaped with no rim on the south.

Stearns suggests, "by analogy with other similar basaltic volcanoes the collapse probably proceeded with the growth of the upper part of the cone, the caldera widening and deepening progressively even though lavas were erupted simultaneously on the floor. Subsidence finally stopped as the volcano approached old age and the lavas began to differentiate. Filling gained on subsidence, and at the cessation of activity the caldera was filled to the brim in the northwestern sector, overtopped and buried by perhaps as much as 150 m of lava in the southern and southeastern sectors, but lacked about 180 m of being filled in the northeastern sector . . . "

Trachyte Plugs

Several trachyte plugs and domes can be seen associated with the Pago volcano. The plugs fed bulbous domes, like Pioa which looms over Pago Pago Harbor. Some volcanic plugs, like Vatia and Pioa, are so little eroded that the upper bulbous parts of the domes remain. In other places, so much structure has been eroded that the remnants have a transitional appearance between domes and plugs.

The magnificent harbor of Pago Pago is dominanted by the trachyte dome, called Pioa, or more commonly the Rainmaker. The Rainmaker is a quartz trachyte dome 300 m wide, 720 m long and 515 m high. The landmark was described by Daly (1924) as reported by Stearns (1944), p. 1300-1301,

> "a monolithic rhyolite with quartz-poor and quartz-free phases in contact with basaltic flows on the north corner, but elsewhere with tuffs and breccias composed almost entirely of basaltic fragments but containing a few blocks of trachyte or rhyolite. He regarded the Pioa plug as an endogenous dome pushed through a funnel 1.25 mile (2 km) in diameter filled with explosion breccia. This is an excellent description of the mass except that the diameter of the funnel is less than half a mile (0.8 km) across and the breccia is almost entirely pre-funnel in age."

Stearns suggests,

> "The trachyte evidently rose to a point nearly 900 ft (270 m) above sea level where it frothed violently producing red, white and black biotite rich pumice and cinders about the vent, at the same time blasting out a few blocks of older basaltic tuff, breccia and lava. After a pumice cone indented with a crater had been

made, the pasty lava squeezed up and partly filled it . . . Columnar jointing developed at right angles to the base of the cooling lava."

A pumice cone was built on slightly weathered tuff covering the floor of the Pago caldera, near the foot of the eastern wall. Thus, the Pioa eruption was a late event in the development of the Pago volcano. Stearns suggests the other trachyte plugs associated with the Pago caldera were formed in a similar way.

Extra-Caldera Volcanics

The extra-caldera volcanics include two units, the pre-caldera basalts and the post-caldera basalts. The pre-caldera lavas include flows, dikes and pyroclastics. The post-caldera lavas unconformably overlie the pre-caldera lavas, with only limited associated gravels and hillwash deposits observed. The best exposures of the extra-caldera rocks are seen along the nearly continuous sea cliffs of the north shore. Stearns mapped them from a boat. Many dikes and faults are exposed in the long bays that cut the volcano transversely. The widest dike is 18 m wide. At Point Nelson (now called Vaisa Point) a dike 9 m wide is seen not far below the level at which its lava erupted (Stearns, 1944). These wide dikes are rare.

Pre-caldera olivine basalts, tuffs, and a dike complex and associated faults can be seen at Fagasa Bay. Thirty dikes cutting thin aa flows are exposed on the coast at the head of the bay. Sixty-eight dikes and several faults are exposed in the shore on the west side of the bay. The dikes are more numerous in the southeastern part of the bay. The number of dikes decrease however toward the promontory at the entrance to the bay. Stearns believes this is due to its location stratigraphically close to the original surface of the volcano.

Numerous faults are seen around the bay. Many are associated with breccias. Stearns reports many faults are

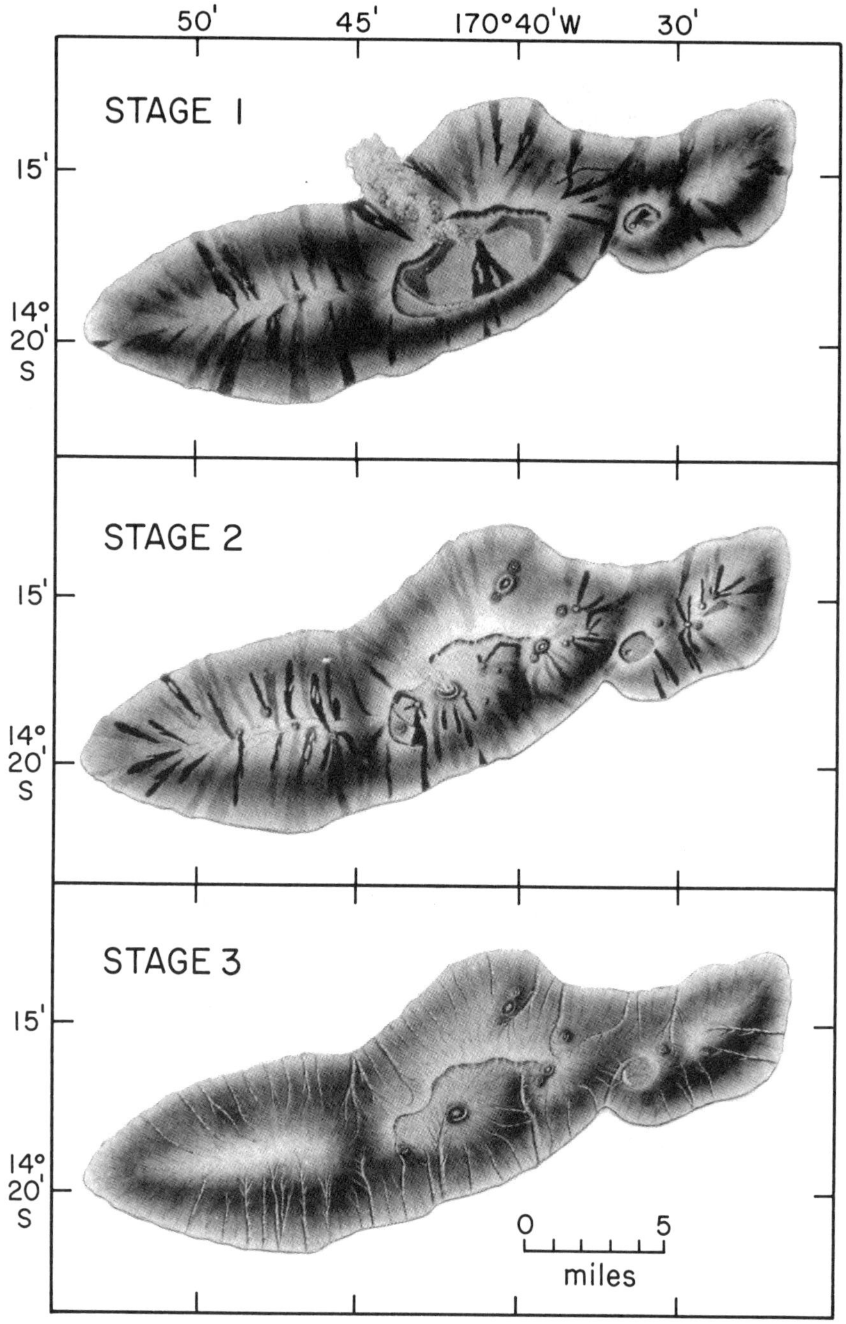

Figure 21. Illustration of the stages of geologic development of the island of Tutuila proposed by Stearns (1944). Stage 1 (top) illustrates the general configuration of the island at the close of the shield-building volcanism. Stage 2 illustrates the configuration of the island at the end of alkalic volcanism. Stage 3 illustrates the configuration of Tutuila at the cessation of volcanism and beginning of stream erosion. Notice the absence of the Leone Peninsula in these figures by Stearns (1944).

bound by cracks which are filled with blocks of basalt up to 1 m across, vitric ash and tuff. Stearns suggests a correlative relation with those on the southwest rift zone of Hawaii, where the cracks fill with blocks falling from thc wall as well as hillwash and ash which infiltrate the cracks.

Vitric tuff beds a few centimeters to a few meters thick are intercalcalated with lava flows. At the tip of the Fagasa Valley promontory, a lava tube can be found with a core of horizontally bedded vitric ash. Likewise, within a platy basalt dike a half m wide, cracked by cooling, vitric-lithic tuff has infiltrated and filled the void. Stearns suggests fire fountains were repeatedly active in this area.

Thirty dikes (mostly less than 1 m wide) are exposed on the northern coast of Fagasa Bay. Some of these appear to be continuations of those exposed on the south side of the bay. An olivine dike just beyond the village contains dunite xenoliths oriented with their vertical axes parallel to the side of the dike and their heavier, larger ends down. The core on the dike appears to have remained liquid longer allowing the heavy xenoliths to sink in the liquid at the close of the eruption. Stearns suggests the concentration occurred 150 m or less below the surface. Paleomagnetic studies have been completed on these dikes. Dikes with xenoliths are common in Tutuila.

Many dikes are observed at Afono Bay. Stearns suggests the series of lava, tuff, and breccia beds exposed there plunge over the buried escarpment of the northern side of the Masefau dike complex. The dike complex associated with the west rift of the Pago volcano is exposed on a road on the west wall of Mormon Valley (Malaeimi Valley), and in a gorge northeast of Tau Peak.

Taputapu Volcano

The Taputapu volcanics form the western end of Tutuila Island. The volcano is roughly 10 km long and 5 km wide, and 450 m above sea level. The flows are dominantly olivine basalts from 2–15 m thick. Locally, they are capped by thicker flows of porphyritic and nonporphyritic olivine poor basalts. The flows dip at angles of 5 to 10 degrees. The lavas where extruded from a rift zone extending N 75° E. Beds of red vitric tuff and cinder are common between flows a few centimeters to a couple meters thick, indicating cinder cones were very common along the rift zone.

Exposures of the dipping beds of the Taputapu volcanics in the streams flowing into Massacre Bay led Stearns to separate the Taputapu volcanics from the Pago volcanics. An unconformity was not found between the two volcanoes. Based upon the relatively small erosion and dissection of the Taputapu lavas compared to the Pago volcanics, Stearns concluded the Taputapu volcanics overlap the Pago lavas, and are younger. Mantling by Recent pyroclastics may explain this apparently youthful appearence according to P. Eyre (personal communications, 1990).

Leone Volcano

The Leone volcanics are the least weathered volcanics on the island and appear very young. They form a large peninsula on the south side of the island (referred to in some publications as the Tafuna-Leone plain) and are associated with several cinder and ash cones. The top of the pahoehoe lavas on the Leone peninsula at Steps Point are only about 2 m above sea level. Two thick tuff units were identified by Stearns but could not be mapped due to the thick vegetation.

Stearns (1944) suggests a crack, 6.4 km long, opened from Olotele Peak across the present fringing reef and the submerged 72 m shelf in Recent times. Eruptions began at the upper end of the crack with fire fountains. Where the crack reached the sea, gas-charged lava rose through the water and produced violent explosions. The ocean water repeatedly entered the vent producing many explosions, with sufficient intensity to hurl material 1.5 km in the air, according to Stearns. The strong trade winds blew much of this material northwestward across the island and partly filled the bay and valley at Leone (P. Eyre, personal communications, 1990). Fragments of drowned coral reef were involved in the eruptions (Daly, 1924).

Relative Ages

The Samoan Islands are much like the Hawaiian islands. Thomson (1921) remarked on similarities and differences between the Samoan and Hawaiian islands which are worth reiterating. First, like Hawaii, with the exception of coral reefs along the coast, the rocks of the Samoan island chain are volcanic. Second, there does seem to be a progressive shifting of volcanic centers in Samoa. In Hawaii, the shifting of volcanic activity is from northwest to southeast. Thomson (1921) points out that Savaii has experienced three eruptions in historic times and is comparable in form and state of erosion to the island of Hawaii. Upolu appears more dissected and is comparable to Maui. Tutuila, he points out, is still more dissected and is comparable to Oahu. Recent radiometric dating by Natland and Turner (1985) has yielded a date near the base of the volcanic pile on Tutuila island of 1.5 Ma. Thus, recent results confirm Thomson's geologic interpretation.

Dana (1849) pointed out that the character of the rocks and general topography of the central district of Upolu were similar to that of Tahiti and Kauai and thus suggest a similar age. The ages inferred by these early investigators are compared to the recent radiometric dates for these islands in Table 3.

Table 2. Rock Units of American Samoa (from Stearns, 1944, Table 1).

Geologic age	Formation	Thickness	General description
Recent	Sediments	60 m	Brown, silty, poorly unconsolidated alluvium in the valley floors; coarse talus debris at the foot of cliffs; and loose calcareous sand and coral gravel with small amounts of cemented beach rock along the coast.
Recent	Leone volcanics	60 m	Olivine (picritic) pahoehoe basalt spread from a fissure reaching Olotele Peak to Fagamaa Crater which is marked by a cone chain. The lava is veneered with tuff over a large area. Cinder member forms cones about 75 m high at the source of the Tafuna flow. Stony ash member forms a cone about 60 m high composed chiefly of nodular stony ejecta and a small amount of accessary basaltic ejecta in a matrix of black volcanic sand and lapilli. The ash is overlain by the tuff member ejected from three craters near Steps Point. The tuff from Fagatele Crater is separated by an erosional unconformity from the overlying Fagamaa tuff, but the tuffs can only be separated along the coast near Steps Point.
Recent	Aunuu tuff	200+	Lithic-vitric tuff forming Aunuu Island.
Great erosional unconformity			
Pliocene and early Pleistocene	trachytes	642 m	Dense, jointed, cream colored, trachyte dikes, plugs, and crater fills later than most of the Pliocene volcanics.
	Taputapu volcanics	492 m	Thin bedded olivine basalts capped in places with a few somewhat thicker andesitic basalts forming a shield-shaped dome over a N 75° E rift. Vitric tuff beds and cinder cones are interbedded with the lavas. The lavas appear to overlap the Pago Volcano.
	Pago extra-caldera volcanics	321 m	The lower member is composed of thin-bedded primitive olivine basalts, associated with thin dikes, extra-caldera volcanics and thin beds of vitric and lithic-vitric tuff having an aggregate thickness of more than 300 m, all dipping (to 25°) away from the caldera. The upper member is composed of basaltic andesites, and andesites, associated with trachyte plugs (Tau, Matafao, Pioa, Vatia, and Afono plugs.) The flows are massive and contain much interbedded vitric tuff and local thick cinder deposits. They have a maximum thickness of 150 m and form a conformable cap on the lower member. In places erosion exposes the lower member.
	Pago intra -caldera volcanics		Massive aphanitic porphyritic basaltic and andesitic lava flows, and cinder cones, associated with three trachyte plugs (Matafao, Papatele, and Pioa Plugs) partly fill a broad caldera. A few thin lenses of gravel are included. The rocks are similar in character and age to the upper member of the extra- caldera volcanics. The interbedded lithic-vitric tuff member is 15-150 m thick and is composed chiefly of thin- and thick- bedded, fine- and coarse-grained tuffs and breccias.
	Alofau volcanics	962+ m	A shield-shaped dome built over a N 70° E trending rift zone and composed almost entirely of thin-bedded primitive olivine basalts associated with one plug of trachyte (Leila plug). The summit was indented by a caldera 1.5 km across in the walls of which are now exposed several hundred basaltic dikes, a few of which are porphyritic. The latest extra-caldera lavas of the Pago Volcano apparently overlap the Alofau dome on the northwest side.
	Olomoana volcanics	322 m	Composed largely of olivine basalts but capped with andesitic basalts and perforated by one plug of trachyte (Lefulufulua plug). Several large cinder cones and numerous beds of vitric tuff are exposed in the volcano. The latest lavas from Alofau Volcano apparently overlap the Olomoana cone.
Erosional (?) Unconformity			
	Masefau dike complex	60 m	Thin basaltic flows dipping 10 to 20° NW cut by hundreds of basaltic dikes a few cm to 2 m wide striking N 70° E and dipping slightly southward. Much of the lava rock between the dike is shattered by faulting. The complex is truncated on the top apparently by an erosional unconformity and on the northern side by a fault (?) scarp. Above the fault (?) plane lies 6-30 m of talus and firefountain debris and 15 m of lavas of the Pago volcanic series.

Paleomagnetism

Paleomagnetic studies of lavas from Samoa were first carried out by Tarling (1962, 1965). These early studies showed that normal polarity is present on Savaii. On the islands of Tutuila and Upolu, the stratigraphically higher units are normal polarity and the stratigraphically lower units are reversed polarity. Tarling compared the results directly with the Hawaiian Islands and concluded the rocks sampled did not exceed 2.6 million years in age. Furthermore, he suggests the overall migration of volcanic activity from East to West "is at an average rate not less than 7.7 cm/year and probably similar to the migration rate of 10 cm/yr observed in Hawaii (McDougall, 1964)."

More recent paleomagnetic studies have been conducted by Keating and Tarling. Care was taken to collect the freshest outcrops, often along the shoreline, in order to avoid the problems of secondary overprints experienced by Tarling (1965). In general, the sampling was very successful and well-defined paleomagnetic directions resulted from the work. The poles from the island of Savaii fall into seven groups (Keating, 1985a). With few exceptions, each of the volcanic units of Savaii yields a tight cluster of paleomagnetic directions; the sites do not appear to display substantial rotation and all appear to be normally magnetized. The mean site directions are plotted in Figure 22 using the geologic names for volcanic units mapped by Kear and Wood (1959). While a few sites appear to be incorrectly assigned to younger mapping units (Keating, 1985a) most form tight clusters of direction which can be isolated from the rock units.

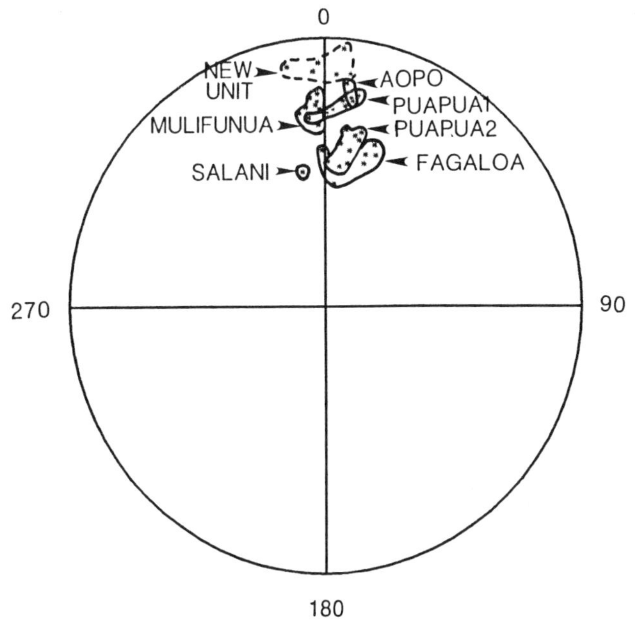

Figure 22. Mean site directions for paleomagnetic sites from the island of Savaii. The site directions from individual volcanic "formations" appear to occur in three clusters. The ovals of confidence have been omitted for clarity. The α^{95} is less than 2° for most mean site determinations.

Table 3. Age comparison between Samoan and Hawaiian islands.

Island	Observer	Comparable to:	Inferred Age:	Observed age: (Ma)
Savaii	Thomson	Hawaii	0.1	0-0.45
Savaii	Dana	Mauna Kea		0.4-0.6
Upolu	Thomson	Maui	0.4	1.63
Upolu	Dana	Kauai	1.4	5.72
Upolu	Dana	Tahiti	0.3	1.2-2.5
Tutuila	Thomson	Oahu	1.3	3.8-15
Manua	McDougall	Hawaii	0	0.1

On Savaii, the rock units are all normally magnetized (Fig. 23) and display a systematic change in inclination. The oldest (stratigraphically lowest) units have the highest inclinations, intermediate units have moderate inclinations, and the youngest (stratigraphically highest) units have the shallowest inclinations. Because the paleoinclinations from these various rock units fall into these groupings, it would appear that at least three major volcanic episodes are present on Savaii.

A comparison of the paleomagnetic results from rock units mapped using the same geologic names proposed by Kear and Wood (1959) indicates that the use of the same names on both islands is incorrect (Keating, 1985a). Units sampled on Savaii have normal polarity (Fig. 24) whereas on Upolu these units have reverse polarity (Tarling, 1965). The paleomagnetic results show that volcanism is not wholly contemporaneous on these islands. A revision of the geologic mapping on these islands would be very useful.

Mixed polarity was found on the island of Upolu (Fig. 23). It appears that the rock units of western Upolu are characterized by normal polarity while the stratigraphically lower units on eastern Upolu are reversely magnetized.

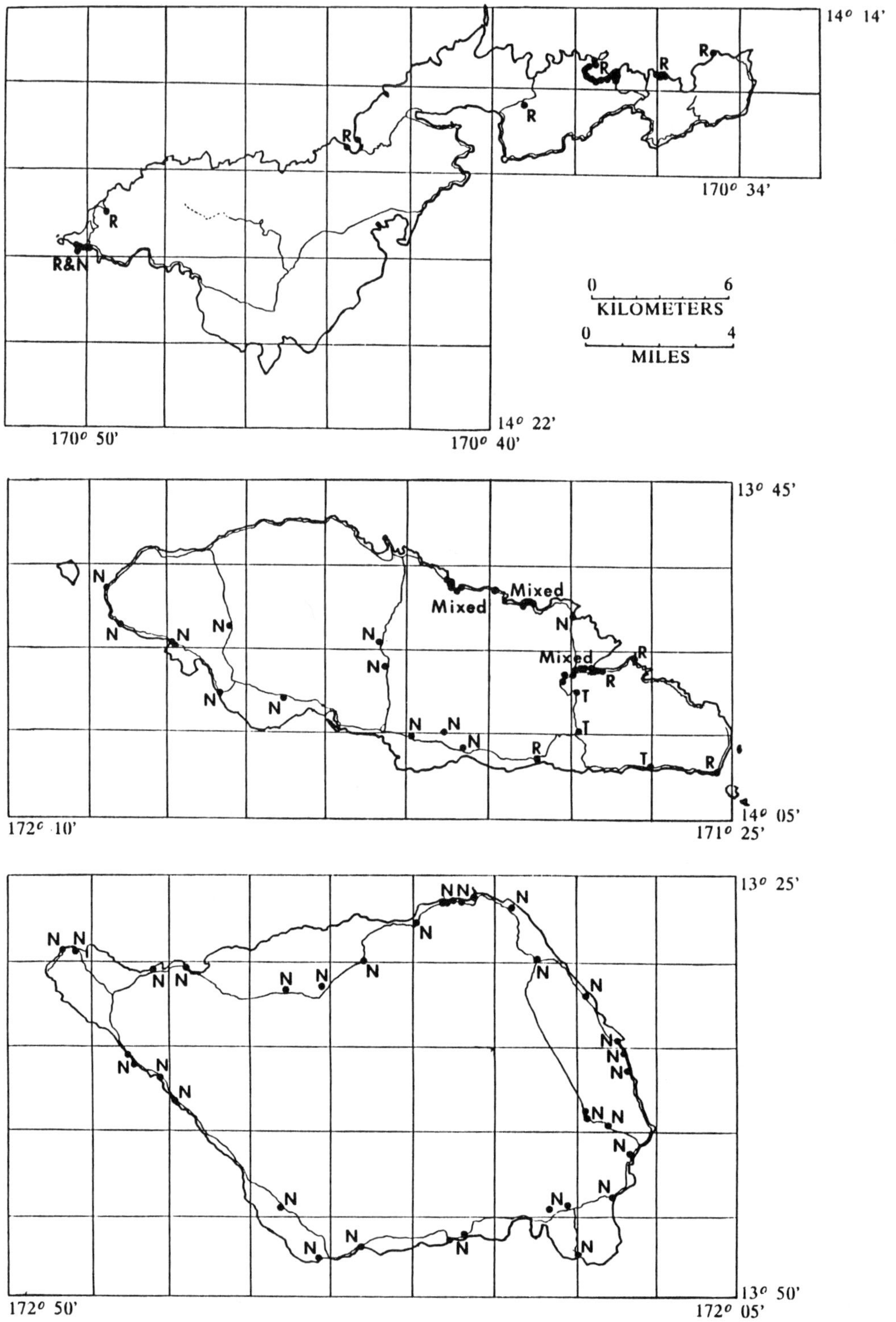

Figure 23. Maps of the main islands of the Samoan chain illustrating the magnetic polarity of sites studied by Keating and Tarling and summarized in Keating (1985a). (N = Normal polarity, R = reversed polarity; T = transitional directions of undetermined polarity, M = both normal and reversed polarity magnetization is present in the lavas and dikes at these sites). The kilometer scale refers only to the island of Tutuila.

On Tutuila, all but two of the sites sampled are reversely magnetized (Fig. 23). This observation is consistent with magnetization of much of the Pago volcano (the oldest volcano) during the Matuyama (Reversed) Polarity Epoch, and ages of 1–1.5 Ma (determined by I. Mcdougall, personal communication, 1984). Rocks of the Taputapu volcano are characterized by mixed polarity with normally magnetized dikes intruding reversely magnetized basalts. If the reversely magnetized pillows are associated with the Matuyama Epoch, which is indeed consistent with the radiometric ages of Taputapu volcano (I. Mcdougall, personal communication, 1984), then the intruding younger dykes are of Brunhes Normal age (0.7 Ma).

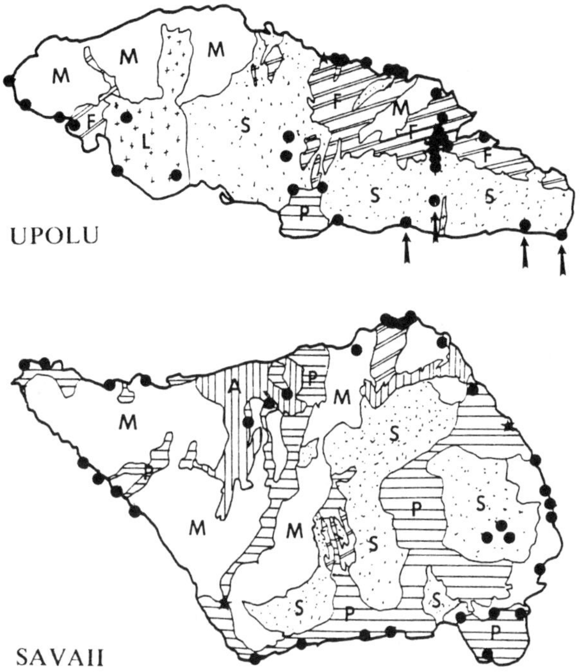

Figure 24. Geologic map of the islands of Western Samoa showing the paleomagnetic sites reported by Keating (1985a). The arrows mark sites on Upolu having reversed polarity or transitional direction within the Salani volcanic unit. On the island of Savaii all of the Salani volcanics are normally magnetized.

Petrography and Isotope Geochemistry

Western Samoa

Petrographic studies of rocks from Samoa are numerous. The descriptions of volcanic units from Western Samoa used in Kear and Wood (1959) were made by a petrologist, R.N. Brothers. Similarly, petrographic descriptions for volcanic units in American Samoa made by G. Macdonald were included with the geologic descriptions of H. Stearns (1944) and a later publication by Macdonald (1968). Many of the early detailed descriptions of the rocks from Samoa were made at the time of the volcanic eruptions early in the century. Mohle (1902) also described rocks collected from Upolu, Savaii, Apolima and Fanuatapu Islands. Kaiser (1904) as well as Wegener (1902, 1903a,b) described rock from the 1902 lavas of Mauga Afi, Savaii. Rocks from the 1905 Matavanu lava flow were analyzed by Heuseler and described by Klautsch (1907). The 1905 flow was also described by Jensen (1907). Analysis of Matavanu lavas (Jensen, 1908) differ somewhat from those of Heuseler, reported by Kear and Wood (1959). Other descriptions of the Matavanu eruption and its lavas, are reported by Angenheister (1909), Grevel (1911), von Bulow (1906), Friederici (1910), Reinecke (1905, 1906), Sapper (1906, 1909, 1911a, 1911b, 1912, 1915), and Schmittmann (1911).

Friedlander (1910) examined the rift zones of Savaii and his collection of rocks was subsequently examined by Weber (1909) and reviewed by Macdonald (1944). Thomson (1921) summarized previous rock descriptions. Bartrum (1927) described the petrology of rocks collected by members of the New Zealand Geological Survey and included a few analyses by F. T. Seelye.

Hedge and others (1972) analyzed the major element chemistry and the minor elements Rb, Sr and Ni. The basalts yielded the highest initial $^{87}Sr/^{86}Sr$ ratios found from ocean basins. They concluded the high potash content and high Sr ratios suggest that these rocks from Upolu and Savaii were derived from a mantle source which was less depleted by previous magmatism than much of the suboceanic mantle.

American Samoa

Friedlander (1907) produced an early account of the geology of Samoa, which included petrographic descriptions by Weber. Comparisons are made with rocks collected from American Samoa by Daly (1924). Stearns (1944) describes the geology of American Samoa, including petrographic descriptions by Macdonald (1944) of Tutuila (American Samoa), Upolu, and Savaii (Western Samoa). Macdonald (1968) describes petrographic analyses of rocks from Tutuila and the Manua group. Macdonald concluded that the "differentiation trend of Samoan rocks" for the most part closely parallels that of the Hawaiian alkali suite.

Where the silica ratios lie very close to the boundary between the tholeiitic and alkalic fields (Upton and Wadsworth, 1965) with some rocks on each side of it.

Stice and McCoy (1968) discuss the geology of the Manua islands, with Stice (1968) reporting the major element petrology. Hubbard (1971) utilizes rocks from Stice (1968) and reports that lavas from the Manua islands are typical oceanic island alkali lavas; and also reports the rare earth abundances. He suggests the lavas were segregated from a normal oceanic upper mantle at greater than 40 km depth.

Dredged Rocks

Hawkins and Natland (1975) examined the major element chemistry and trace element geochemistry of rocks dredged from the flanks of Savaii, Upolu, and Tutuila as well as nearby seamounts (including Machias) within the Samoan seamount chain. The post-erosional lavas sampled are strongly undersaturated with silica. In order to explain the Si and Al-poor, and Ca, Ti, Mg, Fe-rich basanites and olivine nephelinites, the authors suggest that magma generation took place close to the probable base of the lithosphere. Magma generation was due to moderately extensive melting (10-15%) of an essentially anhydrous and alkali-poor mantle at depths of about 85 km. Hawkins and Natland suggest a hot spot mechanism is uncertain and a more likely explanation is a combination of viscous shear melting and lithosphere dilation due to the deformation of the Pacific plate at the nearby Tonga Trench subduction zone.

Natland (1980) suggests that at a magmatic lineage can be identified on Machias Seamount producing trachytes and phonolites. Natland suggests the compositional differences observed result from differences in the extent and depth of melting of the lavas. All Samoan lavas were found to have an undepleted radiogenic mantle source. The post-erosional lavas have higher $^{87}Sr/^{86}Sr$ and Rb/Sr ratios and a lower K/Rb ratio than shield basalts. Natland suggests the chemistry is indicative of thermal-convective disturbances in the mantle, caused by deformation of the Pacific plate where the Tonga Trench intercepts the plate in the vicinity of the Samoan chain.

Source Material

Natland and Turner (1985) reported on the petrology of lavas from American Samoa and Western Samoa. They suggest the shield volcanoes evolved from a dominantly tholeiitic stage to a transitional stage with tholeiitic and alkalic basalts interbedded, to a dominantly alkalic stage. The Samoan lava, including the tholeiites, are enriched in alkalis and TiO_2. Trace element geochemistry indicates that mantle sources are less depleted than those in Hawaiian volcanoes. The mantle sources of Samoan basalts on each island are less depleted sources, tapped through time. This pattern of depletion is the opposite sequence from that observed on any Hawaiian volcanoes. Natland and Turner suggest a pronounced lateral isotopic heterogeneity beneath the island chain gives rise to the trace-element compositions of the post-erosional lavas.

White and Hofmann (1982) examined the Sr and Nd isotope geochemistry of lavas from Samoa. They concluded that the ratios from Tutuila and Upolu diverge significantly from the mantle array (Davies, 1981). (The results from the Manua islands, at the eastern end of the chain plot within the mantle array). The results cannot be explained by binary mixing of depleted and undepleted mantle reservoirs or by variable magmatic depletion of a planetary reservoir. White and Hofmann (1982) suggest a reinjection of crustal material into the mantle. Newman and others (1984), however, point out that this recycling model is difficult to reconcile with values of $^{3}He/^{4}He$ which are 18 times atmospheric (Rison and Craig, 1982). Newman and others suggests that deviations in ($^{230}Th/^{232}Th$) - $^{87}Sr/^{86}Sr$ correlations exist in Samoa. These could be due to modification of the mantle source or by a significant residence time at depth prior to eruption.

Matsuda and others (1984) examined Sr isotopes from Samoan rocks, and concluded that the correlation in ($^{87}Sr/^{86}Sr$) and Rb/Sr ratios reflect multi-component mixing. They concluded mantle heterogeneity exists over a large scale within the south Pacific.

Wright and White (1987) examined Sr, Nd, and Pb isotope ratios and daughter element concentrations from highly undersaturated post-erosional volcanics erupted in Recent to Historic time. They suggest the post-erosional lavas are derived from mixtures of the shield-building volcano source and a post erosional volcanic source which is characterized by high - $^{207}Pb/^{204}Pb$, $^{87}Sr/^{86}Sr$ and low $^{206}Pb/^{204}Pb$ and $^{143}Nd/^{144}Nd$ ratios. They conclude the source of the post-erosional volcanics contains recycled ancient sediment.

Studies of two types of xenoliths from Samoan post-erosional lavas are reported by Wright (1987). Wright concludes the xenoliths were probably formed during diapirism associated with the melting which produced the Samoan shield volcanoes (1-3 Ma) before their eruption as part of the post-erosional lava flows.

The studies of the isotopic composition of rocks from Samoa has contributed to a better understanding of the nature of the earth's mantle. Strontium, Nd and Pb isotopic abundance data shows that important isotopic mantle anomalies exist (Hart, 1984, Dupre and Allegre, 1983). The isotopic anomalies require a mantle source which may contain recycled ancient sediments (Wright and White, 1986/87). Arguments have been made for a compositionally layered mantle or a heterogeneous mantle mixed by large scale mantle convection (Davies, 1981; Zindler and Hart, 1986). Discussions of mantle mixing are numerous, e.g., Anderson (1982 a and b, 1984, 1985), Davies (1981, 1983, 1984).

Zindler and Hart (1986) conclude "the evidence for a chemically heterogeneous mantle is, today, unequivocal." Arguments continue however about the nature, development, and scale of the heterogeneity. Most authors agree that the heterogeneities are long-lived, being preserved in the mantle for times of approximately 1-4 AE (DePaolo and Wasserburg, 1976). Seismic tomography studies (Dziewonski and Anderson, 1981 and 1983; Dziewonski, 1984; Morelli and Dziewonski, 1987; Doornobs and Hilton, 1989) indicate the heterogeneity is large scale. The isotopic studies of rocks from Samoa summarized here have played an important part in developing new models of the earth's mantle.

Radiometric Age Dating

Radiometric age dates for the Samoan Islands were summarized by Keating (1987). By combining the radiometric results from the Samoan Islands with those from the seamounts and banks of the Melanesian Borderland (Duncan, 1985), a progression in volcanism is observed. A comparison of ages for the Samoan chain and the Hawaiian chain (Fig. 25), shows a progression in ages along the chain.

A comprehensive summary of the radiometric ages for Tutuila, American Samoa has been made by McDougall (1985). McDougall reports that the subaerial portion of the Pago shield volcano was constructed between 1.54 and 1.28 Ma. The Pago caldera was formed at roughly 1.27± 0.02 Ma. The emplacement of trachyte bodies occurred at 1.3± 0.01 Ma. The Olomoana and Taputapu volcanoes formed over the same interval as the Pago volcanism. McDougall (1985) suggests that the rate of migration within the island chain is about 9 cm/yr. Modern post-erosional volcanism is present on the main Samoan Islands and yields ages less than 1 Ma.

Natland and Turner (1985) published radiometric dates for the subaerial portion of the Fagaloa shield volcano in Western Samoa, and suggests it was active between 2.7 and 1.5 Ma as well as several dates from Tutuila.

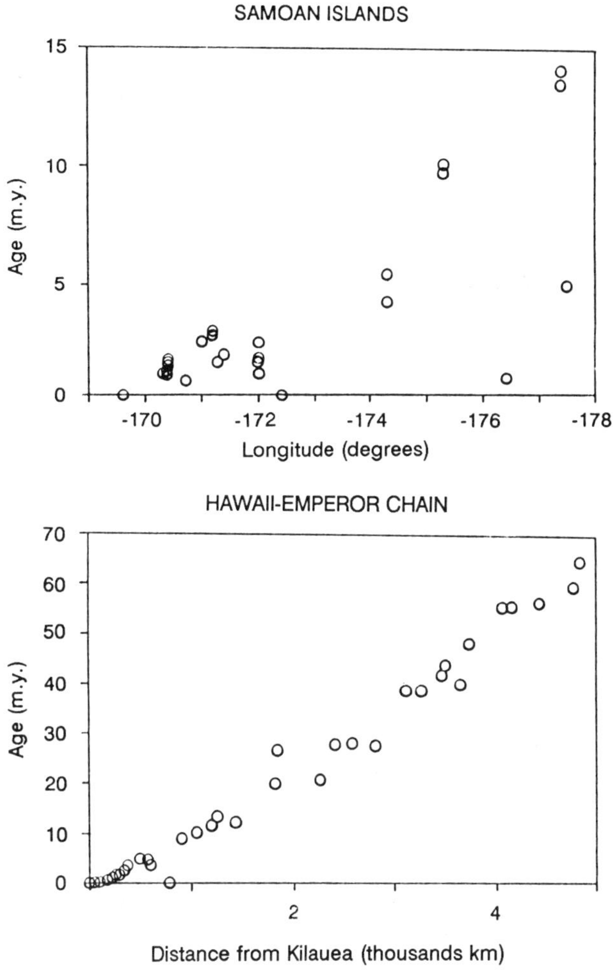

Figure 25. Comparison of the radiometric ages from the Samoan chain and Melanesian Borderland with those of the Hawaiian Islands from Keating (1985a). Both island chains appear to display a linear progression of the oldest ages of volcanism westward, consistent with a tectonic origin as hot spot traces.

Hydrology

Water stored within rocks is a valuable resource. The water is derived from precipitation. Some of the rainwater evaporates from the land surface and returns to the atmosphere; some is utilized in transpiration by plants and also returns to the atmosphere; most of the rain water either runs off the land in streams and eventually is discharged to the sea or is absorbed into the soil or rocks and becomes ground water recharge. Water that percolates past the root zones of plants descends to a level where the voids in the rocks are filled with water; this is referred to as ground water. The upper surface of the water is referred to as the main "water table." The water table slopes gradually from the interior of the island to near sea level at the coast (Fig. 26). Where sufficient fractures or cavities exist near the coast line, springs and fresh water ponds exist. The ground water escapes to the sea at a relatively constant rate. At times of drought, the water table may drop below the base of a stream or pond. As a result, surface waters may dry up. The

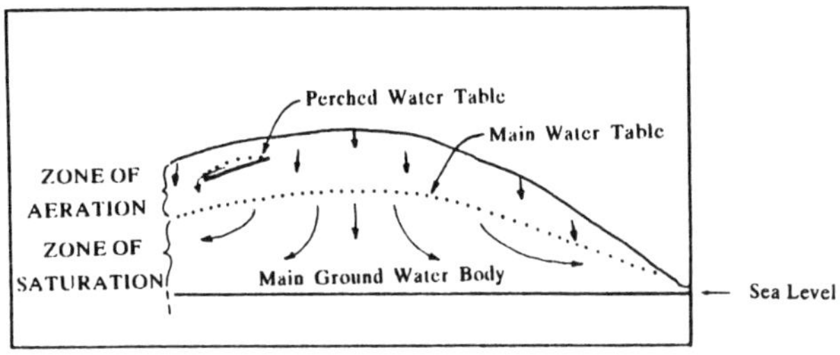

Figure 26. Illustration of the main water table and perched water table. (Figure modified from Macdonald and Abbott, 1970).

slope of the water table is referred to as the ground water gradient. In highly permeable rocks, e.g., fractured and loose basalts, the gradients are low (Stearns, 1941). Clays and unjointed lava flows and dikes are relatively impermeable and have very steep gradients. In dike zones in Hawaii, the gradients are nearly vertical indicating that these dikes act as a barrier to water movement (Fig. 27). In this case, high water tables can be found on the landward side of the dike units (Stearns, 1941).

Rain water is slightly lighter than salt water. This property allows the fresh water to float on the salt water, forming a lens-shaped body of fresh water referred to as the Ghyben-Herzberg lens. The theory developed by Ghyben and Herzberg (Ghyben, 1888 and Herzberg, 1901) and elaborated upon by Hubbert (1940), estimates that for every foot of fresh water above sea level, there will be 40 feet (12 m) of fresh water below it before the salt water is encountered. The division between the fresh and salt water, however, is not sharp; a zone of mixing is present in which the water is not desirable for human consumption. Tidal flushing at the coast line, particularly where permeable rocks are present, influences the size of the zone of mixing.

Discussion of the Hydrology of American Samoa

Eyre (1990) like Stearns (1941) and Bentley (1975) separates rock units into older volcanics (equivalent to shield-building lavas described by other authors), younger volcanics of the Leone plain (which are equivalent to post-erosional lavas as observed in Hawaii) and alluvium. The young, post-erosional lavas are the most permeable, sustaining yields of up to 300 gallons per minute to many wells with little drawdown of the water table. By contrast, the permeability of the alluvium and older volcanics is highly variable. Much of the intracaldera volcanics are not productive aquifers while any particular flow unit can have high permeability, it is generally of limited areal extent thus the unit overall has generally low permeability (Fig. 28). Locally productive aquifers have been discovered in the intercaldera volcanics. These aquifers are associated with fractures, faults, occurrences of volcanic ejecta, and permeable contacts between separate flow units. These zones may sustain yields to a well of 200 gallons per minute or more. However, if such a yield is in excess of surrounding ground water recharge rates, water levels will steadily decline and salt water intrusion will eventually ensue (P. Eyre, personal communication, 1990).

Eyre (personal communications, 1990) states that "even the flank flows of the older volcanoes may provide aquifers of only limited productivity, relative to the post-erosional lavas of the Leone plain, because they are frequently dense

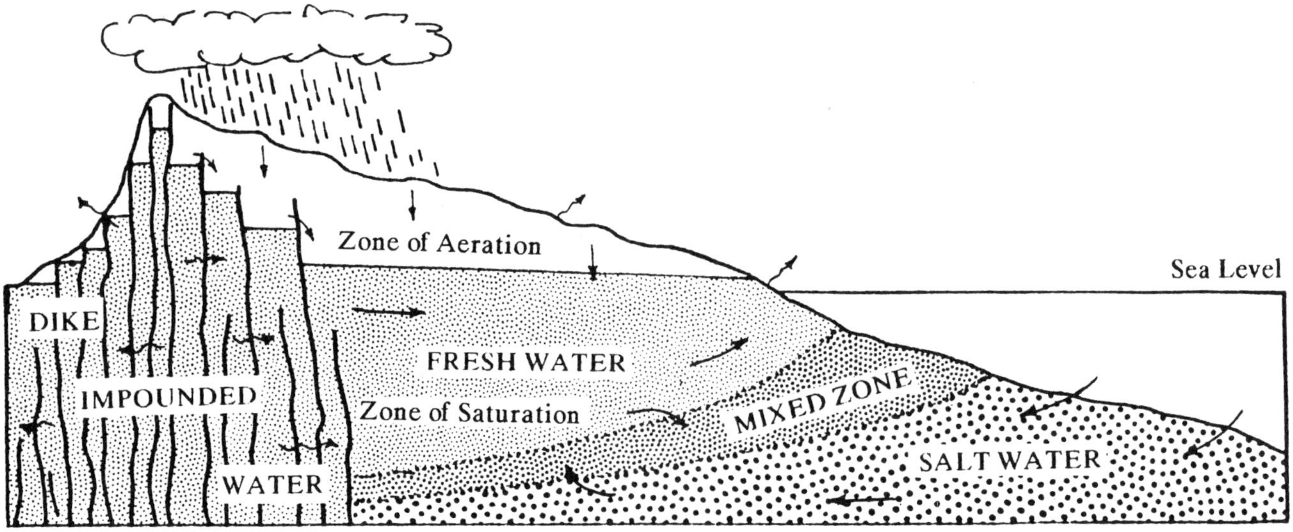

Figure 27. Illustration of the relationship between the fresh water aquifer and the underlying salt water observed in volcanic islands. Areas of dike-impounded waters are also illustrated. (Figure modified from Macdonald and Abbott, 1970).

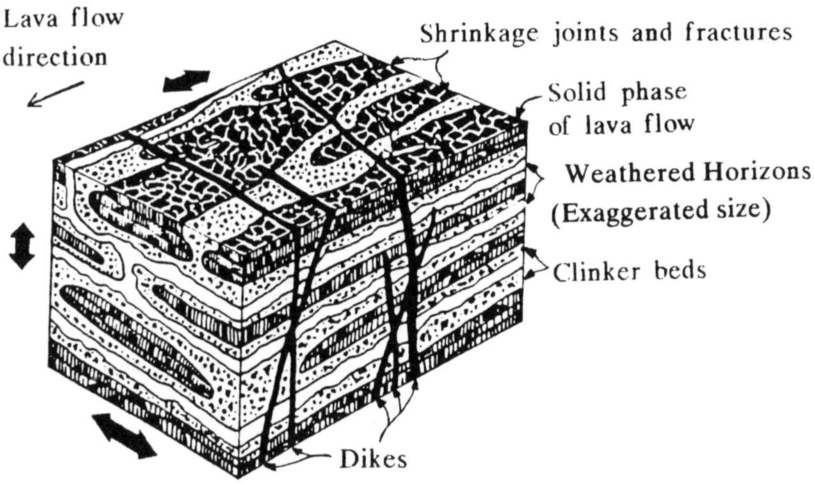

Figure 28. Block diagram illustrating the complex geologic relationships in volcanic landscapes. A complex interfingering of flow units and volcaniclastic units can be present. Dikes, intruding the edifice, can complicate the hydrology by acting as vertical barriers to water movement. (Figure modified from Hunt and others, 1988).

and thick bedded and of limited areal extent. This may be explained by the predominance of viscous alkalic basalt and andesite in Samoa, as well as by the compactness and intensity of eruptive activity on this small island." Eyre also states that "the poor performance of many wells drilled in the older volcanics attests to the difficulty of finding good aquifers." The older volcanics do become an important water supply at high elevations where gravity flow is stopped by dikes, or near-sea level where tuffs and alluvium can be a cap rock which prevents discharge to the sea (Stearns, 1941).

Unweathered tuffs are characterized by moderate permeability. Weathered tuffs are often impermeable. Breccias tend to be dense and have low permeability, when weathered. Fresh cinders have high permeability. Weathered cinders have low permeability. Dikes, interbedded tuffs, and breccias reduce the effective areal extent of lava flows, having a profound effect on ground water movement.

Post-Erosion Volcanics – Leone Peninsula

The eruption of post-erosional volcanics in American Samoa was voluminous and intense. Major centers include the craters of Fagatele, Fogamaa, Olovalu, and the craters near Olotele Peak (P. Eyre, personal communications, 1990). These eruptions produced several areally extensive flows; most are permeable. A large part of the flows covered existing coasts and reefs. The Leone peninsula is a major geographic feature (Fig. 16) formed as the flows built out 3.2 km from the former coastline. Ground water recharge to the lavas of the Leone plain is derived from mountain stream flow as well as direct infiltration of rainfall. Thus, the Leone peninsula is the major aquifer supplying dependable water for Tutuila. Because saltwater underlies the freshwater at a relatively shallow depth, excessive pumping can cause saltwater intrusion when wells reach more than 6 to 9 m below sea level (Stearns, 1941; Davis, 1963, Bentley, 1975; P. Eyre, Personal Communication, 1990). Wells drilled deeper than 18 m below seawater should yield salty water (Bentley, 1975). Saltwater intrusion has caused wells to be shut off during drought years, resulting in water shortages and rationing. Salt water intrusion is exacerbated by the close spacing of wells.

Eyre (personal communication, 1990) estimates that 25 million gallons per day (Mgal/d or 1 m^3/s) of ground water recharges Leone peninsula. Most of the runoff from the mountains adjacent to the plain appears to infiltrate the aquifer, and very little appears to reach the ocean as surface runoff.

Bentley (1975) reviews drillers' logs for wells in the Leone peninsula. The logs indicate that the dense interior portions of lava flows forming the volcanic peninsula are up to 9–15 m thick. The well logs show little uniformity from one well to another, other than the presence of alternating layers of hard basalt and softer units of ash, rubble, and fractured rock. One well at PuaPua drilled through two hard basalt units (29 m and 10 m thick) which were separated by soft material (7 m thick), interpreted as tuff.

Eyre (personal communication, 1990) reports that springs issue at high elevations from the Pago Pago volcano into Taumata Stream in Mormon Valley. The mapping of dikes in the area by Stearns (1941) suggests that high level dike water may occur here. Logs from wells drilled on the floor of Mormon Valley indicate the original valley was filled by Leone volcanics overlying the weathered surface of the older volcanics. The older volcanics have an "impermeable weathered surface," which perches the water to several tens of feet above sea level. As the weathered slope dips seaward beneath the Leone plain, saltwater supports the bottom of the freshwater lens in accordance to the Ghyben-Herzberg principle. Bentley (1975) mentions that reports by consultants (Austin, Smith, and associates, 1963 and 1972) suggest drilling near the boundary of the Leone plain and the Mormoan Valley basalts, to determine if trapped waters are present.

Stearns (1941) reports that the Naval Station in Pago Pago had for years used water from the Pago reservoir near the head of Fagaalu Valley. Stearns found that the reservoir was fed by a spring, 4 m below the top of the dam in the head of the reservoir (234 m above sea level). The reservoir is situated in intra-caldera volcanics according to Stearn's 1944 geologic map, below the steep sides of the Matafao trachyte plug. The spring that feeds the Pago reservoir issues from the fault associated with the Matafao plug.

The region from Vaitele to Fagaalu has no developed ground-water resources (P. Eyre, personal communication, 1990). Outcrops of a'a flows with "alternating layers of high and low permeability" are exposed in the walls of several valleys. Exploration for ground water may prove fruitful in such areas. Several wells have been drilled into the alluvium in the Pago Pago harbor area. These shallow wells yielded fresh water at modest rates, but the wells are susceptible to pollution, and have not been put into production. A private well drilled at the south end of the head of Pago Pago Harbor tapped a thick zone of fresh water under artesian pressure deep in the intra-caldera volcanics (P. Eyre, personal communication, 1990). This successful well shows the benefits possible from artesian aquifers which exist deep in the older volcanics.

At Utulei, a productive well was dug into coastal sediments, 4 m deep (Bentley, 1974) and only 35 m from the shore. The well is in-filled marshland, and appears to tap an artesian discharge from underlying intra-caldera volcanics. There is insufficient surface runoff or rainfall in this area to supply the flow from the well and chloride concentrations increase only slightly during drought (P. Eyre personal communication, 1990).

A well in Aua also tapped an artesian aquifer deep in the intra-caldera volcanics. However, the well was over-pumped and its water became salty. Sparse data from a shallow well on the valley floor indicates that, similar to Utulei, water discharges from the older volcanics into the valley floor sediments.

The lower ends of most stream valleys in Samoa contain thick deposits of alluvium. Often the alluvium contains clays which are impermeable and trap water behind coastal alluvium. Shallow wells dug into these deposits are common and produce brackish water suitable for bathing and laundry. Wells in the alluvium are present in Fagaalu, Pago Pago, Aua, Autu, Alofau, Amouli, and Tula (Bentley, 1975).

In eastern Tutuila, from Lauliituai to Tula the "drainage basins are small and rugged, most wells have low capacity and deliver brackish water," (P. Eyre, personal communications, 1990).

In west and north Tutuila, no wells have been drilled because the stream flow is adequate for the population. The geology is favorable for ground water. A'a flows are common as flank flows on the Olomoana, Alofau and Pago Pago volcanoes. These flows form sequences of dense lavas alternating with permeable, interflow layers. The catchment area is also large, therefore, important aquifers are likely (P. Eyre, personal communications, 1990).

Analyses of water samples reported by Bentley (1975) indicate that the waters from volcanic rocks and alluvium show low concentrations of dissolved solids, except those samples from deep wells and wells near the ocean which show effects of salt water intrusion. Chloride concentrations ranging from 7 to 1,200 mg/L during 1975 to 1983 have been reported by the U. S. Geological Survey (Yee, 1987; Yee, 1988). A level of less than 250 mg/L is the recommended limit for drinking water by the U. S. Environmental Protection Agency (1982).

In modern times, American Samoa has experienced droughts, which have greatly reduced the surface water supplies. In the 1970's, water rationing programs were necessary, and the operations of two tuna canneries were interrupted (Matsuoka, 1978). During the 1980's, powerful hurricanes have hit both American Samoa and Western Samoa, contaminating surface water sources. In addition, frequent landslides occur in these islands (Buchanan-Banks, 1981) which fill and destroy surface water supplies and reservoirs (Stearns, 1941). Further development of ground water supplies are needed to meet the needs of the populations of American Samoa and Western Samoa.

Soils in American Samoa

Wingert (1981) illustrates the soil formations designated in American Samoa. The basic igneous rocks weather to form clayey soils which are nearly impermeable. The volcanic ash and cinders weather to loamy soils. Colluvium forms at the base of the steeper slopes, consisting of silt, clay loam, and silty clay, it is poorly sorted, containing large boulders, and gravels, which constitute up to 35% of the material. Alluvium is deposited by water and ranges from silty clay to fine sand. Thirty soils have been mapped in American Samoa.

GEOLOGIC DEVELOPMENT OF THE SAMOAN ISLAND CHAIN

In discussing the stages of Hawaiian volcanism, Macdonald and Abbott (1970) state that, "volcanoes, like people, pass through a succession of stages in their development." Stearns (1946) outlined eight stages of volcanism (Fig. 29). Macdonald and Abbott (1970) outline nine stages (Fig. 30). The models are similar and outline a history in which the volcanoes build from the sea floor in a phase referred to as the youthful stage (Macdonald and Abbott, 1970) or the shield-building phase (Stages 1-3, Fig. 30). Most of the volcano is formed during this shield-building phase. The next phase of development is referred to as

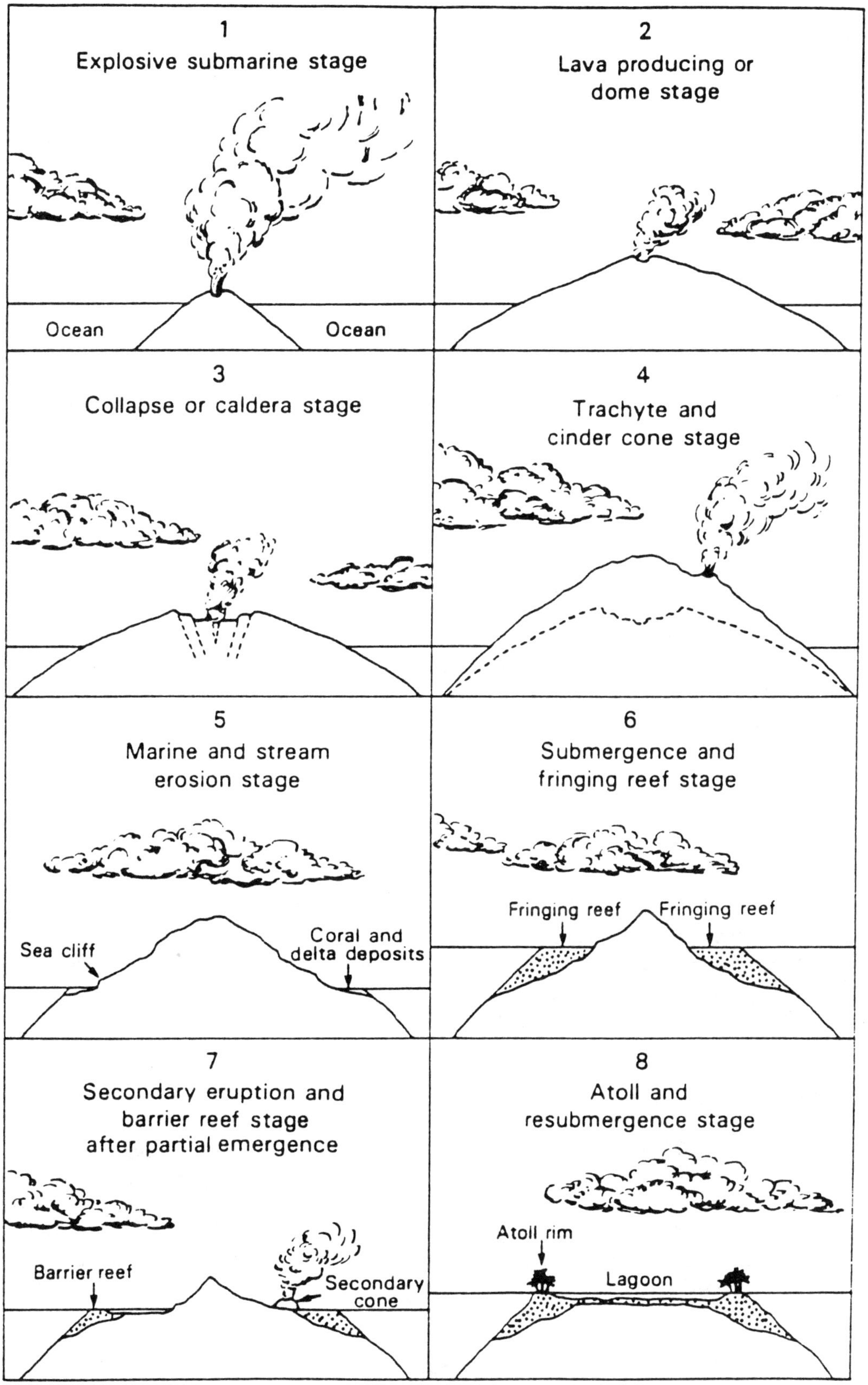

Figure 29. Stages of volcanism proposed by Stearns (1946).

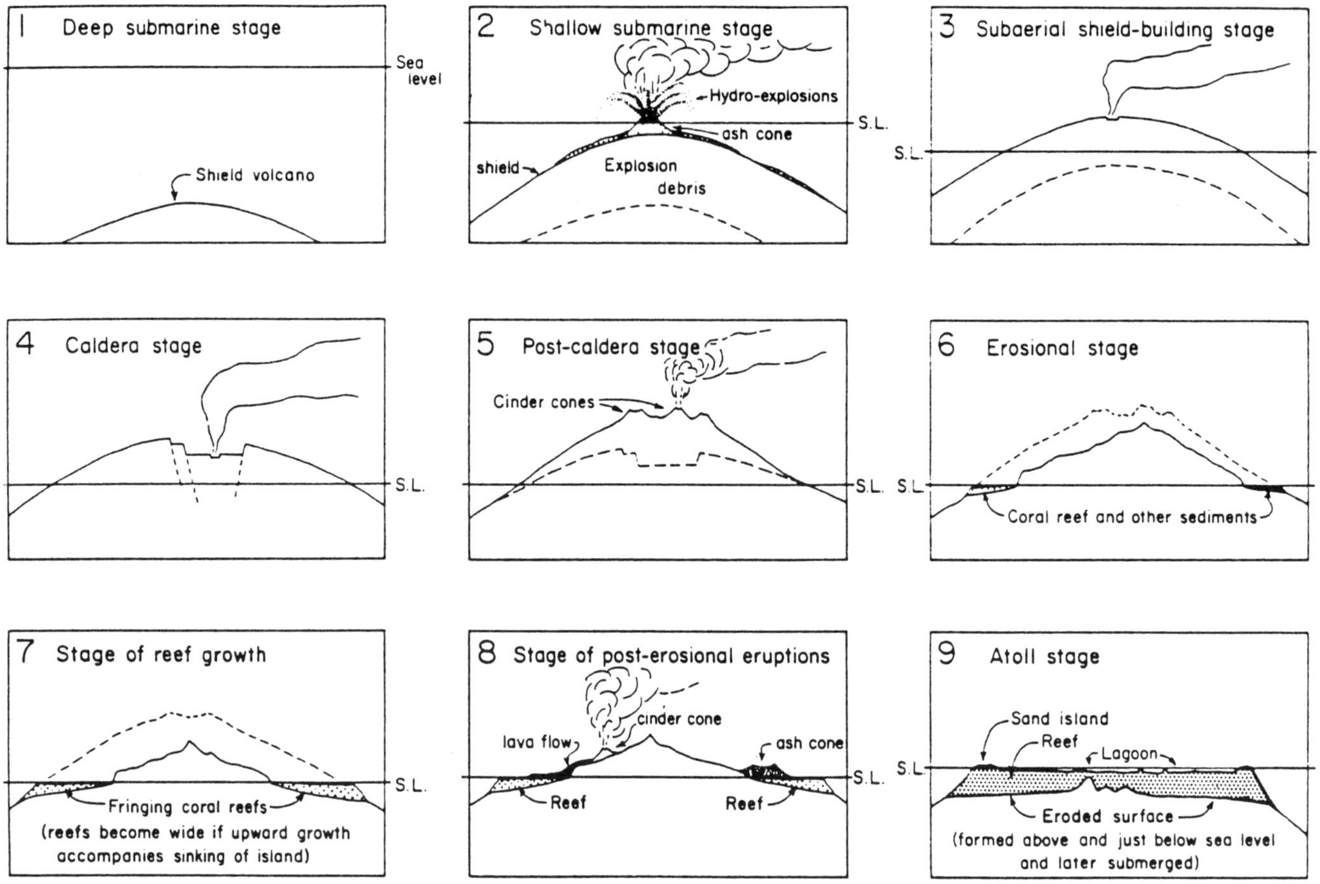

Figure 30. Stages of volcanism observed from the Hawaiian chain (from Macdonald and Abbott, 1970).

the mature stage (Macdonald and Abbott, 1970). Stearns (1946) suggests the formation and collapse of the caldera characterizes the end of this major episode in the development (Stages 4-5), followed by trachytic intrusions and formation of abundant cinder cones (Stage 5). An extensive period of erosion follows (Stage 6). Coral reefs form on the slopes of the volcanoes during the erosional phase (Stages 7 and 8). Late in the history of the volcano, post-erosional volcanism occurs, often concentrated on the slopes of the volcanoes rather than concentrating in the caldera area (Stage 8). This is referred to as the old age phase by Macdonald and Abbott (1970). The volcano continues to subside throughout its history, eventually sinking below sea level until coral reefs grow to form an atoll.

Samoan volcanic development

Shield-Building Phase and Caldera Collapse

The shield building phase of volcanism on Upolu is represented by the Fagaloa volcanics. Kear and Wood (1959) and Kear (1967) suggest the Fagaloa volcanics include two somewhat differently weathered landforms. Tarling (1966) found that the Fagaloa volcanics in western Upolu were normally magnetized while those in eastern Upolu were dominantly reversely magnetized. The paleomagnetic data verifies the Kear and Wood (1959) and Kear (1967) notion that the Fagaloa Volcanics could be divided into an upper and lower series.

Natland and Turner (1985) argue that Fagaloa Bay on Upolu represents the collapsed caldera of the Fagaloa Volcano. They argue that the more alkalic basalts and all of the silica differentiates are centered on Fagaloa Bay. Natland and Turner (1980) however point out that the difference between tholeiites and alkalic basalts in Samoa is arguable. In the case of Samoa, the influence of mantle sources is important, "since they very probably have enhanced the alkalicity of all Samoan basaltic lavas compared with those of Hawaii." As one reviewer pointed out, "alkalicity and silica saturation are not completely interchangeable concepts," according to Natland and Turner (1985).

Regardless of the petrologic arguments, Natland and Turner make a strong case for the existence of a caldera at Fagaloa Bay but acknowledge the lack of "the critical evidence of a ring fault." More mapping is needed in order to verify the geologic relationships and verify the existence of the proposed caldera on Upolu.

On Tutuila, the dikes of Masefau Bay, and the Pago, Alofau, Olomoana, and Taputapu volcanic series represent the shield-building lavas. These four volcanoes formed nearly contemporaneously at about 1.4 Ma. Pago Pago Harbor marks the center of the collapsed Pago caldera.

On Savaii, the oldest exposed volcanics are the Salani volcanics. The island of Savaii is thickly mantled by post-erosional rocks, leading most geologists to believe the shield-building stage of volcanism is buried by later volcanic units on this island. Recently a SeaMARC II side-scan sonar survey was made on the southern flank of Savaii (Figs. 31–34). Despite surveying areas of the flank down slope from extensive cinder cones, little indication of young volcanism is present. The talus-covered slopes appear consistent with the view that the base of Savaii is indeed old and that the younger volcanics have buried the older shield building volcanics. Stice and McCoy (personal communications, 1990) point out that this appears to be the case on Ofu, Olosega and Tau based upon their work in the islands (Stice and McCoy, 1968).

Post-caldera Stage Volcanism

Post-caldera stage volcanism is clearly present on Upolu and Tutuila. Trachytic plugs and numerous cinder cones are evident, along the axis of both islands.

First Erosional Stage

Upolu and Tutuila have experienced considerable erosion. Both islands display extensive dissection by streams.

Post-erosional Volcanism

On Upolu, the Salani volcanics represent post-erosional volcanism. The lavas fill pre-existing valleys. On Tutuila, post-erosional volcanism built the Leone peninsula. The Aunuu tuff covered parts of Tutuila, Ofu and Tau Islands.

Second Erosional Phase

A second erosional phase is evident in Western Samoa. On Upolu, the surface of the Salani volcanics are deeply weathered and deep soils have formed. Deep canyons cut the post-erosional Salani volcanics indicating a long period of erosion took place. On Savaii, the Salani volcanics are moderately weathered and a thick soil cover is present. Numerous deeply incised rivers drain these volcanics.

On Tutuila, the youngest of the high volcanic islands in the chain, the post-erosional volcanics which form the Leone peninsula show little evidence of erosion.

Reef Growth Stage

Reefs are present around Upolu, Savaii, Upolu, Tutuila, Ofu, Olosega and Tau to varying extents. Many reefs have been buried by lava flows. The coral reefs in the Samoan chain have probably not fully recovered from the oscillations of sea level associated with the ice ages. The planation of the flanks of the volcanoes by wave action has provided a stable base for the reefs to grow. Barrier reefs are present where stable platforms remain on the slopes of the islands. Mass wasting of the flanks of the islands, particularly on Savaii, however, has contributed to the loss of coral reefs on the flanks of the volcanoes as portions of the volcanoes slump toward the sea floor.

Continued Volcanic Rejuvenation

On the islands of Savaii and Upolu, volcanic activity has continued until Recent times. On both islands, the relatively unweathered lavas (from the Aopa, PuaPua and Lefaga volcanics) are widely distributed (Fig. 7). Historic volcanism has covered much of Savaii. The volcanic activity on these islands has been repeatedly rejuvenated, producing the Lefaga volcanics, then the PuaPua volcanics, and finally the Aopa volcanics, as well as submarine volcanism near Manua.

Western Samoa lies just north of the Tonga Trench (Fig. 35). Hawkins and Natland (1975) and Natland (1980) suggest that plate bending associated with subduction of the Pacific Plate in the Tonga Trench is responsible for the rejuvenated (post-erosional) volcanism so common in Western Samoa. Natland (1980) suggests shear-melting beneath a zone of lithospheric dilation is responsible for the continued volcanism found in Western Samoa.

Petrogenesis of the Samoan Lavas

Three manuscripts (Hawkins and Natland, 1975; Natland, 1980; and Natland and Turner, 1985) compare the petrologic development of the Samoan volcanic chain to that of the Hawaiian chain. The recent studies of the isotope geochemistry indicate that the mantle geochemistry under the Samoan chain is anomalous. The mantle geochemistry complicates the general interpretation of tholeiitic, transitional, and alkalic volcanism. Samoan lavas have probably all been enhanced in alkalicity compared with Hawaiian lavas. As stated previously, "the alkalicity and degree of silica saturation are not completely interchangeable concepts." Despite the problems produced by enhanced alkalicity, Natland and Turner (1985) believe that "ranges in basalt composition exist encompassing at least the Samoan equivalent of the tholeiite-alkalic basalt transi-

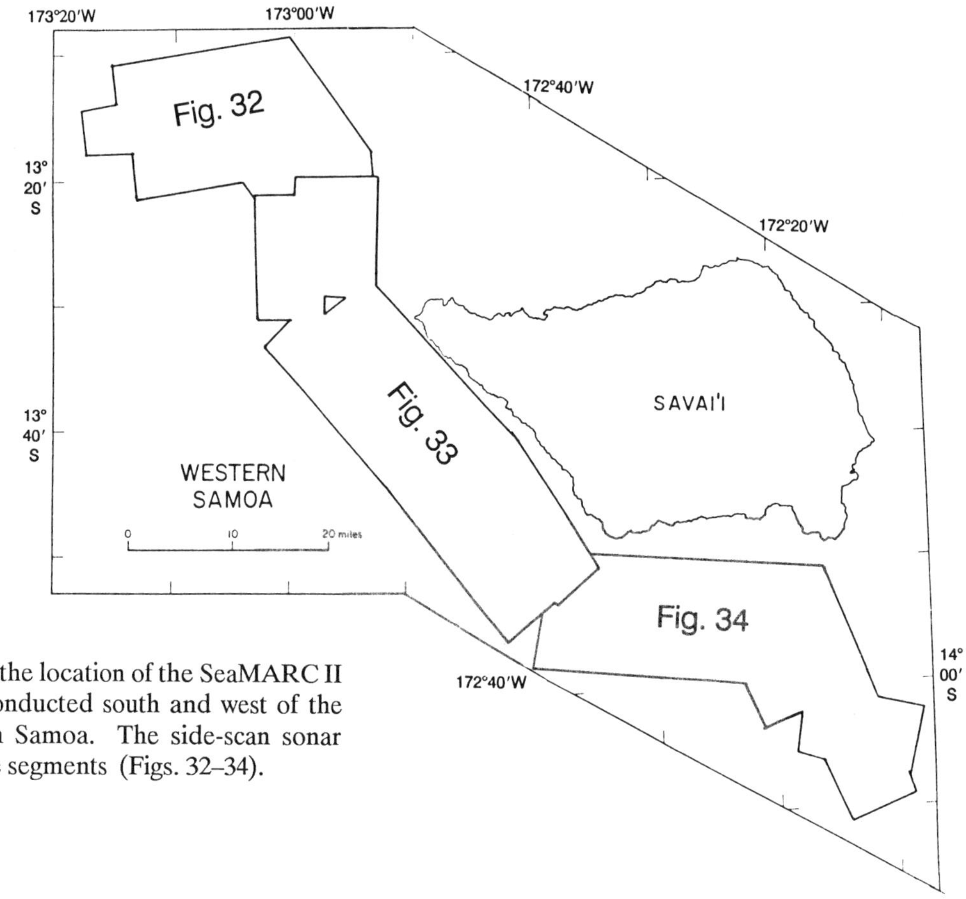

Figure 31. Map showing the location of the SeaMARC II side-scan sonar survey conducted south and west of the island of Savaii, Western Samoa. The side-scan sonar images are shown in three segments (Figs. 32–34).

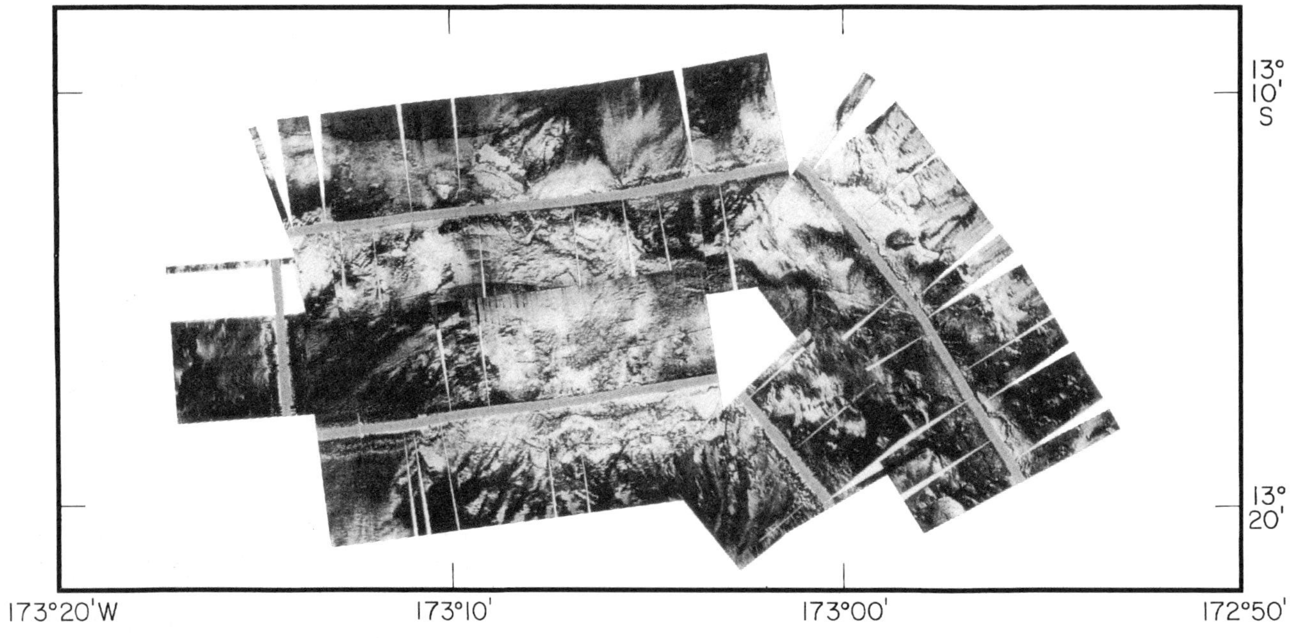

Figure 32. SeaMARC II side-scan sonar image of a seamount immediately west of Savaii. The seamount is capped by low reflectivity material, probably sediments, which appears white in the image. Several faults that parallel the trend of the island chain can be seen on the sea floor east of the seamount. Four small cones can be identified in the image.

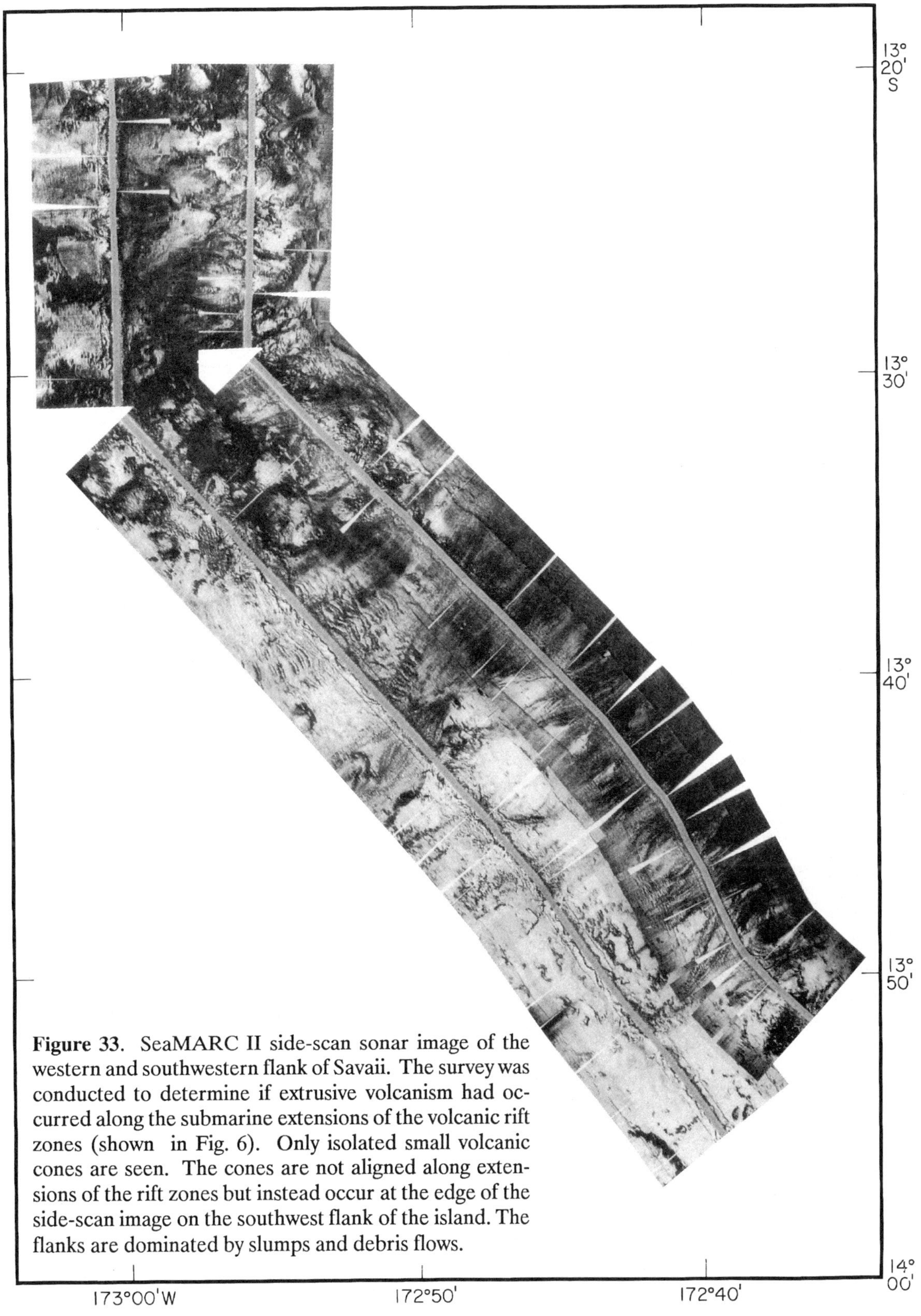

Figure 33. SeaMARC II side-scan sonar image of the western and southwestern flank of Savaii. The survey was conducted to determine if extrusive volcanism had occurred along the submarine extensions of the volcanic rift zones (shown in Fig. 6). Only isolated small volcanic cones are seen. The cones are not aligned along extensions of the rift zones but instead occur at the edge of the side-scan image on the southwest flank of the island. The flanks are dominated by slumps and debris flows.

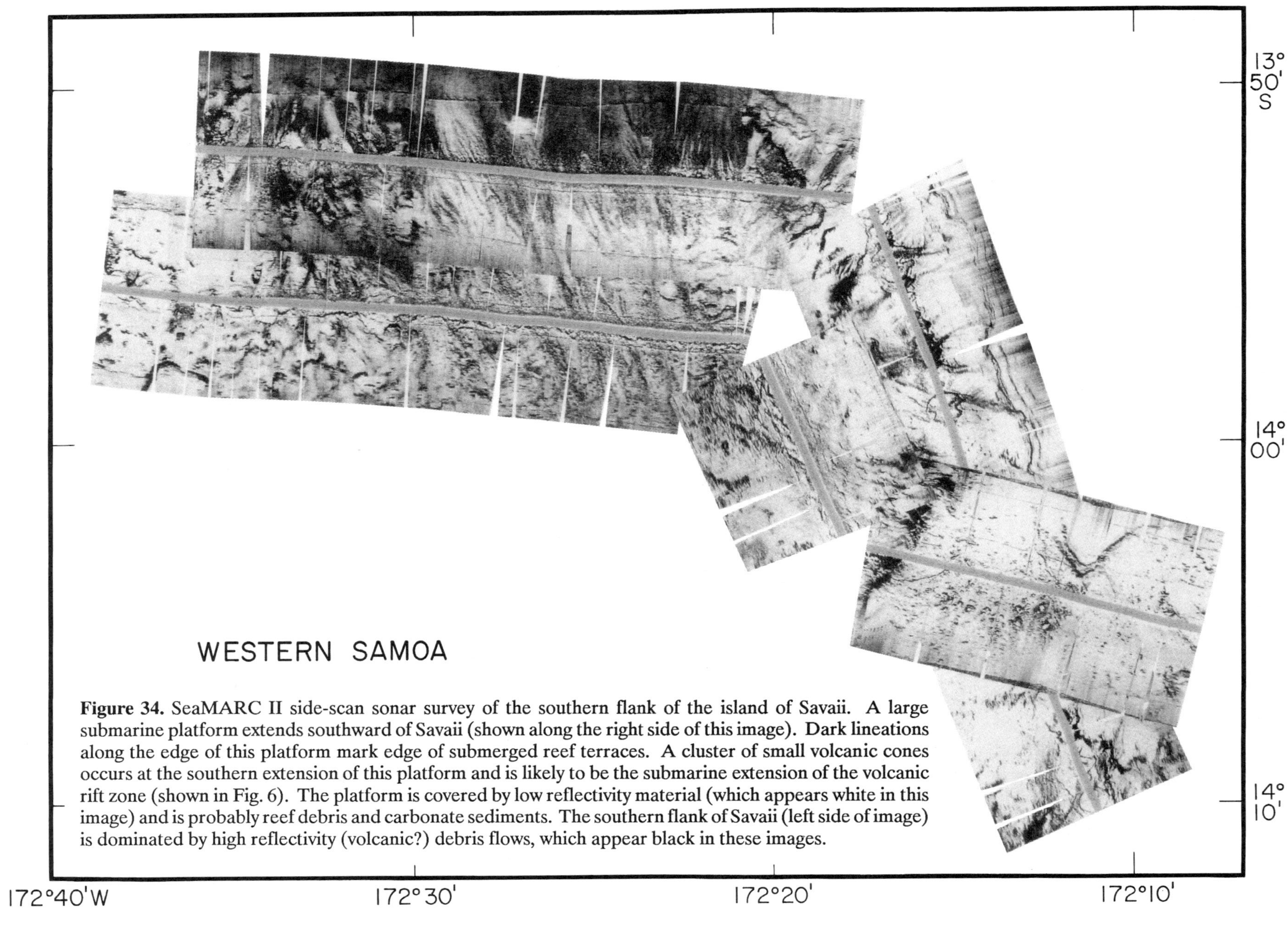

Figure 34. SeaMARC II side-scan sonar survey of the southern flank of the island of Savaii. A large submarine platform extends southward of Savaii (shown along the right side of this image). Dark lineations along the edge of this platform mark edge of submerged reef terraces. A cluster of small volcanic cones occurs at the southern extension of this platform and is likely to be the submarine extension of the volcanic rift zone (shown in Fig. 6). The platform is covered by low reflectivity material (which appears white in this image) and is probably reef debris and carbonate sediments. The southern flank of Savaii (left side of image) is dominated by high reflectivity (volcanic?) debris flows, which appear black in these images.

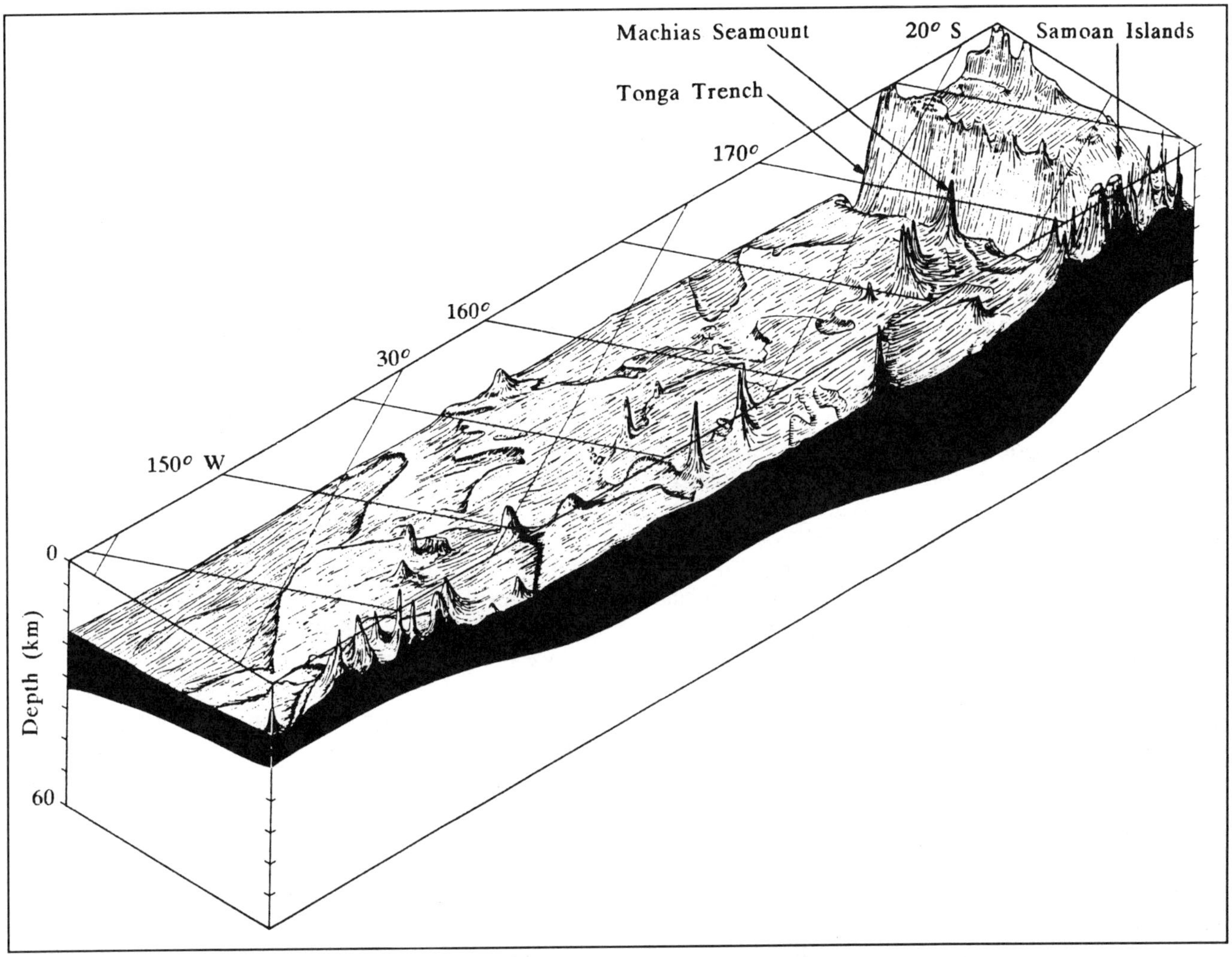

Figure 35. Cross-sectional view of the Samoan and Austral-Cook seamount and island chains, showing their proximity to the Tongan Trench. The view point is near the equator looking south toward the Tongan Trench. The black shading was added by the artist and does not reflect crustal structure. Drawing by Dick Rhodes.

tion. As in the Hawaiian case, this represents a temporal succession, but unlike Hawaii, the transition is from more to less depleted compositions through time." Readers desiring more information of these petrologic arguments are referred to the references previously cited.

THE CORAL REEF PROBLEM

In the early 1800's the descriptions of atolls as great rings of coral around calm lagoons, within the deep Pacific captured the interest of many. Charles Lyell (who was mentor to Charles Darwin) suggested the atolls were the crests of submarine volcanoes overgrown by coral (Darwin, 1842). Darwin conceived the idea that volcanoes grow from the sea floor, form high volcanic islands, then die, subsiding into the sea. Corals grow on the shorelines rimming the volcanic islands; as the volcanoes subside, the corals grow upward, leaving a gap between the barrier reef and the central volcanic island. As subsidence continues and the volcano subsides below sea level, the coral atoll remains. Darwin used direct observations from the fouling of ship bottoms to estimate coral growth rates, concluding corals would have no difficulty in growing fast enough to keep up with subsidence. Darwin's observations were followed by those of James Dana (1875), whose book "Corals and Coral Islands," increased the acceptance of Darwin's ideas on island subsidence.

As a consequence of the Challenger Expedition, John Murray rejected the subsidence hypothesis (Menard, 1986). As a result of the Challenger Expedition, Murray was aware of the deep sea sedimentation of skeletons of microorganisms in the sea. He concluded atolls formed because sediments accumulated on the summit of seamounts, eventually producing shallow banks from which corals could grow. The Murray hypothesis gained considerable support in the late 1880's. Public interest in the Challenger Expedition contributed to the favorable reception of Murray's hypothesis. Darwin's reaction to the hypothesis was that he would like to see some "doubly rich millionaire" pay for borings of atolls to test the hypotheses. The text by Stoddart (this volume) outlines the results of the scientific drilling of atolls, including the early drilling by the Royal Society of London on Funafuti Atoll. Because the early drilling did not reach volcanic rock, arguments regarding the origin of atolls continued. The ideas of Murray and Daly have been referred to as the "antecedent platform theory"; simply stated, reefs grew on pre-existing platforms. Daly (1910) analyzed atolls and concluded that the larger atolls had lagoons of uniform depth (70–90 m). He concluded that during the ice ages the cold waters killed reef building organisms, exposing coastlines to erosion. The oceanic platforms were planed off by waves during periods of low sea level (when water was locked in ice caps). After the ice ages, the seas warmed, coral thrived again, and corals grew to the sea surface, forming barrier reefs on the outer edges of the islands which had been partially truncated. This theory is referred to as the glacial-control hypothesis. Many important observations were made in Samoa early in the 1900's in order to evaluate these hypotheses explaining the nature and occurrence of Pacific coral reefs.

In 1918, Mayor observed wave cut benches around Tutuila at an elevation of 2–3 m above high tide. Daly (1924) observed similar wave cut benches and suggested that sea level was 2–4 m higher than present sea level. Observations were also recorded on the emergent wave-cut benches exposed on the eastern shore of Aunuu Island, Rose Atoll, Ofu and Tau.

Daly (1924) states, "the prolonged denudation of Tutuila was naturally accompanied by the offshore deposition of much sediment." The sediment built broad shelves on the flanks of the seamount. Daly estimated that the total area of shelf surrounding Tutuila is 330 square kilometers, within the 100-fathom (183 m) line. The shelf area is twice that of the island. These observations lend credence to the antecedent platform theory. Davis (1921) examined U. S. Hydrographic maps of Tutuila, and stated, "the shallower parts of the bank are interpreted as submerged fringing and barrier reefs, which are supposed to rest on a wave-cut platform now lying between 60 and 70 fathoms (109–128 m) below sea level by reason of island subsidence."

Chamberlin (1924) attempted to examine the coral reefs of Tutuila in terms of Daly's (1910, 1915) glacial control theory. Chamberlin states, "as far back as 1868, Alfred Tylor suggested that the upper 600 feet (183 m) of coral deposits in the Pacific Ocean might be explained as well by oscillation of sea-level due to the ice-caps of the glacial period as by the accepted hypothesis of sea-bottom subsidence" (Alfred Tylor, 1868).

Penck (1894), von Drygalski (1887), and Daly (1915) estimated the lowering of sea level by withdrawal of water into Pleistocene ice-caps, between 150 and 55 m below present sea level.

Chamberlin (1924) examined the maps constructed by Mayor (1920). These maps indicate that submerged barrier and fringing reefs are present and are well highlighted by the 32-fathom (58 m) or 40-fathom (73 m) contours. Chamberlin also provided additional evidence of planation of Tutuila by pointing out the presence of eroded extrusive and intrusive bodies (e.g., cockscomb trachyte body) situated offshore Tutuila. He concludes the slopes of Tutuila extended at least halfway and likely much further, toward the margin of the existing platform. Chamberlin concluded the broad shelf was partly the result of erosion of the land surface and the buildup of detritus offshore and "partly from the work of the sea." He indicated that the extent of reef growth on the platform was uncertain. During the formation of the reefs, "the sea must have been creeping higher and higher upon Tutuila, as indicated by the vertical thickness of coral. Further subsidence followed the building of the reefs, for they are now deeply submerged." Chamberlin believed that the reef growth on a wave-cut platform rather than on the slopes of a sinking island represented a significant departure from the Darwin-Daly coral reef hypotheses. Davies (1921) however argues very convincingly that this idea did not represent a significant departure from Darwin's writings.

Daly (1924) raised a question about the absence of protecting reefs around Tutuila. Daly stated, "that the island was long devoid of protecting reefs, in spite of the existence of a shelf, is shown by the great height of the sea cliffs, cut before the 6 m shift of ocean level. . . . This reefless condition of the island may thus conceivably have been continued from the last glacial stage. But, the factor leading to the special prolongation of the reefless condition is not easy to discover." Daly (1924) concludes that "subsidence, probably differential, [is required] in order to explain the drowned barrier reef around Tutuila."

Mayor (1924) reported on studies of the coral reefs of Samoa. His in situ investigations of sediment accumulation, dissolution, and so forth, are very insightful. He placed 5 lb (2.2 kg) fragments of tagged reef coral in the lagoon landward of the barrier reef. He returned nine months later and retrieved the stones. He found an average of 115 g weight had been lost. He estimated it would require only 14.5 years to "wholly disintegrate" rocks of this size. Mayor concluded that organic as well as physical abrasion of corals, not dissolution, accounts for lagoon formation around the Samoan Islands.

Davies (1928) argues that the glacial-control theory is largely invalidated by abundant evidence of island instability. Davies views Darwin's theory "modified to advantage," as preferable. He states, "subsidence, therefore, appears to be essential not alone in disposing of outwashed detritus, but also in introducing the special conditions which permit the first successful establishment of young reefs and their further growth."

Stearns (1941) states that both Tutuila and Upolu are characterized by drowned barrier reefs. Stearns (1944) reports that soundings off Upolu indicate a submarine shelf exists 46–55 m along the shore and 73–91 m at a distance of 1.5–5 km offshore, which is similar in size and width to the drowned barrier reef on Tutuila. Stearns also correlates the submarine slope break to the shore line at the time of the "great erosional period" at least 182 m below the present shore line. Stearns concludes the drowned barrier reef rests on a thick section of marine and land deposits on an older platform. The studies of the past century indicate that large submarine platforms are present off the coastlines of the Samoan Islands. Modern studies of reef platforms in the Hawaiian chain indicate that similar platforms and wave-cut benches are extremely common on the slopes of the islands and seamounts down to 3 km. These observations by Moore and Fornari (1984) and Keating (1989) indicate that subsidence is the most important factor in the formation and drowning of reefs on seamounts.

Modifications to the Barrier Reefs

Maps showing the extent of reefs off the Samoan Islands indicate that they are very discontinuous. The discontinuity is caused by mass wasting of the flanks of the volcanoes and modern volcanism which buries the reefs. Thomson (1921) noted that barrier reefs are well-developed in Samoa but "by no means form continuous girdles to the islands." He pointed out that locally it is assumed that the absence of the offshore reef and presence of an "ironbound" (rocky) coast, is a reflection of recent volcanism. He observed that while this is true in Savaii, it "is by no means always the case in Upolu." He proposed that the drowning of the coasts by depression and down-faulting of parts of the coast was an important factor in removing the barrier reef. Thomson suggests that the mass wasting of the flanks of the volcano produced Fagaloa Bay, precluding a simple history of a drowned valley resulting from a change in sea level.

Modern Sea Level

The presence of drowned reefs, truncated ridges, cliffed coastlines, and wave-cut benches in Samoa, document the power of the sea to modify the coastline. Modern studies of climate predict a rise of sea level. The nature and magnitude of such a rise is hotly debated. Global change and its effect on the Pacific islands has also been a topic of wide attention. Islands throughout the Pacific will be affected by even a small rise in sea level. Nunn (1988) has examined the possible areas of inundation for selected parts of the Cook Islands, Fiji, Kiribati, Tonga, and Western Samoa. For Western Samoa, only the area around the capital city of Apia was examined. Nunn reports that most of the coastal mangrove coastal plain would be inundated by a slight rise. If the rise reached 1.5 m, most of the commercial center of Apia, the government building on the Mulinuu Point, and many of the hotels would not be exposed to the sea. A rise to 3.5 m would double the land area impacted and would affect most of the commercial and residential area of Apia.

General Unresolved Questions About the Geology of Savaii

Most geological studies of Pacific islands have been concentrated on the islands of Hawaii. Thus the geological observations on Hawaii are generally used as examples of island volcanism to which all other island groups are compared. Relative to the Hawaiian group, the Samoan group appears anomalous. First, the islands grow larger from east to west, rather than west to east as in Hawaii. Secondly, active volcanism is recorded on both the eastern and western ends of the chain. In Hawaii, the active volcanism is found only on the eastern end. Thirdly, the southeasternmost island in the Samoan chain is an atoll rather than an active volcano. And finally, the tholeiitic rocks so abundant in Hawaii are nearly absent in Samoa. Alkalic rocks and transitional rocks are dominant in Samoa. These rocks constitute a veneer in the Hawaiian Islands, but in Samoa appear to form nearly the entire mass of the volcanoes.

If Hawaii is the proper corollary, then we should be able to use the age relations observed on Tutuila and Upolu to estimate the age of Savaii, providing the volcanic propagation rates are similar to those in Hawaii. Figure 33 illustrates the available radiometric dates for the Samoan chain. The extrapolation of age from the illustration suggests that the age for Savaii should be approximately 4 Ma. The paleomagnetic data suggest ages from 1 to 3 Ma.

The absence of older rocks on Savaii is a major concern of scientists working on the Samoan islands. The majority of the landforms on Savaii are very young. Most geologists have assumed that the younger volcanics bury the much older shield-building volcanics. Thus, a deeper tholeiitic core should be present and be simply capped by a thick sequence of eroded and weathered debris which is, in turn, covered by the Recent post-erosional basalts. If this is the case, this structure has important implications relative to ground water supplies. In order to examine the possibility, dredging is needed on the lower submarine slopes of Savaii in order to obtain samples of the deeper shield-building lavas forming the lower slopes of the seamount. Recently, a side-scan sonar survey was conducted on the southern flank of Savaii. The side-scan image shows the slopes are relatively free of volcanic cones. This lack of obvious volcanism, is suggestive that Savaii is indeed an old volcano with a thick cap of Recent volcanism.

Tectonic setting

Studies of the bathymetric trends of the linear seamount and island chains in the Pacific (e.g., Clague and Jarrard, 1973; Jarrard and Clague, 1977; Epp, 1978) have yielded a great deal of information regarding the tectonic history of the Pacific plate. Linear seamount chains in the Pacific are numerous and varied. The Hawaiian, Emperor, and Line Islands seamount chains are the best studied. West of these three seamount and island chains, however, the bathymetry becomes extremely complex, and the trends are so numerous that even the linear bathymetric trends become less apparent. Analyses of the bathymetric trends show that most of the Pacific seamount and island chains lie on parallel small circles about Pacific plate hot spot poles. The Hawaiian Islands and Emperor Seamounts have been associated with a melting anomaly (e.g., Morgan, 1972; Shaw and Jackson, 1973) referred to as a hot spot. Very limited radiometric, petrographic or paleomagnetic results are available from the western and southern Pacific seamount chains, particularly those that are old and submerged. Due to the lack of dating of the western Pacific seamount chains it is difficult to determine if these chains display simple age progression patterns like the Hawaiian-Emperor chain, or if they were necessarily produced by passage of the Pacific plate over relatively fixed melting anomalies, or hotspots, in the mantle.

The Line Islands chain, for example, has proven to be a composite of two or more major volcanic episodes (Jarrard and Clague, 1977; Jackson, 1976; Saito and Ozima, 1977; Jackson and Schlanger, 1976; Schlanger et al., 1984; Keating, this volume). Clague and Jarrard (1973) and subsequent workers have proposed that the Samoan Islands are a hot spot trace. Hawkins and Natland (1975) and Natland (1980) suggest that plate deformation associated with subduction in the nearby Tonga Trench has influenced, perhaps even caused, Samoan volcanism. It has been suggested that the deformation has determined the location and orientation of Samoan shield volcanoes, and contributes to the unusual volume of post-erosional volcanism at the western end of the chain. This subduction-related deformation has determined the orientation of the post-erosional volcanic rift zone, and has deformed the sea floor around the Samoan islands.

The results of new studies provide basic observations that constrain the origin, age, and evolution of the Samoan Islands and Melanesian borderland seamounts (Brocher, 1985; Duncan, 1985; Natland and Turner, 1985; Sinton et al., 1985; Keating, 1985). These authors suggest that the results of the individual studies are consistent with a hot spot origin for the Samoan islands. In examining the data collectively, sparsity of data becomes an important concern. At the present time, the paleomagnetic, radiometric, and geochemical data remain very limited. The paleomagnetic data, for example, can be correlated with a hot spot model for the origin of the Samoan islands. But at the same time, the available data can also be used to support a model for progressive volcanism toward Savaii, which is quite inconsistent with a hot spot model. Likewise, the existing radiometric dates on the islands are limited; they allow a positive correlation with a hot spot model in only two islands and two seamounts west of the Samoan chain. Thus, although the data are "broadly consistent with a hot spot mode of origin for the volcanoes," as stated by McDougall (personal communications, 1984), it is important that variations of the hot spot mode of origin be examined in order to explain the anomalous patterns of volcanism observed in the Samoan Islands.

Duncan (1985) shows that recent (0.8 Ma) volcanic activity has occurred on Wallis Island (roughly 180 nm west of Savaii; Fig. 36). On the basis of the present geological knowledge of the Samoan islands, it is likely that the latest volcanism is contemporaneous on Wallis, Savaii, and the Manua islands. Could these three islands be related? The three form a linear trend that cuts across the main trend of the Samoan group. Assigning a deformational history to this recent activity, as suggested by Natland, resolves many of the anomalous features of the volcanic propagation in the Samoan chain. Continued studies of the age, geochemistry, and magnetic history associated with mapping of these islands and seamounts are the key to our knowledge of the evolution of this unusual chain of islands and seamounts.

ACKNOWLEDGMENTS

This review has been many years in the making. Without the encouragement of Harold Stearns, it would never have been completed. The opportunities to visit Harold and his wife, stimulated this work. I also was encouraged by Paul Eyre, whose conversations regarding the hydrology aided me in completing this manuscript. I thank Gary Stice, Floyd McCoy, and Paul Eyre for their excellent reviews of the paper. I also thank Don Tarling and Jim Natland for their encouragement during the final stages of manuscript production. The figures were drafted by Dick Rhodes and Kenneth Helsley. I thank the Agency for International Development for support of this undertaking and

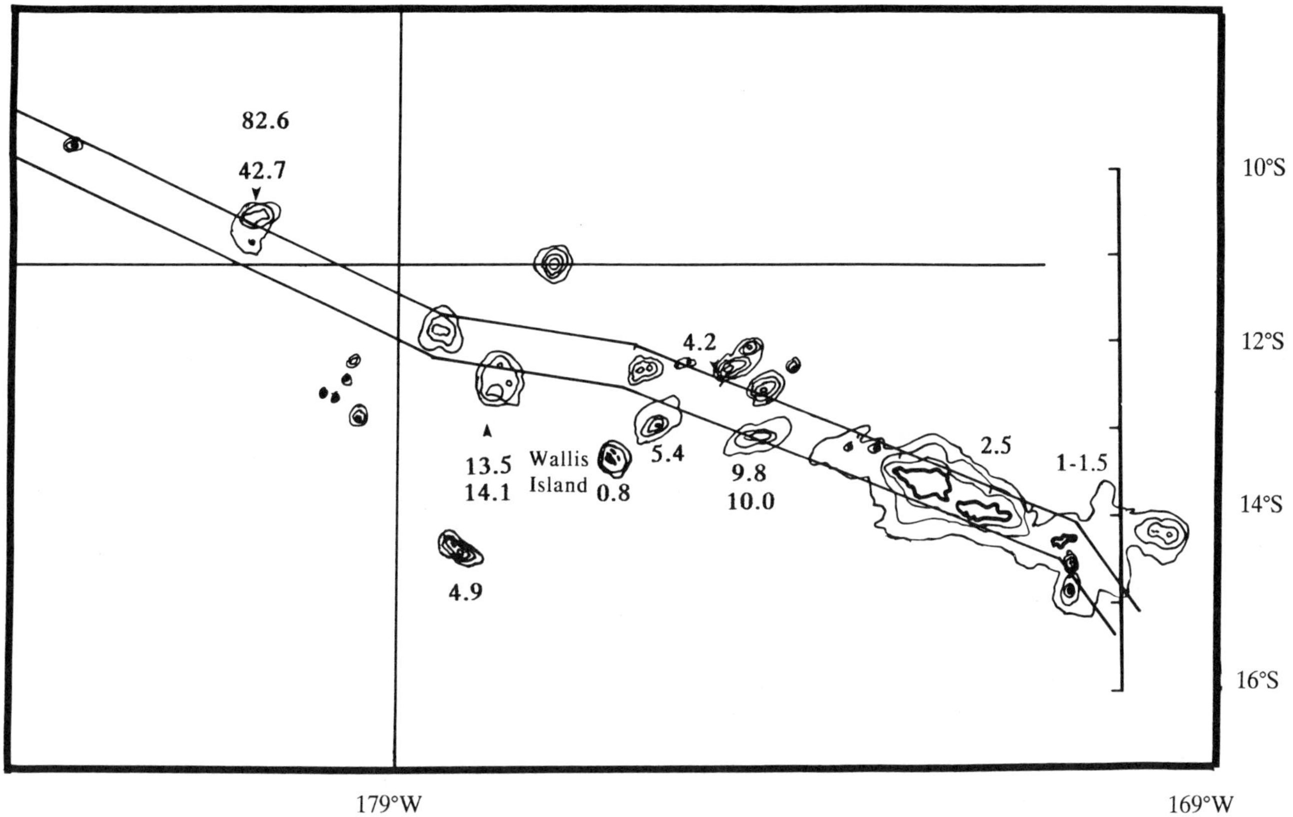

Figure 36. Map of the Samoan island chain and the seamounts and banks of the Melanesian Borderland. Radiometric dates from McDougall (1985) and Duncan (1985) are shown along the projection of the hot spot trace constructed parallel to that of the Hawaiian chain by Epp (1978). Wallis Island is dated as 0.8 Ma.

the National Science Foundation for funding the paleomagnetic studies cited here. Hawaii Institute of Geophysics contribution no. 2317.

REFERENCES

Anderson, T., 1910, The volcano of Matavanu in Savaii: Q. Jour. Geol. Soc. Lond., v. 66, p. 621–639.

Anderson, T., 1911, Matavanu, a new Volcano in Savaii (German Samoa), Rep. Brit. Assoc., v. 80, p. 654.

Anderson, T., 1912, Matavanu: A New Volcano in Savaii (German Samoa), Proc. Roy. Inst. Great Brit., v. 19, p. 856.

Anderson, D.L., 1982(a), Hotspots, polar wander, Mesozoic convection and the geoid: Nature, v. 297, p. 391–393.

Anderson, D.L., 1982(b), Isotopic evolution of the mantle, the role of magma mixing: Earth and Planetary Science Letters, v. 57, p. 1–12.

Anderson, D.L., 1984, The Earth as a planet: paradigms and paradoxes, Science, v. 223, p. 347–355.

Anderson, D.L., 1985, Hotspot magmas can form by fractionation and contamination of mid-ocean ridge basalts: Nature, v. 318, p. 145–149.

Angenheister, G., 1909, Beobachtungen am Vulkan der Insel Savaii'i (Samoa): Globus. II., March 1909, p. 138–42.

Army Corps of Engineers, 1980, American Samoa coral reef inventory: Honolulu, Pacific Ocean Division, 314 pp.

Austin, Smith and Associates, 1963, Report covering a master planned water supply and distribution system for the Pago Pago and Tafuna areas: A report to the government of American Samoa, 48 pp.

Austin, Smith and Associates, 1972, Report updating the 1963 master planned water supply and distribution system for the Pago Pago and Tafuna areas: Honolulu, Hawaii, 48 pp.

Bartrum, J.A., 1927, Igneous Rocks from Western Samoa: N.Z. Inst., Trans, v. 57, p. 253-264.

Bentley, C.B., 1975, Ground water resources of American Samoa with emphasis on the Tafuna - Leone plain, Tutuila Island: U.S. Geological Survey Water Resources Investigation, 29-75, 32 pp.

Bramlette, M.N., 1926, Some marine bottom samples from Pago Pago Harbor, Samoa: Washington, Carnegie Institution Publication, 35 pp.

Brocher, T.M., 1985, Geological Investigations of the Northern Melanesian Borderland, Earth Science Series, v. 3: Houston, TX, Circum-Pacific Council for Energy and Mineral Resources, 199 pp.

Bryan, E.H., Jr., 1939, Rose Atoll, U.S.A., Paradise of the Pacific, v. 51, p. 9, 25–26.

Bryan, E.H., 1941, American Polynesia, Coral Islands of the Central Pacific: Tongg Printing, Honolulu.

Buchanan-Banks, J.M., 1981, The Oct 28, 1979, Landsliding on Tutuila, American Samoa: U.S. Geological Survey Open File Report No. 81-0081, 21 pp.

Chamberlin, R.T., 1924, The geologic interpretation of the coral reefs of Tutuila, American Samoa: Washington, Carnegie Institution Publication 340, p. 147–172.

Clague, D.A., and Jarrard, R.D., 1973, Tertiary Pacific plate motion deduced from the Hawaiian-Emperor Chain: Geological Society of America Bulletin, v. 84, p. 1135–1154.

Cline, M.G., 1955, Soil Survey: Territory of Hawaii, USDA Series 1939, No. 25.
Couthouy, J.P., 1842, Remarks upon coral formations in the Pacific: Boston Journal of Natural History, v. 4, p. 67–105; 137–162.
Coulbourn W., Hill, P.J. and Bergersen, D.D., 1989, Machias Seamount, Western Samoa: sediment remobilization, tectonic dismemberment and subduction of a guyot: Geo-Marine Letters v. 9, p.119–125.
Daly, R.A., 1924, The geology of American Samoa, Carnegie Institution Publication 340, p. 95-145.
Dana, J.D., 1849, U.S. Exploring expedition during the years 1838-1842 under command of Charles Wilkes, U.S.N., 10, Geology; New York.
Dana, J.D., 1875, Corals and Coral Islands: London, Sampson Low, Marston, Low and Searle, 348 pp.
Dana, E.S., 1880, Characteristics of Volcanoes: London, Low.
Darwin, C., 1842, The structure and distribution of coral reefs, reprinted by the University of Arizona Press in 1984, Tucson, 205 pp.
Davies, G.F., 1983, Geophysical and isotopic constraints on mantle convection (abs.): Eos, Transactions of the American Geophysical Union, v. 63, p. 612.
Davies, G.F., 1984, Lagging mantle convection, the geoid and mantle structure: Earth and Planetary Science Letters, v. 69, p. 187–194.
Davies, G.F., 1981, Regional compensation of subducted sphere: effects on geoid, gravity and topography from a preliminary model: Earth and Planetary Science Letters, v. 54, p. 431-441.
Davis, W.M., 1921, The coral reefs of Tutuila, Samoa: Science, v. 53, p. 559-565.
Davis, D.A., 1963, Ground-water reconnaissance of American Samoa: U.S. Geological Survey Water-Supply Paper 1608-C, 21 pp.
DePaolo, D.J. and Wasserberg, G.J., 1976, Nd isotopic variations and petrogenetic models: Geophysical Research Letters, v. 3, p. 249–252.
Dingler, J., R., Wolf, S. C, and Carlson, D., 1986, Offshore sand resources along the south shore of Tutuila Island, American Samoa (1978) U.S. Geological Survey Open-File Report No. OF 86-0112, 43 pp.
Dingler, J. R., Carlson, D.V.; and Sallenger, A.H., Jr., 1987, Map showing locations and statistical parameters of beach and offshore sand samples, Tutuila Island, American Samoa: U.S. Geological Survey Miscellaneous Field Studies Map . No. MF-1875, 1 sheet.
Doornobs, D.J. and T. Hilton, 1989, Models of the core-mantle boundary and the travel times of internally reflected core phases: Journal of Geophysical Research, v. 94, p. 15, 741–15, 751.
Duncan, R.A., 1985, Radiometric ages from volcanic rocks along the New Hebrides-Samoa Lineament, in T. Brocher, ed., Geological Investigations of the Melanesian Borderland, Earth Science Series, v. 3: Houston, TX, Circum Pacific Council for Energy and Mineral Resources , p. 67-76.
Dupre, B. and Allegre, C.J., 1983, Pb-Sr isotope variation in Indian Ocean basalts and mixing phenomena: Nature, v. 303, p. 142–146.
Dziewonski, A., 1984, Mapping the lower mantle: determinations of lateral heterogeneity in P velocity up to degree and order 6: Journal of Geophysical Research, v., 89, p. 5929-5952.
Dziewonski, A.M. and D.L. Anderson, 1981, Preliminary reference earth model (PREM): Physics of the Earth and Planetary Interiors, v. 25, p. 297-356.
Dziewonski, A.M and D.L. Anderson, 1983, Travel times and station corrections for P waves at teleseismic distances: Journal of Geophysical Research, v. 88, p. 3295-3314.
Dziewonski, A.M. and Gilbert, F., 1976, The effect of small, aspherical perturbations on travel times and a re-examination of the corrections for ellipticity: Geophysical Journal of the Royal Astronomical Society, v. 44, p. 7-17.
Edwards, E., and Hamilton, G., 1915, Voyage of **H.M.S. Pandora** dispatched to arrest the mutineers of the **Bounty** in the South Seas, 1790-91: London, Edwards, 117 pp.
Epp, D., 1978, Age and tectonic significance among volcanic chains on the Pacific Plate, Ph.D. Dissertation, University of Hawaii, Honolulu, HI.
Exon, N.F., 1982, Offshore sediments, phosphorite and manganese nodules in the Samoan region, southwest Pacific: South Pacific Marine Geological Notes, v. 2, p. 103-120.
Eyre, Paul, R., 1990, Ground water resources of Tutuila, American Samoa: U.S. Geological Survey Water Resources Investigation Report, 55 pp.
Friederici, G., 1910, Ein besuch des vulkans von Savai' i (Samoa): Globus, v. 97, p. 14.
Friedlander, I., 1910, Beitrage zur geologie der Samoa - Inseln: Abh. Bayer Akad d. Wissensch II kl. Bd. 24, pp 507-541.
Gyyben, W.B., 1888, Nota in verband net de voorgenoman putboring nabi, Amsterdam. Tidschrift van Let Koninklijk Inst. van Ing.
Grevel, W., 1911, Der Matavanu auf Savai'i am 9 April 1, 1911, Z. Ges. Erdk. Berl., p. 485 - 491.
Hamilton, W.M. and Grange, L.I., 1938, The soils and agriculture of Western Samoa: New Zealand DSIR Bulletin 61, Wellington.
Hart, S.R., 1984, A large scale isotope anomaly in the Southern Hemisphere mantle: Nature, v. 309, p. 753-757.
Hawkins, J.W. and Natland, J.H., 1975, Nephelinites and basanites of the Samoa linear volcanic chain: the possible tectonic significance: Earth and Planetary Science Letters, v. 24, p. 427-439.
Hedge, G.E., 1966, Variations in radiogenic strontium found in volcanic rocks: Journal of Geophysical Research, v. 71, p. 6119-6120.
Hedge, C.E., Peterson, Z.E. and Dickinson, W.R., 1972, Petrogenesis of Lavas from Western Samoa: Geological Society of America Bulletin, v. 83, p. 2709–2714.
Herzberg, A., 1901. Wasserversorgung einger Nordseebader. J. Gasbeleucht. Wassenversorg., v. 44, p. 815-819.
Holmes, A., 1945, Principles of Physical Geology, 2nd Ed.: New York, Ronald, 532 pp.
Hubbard, N.J., 1971, Some chemical features of lavas from the Manua Islands Samoa: Pacific Science, v. 25, p. 178-187.
Hubbert, M. K., 1940, The theory of ground-water motion: Journal of Geology, v. 45, p. 785-944.
Hunt, C.D., Jr., Ewart, C.J., and Voss, C.I., 1988, Region 27, Hawaiian Islands, in Back, W., Rosenshin, J.S., and Seaber, P.R., eds., Hydrogeology, Geology of North America, v. O-2: Boulder, CO, Geological Society of America.
Hutchinson, G.E., 1950, The Biochemistry of Vertebrate Excretion: Bulletin American. Museum Natural History, v. 96, 554 pp.
Jackson,E.D. and Schlanger, S.O., 1976, Regional synthesis, Line Islands, Tuamotu chain and Manihiki Plateau, Central Pacific Basin:Initial Reports of the Deep Sea Drilling Project, v. 33, p. 915-1027.
Jarrard, R.D. and Clague, D.A., 1977, Implications of Pacific island and seamount ages for the origin of volcanic chains: Reviews of Geophysics and Space Physics, v. 15, p. 57-76.
Jensen, H.I., 1907, The geology of Samoa and the eruption in Savaii: Proceedings of the Linnean Society of New South Wales, v. 31, p. 641-672.
Jensen, H.P., 1908, Chemical note on a recent lava from Savaii: Proceedings of the Linnean Society of New South Wales, v. 32, p. 706-707.
Kaiser, E., 1904, Beitrage zur petrographic und geologie der Deutschen Sudsee-Inseln. Jb. Preuss. Geol. Landesanst. v. 24, p. 91-121.
Klautsch, 1907, Der jungste vulkanausbruch auf Savaii, Samoa Jahrb. K. Preuss. Geol. Landesanst., v. 28, p. 169-182.
Kear, D. and Wood, B.L., 1959, The geology and hydrology of Western Samoa: New Zealand Geological Survey Bulletin, v. 63, p. 1-90.
Kear, D., 1967, Geological notes on Western Samoa: New Zealand Journal of Geology and Geophysics, v. 10, p. 1446-1551.
Keating, B., 1985a, Paleomagnetic studies of the Samoa Islands: results from the islands of Tutuila and Savaii, **in** T. Brocher, ed., Geological Investigations of the Melanesian Borderland, Earth Science Series, v. 3: Houston, TX, Circum-Pacific Council for Energy and Mineral Resources, p. 187-199.
Keating, B.H., 1985b, Structural failure and drowning of Johnston Atoll, Central Pacific Basin, in Keating, B.H. Fryer, P., Batiza, R., and Boehlert, G.W., (eds) Seamounts, Islands, and Atolls: Washington, DC, American Geophysical Union, Geophysical Monograph 43, p. 49-60.

Keating, B.H., 1987, Summary of the radiometric ages from the Pacific, Intergovernmental Oceanographic Commission, Technical Series, 32, 67 pp.

Keating, B.H., 1989, Subsidence records from Pacific atolls and islands (abst.): Eos, Transactions of the American Geophysical Union, p. 1370.

Keating, B.H., Insular geology of the Line Islands, this volume.

Lipman, C.B. and Taylor, J.K., 1924, Bacteriological studies on Rose Islet soils: Carnegie Institution Publication 340, p. 209-217.

Lonsdale, P., 1975, Sedimentation and tectonic modification of the Samoa archipelagic apron: Bulletin, American Association of Petroleum Geologists, v. 59, p. 1333-1362.

Machesky, L.F., 1965, Gravity relations in American Samoa and the Society Islands: Pacific Science, v. 19, p. 367-373.

Macdonald, G.A., 1944, Petrography of the Samoan Islands: Bulletin of the Geological Society of America, v. 55, p. 1333-1362.

Macdonald, G.A., 1968, A contribution to the petrology of Tutuila, American Samoa: Sonoerdruck Aus der Geologischen Rundscha Band, v. 57, p. 821-837.

Macdonald, G.A. and Abbott, A.T., 1970, Volcanoes in the Sea, the Geology of Hawaii: Honolulu, The University Press of Hawaii, 282, 398 (reprinted 1974).

Manheim, F.T. and Lane-Bostwick, C.M., 1989, Chemical composition of ferromanganese crusts in the world ocean: a review and comprehensive database, U.S. Geological Survey Open-File Report 89-020, 201 pp.

Matsuda, J., Notsu, K., OKano, J., Yaskawa, K and Chungue, L., 1984, Geochemical implications from Sr isotopes and K-ar age determinations for the Cook-Austral Islands Chain: Tectonophysics, v. 104, p. 145–154.

Matsuoka, I., 1978, Flow characteristics of streams in Tutuila, American Samoa, U.S. Geological Survey Water Resources Investigations, 78-103, 34 pp.

Matsushima, Y., Sugimura, A., Berryman, K., Ishii, T., Maeda, Y., Matsumoto, E. and Yonekura, N., 1984, B. Research report of B-party: Holocene sea-level changes in Fiji and Western Samoa: sea-level changes and tectonics in the Middle Pacific: Report of the HIPAC project in 1981, 1982, 1983, Kobe University, Japan.

Major, A.G., 1920, The reefs of Tutuila, Samoa, in relation to coral-reef theories, Proceedings of the American Philosophical Society, v. 19, p. 3-4.

Mayor, A.G., 1924, Rose Atoll, American Samoa, Carnegie Institution Publication 340, Washington, D.C., p. 73-79.

Mayor, A.G., 1924, Structure and ecology of Samoan reefs, Carnegie Institution Publication 340, Washington, D.C., p. 1-25.

McDougall, I., 1964, Potassium-argon ages from lavas of the Hawaiian Islands: Bulletin of the Geological Society of America, v. 75, p. 107-128.

McDougall, I., 1985, Age and evolution of volcanoes of Tutuila, American Samoa: Pacific Science, v. 39, p. 311-320.

McMurtry, G.M., De Carlo, E.H., Kim, K.H. and Vondehaar, D., 1986, Geochemical investigations of Co-rich ferromanganese crusts in the Hawaiian Exclusive Economic Zone: Addendum to the Final Report - Resource Assessment of Cobalt-Rich Ferromanganese Crusts within the Hawaiian Exclusive Economic Zone, U.S. Department of the Interior, Mineral Management Service, Cooperative Agreement 14-12-001-30177, 92 pp.

Menard, H.W., 1986, Islands: New York, Scientific American Books, 230 pp.

Mohle, F., 1902, Beitrag zur petrographic der Sandwich - und Samoa - Insein: N. Jahrb. f. Min. u. Petr., Suppl., 15, p. 66-104.

Moore, J.G. and Fornari, D.J., 1984, Drowned reefs as indicators of the rate of subsidence of the Island of Hawaii: Journal of Geology, v. 92, p. 752-759.

Morgan, W.J., 1972, Deep mantle convection plumes and plate motion, Bulletin, American Association of Petroleum Geologists, v. 56, p. 203-213.

Morelli, A., and Dziewonski, A.M., 1987, Topography of the core-mantle boundary and lateral homogeneity of the liquid core: Nature, v. 325, p. 678-683.

Morrison, R.J., Regina, A., Prasad, R.A., and Asghan, A., 1986, Taxonomy of some Western Samoa benchmark soils: University of the South Pacific, Environmental Studies Report 28, 65 pp.

Natland, J.H., 1975, Geology and tectonics of the Fagaloa shield volcano, island of Upolu, Western Samoa: plate bending, faulting and possible shear melting north of the Tonga Trench: Geological Society-America, Abstracts, Program, v. 6, p. 226.

Natland, J.H., 1980, The progression of volcanism in the Samoan Island volcanic chain: American Journal of Science, v. 280-A, p. 709-735.

Natland, J.H. and Turner, D.L., 1985, Age progression and petrological development of Samoan shield volcanoes: evidence from K-Ar Ages, lava compositions, and mineral studies, in T.M. Brocher, ed., 1985, Geological Investigations of the Northern Melanesia Borderland, Earth Science Series, v. 3: Houston, TX, Circum - Pacific Council for Energy and Mineral Resources, p. 139-172.

Nakamura, S., 1984, Soil survey of American Samoa, 1984: Washington, Department of Agriculture, Soil Conservation Service, 95 pp.

Needham, R.E. Buland, R.P., Choy, G.L, Dewey, J.W., Engdahl, E.R., Sipkin, S.A., Spence, W., and Zirbes, M.D., 1982, Special earthquake report for the Sept. 1, 1981, Samoa Islands region earthquake: U.S. Geological Survey Open-File Report, 74 pp.

Newman, S., Finkel, R.C., and Macdougall, J.D., 1984, Comparison of Th-U disequilibrium systematics in lavas from three hot spot regions; Hawaii, Prince Edward and Samoa: Geochimica et Cosmochimica Acta, v. 48, p. 315-324.

Nunn, P. D., 1988, Recent environmental changes along southwest Pacific coasts and the prehistory of Oceania: developments of the work of the late John Gibbons: Journal of Pacific Studies, v. 14, p. 42-58.

Ollier, C.D. and Zarriello, P., 1979, Peapea lava cave, Western Samoa: British Cave Research Association, Transactions, v. 16, p. 133-142.

Palfreyman, W.D, et al., 1988, Geologic Map of the Circum-Pacific Region, Southwest Quadrant: Circum-Pacific Council for Energy and Mineral Resources, Tulsa, OK.

Powers, S., 1917: Tectonic lines in the Hawaiian Islands, Bulletin of the Geological Society of America, v. 28, p. 501-514.

Penck, A, 1894, Morphologie der Erdoberflache, Vol. 2.

Pickering, C., 1876, Geographical distribution of animals and plants, [Part II]. U.S. Exploring Expedition 19, 2, 524 pp., Philadelphia.

Reinecke, F., 1905, Der neue vulkanische Ausbruch auf Savaii, Petermanns Mitt., v. 51, p. 255.

Reinecke, F., 1906, Der neue vulkenische ausbruch auf Savai' i, Petermanns Mitt., v. 52, p. 86-88.

Richard, J.J., 1962, Catalogue of the active volcanoes of the world including Solfatara fields; 13, Kermadel, Tonga and Samoa, 38, Inter. Volcano. Assoc. Naples.

Rison, W. and Craig, H., 1982, Helium 3; coming of age in Samoa. [abstr.]: Eos, Transactions, American Geophysical Union, v. 63, p. 1144.

Robertson, E.I., 1987, Gravity surveys of the Samoan Islands, Southwest Pacific Ocean: New Zealand Journal of Geology and Geophysics, v. 30, p. 233-240.

Sachet, M.H., 1954, A summary of information on Rose Atoll: Atoll Research Bulletin, v. 5, p. 1-56.

Saito, K.M., and Ozima, M., 1977, $^{40}Ar/^{39}Ar$ geochronological studies of submarine rocks from the western Pacific area, Earth and Planetary Science Letters, v. 33, p. 353-369.

Sapper, K., 1906, Der Matavanu-Ausbruch auf Savaii 1905-6. Nach Aufzeichnungen von Peter Mennel, Mitteilungen von Dr. B. Funk and gedruckten Berichten dargestellt Z. Ges. Erdk. Berl., p. 686-709.

Sapper, K., 1909, Neuere Mitteitungen uber den Matavanu - Ausbruch auf Savai'i, Nach Aufzeichnungen von Pater Mennel and Dr. W. Grevel un nach gedruckten Berichten zusammengestellt, Z. Ges. Drdk. Berl, p. 172-180.

Sapper, K., 1911a, Nachrichten uber den ausbruch des mata auf Savai'i vanu aus dem jahren 1909 and 1910, Z. Ges. Erdk. Berl., p. 702 - 786.

Sapper, K., 1911b, Das ende des matavanu - ausbruchs auf Savai'i, nach mitteilungen von Dr. W. Grevel und Pater Mennel, Z. Ges. Erdk. Berl., p. 701-704.

Sapper, 1912, Nachklange zum Matavanu - Ausbruch (Savai'i) Aus Breifen Von Pater Mennel und Dr. W. Grevel zusammenges - tellt, Z. Ges. Erdk. Berl., p. 445-451.

Sapper, K., 1915, Bericht uber die vulkanischen Ereignisse der Jahre 1895-1913, Beitr. Geophys, 14, Heft I, p. 35-97; Heft II, p. 99-100.

Schlanger, S.O., Garcia, M.O., Haggerty, J.A., Keating, B.H., Naughton, J.J., Philpotts, J.A., Sager, W.W. and Duncan, R.A., 1984, Geologic evolution of the Line Islands, Journal of Geophysical Research, v. 89, p. 11261-11272.

Schmittmann, J., 1911, Der feuersee auf Savai'i (Deutsch - Samoa), Petermann's Mitt., 57, (1), 77-78, pl. 15 and 16.

Schroth, C.L., 1970, Analysis and prediction of the properties of Western Samoa soils. Ph.D. thesis, Department of Agronomy and Soil Science, University of Hawaii, 259 pp.

Schultz, L.P., 1940, The Navy surveying expedition to the Phoenix and Samoa Islands, 1939. Smiths. Explor. 1939, p. 45-50.

Seelye, C.J., 1943, The climate of Pukapuka, - Danger Islands, and, The climate of Aitutaki. N.Z. Meteor. Off. Series C. nos. 3: 1-8, 4: 1-8.

Seelye, F.T., and Grange, L.I., Davies, L.H., 1938, The laterites of Western Samoa: Soil Science, v. 46, p. 23-31.

Setchell, W.A., 1924, Vegetation of Rose Atoll: Carnegie Institution Publication 341, p. 255-261.

Shaw, H.P and Jackson, E.D., 1973, Linear island chains in the Pacific: results of thermal plumes or graviational anchors?: Journal of Geophysical Research, v. 78, p. 8634-8652.

Sinton, J.M., Johnson, K.T., and Price, R.C., 1985, Petrology and geochemistry of volcanic rocks from the Northern Melanesian Borderland, **in** T.M. Brocher, ed., 1985, Geological Investigations of the Northern Melanesian Borderland, Earth Science Series, v. 3: Houston, TX, Circum-Pacific Council for Energy and Mineral Resources, p. 35-65.

Stearns, H.T., 1941, Ground water supplies and construction rock for United States Naval projects, Tutuila, Samoa. Unpublished Navy report, 1-24.

Stearns, H.T., 1944, Geology of the Samoan Islands: Geological Society of America Bulletin, v. 55, p. 1279-1332.

Stearns, H.T., 1946, Geology of the Hawaiian Islands, Hawaii Division of Hydrography, Bulletin. 8, 105 pp.

Stice, G.D., 1968, Petrology of the Manua Islands, Samoa: Contributions to. Mineral Petrology, v. 19, p. 343-357.

Stice, G.D. and McCoy, F.W. Jr., 1968, The geology of the Manua Islands, Samoa: Pacific Science, v. 22, p. 426-457.

Stone, E.L., Jr., 1951, The soils of Arno Atoll, Marshall Islands: Atoll Research. Bulletin, v. 5, p. 1-56.

Sugimura, A., Maeda, Y., Matsushima, Y. and Rodda, P., 1988, 8 Further report on sea-level investigation in Western Samoa: sea-level changes and tectonics in the middle Pacific, Report of the HIPAC Project in 1986 and 1987, Kobe University, Japan, p. 77-90.

Tarling, D.H., 1962, Tentative correlation of Samoan and Hawaiian Islands using "reversals" of magnetization: Nature, v. 196, p. 882-883.

Tarling, D.H., 1965, The paleomagnetism of the Samoan and Tongan Islands:: Geophysical Journal of the Roy Astronomical Society, v. 10, p. 497-513.

Thomson, J.A., 1921, The geology of Western Samoa: New Zealand Journal of Science and Technology, v. 4, p. 49-66.

Turner, G., 1979, Samoa, a hundred years ago and long before: together with notes on the islands of the Pacific: AMS Press (New York) 395 pp. (Reprint of 1884 ed. pub. by Macmillan, London).

U.S. Environmental Protection, Agency, 1982, Secondary maximum contaminant levels (section 143.3 of part 143, National secondary drinking water regulations): U.S. Code of Federal regulations, Title 40, Parts 100 to 149, revised as of July 1, 1982, 374.

U.S. Geological Survey, Water resources data for Hawaii and other Pacific areas, water year 1979; volume 2, Trust Territory of the Pacific Islands, Guam, American Samoa, and northern Mariana Islands, U.S. Geological Survey.

U.S. Geological Survey, 1981, Water resources data for Hawaii and other Pacific areas, water year 1980; Volume 2, Trust Territory of the Pacific islands, Guam, American Samoa, and Northern Mariana Islands, 171 pp.

U.S. Geological Survey, 1979, Water resources data for Hawaii and other Pacific areas, water year 1978; Vol.2, Trust Territory of the Pacific Islands, Guam, American Samoa and northern Mariana Islands: 121 pp.

Upton, B.G.J. and Wadsworth, W.J., 1965, Geology of Reunion Islands, Indian Ocean: Nature, v. 207, p. 151-154.

Valenciano, Santos, 1985, Trust Territory of the Pacific Islands, Saipan, Guam, and American Samoa, ground water resources: in National water summary 1984; hydrologic events, selected water- quality trends, and ground-water resources, Geological Survey Water-Supply Paper (Washington), 2275, 403-407.

Upton, B.G.J. and Wadsworth, W.J., 1965, Geology of Reunion Islands, Indian Ocean: Nature, v. 207, p. 151-154.

Von Bulow, W., 1906, De volkanische Tatigkeit auf Savaii und deren, Einwirkung auf die Wirstchaftlichen Verhattnisse der eigneboremen, Globus, p. 21-24.

Weber, M., 1909, Eur petrogrpahic der Samoa-Inseln. Abh. K. Buyer. Akad. A. Wiss, II Kl., Bd. 24. Abth. 2, p. 290-310.

Wegener, G., 1902, Vulklanausbruch in Savai'i, Dtsch. Kolon. Ztg., 19, p. 491-492.

Wegener, G., 1903a, Die vulkanischen Ausbruche auf Savai'i, Z. Ges. Erdk. Berl., p. 208-219.

Wegener, G., 1903b, Der vulkanausbruche in Samoa, Weltall., 3, p. 106-111.

White, W.M. and Hoffman, A.W., 1982, Sr and Nd isotope geochemistry of oceanic basalts and mantle evolution. Nature, v. 299,p. 821-825.

Wilkes, H.C., 1852, Narrative of the U.S. Exploring Expedition, Vol. 1, 1557 pp.

Wingert, E.A., 1981, A coastal zone management atlas of American Samoa, University of Hawaii Cartographic Laboratory, Department of Geography, Honolulu.

Wright, A.C.S., 1963, Soils and land use of Western Samoa: New Zealand Soil Bureau Bulletin, Wellington, 191 pp.

Wright, Elizabeth, 1986, Petrology and geochemistry of shield-building and post-erosional lava series of Samoa; implications for mantle heterogeneity and magma genesis, Doctoral Dissertation, Univ. of California, San Diego, CA, 305 pp.

Wright, Elizabeth, 1987, Mineralogical studies of Samoan ultramafic xenoliths; implications for upper mantle processes:, in Keating, B.H. Fryer, P., Batiga, R., and Boehlert, G.W., (eds) Seamounts, Islands, and Atolls: Washington, DC, American Geophysical Union, Geophysical Monograph 43, p. 221-234.

Wright, Elizabeth, and White, William M., 1987, The origin of Samoa; new evidence from Sr. Nd, and PB isotopes: Earth and Planetary Science Letters, v. 81, p. 151-162.

Yee, Johnson J., 1988, Trust Territory of the Pacific Islands, Saipan, Guam, and American Samoa, ground water quality: in National water summary 1986: hydrologic events and ground water quality (Moody, David W., compiler; et al.), U.S. Geological Survey Water supply paper, Rep No. W2325, p. 483-488.

Yee, Johnson J., 1987, Trust Territory of the Pacific Islands, Saipan, Guam, and American Samoa ground water quality: U.S. Geological Survey Open-File Report No. OF 87-0755, p. 1-7.

Zindler, A. and Hart, S.R., 1986, Chemical geodynamics, Annual Review of Earth and Planetary Sciences. p. 493-571.

Keating, B.H., and Bolton, B.R., editors, 1992, Geology and offshore mineral resources of the central Pacific basin, Circum-Pacific Council for Energy and Mineral Resources Earth Science Series, V. 14. New York, Springer-Verlag.

SEDIMENTATION AND HIATUSES IN THE CENTRAL PACIFIC BASIN: THEIR RELATIONSHIP TO MANGANESE NODULE FORMATION

Akira Nishimura
Marine Geology Department, Geological Survey of Japan, 1-1-3 Higashi, Tsukuba City, Ibaraki 305, JAPAN

ABSTRACT

Sediment cores from the Central Pacific Basin were studied to determine (a) stratigraphic hiatuses, (b) sedimentation rates, and (c) their relationship to topography and the occurrence of deep-sea manganese nodules. The results of core studies and of detailed examination of two selected areas in the Central Pacific Basin led to the following conclusions.

(1) Hiatuses of Neogene through Quaternary age occur widely in the Central Pacific Basin. A close relationship exists between the distribution pattern of the hiatuses and areas of low sedimentation rate, indicating the strong influence of a bottom current on sedimentation processes.

(2) The hiatuses within the cores are dated as late Miocene, late Pliocene, and latest Pliocene to early Pleistocene.

(3) The areas where the hiatuses are most common, and where sedimentation rate varies locally, are divided into two types: topographic highs in flat basin plains ("abyssal hill type"), and narrow, elongated topographic lows ("valley type"). They show different styles of sedimentation and hiatus formation related to the intensity of bottom current.

(4) Local variation in the style of occurrence of manganese nodules depends on the relative importance of the bottom current. The differences in the manganese nodule types (r- and s-types) reflect the chemical environment where nodules have been formed. The r- and s-type nodules are related to poorly oxygenated and well oxygenated environments, respectively. Sedimentation rate is the main control on the chemical environment and nodule characteristics, determining the positions of nodule growth in the vicinity of the water-sediment interface; sedimentation rate varies inversely with the support strength of the surface sediments.

INTRODUCTION

The Central Pacific Basin (CPB) is situated far from surrounding continents, with the only land being isolated islands. Thus, sedimentation in the CPB is controlled mainly by pelagic processes. The supply of pelagic material is closely related to surface organic productivity, and its preservation potential is controlled by bottom conditions like depth and the nature of bottom waters.

The CPB is well known as the main passageway for northward flowing bottom water of the Pacific Ocean. This deep circulation connects the southwest Pacific Ocean to the northwest and the northeast Pacific areas. Because of interest in this bottom water flow, the geology of the area beneath the current as well as the current's inflow and outflow zones in the CPB have been studied intensively (Hollister et al., 1974; Lonsdale, 1981). The main part of the CPB, however, has not been studied in detail sedimentologically.

The Geological Survey of Japan (GSJ) conducted a detailed sedimentological study of the CPB during the course of its research program on "Geological Investigation of Manganese Nodules" from 1980 to 1984. Several areas within the CPB were studied in detail and considerable new data on sedimentation and manganese nodule occurrences were obtained.

The manganese nodules in the CPB can be classified into two main types with different textures, compositions, and processes of formation. The distribution and character of manganese nodules vary regionally and locally. Moreover, hiatuses, sediment types, and sedimentation rates recognized in cores, and the thickness of seismic units, can be related to genesis of the manganese nodules (Usui et al., 1987b).

The objectives of this paper are first to clarify the sedimentary history and processes with particular reference to hiatus distribution in the CPB, and then to clarify the relationship between sedimentation and manganese nodule occurrences through the detailed study of two selected areas.

All available geologic data regarding the general geologic setting, sediments, hiatuses, and manganese nodules in the CPB were compiled, including details from cruise reports concerning the two contrasting survey areas of the CPB (Nakao and Moritani, 1984; Nakao, 1986). Results presented in this paper clarify the relationship between local sedimentary processes and the formation of manganese nodules.

THE CENTRAL PACIFIC BASIN

Bathymetry

The Central Pacific Basin (CPB) is located in the western equatorial Pacific, bordered approximately by latitudes 15°N and 10°S, and by longitudes 170°E and 165°W (Fig. 1). The basin is surrounded by the following topographic highs: the Mid-Pacific Mountains, the Line Islands ridge, the Marshall-Gilbert Islands ridge, the Robbie Ridge, and the Manihiki Plateau. The basin floor is deeper than 5,000 m in most parts and exceeds 6,000 m in the northwestern part of the CPB and on the floors of troughs. In the southern part of the CPB, many scattered islands of the Tokelau and the Phoenix islands are found.

The Samoan Passage lies between the Robbie Ridge and the Manihiki Plateau. The Wake Island Passage is present between the Mid-Pacific Mountains and the Wake Island Ridge to the west. Three other passages, the Horizon, the Clarion, and the Kingman passages, cross the Line Island ridge. These passages are narrow and deeper than 5,000 m, connecting the CPB to adjacent basins.

Geologic Setting

Important background knowledge of the geologic, sedimentologic, and geophysical features of the CPB comes from results of Deep Sea Drilling Project (DSDP) Legs 7, 17, and 33 (Winterer et al., 1971,1973; Schlanger et al., 1976) and the other geoscientific studies of various areas in the CPB.

The basement and the surrounding ridges of the CPB were formed in the Cretaceous, with the exception of the Tokelau and the Phoenix islands and the southern part of the Gilbert ridge, which were formed during the Miocene. Based on hot spot theory and plate stratigraphy, the Pacific Plate and the CPB have moved in a WNW direction since the Eocene (van Andel et al., 1975).

Seismic Stratigraphy

The acoustic basement of the CPB is probably composed of basalt. The sediment layers overlying this basement are generally divided into two units, the upper Unit I and the lower Unit II. These have been correlated to sediments cored at DSDP sites 165, 166, and 170 (Tamaki and Tanahashi, 1981).

The upper layer (Unit I) is composed of transparent to semi-opaque material, locally containing very fine stratification of variable thicknesses. The lithology of this unit varies considerably (calcareous ooze, siliceous ooze, zeolitic mud, turbidites) reflecting different sedimentary environments (Mizuno and Okuda, 1982).

Between Units I and II, there is a distinct reflector, designated as "reflector A", which is correlated to a middle Eocene to lower Oligocene chert (Tamaki and Tanahashi, 1981).

Unit II is an acoustically opaque unit which rarely contains thin layering. The lower part of the unit (Unit IIB) is composed mainly of volcanogenic sediments and the upper part (Unit IIA) of chert, calcareous ooze, limestone, and volcanogenic sediments of various ages (Early Cretaceous to Oligocene).

Surface Sediment Distribution

The surface sediment of the CPB is composed of pelagic sediment types classified mainly on character and abundances of biogenic and authigenic components (Nishimura, 1981) (Table 1). The surface sediment distribution in the CPB shows the following four zones which are arranged more or less parallel to latitude with some exceptions around topographic highs (Nakao and Mizuno, 1982).

- Pelagic clay zone: in the northern area, Mid-Pacific Mountains.
- Transitional zone (pelagic clay to siliceous clay): in the northern area.
- Equatorial siliceous biogenic zone: vicinity of the equator.
- Equatorial calcareous biogenic zone: Manihiki Plateau.

In general, the surface sediment lithology in the CPB is controlled by surface organic productivity and depositional depth. Thus, siliceous sediments characterized by large numbers of tests of siliceous organisms, mainly radiolarians,

Table 1. Classification of sediments of the Central Pacific Basin.

Pelagic clay		
Zeolite rich clay 5% < zeolite < 10%	Siliceous fossil rich clay 5% < siliceous fossils < 10%	Calcareous fossil rich clay 5% < calcareous fossils < 10%
Zeolitic clay 10% < zeolite	Siliceous clay 10% < siliceous fossils < 30%	Calcareous clay 10% < calcareous fossils < 30%
	Siliceous ooze 30% < siliceous fossils	Calcareous ooze 30% < calcareous fossils

Siliceous fossils: radiolarians, diatoms, silicoflagellates, and spines.
Calcareous fossils: foraminifers and calcareous nannoplanktons.

of turbidity currents to the present-day sedimentation of the CPB is minor.

Hiatuses

Widespread deep-sea hiatuses have been recognized from DSDP data and numbered through the upper Paleogene and Neogene cored sediments (Keller and Barron, 1983,1987; Keller et al., 1987). The distribution of the hiatuses (Figs. 2 and 3) is based on DSDP and other geologic studies. The hiatuses are scattered throughout the whole CPB, and are attributed to periodic intensification of the bottom current.

Because of the short penetration of piston corers, most cores contain at most one hiatus. In general, the time gap represented by the uppermost hiatus shows the following variation:

(1) In the northern and northeastern part of the CPB, ages of the sediments below this hiatus are older, Oligocene to early Miocene, than those in the central western area of the CPB.

(2) In the same area, the oldest ages of the sediments immediately above the hiatus are also older, Pliocene to early Pleistocene, than those in the central and western areas of the CPB.

(3) In the western part of the CPB, the duration of the hiatus is longer, approximately middle Miocene to latest Quaternary, although core data are limited.

Sedimentation Rates

The sedimentation rates in the Central Pacific Basin are generally low, with sedimentation being dominantly pelagic in character. Estimates of sedimentation rate for the whole of the Pacific Ocean are given by Piper et al. (1985) and are summarized in Figure 4. Rates have also been calculated from age determinations made from cores used in this study. Within the detailed survey areas, these data typically show small local variations as in the detailed survey areas which will be reported later.

Most of the CPB is characterized by a sedimentation rate of 1 to 5 mm/1000 y. Areas along the equatorial high productivity zone have rates greater than 5 mm/1000 y. The Mid-Pacific Mountains area and the southwestern part of the CPB have rates less than 1 mm/1000 y.

Manganese Nodules

The distribution of manganese nodules in the CPB is shown on a map of the whole Pacific Ocean compiled by Piper et al. (1985). The areas with abundant nodules are distributed in the northern part of the CPB, extending westward from the Clarion-Clipperton Fracture Zone with its very high abundance of manganese nodules. The detailed distribution and occurrence of manganese nodules in the northern part of the CPB is reported by Usui et al. (1987a).

The manganese nodules of the CPB can be divided into two types, r-type (rough surface) and s-type (smooth surface). These two types were first classified on the basis of surface textures alone (Moritani et al., 1977), but now it has been shown that these types reflect different mineral compositions with different origins: 10Å manganate or δ-MnO_2 (Table 2) (Usui et al., 1987a). The occurrences as well as the mineral and chemical compositions of these types have been considered to imply that the r-type nodules were formed through early diagenesis, and that the s-type

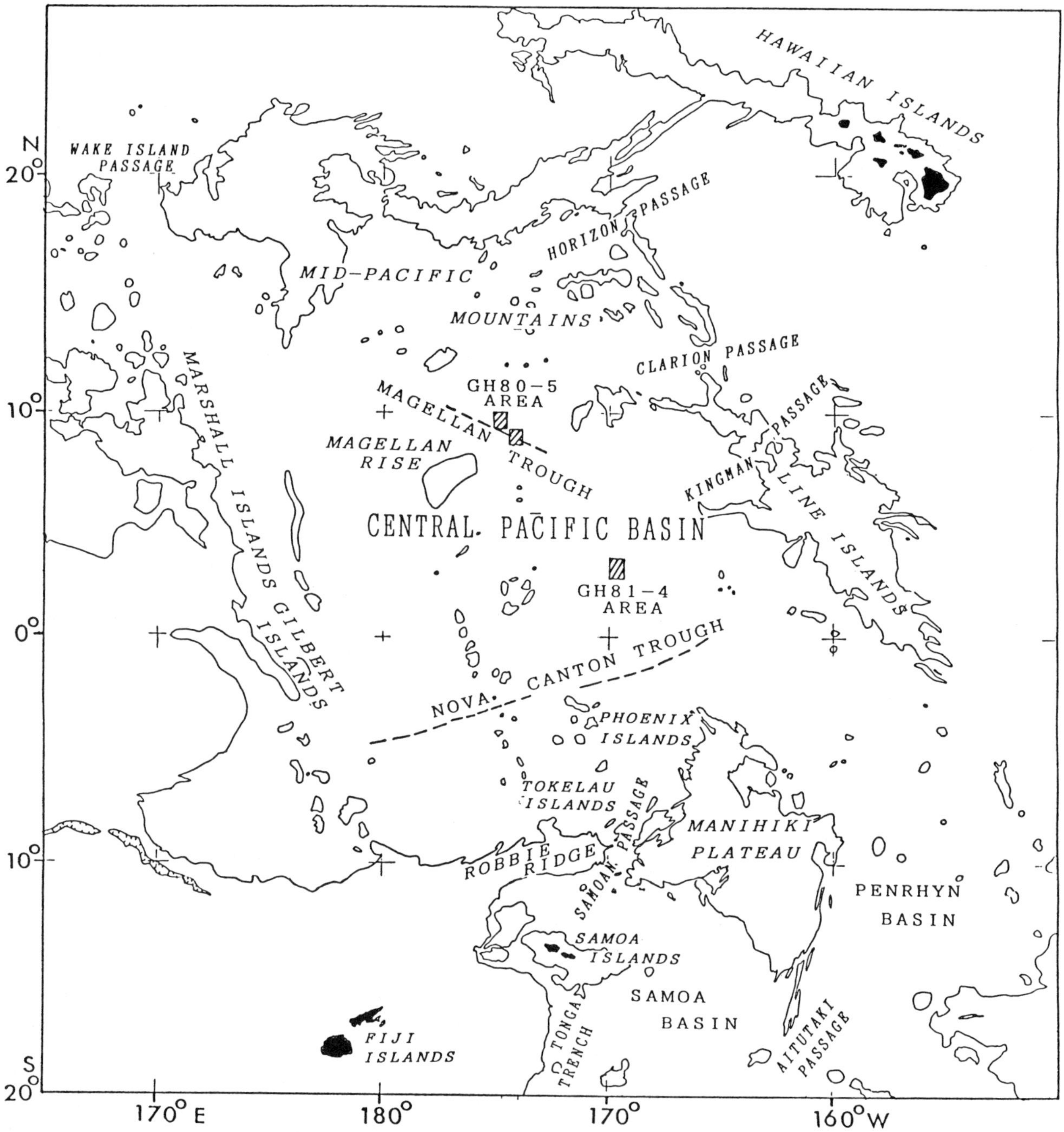

Figure 1. Topography of the Central Pacific Basin (Chase et al., 1971; Mammerickx et al., 1975). Contours show 2400 fathoms.

are deposited around the equatorial high productivity zone, whereas pelagic clay without siliceous tests is deposited elsewhere. The equatorial calcareous biogenic zone is related to a large topographic high, the Manihiki Plateau. This area is shallower than the carbonate compensation depth (CCD), which is located around 5,000 m (Takayanagi et al., 1982). The lithology observed in the core sequences is also dominantly controlled by paleodepositional depths and paleodistance from the equator.

Calcareous biogenic turbidites were derived from topographic highs extending above the CCD. Their distribution is now restricted to the eastern part of the CPB, west of the Line Islands Ridge and around the Manihiki Plateau (Orwig, 1981; Nishimura et al., 1985). Contribution

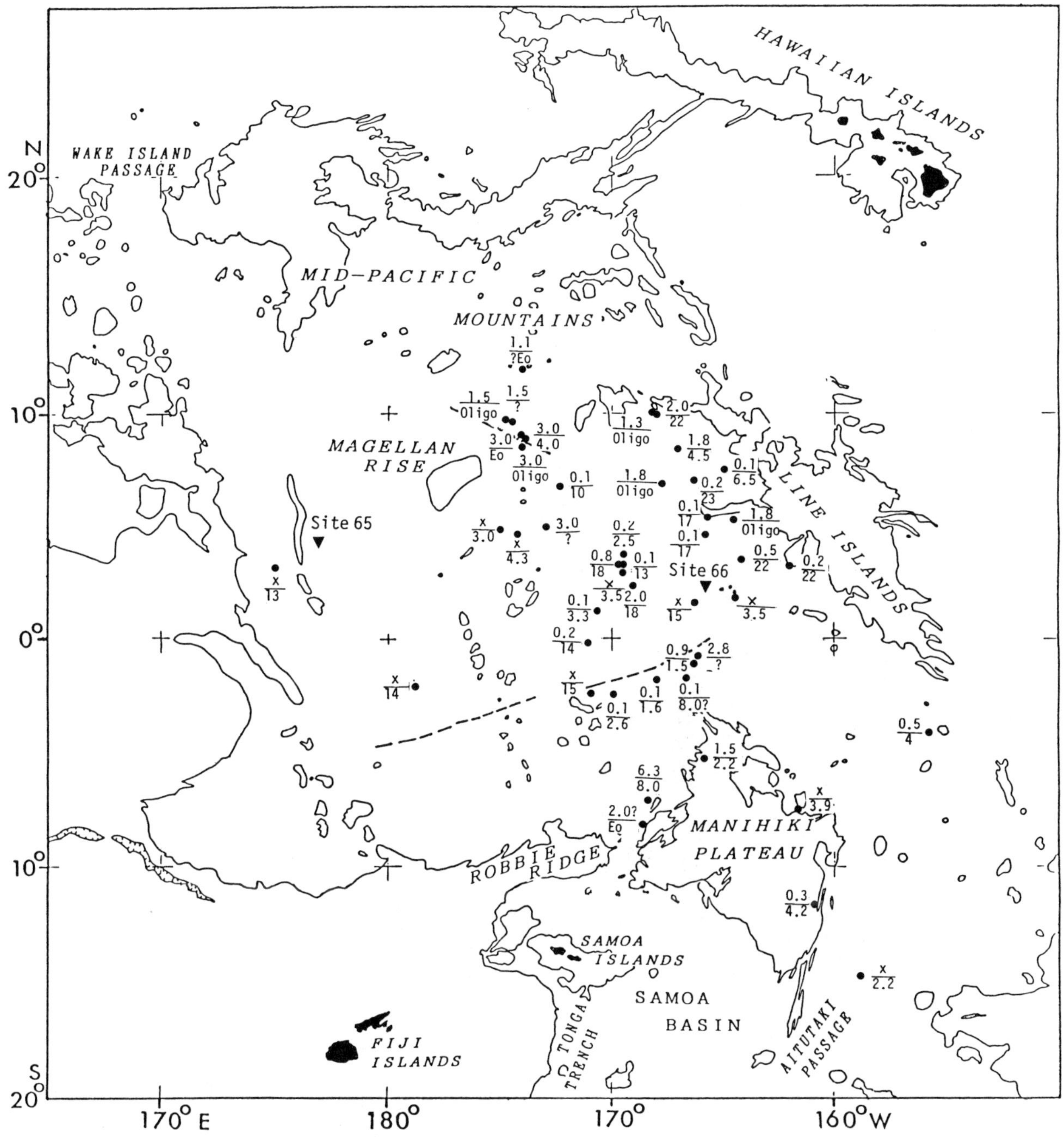

Figure 2. Uppermost hiatus distribution in the Central Pacific Basin. The oldest age above the hiatus and the youngest age below the hiatus (in Ma) are shown above and below a bar, respectively. The symbol x shows absence or near absence of sediments above the hiatus. Ages less than 0.7 Ma are estimated from the sedimentation rate of the nearest cored sequence. Data sources are Kobayashi et al., 1971; Lineberger, 1975; Theyer, Mato, and Hammond, 1978, cruise reports of the GSJ, and unpublished data of the GSJ.

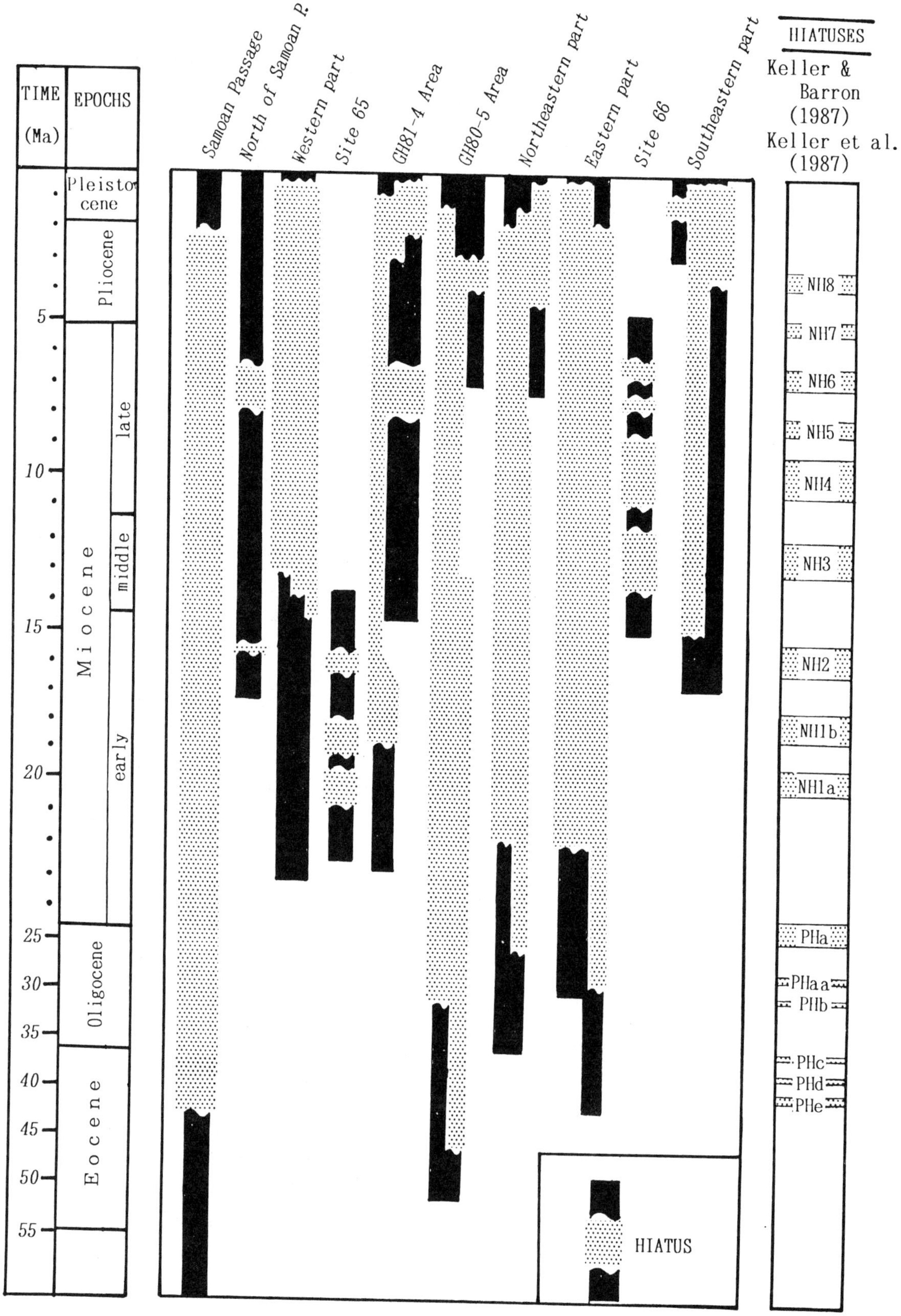

Figure 3. Occurrence of hiatuses in selected core sequences of the Central Pacific Basin. Data are those of Figure 2 plus data from DSDP sites drawn from Keller and Barron (1983).

Table 2. Manganese nodule types of the Central Pacific Basin.

Features	Nodule type r-type	s-type
Surface texture	rough	smooth
Occurrence relative to sediment surface	buried	exposed
Mode of size fraction	1–2 cm	2-4 cm
Internal structure	laminated, concentric layers	massive and compact, surrounding layer slightly laminated
Polynucleation	rare	common
Internal cracks	rare	dominant
Dominant surface sediment	siliceous ooze siliceous clay	zeolitic clay pelagic clay
Mineral composition	10Å manganate	δ-MnO_2
Chemical composition		
(average) Mn/Fe	4.1	1.4
Cu + Ni	2.4 %	0.8 %

Revised from Usui (1982)

nodules are the product of hydrogenetic process, namely the direct precipitation of manganese oxides from seawater colloids. The recognition of these two types is of major significance in considering the genesis of manganese nodules.

Bottom Current

The Samoan Passage, located at the southwestern margin of the CPB, provides a route for northward flow of Antarctic Bottom Water (AABW). In and around the passage area, extensive studies have clarified the nature of the present and past strong bottom current (Hollister et al., 1974 ; Lonsdale, 1981).

A cold and dense water mass with high dissolved-oxygen content and high salinity originates in the Ross and the Weddell seas of the Antarctic and sinks into the deep sea, forming the Circum Antarctic Water Current. A branch of this current, AABW, flows into the western part of the South Pacific Basin and along the Tonga-Kermadec Trench and then reaches the Samoan Basin (Hollister et al., 1984). The AABW, after reaching the Samoan Basin, flows into the CPB through the Samoan Passage (Reid and Lonsdale, 1974) and flows out through several passages – the Wake Island, the Horizon, the Clarion, and the Kingman Passages – at the northern to eastern margin of the CPB (Mantyla, 1975; Edmond et al., 1971; Normark and Spiess, 1976; Johnson, 1972b) (Fig. 1). Topographic features strongly affected by the bottom current along the Gilbert Islands Ridge suggest the presence of a western boundary current along the western margin of the CPB (Lonsdale and Smith, 1980). As for the main CPB area, an eastward flow along the Nova Canton Trough (Matsubayashi and Mizuno, 1982) and a very slow northward flow have been suggested based on the distribution pattern of bottom potential temperatures (Gordon and Gerard, 1970).

SELECTED SURVEY AREAS

The Northern Equatorial Central Part of the Central Pacific Basin Area (GH81-4 Area)

Bathymetry and Geologic Setting

This area is located in the central part of the CPB within latitudes 3°25'N and 2°45'N and longitudes 169°55'W and 169°20'W (Fig. 1). The topography (Fig. 5) is almost flat at depths between 5,600–5,400 m on the basin floor; there are some topographic highs, abyssal hills, a few hundred meters above the basin floor. The three most prominent hills are arranged in a N-S direction. The western slopes of the hills are steeper than the eastern slopes and there are narrow and deep moat-like depressions along the western foot of the central and southern hills.

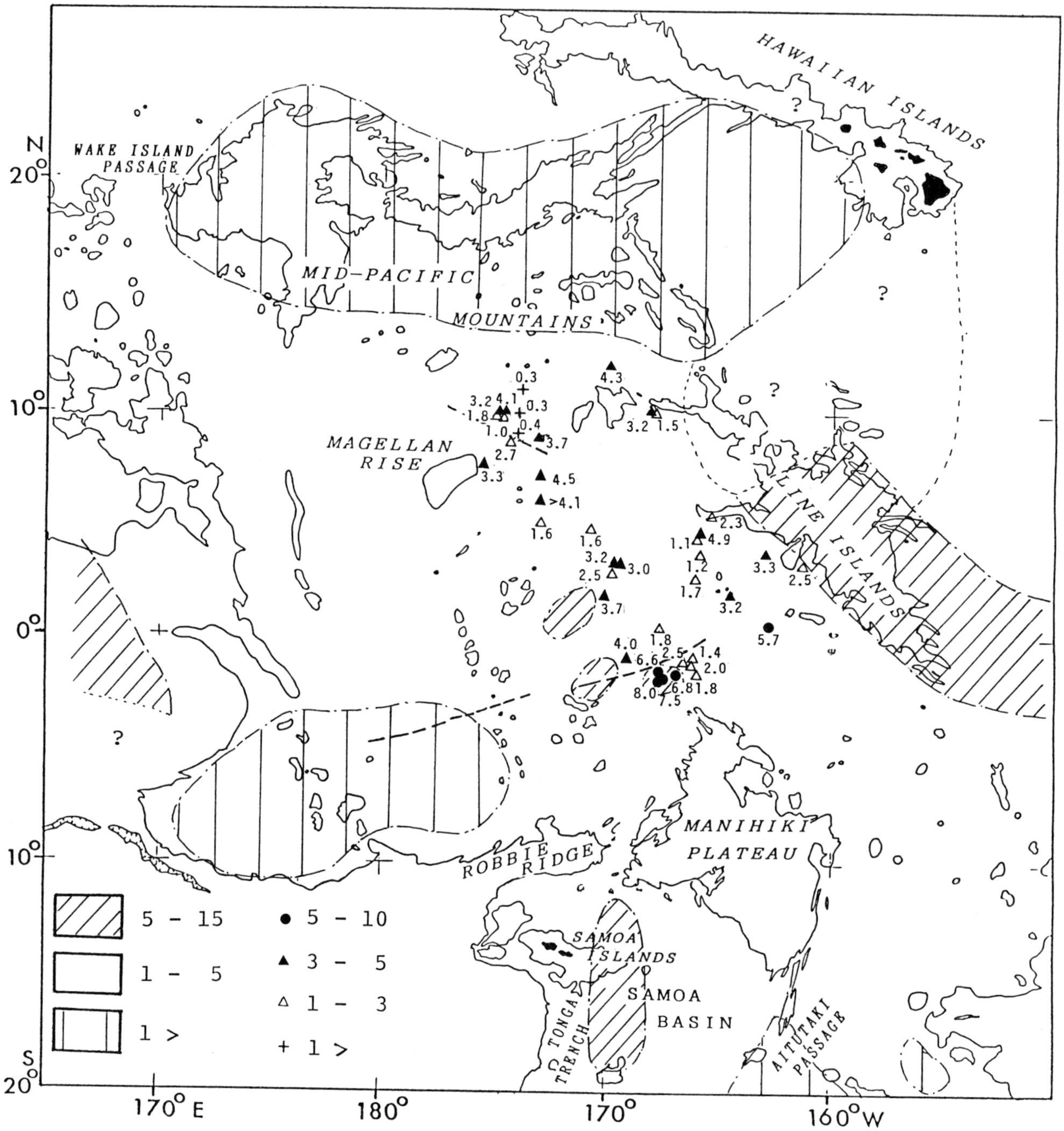

Figure 4. Sedimentation rate distribution in the Central Pacific Basin. Numerals show the average sedimentation rates (mm/1000 y) during the Quaternary or the Brunhes Epoch. Core data base is the same as in Figure 2 and is represented by single points. Areal distributions of average sedimentation rates are partly modified from Piper et al. (1985).

Unit I on seismic records is composed of a transparent layer without distinct internal reflectors and corresponds to the transparent layer in the 3.5 kHz subbottom profiling (SBP) records. Unit I on the western slope of the hills is thinner than on the eastern slope of the hills. There are depressions formed by thinning of Unit I on the basin floor along the western side of the hills. Thin laminated subsurface reflectors are found in Unit I of the SBP records and also mud wave features are seen on these records (Tanahashi, 1986) (Fig. 6). An acoustically semi-opaque layer which belongs to Unit II crops out as small exposures on the top of the abyssal hills, where thick manganese-

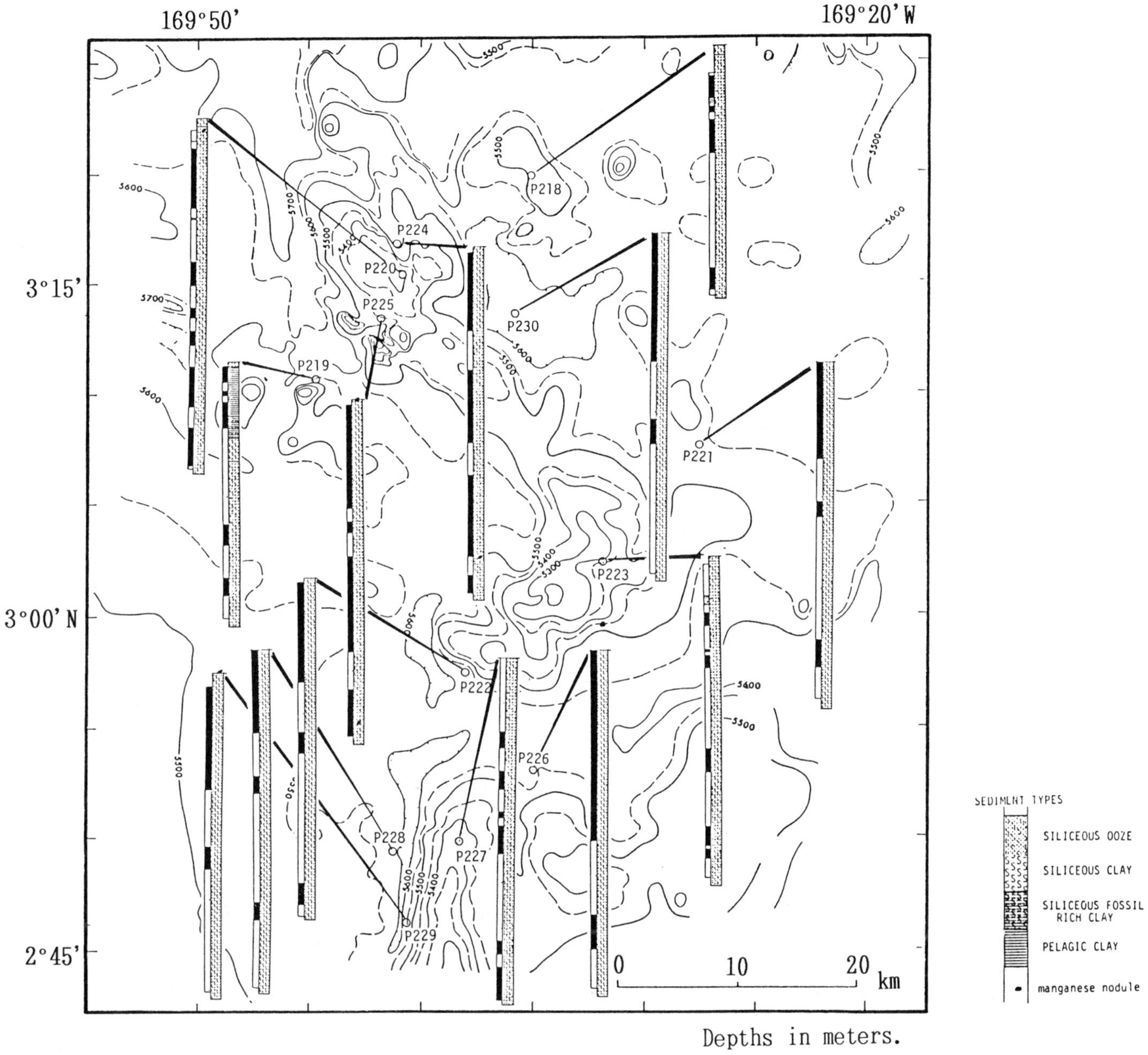

Figure 5. Core samples on the topographic map of the GH81-4 area. The left columns show magnetic polarity of sediments; the right columns show sediment type.

encrusted claystone was dredged. The claystone yields ichthyoliths which have been assigned an Eocene to Oligocene age (Nishimura, 1986).

All the features mentioned above show that sedimentation was strongly affected by a strong bottom current, with erosion and redeposition controlled by the eastward/northward-flowing bottom current.

The top of the bottom water flow (AABW) occurs well above the sea floor at 4500 m depth, based on temperature-depth profiles (Yamazaki, 1986a).

Core Samples

The surface sediment lithology in this area is siliceous clay with abundant radiolarian tests. The Quaternary sequences of this area have the same lithology. Thirteen piston cores are available in this area (Nishimura, 1986) (Figs. 5 and 7). The cores from the basin floor provide evidence for continuous sedimentation through the late Neogene and Quaternary. They are lithologically uniform throughout the sequences and characterized by remanent magnetization

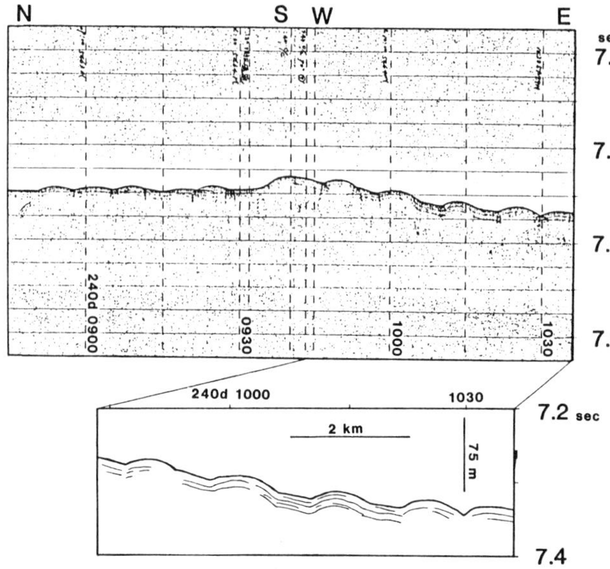

Figure 6. Mud waves in the GH81-4 area of the Central Pacific Basin (Tanahashi, 1986). Profile shows the effect of the eastward-flowing bottom current. The surveyed line is located east of the southern hill of the GH81-4 area.

patterns in good agreement with the standard polarity change patterns for the late Neogene-Quaternary interval. The characteristic occurrence of silicoflagellates, *Mesocena quandrangula* Ehrenberg, the age of which is defined from 1.3 to 0.79 Ma (Berggren et al., 1980), enables a reliable age determination (Fig. 7).

The cores from the hills show lithological changes suggestive of the presence of at least one local hiatus. Most of the cores are composed of siliceous clay in the uppermost part and siliceous ooze below. The radiolarian assemblages in the core sequences and the magnetostratigraphy allow estimation of the duration of the hiatus between the two different lithologic facies. Biostratigraphy and magnetostratigraphy suggest the presence of other hiatuses in the lower siliceous ooze.

The time intervals, determined from the results of the core study, are represented by hiatuses grouped into two time spans: late Miocene (8.5–6.2 Ma) and late Pliocene to middle Quaternary (2.6–0.9 Ma). The sedimentation rate calculated from the magnetostratigraphy is, for the basin floor, 3–6 mm/1000 y through the Brunhes Normal Epoch and 5–10 mm/1000 y around the Jaramillo Event of the Matuyama Reversed Epoch, whereas on the abyssal hills, it is 3.5 mm/1000 y through the Brunhes Epoch determined from the only available core (Yamazaki, 1986b).

Manganese Nodules

The distribution of manganese nodules in this area is rather restricted (Fig. 8). The characteristics of the distribution are listed below (Usui, 1986; Usui et al., 1987b).

(1) Manganese nodules occur on the hills, but not on the basin floor.

(2) The hill areas with thin acoustically transparent layers contain highly concentrated occurrences of manganese nodules.

(3) The variation in the occurrence of the two types of nodules is observed even within small areas on the hills. The mode of occurrence is the same as those in other areas of the CPB; that is, the r-type nodules are typically buried whereas the s-type nodules are generally exposed (Fig. 9).

(4) The distribution of s-type nodules corresponds to the area with outcrop of Unit II on the seismic records.

Relationship Between Sediments and Nodules

The area of abundant manganese nodules corresponds to those elevated hill areas where hiatuses are recognized in cores. The thinness of the acoustically transparent layer on the hills suggests that the sedimentation rate on the hills is lower than that of the basin floor. The sedimentation rate of s-type nodule areas is possibly lower than that of r-type nodule areas, though the acoustically transparent layers are sometimes absent from those areas characterized by s-type nodules.

The sedimentation rate of 3.5 mm/1000 y presented earlier is the average rate above the boundary between Units I and II. The exact recent (for example through the Brunhes Epoch) sedimentation rate in the s-type nodule area could not be determined. Two piston cores (Cores P229 and P225) with r-type nodules on the top, both on the basin floor and on the hill areas, show sedimentation rate of 3.5 mm/1000 y. These values are rather low compared to the other piston cores without nodules (Fig. 7).

The Area North of the Magellan Trough (GH80-5 Area)

Bathymetry and Geologic Setting

Two small areas were surveyed in detail, one bordered by latitudes 9°30'N and 10°00'N and longitudes 174°30'W and 174°50'W, and the other by latitudes 8°40'N and 9°10'N and longitudes 173°50'W and 174°100'W (Figs. 1 and 10). The topography of the surveyed area (Fig. 11) is characterized by rows of hills and troughs which are parallel or subparallel to the WNW-ESE trending axis of the Magellan Trough. The water depth varies from 5,400 to 6,000 m.

The sediment layers which appear on seismic records are again divided into two units, Units I and II. The upper

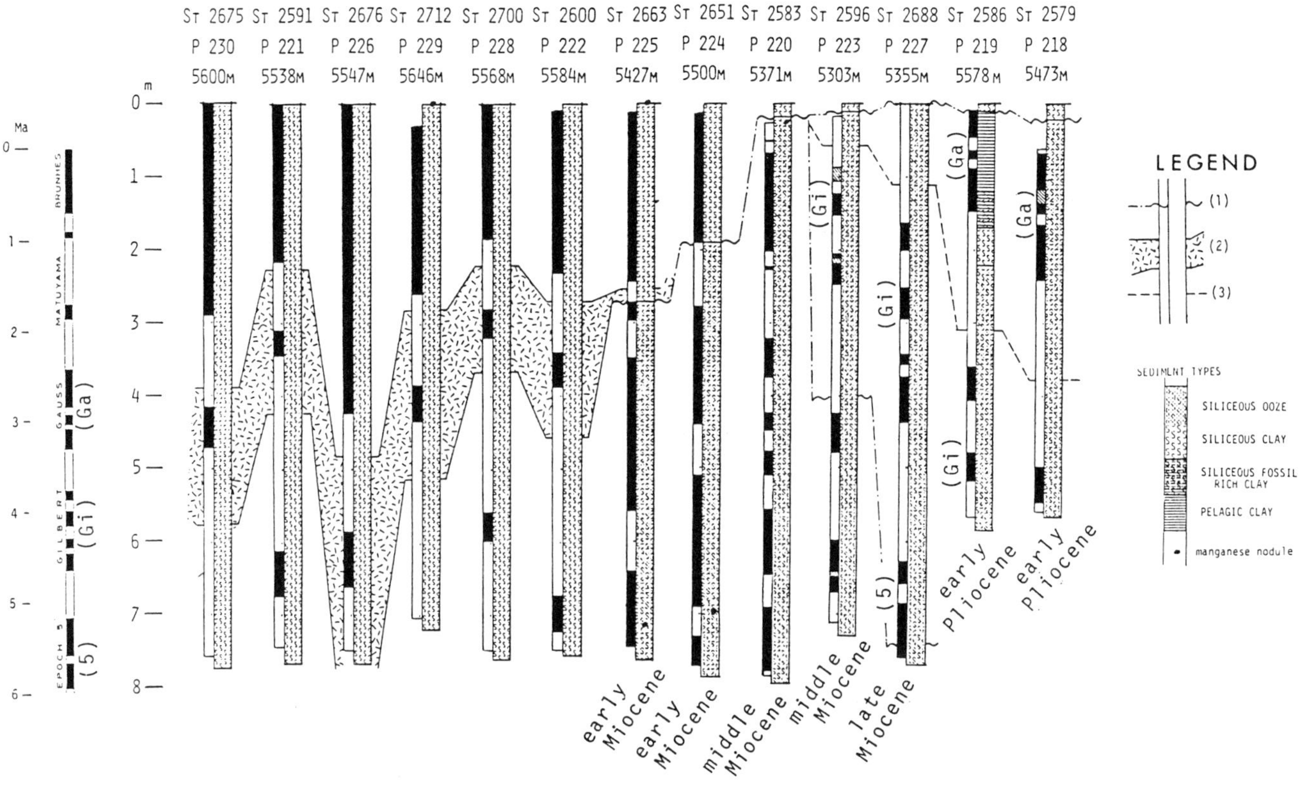

Figure 7. Core profiles of the GH81-4 area of the Central Pacific Basin. Core sequences were described in detail by Nishimura (1986). Right columns show the sediment types and left ones the magnetic polarity of sediments. Radiolarian ages of core bottom sediments are shown below the core. (1) hiatus (2) *Mesocene quandrangula* zone (0.79-1.3 Ma) (3) top horizon of *Spongaster pentas* Riedel and Sanffippo (ca. 3.4 Ma).

transparent unit (Unit I) is thin and varies from 20 to 40 m thick in this area (Tamaki, 1984). Unit I has no internal reflector corresponding to the hiatus detected in the core sequences of this area. Unit I coincides with the acoustically transparent layer on 3.5 kHz records.

Core Samples

The surface sediment lithology of this area is composed of siliceous fossil-rich clay and it changes to pelagic clay within the order of several tens of centimeters below the sea bed. The pelagic clay was possibly formed by the dissolution of biogenic opal during early diagenesis (Nishimura, 1981).

Seventeen piston cores were collected in this area (Nishimura, 1984) (Figs. 11 and 12). Some of these cores show evidence of continuous sedimentation, such as the continuity of the lithology and the good agreement of remanent magnetization patterns with the standard polarity change patterns. These cores were taken from the ridges. Cores collected from the troughs, however, show lithologic changes which suggest the presence of hiatuses. The upper parts of the cores are composed mainly of pelagic clay and the rest mainly of zeolitic (clinoptilolite) clay.

Most of these cores do not contain good age indicators because of dissolution of both calcareous and siliceous microfossils, and because of unstable remanent magnetization in the zeolitic clay.

Core P201, however, provides important information for establishing the sedimentary history of this area. This core shows a hiatus between the upper pelagic clay and the lower zeolitic clay parts. It contains siliceous clay at the lowermost part of the sequence, which yields late Miocene radiolarians with high content of collosphaerids and reworked radiolarians suggesting sedimentation under the influence of a bottom current. The age of the upper part of the core above the hiatus corresponds to the period between the Gauss Epoch to the Recent, and the rest of the core below the hiatus is determined to be between 7 to 4 Ma from the radiolarian assemblage of the siliceous clay and magnetostratigraphy. Therefore, the interval of the hiatus is interpreted to be 4.0 to 3.0 Ma.

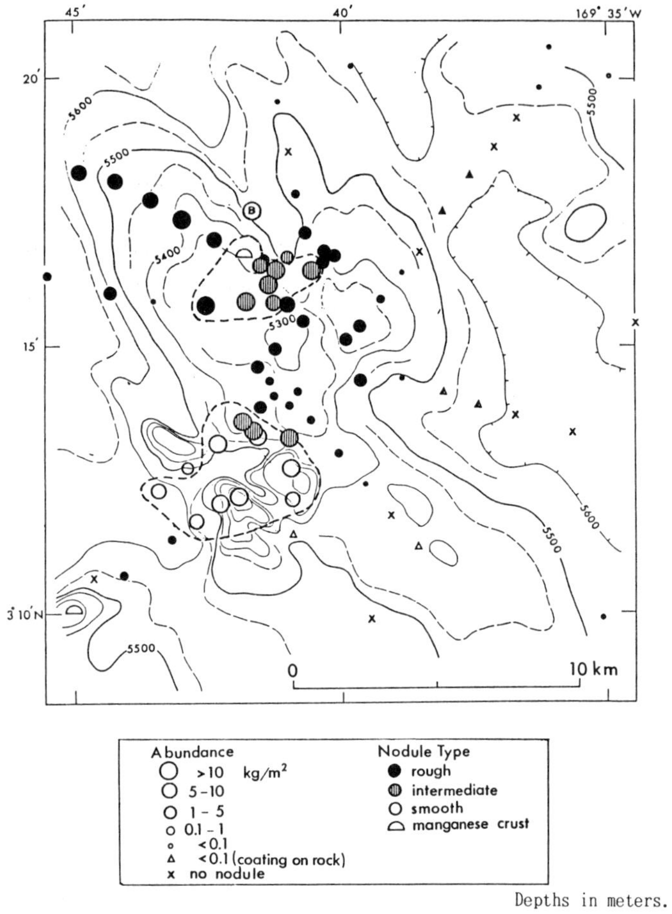

Figure 8. Distribution of manganese nodules around the northern hill in the GH81-4 area (Usui, 1986). Dashed lines indicate areas of nodule coverage greater than 5% on the sea-bed photographs.

The sedimentation rates in this area are summarized below, mainly based on magnetostratigraphy of the core sequences (Joshima and Nishimura, 1984) (Fig. 10).

(1) Cores without hiatuses show a relatively high sedimentation rate (2–4 mm/1000 y) for the Quaternary.

(2) Cores with hiatuses have relatively low sedimentation rates (0.7–2 mm/1000 y) in the Quaternary.

(3) In spite of the generally low sedimentation rates, most of the cores have a distinct normally magnetized part which is correlated to the Jaramillo Event. The sedimentation rate around the Jaramillo Event is about 1.5 times that corresponding to the Brunhes Normal Epoch.

(4) The sedimentation rate of the lower part of the cores below the hiatuses is not clear, except for Core P201, for which a low rate (1.5 mm/1000 y) for the zeolitic clay has been determined.

Manganese Nodules

Most of the sampling sites (box cores, piston cores, and freefall grabs) in the area yielded manganese nodules. The characteristics of the manganese nodules in this area are summarized below (Usui and Nakao, 1984) (Fig. 13).

(1) High to moderate concentrations of manganese nodules occur throughout the area with the exception of the southern part where nodules are absent. Seismic records indicate the presence of "turbidite layers" in this southern region.

(2) The distribution of the two types of manganese nodules varies considerably on a local scale.

(3) Bottom photography and box corer samples show exposed occurrences of r-type nodules (Fig. 14). The exposed r-type nodules in this area are larger and more abundant than the buried r-type nodules in other areas. Cross sections of the nodules sometimes show that they have cores of older s-type nodules.

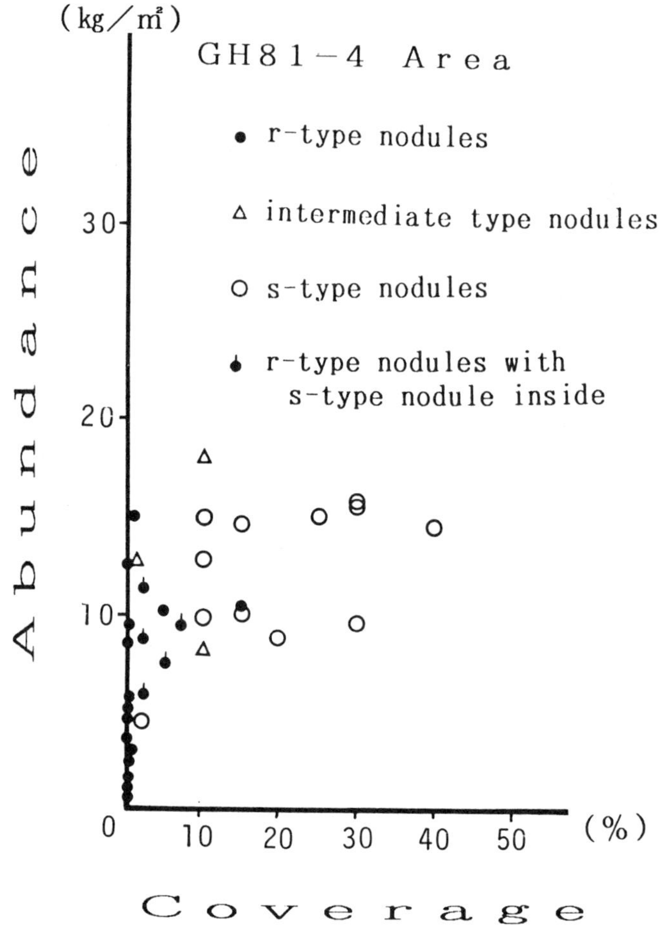

Figure 9. Relationship between nodule abundance and coverage revealed by sea-bed photography in the GH81-4 area (Usui, 1986).

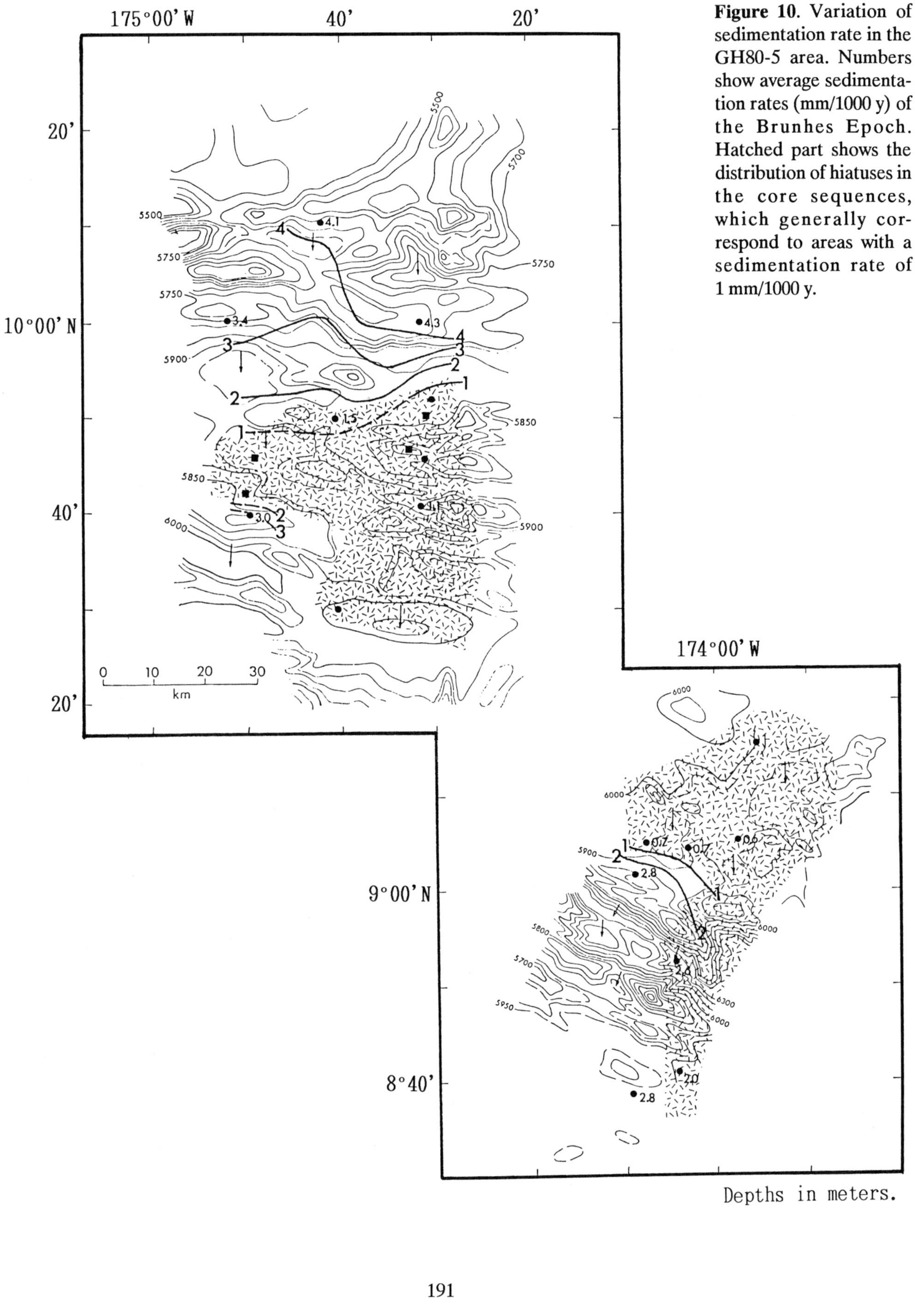

Figure 10. Variation of sedimentation rate in the GH80-5 area. Numbers show average sedimentation rates (mm/1000 y) of the Brunhes Epoch. Hatched part shows the distribution of hiatuses in the core sequences, which generally correspond to areas with a sedimentation rate of 1 mm/1000 y.

Depths in meters.

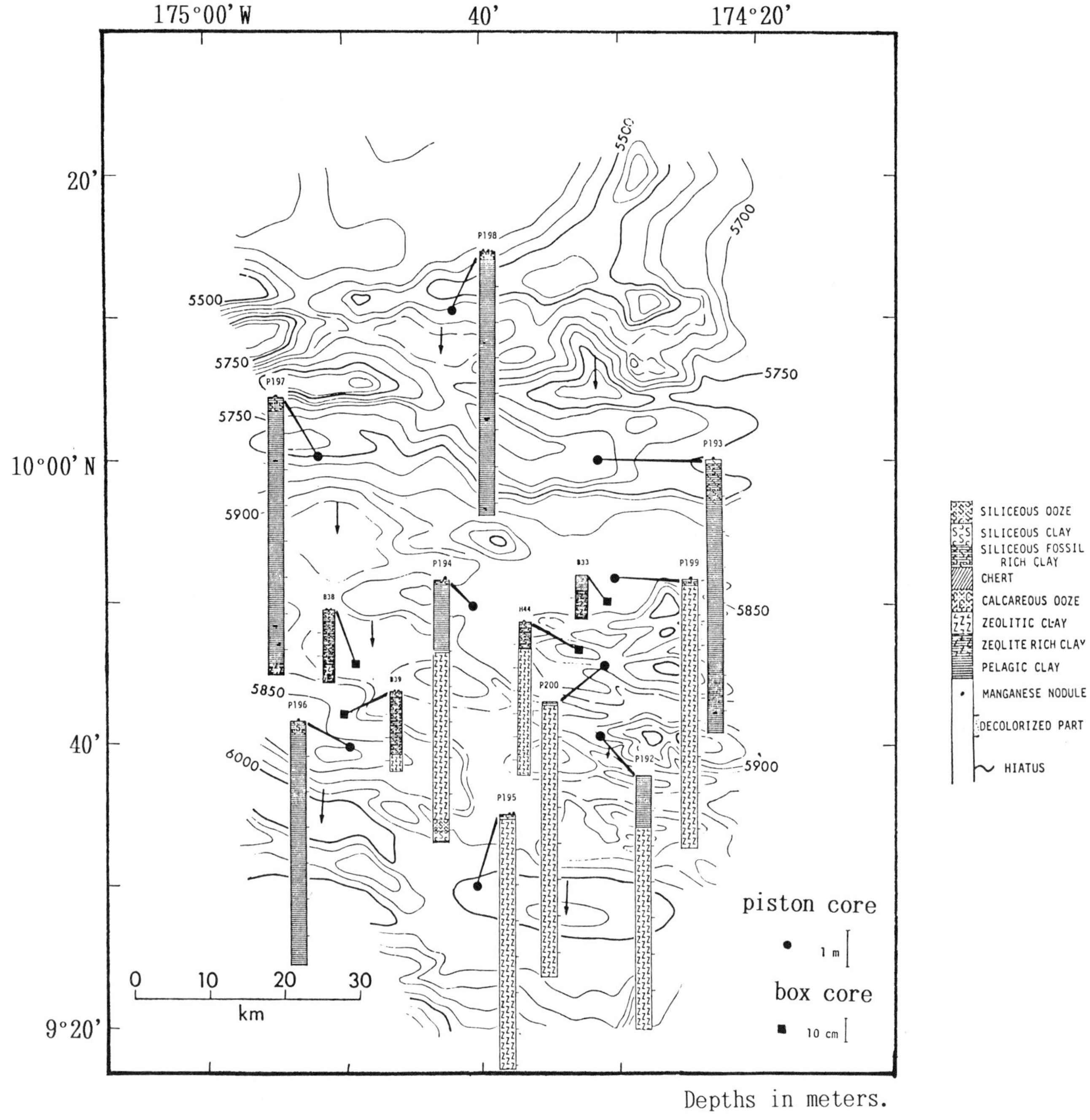

Figure 11-1. Core samples on the topographic map of the western part of the GH80-5 area.

Relation between Sedimentation and Nodules

The occurrences of the two types of nodules appear to have significant relationships, listed below, to the existence of hiatuses, the sedimentation rate, and the topography.

(1) The r-type nodules are distributed mainly on the sea floor in water depth below 5,850 m (Fig. 13).

(2) The distribution of the r-type nodules can be roughly correlated to those areas in which sediment hiatuses are common. These areas often have very thin Quaternary sediments and the sedimentation rates above the hiatuses are less than 2 mm/1000 y, based on magnetostratigraphy.

(3) The s-type nodules are distributed in areas without r-type nodules. Such areas had rather higher sedimentation rates during the Quaternary than areas with r-type nodules.

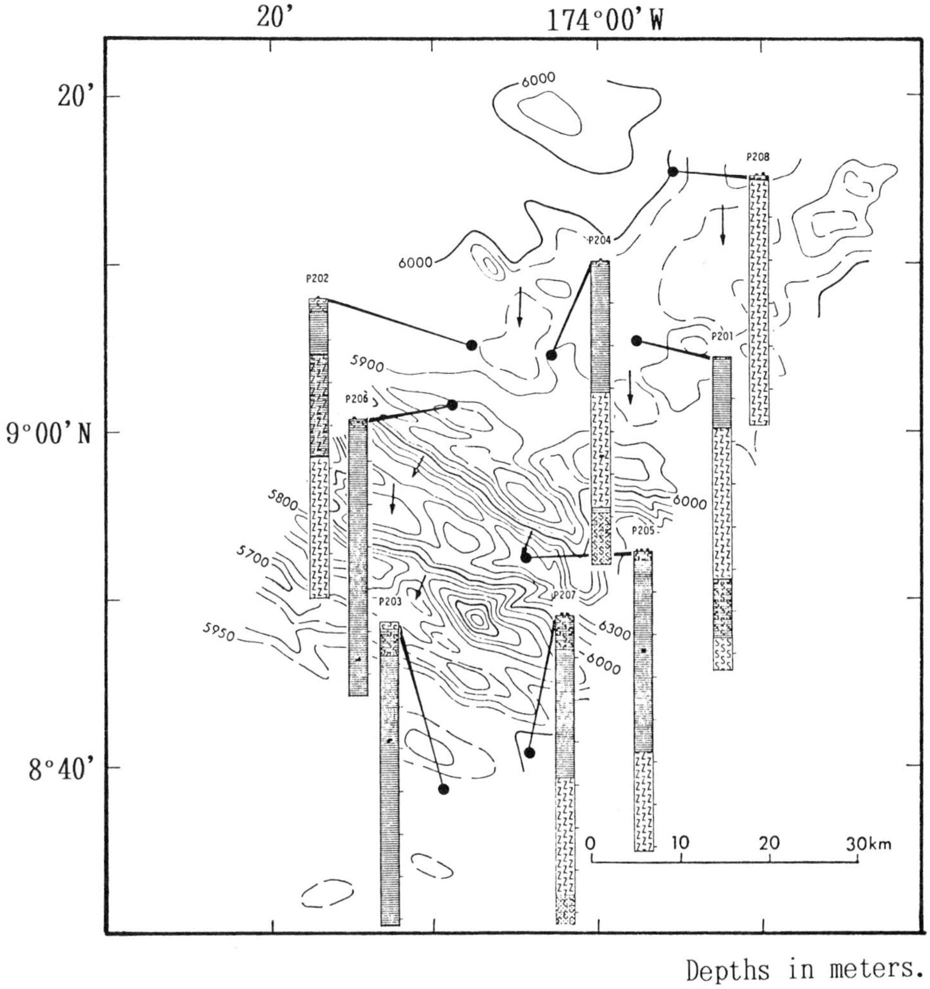

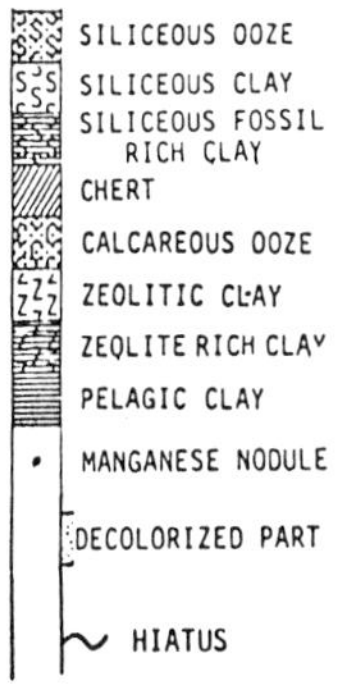

Figure 11-2. Core samples on the topographic map of the eastern part of the GH80-5 area.

Also, the sedimentary sequences in areas characterized by s-type nodules have no hiatuses throughout the late Pliocene to Quaternary time interval.

FORMATION OF HIATUSES – MECHANISM AND AGE

Erosion/sedimentation on the sea floor is determined by the dynamic balance between the rate at which sediments are supplied to the bottom and the rate at which they are removed. The principal cause of erosion is generally believed to be an intensification of bottom current flow that corrodes siliceous and/or calcareous material and transports sediments (Moore et al., 1978).

Hiatuses observed in the sedimentary sequences of the CPB are likewise believed to have been formed by bottom-current erosion (AABW), for the reasons listed below.

(1) The water depths of the core sites with hiatuses are below the top of the bottom water layer (Yamazaki, 1986a). The top of the bottom water during periods of intensified flow is inferred to have been shallower than that of the present bottom water, as in the Vema Channel area of the South Atlantic where the top of the bottom water layer during the last glaciation, with more vigorous circulation, was shallower than at present (Ledbetter and Ellwood, 1980).

(2) Topographic features such as moats and mud waves around the hills suggest the strong influence of a bottom current on sedimentation.

(3) Hiatuses are often associated with specific topographic features such as abyssal hills, or troughs between ridges, and are associated with thinner, acoustically transparent layers on seismic records. These features suggest that hiatuses and low sedimentation rates are both caused by factors related to the sea bottom topography.

(4) Sediments of the CPB are mainly siliceous clay, siliceous ooze, and pelagic clay deposited below the CCD. These sediments show little effect of chemical corrosion by bottom water.

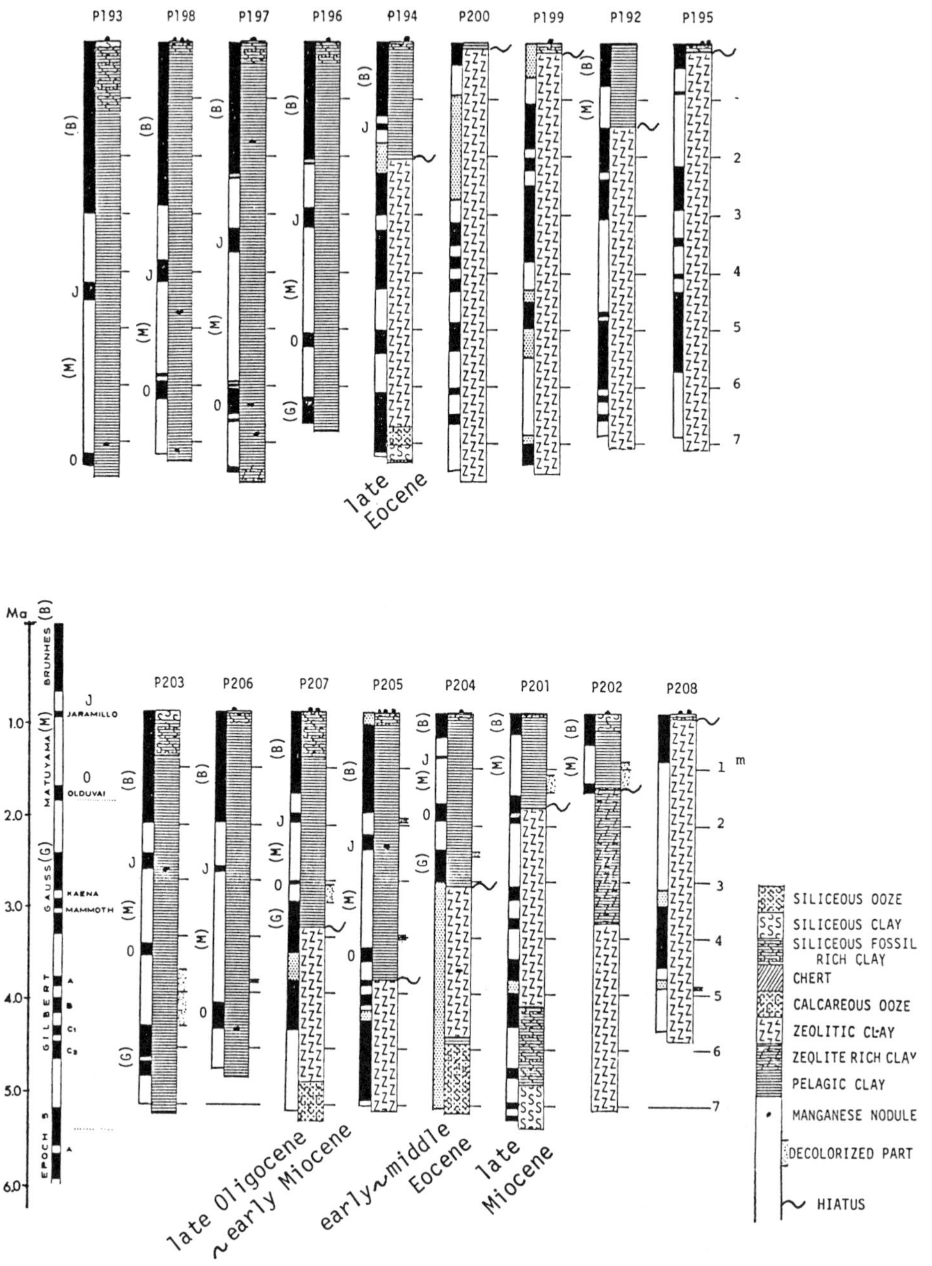

Figure 12. Core profiles of the GH80-5 area of the Central Pacific Basin. Core sequences were described in detail by Nishimura (1984). The ages below the columns show the ages of the core bottom sediments based on microfossil data.

(5) The ages of the hiatuses in this area coincide with globally widespread hiatuses, formed during global cooling events (Barron and Keller, 1982).

The durations of the hiatuses in the core sequences are often long, because successive erosional events removed intervening sediments, leaving no sedimentary record. This is to be expected in areas of very slow deposition.

Precise age determination of the short periods of hiatus formation comes from areas with high sedimentation rate such as in a high productivity zone (Keller and Barron, 1983). It is usually very difficult to determine the precise age of hiatus formation solely on the basis of lack of sediments. The age of hiatus formation is defined from the youngest age below the hiatus and the oldest age above the hiatus in

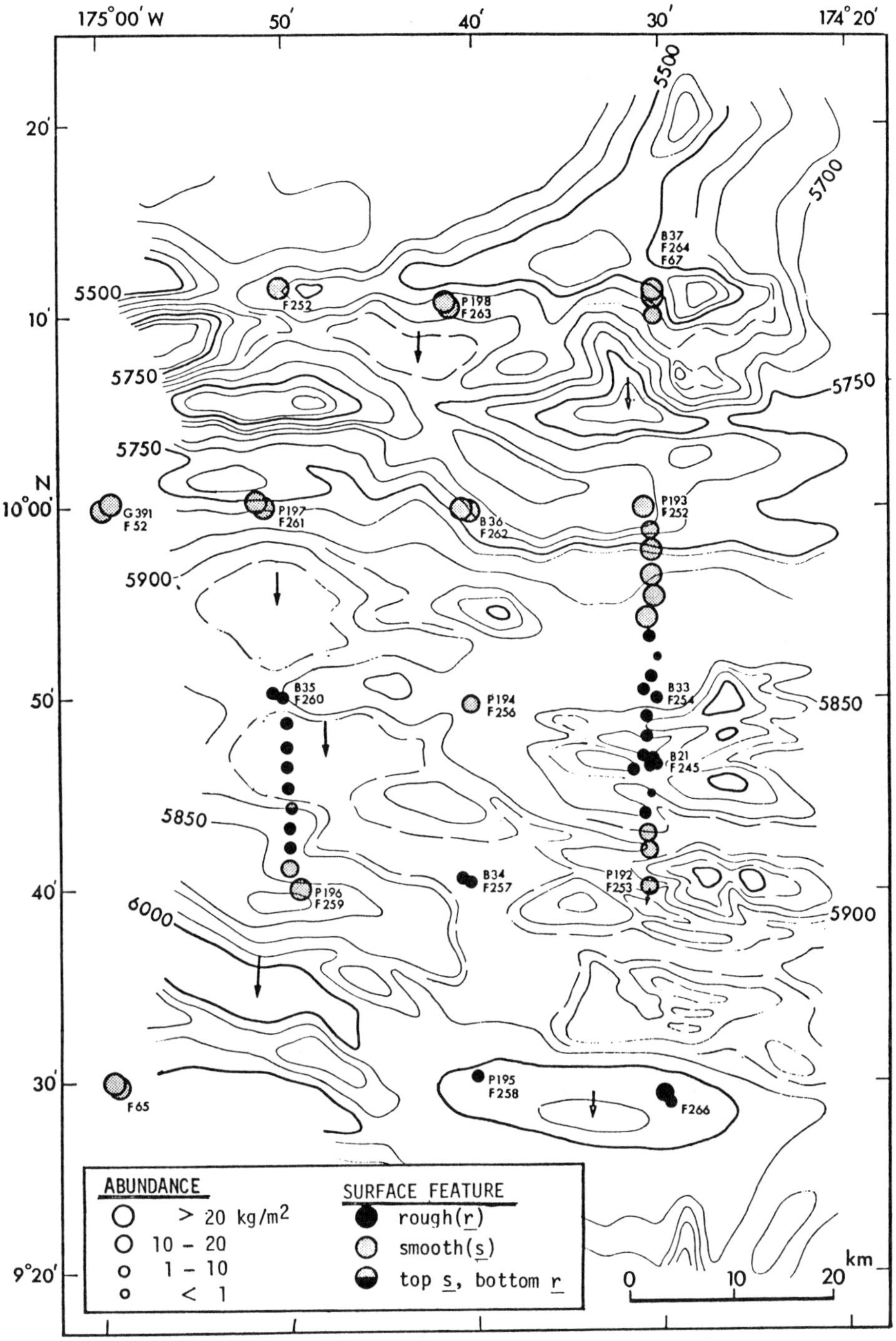

Depths in meters.

Figure 13-1. Distribution of manganese nodules in the western part of the GH80-5 area (Usui and Nakao, 1984).

the sequences of the cores. The time of peak hiatus formation corresponds to the time when the greatest number of cores lack a sedimentary record. In the CPB area, core data allow discrimination of three certain hiatuses: late Miocene, late Pliocene; and latest Pliocene to early Pleistocene, and one possible hiatus period, middle to late Pleistocene (Fig. 15).

(1) Late Miocene: Two cores P223 and P227 from the northern equatorial central part of the CPB (GH81-4 area) clearly indicate the existence of a late Miocene hiatus (9.0-6.2 Ma). The age of the sediment just below the hiatus of Core P220 restricts the period of hiatus formation to 8.5-6.2 Ma. This hiatus period is correlative to the widespread hiatus NH6 (7.0-6.3 Ma) defined by Keller and Barron

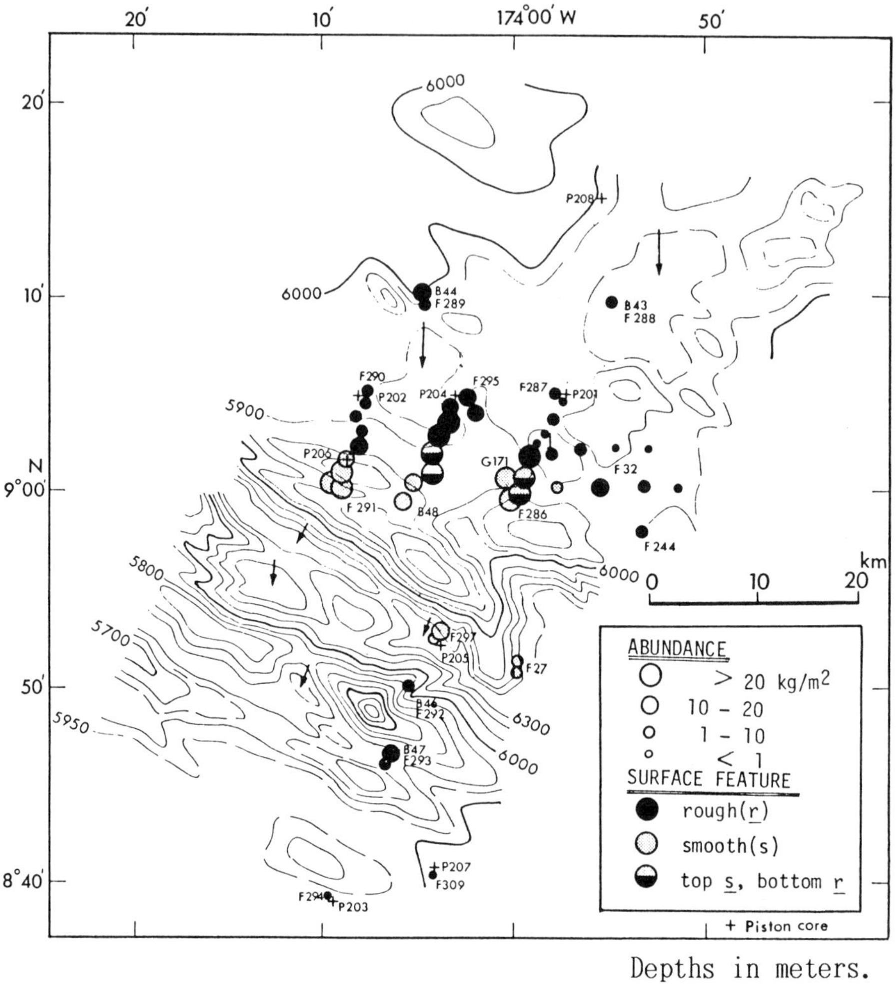

Figure 13-2. Distribution of manganese nodules in the eastern part of the GH80-5 area (Usui and Nakao, 1984).

(1987) and also found at DSDP Site 66 on the basin floor southwest of the GH81-4 area (Keller and Barron, 1983).

Many core sequences have been studied in the Samoan Passage area which is the only inflow passage of bottom water into the CPB (Lonsdale, 1981). Although a long period (from Eocene to Quaternary) of hiatus was reported in the passage (Hollister et al., 1974), the cores of the adjacent area north of the passage contain few hiatuses. Two cores with a distinctive late Pliocene hiatus have been reported. They are from the Northern Drift area and lack the radiolarian '*Ommatartus penutimus* Zone'; this hiatus is correlative to NH6.

The NH6 hiatus is probably very widely distributed throughout the CPB. In the north of the Magellan Trough (GH80-5 area), siliceous clay with a high content of collosphaerid radiolarians was deposited under the influence of probably slightly intensified bottom-current flow during this period (in the bottom of Core P201).

(2) Late Pliocene: This hiatus period is indicated by the core data from the GH80-5 area, especially Core P201. The age is 4.0 to 3.0 Ma and correlates with the NH8 hiatus (3.7–3.1 Ma) defined by Keller and Barron (1987). In the Southern Ocean, there are many cores with a hiatus of this age and it is inferred that there was an intensified circulation around the Antarctic at that time (Ledbetter and Ciesielski, 1986).

(3) Latest Pliocene to early Pleistocene: This period is indicated from the age data of cores P225 and P218. The period is 2.6-0.9 Ma. Among the cores from the GH81-4 and GH80-5 areas, core sequences without a hiatus of this age have distinct and longer normally magnetized parts which are correlated to the Jaramillo Event. This suggests a higher sedimentation rate around the Jaramillo Event. This is thought to be related to the redeposition of reworked sediments generated elsewhere by hiatus formation. In the Southern Ocean, deep bottom current flow increased in velocity during the late Matuyama Reversed Epoch (2.0–1.5 Ma) as a result of Antarctic cooling (Ledbetter and Ciesielski, 1986).

(4) Middle to late Pleistocene: The period of this hiatus has not been clarified in the present studied areas. Many studies of sediment cores with hiatuses in the Pacific, includ-

ing this research, show a wide distribution of a very thin layer (sometimes thinner than several centimeters) of Quaternary sediments above the hiatuses. Some researchers believe that a hiatus below the very thin layer of Quaternary sediments was formed during the last glacial (Johnson, 1972a). However, no sediment cores which have hiatuses formed during a short period of late Quaternary has been reported.

Some of the cores in the CPB have distinctly short sediment sequences with normally magnetized sections correlative to the Brunhes Normal Epoch (e.g., Cores P173 and P174) (Joshima, 1982). This suggests the possibility of hiatuses in the Brunhes Normal Epoch. Although we have no direct evidence for depositional breaks, and precise age determination in the Brunhes Normal Epoch is difficult, the existence of a hiatus in the middle to late Pleistocene cannot be ruled out.

TWO DIFFERENT STYLES OF SEDIMENTATION AND HIATUS FORMATION AFFECTED BY BOTTOM WATER CURRENTS

The areas where detailed studies of hiatus distribution have been made can be grouped topographically into two distinct types. One type occurs on areas of abyssal hills and mountains, with distinct hiatuses on the topographically higher parts; these hiatuses are less apparent on the abyssal plain. The other type occurs in narrow troughs along the sides of ridges and in deep passages.

In the northern equatorial central part of the CPB (GH81-4 area), hiatuses are of the "abyssal hill type", whereas in the area north of the Magellan Trough (GH80-5 area), they are of the "valley type". This two-fold classification can be applied to hiatuses described by other workers, with the former type recognized on hills and mountains (Moore and Heath, 1967; Johnson and Johnson, 1970; Moore, 1970), and the latter type both in passage areas (Lonsdale, 1981) and near fracture zones (Piper and Blueford, 1982).

Sedimentation processes can be related to the intensity of the bottom current: periods with a strongly intensified bottom current correspond to hiatus-forming periods in each area; "normal" periods with a weak bottom current correspond to periods of sedimentation over the whole area; periods with a moderately intensified bottom current mark times of sedimentation with clear bottom-current control on facies and sediment distribution, but without hiatus formation.

Different styles of sedimentation and hiatus distribution in the two areas of detailed study are explained below, and are illustrated in Figure 16. The vertical flux of pelagic

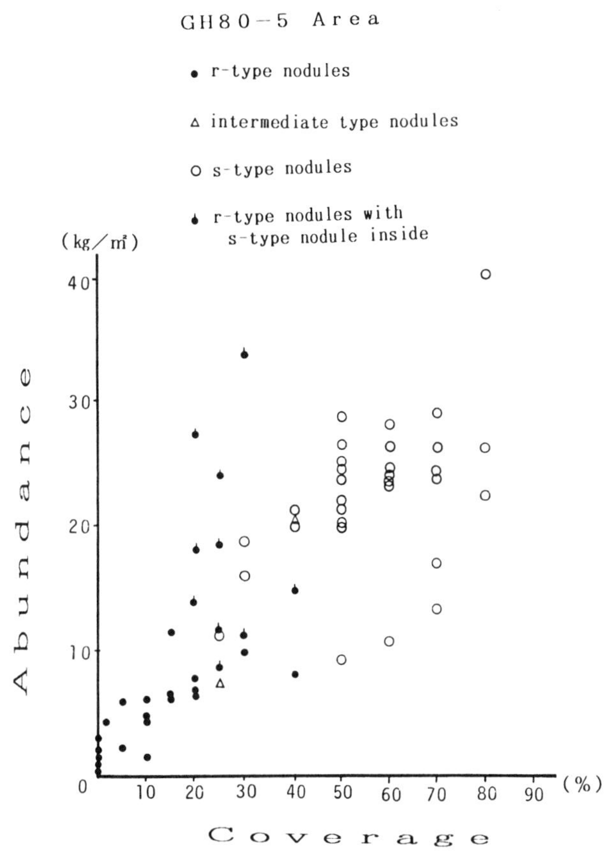

Figure 14. Relationship between nodule abundance and coverage revealed by sea-bed photography in the GH80-5 area (Usui and Nakao, 1984).

grains is assumed to be relatively constant, with small fluctuations mainly related to surface productivity in each survey area.

"Abyssal hill type": During "normal" periods with a weak bottom current, relatively slow sedimentation characterizes topographic highs (hills). During periods of intensified bottom current activity, hiatuses are formed on the hills. The sedimentation rate on the hills decreases from the "normal" period to the hiatus-forming period, and increases from the hiatus-forming period to the "normal" period (Yamazaki, 1986b). Moats and mud waves are formed around the hills by bottom currents.

"Valley type": During "normal" periods with a weak bottom current, relatively slow sedimentation characterizes the valley areas, compared to the adjacent ridge areas. During periods with an intensified bottom current, hiatuses are formed in the valleys. During periods with a moderately intensified bottom current, rapid sedimentation occurs in the valleys, and sediment lithology changes from pelagic/zeolitic clay to siliceous ooze/clay including reworked radiolarians (for example, Lonsdale, 1981; data of Core P201).

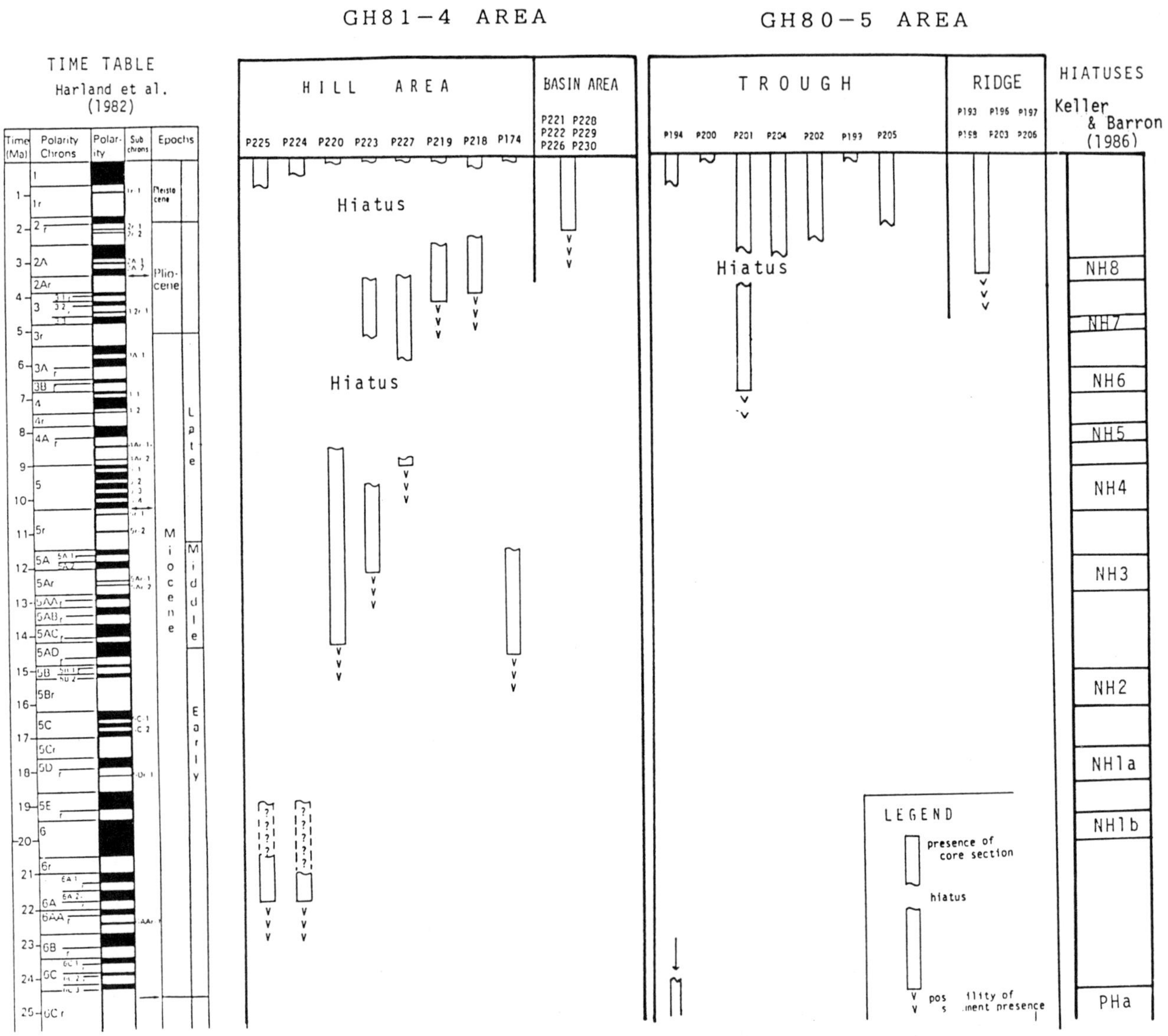

Figure 15. Occurrence of hiatuses in the core samples of the GH81-4 and GH80-5 areas. Ages of hiatuses recognized by Keller and Barron (1986) are based on the correlation of Anomaly 5 to paleomagnetic Chron 11 (Barron et al., 1985).

In the "abyssal hill type" areas, bottom currents always have a stronger influence on the hills than in the basin, and tend to prevent significant pelagic sedimentation. This has caused the difference in sedimentation rates between the hills and basins, even in "normal" periods.

Conversely, in the "valley type" areas, an intense bottom current affects sedimentation in the valleys or the troughs by eroding the seabed (hiatus formation). Bottom currents with moderate intensity cause rapid deposition of reworked sediments in the valleys. These sediments must be supplied laterally from nearby areas of erosion. The details of sedimentation processes during "normal" periods in valley-type areas are still unresolved. Two possible mechanisms are outlined below.

(1) Bottom currents have a stronger impact in the confined valleys than outside of the valleys. The currents prevent significant pelagic sedimentation, and thus result in a marked difference in sedimentation rate between the troughs and the ridges in the "valley type" areas.

(2) The sediment supply to these areas is mainly by lateral supply by bottom currents. The bottom currents deposit more sediment outside of the valleys than in the valleys.

The latter explanation better fits the GH80-5 area, where the vertical supply is thought to be very small because of great distance from the equatorial high productivity zone.

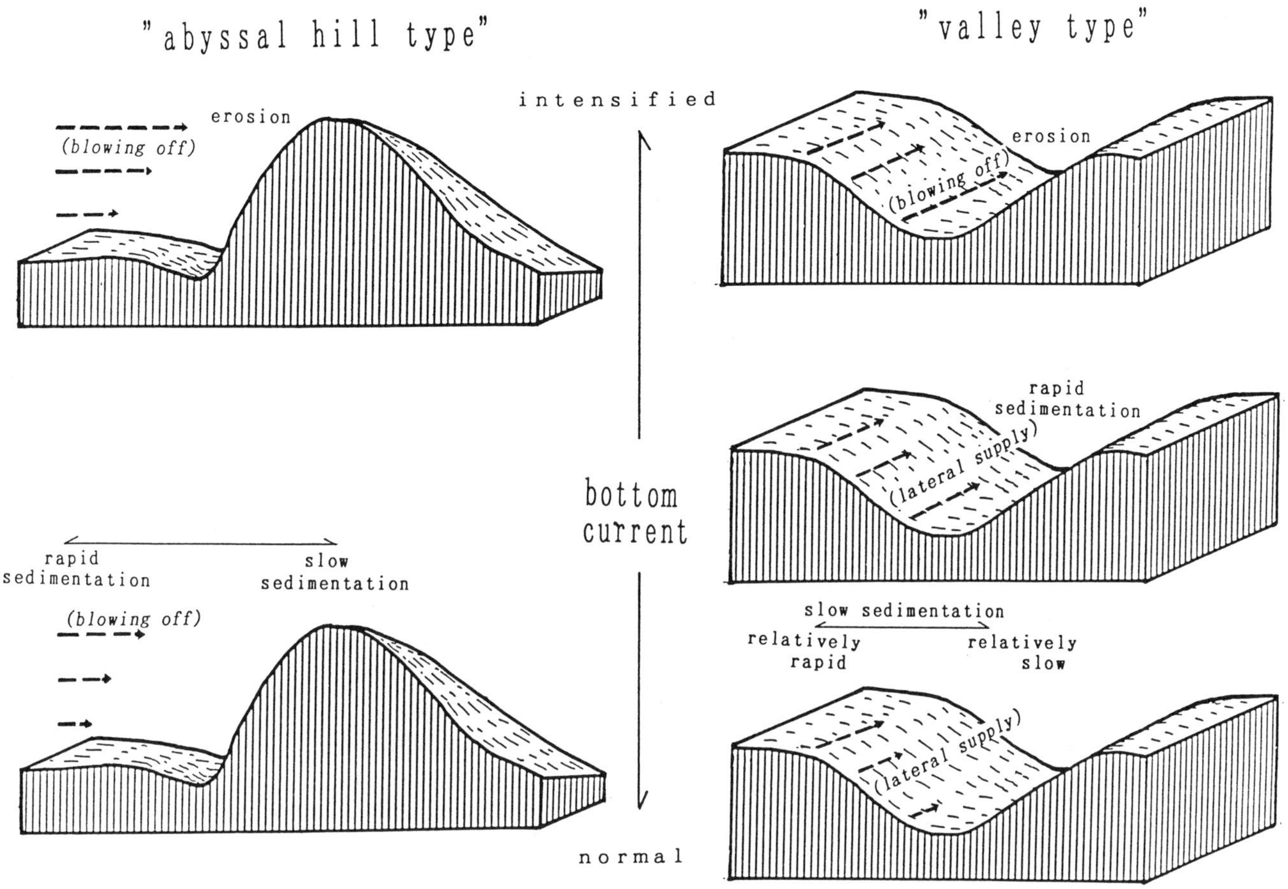

Figure 16. Schematic diagram of sedimentation processes in the two contrasting areas. Thick broken arrows show bottom currents. Their lengths reflect intensities.

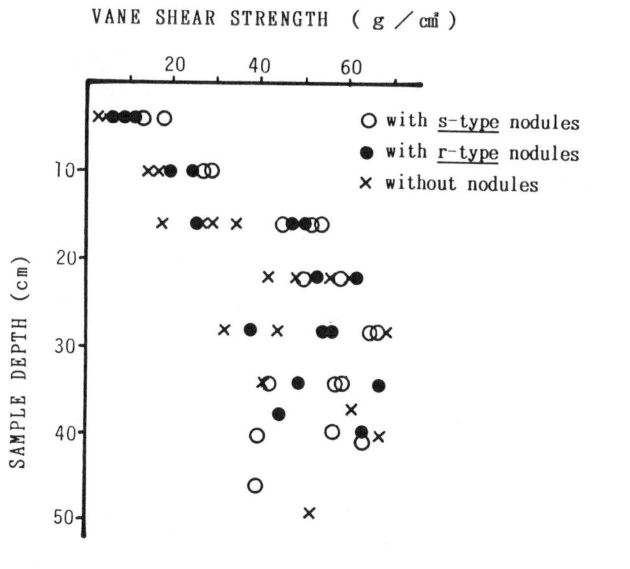

Figure 17. Vane shear strength of box core sediments in the GH81-4 area (Handa and Yamazaki, 1986).

NODULE FORMATION AND SEDIMENTATION

Differences in manganese nodule distribution patterns in the two areas are outlined below.

"Abyssal hill type" (GH81-4 area) : A high concentration (10–20 kg/m^2) of s-type nodules occurs on the surface of the sediment. Sedimentation rate is relatively low (probably less than 3.5 mm/1000 y). A low concentration of r-type nodules occurs below the sediment surface (i.e. buried) in areas with moderately high sedimentation rate (~3.5 mm/1000 y). Areas without nodules have higher sedimentation rates (3–6 mm/1000 y).

"Valley type" (GH80-5 area): A high concentration (10–20 kg/m^2) of s-type nodules is exposed on the sea floor in areas with relatively high sedimentation rate (2–4 mm/1000 y). Higher concentrations (more than 20 kg/m^2) of r-type nodules are exposed on the sea floor in areas with relatively low sedimentation rate (less than 2 mm/1000 y).

In order to relate the present distribution of manganese nodules to present environments, it must be assumed that the manganese nodules in the vicinity of the water-sediment interface are still growing. Although data on the growth rate of manganese nodules in the CPB are limited, the alpha-activity method indicates average growth rates of 4.47 and 3.35 mm/m.y. for r-type and s-type nodules, respectively, in a small area west of the Line Islands Ridge (Piper and Gibson, 1981). Moreover, the growth rate of the exposed r-type nodules in the GH80-5 area was determined to be about 3 mm/m.y. using ^{10}Be dating techniques (Usui, personal communication). The internal structures (growth structures) of the outer several millimeters of the manganese nodules in the GH81-4 and GH80-5 areas are similar to those of the dated nodules and show no evidence of an interruption in growth. These data suggest that both the r- and s- type nodules around the water-sediment interface have been growing at least during Pleistocene, and that they probably continue to grow today.

The r- and s-types have different mineral compositions. The MnO_2 (s-type) nodules indicate highly oxygenated conditions and 10Å manganate (r-type) less well oxygenated conditions.

In almost all parts of the CPB (including GH81-4 area), geochemical conditions correspond exactly to these predicted above for the nodules that are present in the area. That is, s-type nodules are exposed to oxygenated bottom water, while r-type nodules are buried in the surface sediments and are affected by diagenetic processes that result in lower oxygen concentrations.

Is there a dominant factor which determines whether r- or s-type nodules will predominate in a particular area? Also, why are the growing manganese nodules found on or near the sea floor, in spite of their very slow growth rate (order of mm/m.y.) compared to the sedimentation rate of the underlying sediments (order of mm/1000 y)? It has been suggested that nodules are maintained at sediment-water interface by the activity of benthic organisms (Mero, 1965; Piper and Fowler, 1980; Usui et al., 1987b; McCave, 1988). Why, then, are only s-type nodules found at the surface, whereas r-type nodules are generally buried? Variable intensity of benthic organism activity (e.g., Nishimura, 1983; von Stackelberg, 1987) is an unsatisfactory explanation for this observation. Instead, the major factor determining the positions of the nodules around the water-sediment interface is apparently the physical properties of the surface sediments, namely the support strength, or firmness of the material.

In the GH81-4 area, the physical properties of the sediments, particularly vane shear strength, were measured, and the results show that strength increases from areas with no nodules through r-type nodule areas, and finally to s-type nodule areas (Handa and Yamazaki, 1986) (Fig. 17). Larger vane shear strength values were obtained from the sites with lower sedimentation rates, even though the sediment types were the same (Miyata and Tanaka, 1983; core sample data of the GH81-4 area of Handa and Yamazaki, 1986). The ability to support nodules probably varies directly with relative positions of nodules near the sea floor, providing a simple explanation for vane shear strength (Fig. 18).

For example, the low sedimentation rate areas with high support strength contain nodules in an exposed state, and areas with a high sedimentation rate and low support strength contain nodules in a buried state. The areas with a very high sedimentation rate have insufficient support strength to retain nodules in the surface sediments, where nodules they must be to grow. Hence, no nodules are found in such areas.

For the GH80-5 area, the values of vane shear strength obtained on box core samples are as large as those found in the s-type nodule part of the GH81-4 area, except for several stations with buried r-type nodules and without nodules (Tsurusaki, 1984). This reflects the generally low sedimentation rate in the whole GH80-5 area. The r-type nodules in this area have been growing in an exposed state because of slow sedimentation.

Recall, however, that r-type nodules imply relatively low oxygen levels. Why, then, are r-type nodules growing on the sediment surface in this area? Nodules with 10Å manganate are common in the East Pacific and rare in the West Pacific. Nodules with 10Å manganate in the East Pacific are highly concentrated and sometimes occur in an exposed state (Piper and Blueford, 1982). The bottom water in the East Pacific has passed through the CPB, originating from the AABW. The physico-chemical conditions of the sea water change as a result of oxidation of organic matter, providing a lower oxygen content than that in the central part of the CPB.

This is particularly true beneath the slowly flowing bottom current in the semi-enclosed topography of the valley below 5,850 m in the GH80-5 area, allowing growth of exposed r-type nodules. The metals in the nodules are thought to be derived not only from solution in the bottom water, but also from sediment particles temporally resting on the rough surface of the nodules.

Factors which control the abundance of nodules are very difficult to determine. The growth rates of r-type and s-type nodules are similar under similar conditions (Mizuno et al., 1980). Therefore, the total volume, plus total time for growth, of precipitated manganese minerals per unit time in a unit area must determine the abundance. It is reasonable to expect that the nodules can acquire more ions from sea water than can the buried nodules from their pore water, because ion concentrations in the bottom water are maintained by the bottom current. Although investigations are needed to clarify the cause of the differences in nodule abundance, the high concentration of both exposed s- and

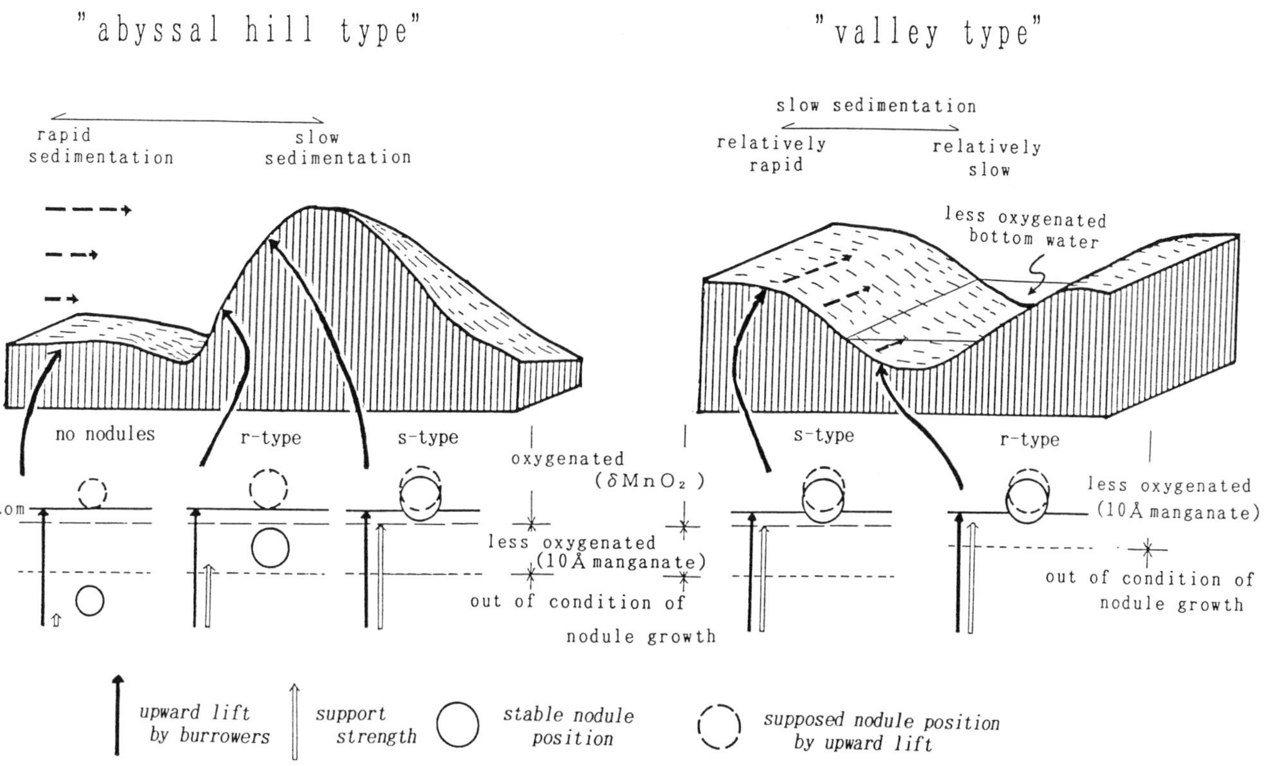

Figure 18. Schematic diagram of manganese nodule formation in the two sites distinguished in this paper.

r- type nodules is possibly a result of this constant renewal of constituent ions by bottom-water flow.

There is a general relationship between nodule occurrence and the thickness of the acoustically transparent layer on seismic records (Tamaki et al., 1977; Usui and Tanahashi, 1986). In the GH81-4 area, the area of abundant nodules has a thin (less than 70 m) transparent layer; s-type nodules are distributed in areas with little or no development of the layer; r-type nodules are in areas of moderate development of the layer (Usui and Tanahashi, 1986). This relationship is clearly explained by the model for nodule formation outlined above. The area of abundant nodules has experienced a low sedimentation rate, favorable for nodule growth, for a long period, resulting in a thinner sediment layer.

In the core sequences, deeply buried nodules which have ceased growing are sometimes found (e.g., Fig. 12). These nodule occurrences indicate a change in the sedimentation rate. Increased sediment supply caused a decrease in the support strength of the surface sediments so that the nodules could not continue to migrate upward as the sea bed accreted. Note that the critical sedimentation rate for stranding and burial of nodules is not the average sedimentation rate over a period of several million years, which is generally all that available age data allow us to calculate. Instead, it is the sedimentation rate corresponding to the several-centimeter-thick surface sediments that must support the nodules on or near the face.

There is a close relationship between high sedimentation rates and the occurrence of deeply buried nodules. In the GH80-5 area, the sedimentation rate around the Jaramillo Event, where deeply buried nodules are found in the three cored sequences, is 1.5 times that determined for other periods of the late Pliocene through Quaternary (Joshima and Nishimura, 1984).

CONCLUSIONS

Study of cores and correlations between the distribution of sediments and manganese nodules in the Central Pacific Basin reveals the following general features.

(1) Hiatuses formed in the Neogene through Quaternary by the strong influence of bottom currents are widely distributed in the sedimentary sequences in the CPB.

(2) The period of the main hiatus of the CPB revealed in the piston core sequences is NH6 (7.0-6.3 Ma) of late Miocene age. Subsequently, hiatuses were formed in various local areas, with age ranges of late Pliocene (NH8, 3.7-3.1 Ma), the latest Pliocene to early Quaternary (2.6-0.9 Ma), and possibly the late Quaternary.

(3) The areas where hiatuses have been frequently formed are grouped into two types. One is the topographic highs, hills and mountains, that rise above the flat basin floor of the CPB ("abyssal hill type"), and the other comprises narrowly elongated topographic lows, channels and troughs ("valley type"). In these areas, the sedimentation rate is lower than that of the adjacent areas.

The presence of hiatuses and the low sedimentation rates can be ascribed to the influence of bottom currents in these areas. The bottom currents cause slow sedimentation in "normal" periods and hiatuses during times of intensified flow.

(4) Local variations in the style of occurrence of manganese nodules are related to local sedimentation processes that are governed by bottom currents. Sedimentation rate determines the position of nodule growth around the water-sediment interface, because the rate controls physical properties of the surface sediments. The differences in the manganese nodule types (r- and s-types) reflect the chemical environment where the nodules have been formed, and the differences in the manganese nodule abundance reflect the duration of growth and the source of materials for nodule formation--from bottom water (faster growth) or from the pore water of sediments deposited on the sea floor (slower growth).

Exposed r-type nodules are rarely observed in the CPB. However, with very slow sedimentation assuring maintenance of nodules on the sediment surface, and with less oxygenated bottom water generated by oxidation of organic matter, exposed and growing r-type nodules can occur.

ACKNOWLEDGMENTS

The author is greatly indebted to the staff of the Geological Survey of Japan and R/V *Hakurei-Maru* crew members for cooperation during research cruises and shore-based studies. Special thanks are due to GSJ colleagues Drs. Y. Shimazaki, T. Moritani, A. Usui, S. Nakao, and Y. Inouchi for their critical review of the manuscript and instructive comments. R. Hiscott (Memorial University, Canada) is thanked for advice on English phrasing. I also wish to thank A. Mizuno, Yamaguchi University; T. Kamei, T. Shiki, and K. Chinzei, all of Kyoto University, for many encouraging discussions.

REFERENCES

Barron, J.A. and G. Keller, 1982, Widespread Miocene hiatuses: Coincidence with periods of global cooling: Geology, v. 10, p. 577-581.

Barron, J.A., G. Keller, and D.A. Dunn, 1985, A multiple microfossil biochronology for the Miocene: Geological Society of America Memoir, no. 163, p. 21-36.

Berggren, W.A., L.H. Burckle, M.B. Cita, H.B.S. Cooke, M.B. Funnel, S. Gartner, J. D. Hays, J. P. Kennet, N .D. Opdyke, L. Pastouret, N.J. Shackleton, and Y. Takayanagi, 1980, Towards a quaternary time scale: Quaternary Research, v. 13, p. 277-302.

Chase, T.E., H.W. Menard, and J. Mammerickx, 1971 , Topography of the North Pacific: IMR Technical Report Series, TR 14.

Edmond, J .M., Y. Chung, and J.G. Sclater, 1971, Pacific bottom water: Penetration east around Hawaii: Journal of Geophysical Research, v. 76, p. 8089-8097.

Gordon, A.L. and R.D. Gerard, 1970, North Pacific bottom potential temperature: Geological Society of America, Memoir, no. 1 26, p. 23-39.

Handa, K. and T. Yamazaki, 1986, Physical and engineering properties of manganese nodules and deep-sea sediments in the GH81-4 area: Geological Survey of Japan Cruise Report, no. 21, p. 84-97.

Harland, W.B., A.V. Cox, P.G. Llewellyn, C.A.G. Pickton, A.G. Smith, and P. Waters, 1982, A geologic time scale: Cambridge University Press (Cambridge), 131 p.

Hollister, C.D., A.R.M. Nowell, and P.A. Jumars, 1984, The dynamic abyss: Scientific American, 250(3), p. 42-53.

Hollister, C.D., S.A. Johnson, and P.F. Lonsdale, 1974, Current-controlled abyssal sedimentation: Samoan Passage, equatorial Pacific: Journal of Geology, v. 82, p. 275-300.

Johnson, D.A., 1972a, Ocean-floor erosion in the equatorial Pacific: Geological Society of America Bulletin, v. 83, p. 3121-3144.

Johnson, D.A., 1972b, Eastward-flowing bottom currents along the Clipperton Fracture Zone: Deep-Sea Research, v. 19, p. 157-169.

Johnson, D.A. and T.C. Johnson, 1970, Sediment redistribution by bottom currents in the central Pacific: Deep-Sea Research, v. 17, p. 157-169.

Joshima, M., 1982, Remanent magnetization of piston cores from the Wake-Tahiti transect in the Central Pacific: Geological Survey of Japan, Cruise Report, no. 18, p. 276-287.

Joshima, M. and A. Nishimura, A., 1984, Remanent magnetization of sediment cores in the GH80-5 survey area: Geological Survey of Japan, Cruise Report, no. 20, p. 165-192.

Keller, G. and J.A. Barron, 1983, Paleoceanographic implications of Miocene deep-sea hiatuses: Geological Society of America Bulletin, v. 94, p. 590-613.

Keller, G. and J.A. Barron, 1987, Paleodepth distribution of Neogene deep-sea hiatuses: Paleoceanography, v. 9, p. 697-713.

Keller, G., T. Herbert, R. Dorsey, S.D'Hondt, M. Johnsson, and W.R. Chi, 1987, Global distribution of late Paleogene hiatuses: Geology, v. 15, p. 199-203.

Kobayashi, K., K. Kitazawa, T. Kanaya, and T. Sakai, 1971, Magnetic and micropaleontological study of deep-sea sediments from the west-central equatorial Pacific: Deep-Sea Research, v. 18, p. 1045-1062.

Ledbetter, M.T. and P.F. Ciesielski, 1986, Post-Miocene disconformities and paleoceanography in the Atlantic sector of the Southern Ocean: Palaeogeography, Palaeoclimatology, Palaeoecology, v. 52, p. 185-214.

Ledbetter, M.T. and B.B. Ellwood, 1980, Spatial and temporal changes in bottom-water velocity and direction from anlysis of particle size and alignment in deep-sea sediment, Marine Geology, 38, p. 245-261.

Lineberger, P.H., 1975, Sedimentary processes and pelagic turbidites in the eastern Central Pacific Basin: Hawaii Inst. Geophys. Techn. Rept., HIG-75-24, 38 p.

Lonsdale, P.F., 1981, Drifts and ponds of reworked pelagic sediment in part of the southwest Pacific: Marine Geology, v. 43, p. 153-193.

Lonsdale, P.F. and S.M. Smith, 1980, "Lower insular rise hills" shaped by a bottom boundary current in the Mid-Pacific: Marine Geology, v. 34, p. M.19-25.

Mammerickx, J., S. M. Smith, I. L. Taylor, and T. E. Chase, 1975, Topography of the South Pacific: IMR Technical Report Series, TR 56.

Mantyla, A., 1975, On the potential temperature in the abyssal Pacific Ocean: Journal of Marine Research, v. 33, p. 341-354.

Matsubayashi, O. and A. Mizuno, 1982, Bottom potential temperature and vertical temperature profiles of near bottom waters in the Central Pacific: Geological Survey of Japan, Cruise Report, no. 18, p. 231-237.

McCave, I.N., 1988, Biological pumping upwards of the coarse fraction of deep-sea sediments: Journal of Sedimentary Petrology, v. 58, p. 148-158.

Mero, J.L., 1965, The mineral resources of the sea, Elsevier, Amsterdam, 312 p.

Miyata, Y. and T. Tanaka, 1983, Physical property (shear strength) of piston core samples: Preliminary Report, Hakuho-Maru Cruise KH82-4, p. 49-55.

Mizuno, A., T. Miyazaki, A. Nishimura, K. Tamaki, and M. Tanahashi, 1980, Central Pacific manganese nodules, and their relation to sedimentary history: Proc. Offshore Technology Conference in Houston, May 5-8, 1980, p. 331-340.

Mizuno, A. and Y. Okuda, 1982, Seismic reflection profiles of the Wake-Tahiti transect in the Central Pacific Ocean: Geological Survey of Japan, Cruise Report, no. 18, p. 46-89.

Moore, T.C., Jr., 1970, Abyssal hills in the Central Equatorial Pacific: Sedimentation and stratigraphy: Deep-Sea Res., v. 17, p. 573-593.

Moore, T.C., Jr. and G.R. Heath, 1967, Abyssal hills in the Central Equatorial Pacific: Detailed structure of the sea floor and subbottom reflectors: Marine Geology, v. 5, p. 161-179.

Moore, T.C.Jr., T..H. van Andel, C. Sancetta, and N. Pisias, 1978, Cenozoic hiatuses in pelagic sediments: Micropaleontology, v. 24, p. 113-138.

Moritani, T., S. Maruyama, M. Nohara, K. Matsumoto, T. Ogitsu, and H. Moriwaki, 1977, Description, classification, and distribution of manganese nodules: Geological Survey of Japan, Cruise Report, no. 8, p. 136-158.

Nakao, S., 1986, ed., Marine geology, geophysics, and manganese nodules around deep-sea hills in the Central Pacific Basin August-October 1981 (GH81-4 Cruise): Geological Survey of Japan, Cruise Report, no. 21, 257 p.

Nakao, S. and A. Mizuno, 1982, Regional sedimentologic data: the Central Pacific Wake-Tahiti transect, GH80-1 cruise: Geological Survey of Japan, Cruise Report, no. 18, p. 95-123.

Nakao, S. and T. Moritani, 1984, eds., Marine geology, geophysics, and manganese nodules in the northern vicinity of the Magellan Trough August-October 1980 (GH80-5 Cruise): Geological Survey of Japan, Cruise Report, no. 20, 272 p.

Nishimura, A., 1981, Deep-sea sediments in the GH79-1 area; Their geological properties: Geological Survey of Japan, Cruise Report, no. 15, p. 110-142.

Nishimura, A., 1983, Sediments in the Central Pacific manganese nodule province: Marine Science, v. 15, p. 421-425.

Nishimura, A., 1984, Deep-sea sediments in the GH80-5 area in the northern vicinity of the Magellan Trough: Geological Survey of Japan, Cruise Report, no. 20, p. 67-89.

Nishimura, A., 1986, Deep-sea sediments in the Central Equatorial Pacific (GH81-4 area): Geological Survey of Japan, Cruise Report, no. 21, p. 56-83.

Nishimura, A., T. Yamazaki, K. Ikehara, and M. Tanahashi, 1985, Calcareous biogenic turbidites in the Central Pacific Basin: Journal of Sedimentological Society of Japan, no. 22/93, p. 65-70.

Normark, W. R. and F. N. Spiess, 1976, Erosion on the Line Islands archipelagic apron: effect of small-scale topographic relief: Geological Society of America Bulletin, v. 87, p. 286-296.

Orwig, J.L., 1981, Channeled turbidites in the Eastern Central Pacific Basin: Marine Geology, v. 39, p. 33-57.

Piper, D.Z. and J.R. Blueford, 1982, Distribution, mineralogy, and texture of manganese nodules and their relation to sedimentation at DOMES Site A in the equatorial Pacific Ocean: Deep-Sea Research, v. 29, p. 927-952.

Piper, D.Z. and B. Fowler, 1980, New constraint on the maintenance of Mn nodules at the sediment surface: Nature, v. 286, p. 880-883.

Piper, D.Z. and C.N. Gibson, 1981, Nodule growth rates in the GH79-1 area: Geological Survey of Japan, Cruise Report, no. 15, p. 265-269.

Piper, D.Z., E.W. MaCoy, and T.R. Swint, 1985, Manganese nodules, seafloor sediment, and sedimentation rates of the Circum-Pacific Region: American Association of Petroleum Geologists, Tulsa, Oklahoma, scale 1:17,000,000.

Reid, J.L. and P.F. Lonsdale, 1974, On the flow of water through the Samoan Passage: Journal of Physical Oceanography, v. 4, p. 58-73.

Schlanger, S.O., E.D. Jackson, et al., 1976, Initial Report of the Deep Sea Drilling Project, v. 33, 973 p.

Takayanagi, Y., T. Sakai, M. Oda, and S. Hasegawa, 1982, Micropaleontology of piston cores, Wake to Tahiti: Geological Survey of Japan, Cruise Report, no. 18, p. 238-263.

Tamaki, K., 1984, Seismic reflection survey in the Central Pacific Basin during GH80-5 cruise: Geological Survey of Japan, Cruise Report, no. 20, p. 43-52.

Tamaki, K., E. Honza, and A. Mizuno, 1977, Relation between manganese nodule distribution and acoustic stratigraphy in the Eastern half of the Central Pacific Basin: Geological Survey of Japan, Cruise Report, no. 8, p. 172-176.

Tamaki, K. and M. Tanahashi, 1981, Seismic reflection survey in the northeastern margin of the Central Pacific Basin: Geological Survey of Japan, Cruise Report, no. 15, p. 77-99.

Tanahashi, M., 1986, Subsurface acoustic layers detected by 3.5 kHz subbottom profiler in the GH81-4 area (eastern part of the Central Pacific Basin): Geological Survey of Japan Cruise Report, no. 21, p. 21-32.

Theyer, F., C.Y. Mato, and S.R. Hammond, 1978, Paleomagnetic and geochronologic calibration of latest Oligocene to Pliocene radiolarian events, Equatorial Pacific: Marine Micropaleontology, v. 3, p. 377-395.

Tsurusaki, K., 1984, Geotechnical properties of deep-sea sediments in the northern part of Central Pacific Basin (GH80-5 area): Geological Survey of Japan, Cruise Report, no. 20, p. 90-105.

Usui, A., 1982, Variability of manganese nodule deposits: the Wake to Tahiti transect: Geological Survey of Japan, Cruise Report, no. 18, p. 138-223.

Usui, A., 1986, Local variability of manganese nodule deposits around the small hills in the GH81-4 area: Geological Survey of Japan, Cruise Report, no. 21, p. 98-159.

Usui, A., A. Mizuno, T. Moritani, and S. Nakao, 1987, Manganese nodules in the northern and central parts of the Central Pacific Basin - Results of the "Basic Study on Exploration of Deep Sea Mineral Resources": Bulletin of the Geological Survey of Japan, v. 38, p. 539-585.

Usui, A. and S. Nakao, 1984, Local variability of manganese nodule deposits in the GH80-5 area: Geological Survey of Japan, Cruise Report, no. 20, p. 106-164.

Usui, A., A. Nishimura, M. Tanahashi, and S. Terashima, 1987, Local variability of the manganese nodule facies on small abyssal hills of the Central Pacific Basin: Marine Geology, v. 74, p. 237-275.

Usui, A. and M. Tanahashi, 1986, Relationship between local variation of nodule facies and acoustic stratigraphy in the GH81-4 area: Geological Survey of Japan, Cruise Report, no. 21, p. 160-170.

Van Andel, T..H., G.R. Heath, and T.C. Moore, Jr., 1975, Cenozoic history and paleoceanography of the central equatorial Pacific Ocean: Geological Society of America Memoir, no. 143, 134 p.

Von Stackelberg, U., 1987, Growth history and variability of manganese nodules of the Equatorial North Pacific, *in* Marine Minerals; Advances in Research and Resource Assessment (P.G. Teleki, ed.), NATO Advanced Study Institutes Series C: Mathematical and Physical Sciences, 194, p. 189-204.

Winterer, E.L. et al., 1971, Initial Report of the Deep Sea Drilling Project, v. 7, 1757 p.

Winterer, E.L., J.I. Ewing, et al., 1973, Initial Report of the Deep Sea Drilling Project, v. 17, 93O p.

Yamazaki, T., 1986a, Bottom water temperature in the Central Pacific Basin (GH81-4 area): Geological Survey of Japan, Cruise Report, no. 21, p. 171-172.

Yamazaki, T., 1986b, Paleomagnetic stratigraphy of deep sea sediments in the Central Equatorial Pacific (GH81-4 area) : Geological Survey of Japan, Cruise Report, no. 21, p. 173-194.

Keating, B.H., and Bolton, B.R., editors, 1992,
Geology and offshore mineral resources of the
central Pacific basin, Circum-Pacific Council for
Energy and Mineral Resources Earth Science
Series, V. 14. New York, Springer-Verlag.

MANGANESE NODULE DEPOSITS IN THE CENTRAL PACIFIC BASIN: DISTRIBUTION, GEOCHEMISTRY, MINERALOGY, AND GENESIS

Akira Usui and Tomoyuki Moritani
Marine Geology Department, Geological Survey of Japan, 1-1-3 Higashi, Tsukuba, Ibaraki, 305, JAPAN

ABSTRACT

The occurrence, morphology, and composition of manganese nodule deposits in the Central Pacific Basin have been characterized from the data of about 1400 bottom samplings, 900 sea-bed photographs, 1000 chemical analyses, and 800 mineral analyses, together with relevant seismic profiling and sediment stratigraphy obtained during a 10-year research program of the Geological Survey of Japan. The manganese nodules of this area are formed through two different depositional processes: 1) hydrogenous: direct deposition of ferromanganese oxide from normal oxygenated bottom water, and 2) diagenetic: redeposition of dissolved metals within the mildly reduced surface siliceous sediments. These two processes form smooth surface nodules (type s) and rough surface ones (type r), respectively. They also strongly control the nodule mineralogy and chemistry. Both types develop where sedimentation rate is low enough, as revealed by thin lower Eocene to Quaternary uppermost transparent layers on seismic records on regional and local scales. Type r nodules require both a moderate sedimentation rate and sufficient supply of organic material to keep the surface sediments reduced, whereas type s nodules require much less or no sedimentation and the consequent oxidized condition probably caused by bottom currents. Abundant type s nodules occur in the higher latitude zones (northwestern part of the Central Pacific Basin and the Penrhyn Basin of the South Pacific) probably as a result of the activity of AABW during the Neogene through the Quaternary. In contrast, the presence (up to moderate abundance) of type r nodules is closely related to the high-productivity zone near the equator, although local variations in nodule type are quite common.

INTRODUCTION

Iron and manganese oxides are common in the marine environment, where they form deep-sea manganese nodules, seamount manganese crusts, and hydrothermal deposits. Manganese nodules are the most widely spread of these types on the world ocean floors. Pacific manganese nodules have been extensively investigated during the past two decades from scientific and economic points of view. The regional distribution of nodules in the Pacific have been shown by Glasby (1974), Rawson and Ryan (1978), Cronan (1980), Exon (1983), Usui et al. (1983), CPCEMR (1984, 1985), Piper et al. (1987), and others. These investigations of nodule abundance, sea-floor coverage, and related geological study give the general distribution patterns of manganese nodules in the Central Pacific Basin. However, only a few surveys have revealed the local small-scale variations (in the range of kilometers to hundreds of meters) in nodule occurrence and characteristics even in the prime nodule provinces (Andrews and Friedrich, 1979; Piper et al., 1979; Fewks et al., 1980; Usui et al., 1987).

Piper et al. (1987) indicated five areas of the Pacific in which nodules are abundant: 1) Clarion-Clipperton fracture zones, 2) northeast Pacific-Musicians Seamount area, 3) central Southwestern Pacific Basin, 4) Southern Ocean around 60°S, and 5) northern Peru Basin. Nodule abundance, composition, and morphology are variable both on regional and local small scales on the sea floor even within some abundant nodule provinces. These variations are sometimes related to bathymetric features (Halbach and Özkara, 1979; Pautot and Melguen, 1979), distance from the equator (Exon, 1983), surface sediment type (Usui, 1983; Usui et al., 1987), and/or sediment thickness (Moore and Heath, 1966; Calvert et al., 1978; Craig, 1979; Mizuno et al., 1980). These nodule characteristics corresponding to various factors suggest a close relationship between nodule growth and sedimentary conditions.

We investigated the Central Pacific Basin (abbreviated as CPB) west of the Clarion-Clipperton Fracture Zone beyond the Line Islands Chain (Fig. 1). A sporadic but moderately abundant distribution of nodules has been reported in the CPB, also called the "Western Basin" (Cronan, 1972; Horn et al., 1972). We describe here the general features of nodule distribution, chemical and physical characteristics of the CPB, and briefly summarize nodule growth history. This represents a compilation of results of a 10-year research program of the Geological Survey of Japan (abbreviated as GSJ). Some of the results of on-site observation and shore-based analyses have been published as GSJ Cruise Reports (Table 1) and other articles (Usui, 1983; Usui et al., 1987; Usui and Mita, 1987).

FRAMEWORK AND METHODS OF GSJ TEN-YEAR RESEARCH PROGRAM

From 1974 to 1983, GSJ conducted ten research cruises on R/V *Hakurei Maru*, with coworkers from the Japanese universities, National Research Institute for Pollution and Resources, Metal Mining Agency of Japan, U. S. Geological Survey, Clausthal University of Technology, Korean Institute of Energy and Resources, and South Pacific governments of Fiji, Tonga, Western Samoa, and Cook Islands. The aim of this program was to map the distribution patterns and to understand depositional environments for deep-sea manganese nodules in the CPB. The survey areas are bounded to the north by the Mid-Pacific Mountains, to the west by the Marshall-Gilbert Islands, to the east by the Line Islands chain, and to the south by the Manihiki Plateau and the Penrhyn Basin (Fig. 1). The CPB includes the Magellan Trough, Magellan Rise, and Nova-Canton Trough.

The program was divided into Phase I and Phase II (Fig. 1, Table 1). Phase I focussed on reconnaissance grid surveys within the wide area enclosed by 5°N-13°N and 175°E–165°W (an area of about 900 x 2000 km), in the central to northern part of the CPB (Fig. 1). Shipboard programs included nodule sampling, sea-bed photography, sediment coring, dredging, TV-monitored continuous sea-bed photography, 3.5-kHz sub-bottom profiling, seismic reflection and refraction surveys, geomagnetic surveys, heat flow measurements, and gravity measurements. One wire-line sediment coring (core length, 0.5 to 8 meters) and a few

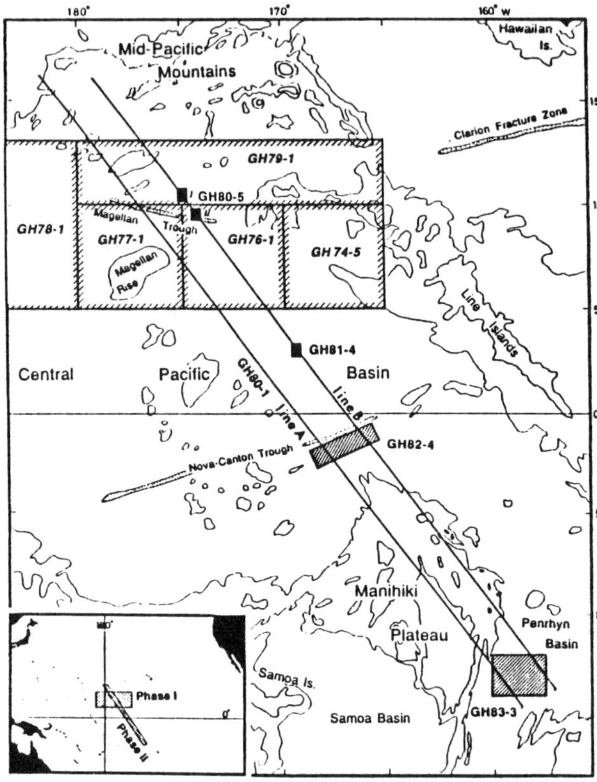

Figure 1. Survey areas of GSJ ten-year research program, 1974-1983. Phase I: grid survey (cruises GH74-5, 76-1, 77-1, 78-1, 79-1). Phase II: two transects (cruise GH80-1) and detailed survey (cruises GH80-5, GH81-4, GH82-4, GH83-3).

adjacent free-fall photograb samplings were planned at each one-degree sampling station. Some supplementary samplings were conducted between these fixed stations. During Phase II, investigation was focussed on small-scale detailed surveys within the four model areas (40 x 40 km or less) selected over the two NW-SE trending 4000-km traverses (the Wake-Tahiti Transect, Cruise GH80-1). Bottom samples and photographs were taken at one-nautical-mile intervals or less along short survey lines.

Selected nodule samples were subject to chemical, mineralogical, radiochemical, and paleontological analyses, after on-board description of morphology (shape and surface features), abundance (wet weight of sampled nodules per square meters), sea-floor coverage (calculated from sea-bed photos), and size analyses (classification, weighing and counting). Chemical analysis was made for 13 elements by atomic absorption spectroscopy according to the GSJ analytical procedures (Terashima, 1978) in reference to Rock Standards USGS-Nod-A-1 and USGS-Nod-P-1. Mineralogical analysis was performed by X-ray powder diffraction and reflecting microscope (Usui, 1984) on the same powder samples selected for chemical analysis.

Table 1. Operation of GJS ten-year program in the Pacific, 1974-1983.

Cruise	Phase I GH74-5	GH76-1	GH77-1	GH78-1	GH79-1	Phase II GH80-1	GH80-5	GH81-4	GH82-4	GH83-3
Year	1974	1976	1977	1978	1979	1980	1980	1981	1982	1983
Area	Line Is.	CPB	CPB	NE of CPB	N of CPB	Wake-Tahiti	CPB	CPB	Nova Trough	Penrhyn Basin
Mode of survey	regional	regional	regional	regional	regional	transect	small-scale	small-scale	small-scale	small-scale
Period (days)	65	60	60	60	60	60	60	60	60	60
Stations	27	31	39	39	45	59	72	138	158	131
box core/grab	15	30	31	36	42	32	24	11	14	16
free-fall grab	0	67	85	71	170	134	70	124	140	146
piston cores	1	6	5	0	5	22	17	13	20	17
dredges	9	2	2	2	1	2	3	1	2	1
sea-bed photos	2	2	77	66	122	142	71	122	140	135
GSJ Cruise Report	no. 4 (1975)	no. 8 (1977)	no. 12 (1979)	no. 17 (1981)	no. 15 (1981)	no. 18 (1982)	no. 20 (1984)	no. 21 (1986)	in press	in press
editor(s)	Mizuno & Chujo	Mizuno & Moritani	Moritani	Moritani & Nakao	Mizuno	Mizuno & Nakao	Nakao & Moritani	Nakao	-	-

Growth rates were radiochemically determined for ^{10}Be methods. Several samples by alpha-track or Be methods.

BATHYMETRY AND GEOLOGY OF THE SURVEY AREA

The general topographic trends of the northern to central parts of the CPB (Phase I) are described on a detailed bathymetric map completed during GSJ cruises (5°N–14°N, 177°E–167°W, 1:2000000; by Onodera and Mizuno, 1981). The basin floors are characterized by abyssal knolls trending WNW-ESE at depths from 5000 to 6000 m in the northern part of the CPB, while the central part of the CPB is a rather flat abyssal plain. Some isolated seamounts are distributed within the basins.

The NW-trending island chains or seamount chains, e.g., Line Islands, Marshall Islands, Gilbert Islands, are considered to have originated by hot spot activity of the Hawaiian-Emperor style (Morgan, 1972), although some recent geophysical findings speculate multi-hot spot origin or fracture zone origin (Sager and Keating, 1984). The volcanic rocks from the Line Islands chain are Late Cretaceous age as old as 85 Ma (Schlanger et al., 1984), and the DSDP core stratigraphy (sites 66, 67, 165) suggests that the age of basement basalt in the CPB is early Cretaceous (Winterer, Ewing et al., 1973). Within the CPB, paleomagnetic evidence for ancient sea floor spreading along the Magellan Trough (Tamaki et al., 1979), the central basin rise of the CPB (Orwig and Kroenke, 1980), and the Nova-Canton Trough (Winterer, 1976) was suggested. These works revealed that most of the basin areas have been below the CCD since Cretaceous, while calcareous sediments predominate on local topographic highs. Thick calcareous sediments on the Line Islands chain and Magellan Rise occasionally served as loci for turbidite flows into abyssal depths. Deposition as channel fills, ponds, and beds was controlled structurally by basement topography (Orwig, 1981; Tamaki, 1984).

The sedimentary history of the CPB has been discussed in terms of bottom currents and tectonic movements by van Andel et al. (1975), Orwig and Kroenke (1980), and others. Active and fluctuating flow of Antarctic Bottom Waters promoted regional and local complex patterns of erosion and resedimentation, resulting in prominent hiatuses in the sediment cores, which are most frequent near the Plio-Pleistocene boundary and within the Paleogene (van Andel et al., 1975; Nishimura, 1986). Local topographic complexities like seamount areas often display distinct sedimentary histories which may vary over short distances (Mizuno et al., 1980; Nishimura, 1986).

Sediment sequence of the CPB has been characterized by Winterer and Ewing et al. (1973), Tamaki (1977), and GSJ workers (Tanahashi, 1986 and others). During Phase I of the program, two prominent continuous reflectors (A' and B') were recognized on seismic reflection profiles over the CPB area and correlated to DSDP core stratigraphy (Tamaki, 1977; Tanahashi, 1986). The uppermost highly transparent layer (called Unit I) above reflector A' consists mainly of deep-sea clay/biogenic oozes of early Eocene to Quaternary age. The thickness of Unit I varies from zero to more than 300 m. The unit contains several high-amplitude flat beds of calcareous turbidites in some areas. The lower layer between reflector A' and the acoustic basement (called Unit II) is subdivided into IIA and IIB. IIA consists of upper high-amplitude layers of middle Eocene chert and a lower transparent layer of Late Cretaceous pelagic clay, and IIB consists of high-amplitude Late Cretaceous limestone or volcanics. The thickness of Unit II ranges from 110 to 190 m, although reflector B' and basement are often obscured.

This sequence is commonly observed throughout the basin area in the CPB, although the thickness and lithology are variable on regional and local scales. Unit I, most closely related to nodule growth history, is classified into three types of seismic character, A, B, and C (Tamaki, 1977): (A) wholly transparent and structureless, (B) transparent and associated with semi-transparent layers undulating in phase with sea-floor topography, and (C) transparent and associated with high-amplitude and flat turbidite beds abutting on Unit II. Unit I is generally thick in basin areas but thin or scarce around topographic highs. However, it is exceptionally thin or scarce in the deeper parts of basins, probably reflecting deep-sea erosion or lack of sedimentation due to strong bottom currents (Nishimura, this volume).

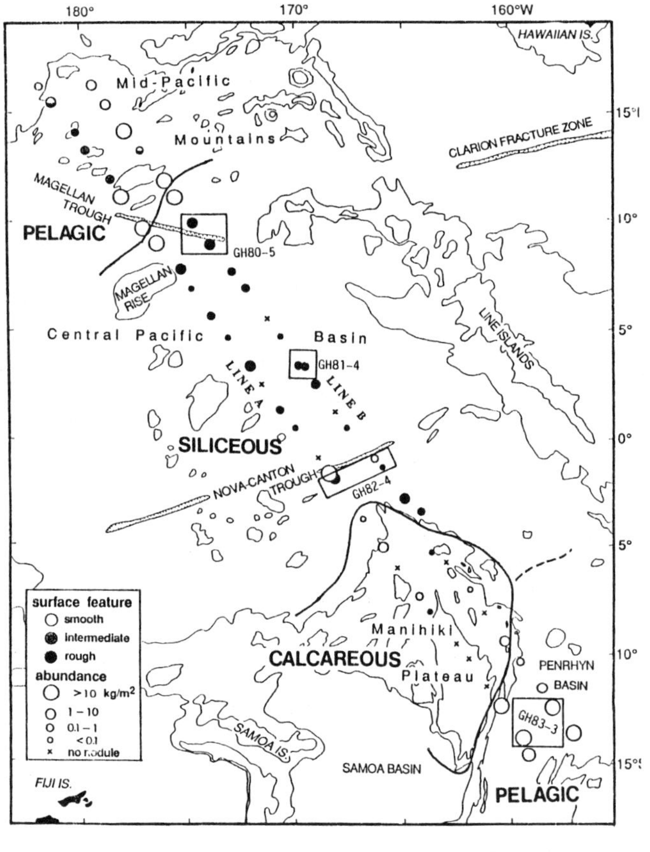

Figure 2. Regional distribution of manganese nodules on the Wake-Tahiti Transect (cruise GH80-1) modified from Usui (1983).

The general distribution pattern of surface sediments was characterized along the Wake-Tahiti Transect (cruise GH80-1) crossing the equator from 16°N to 15°S (Mizuno and Nakao, 1982). Surface sediment type is closely related to the modern topography and distance from the equatorial high productivity zone. Biogenic siliceous sediments are dominant on the basin floor around the equator at depths below the CCD (about 4900 m; Berger et al., 1976). The crest and flanks of seamounts, the Manihiki Plateau, and the Magellan Rise are mostly covered with thick biogenic calcareous sediments. Other areas in higher latitudes, the northern part of the CPB and the Penrhyn Basin of the South Pacific, are predominated typically by brown deep-sea clay (occasionally zeolitic). Manganese nodules are most abundant in the pelagic clay province, are also common in siliceous sediment areas, but are scarce on calcareous sediments. (Fig. 2 and Usui, 1983). Nishimura (1986) described the changes in sedimentary conditions from the Paleogene to Quaternary age in these nodule fields. He indicated that the central part of the CPB floor encountered two or three periods of intense deep-sea erosion or no sedimentation caused by bottom currents (Nishimura, this volume). During the long geological time, sedimentation has changed from pelagic to biosiliceous, according to the northwestern motion of the Pacific plate crossing the equator. Thus, the sedimentary history of the CPB has been strongly controlled by the CCD, bottom current, equatorial high productivity, plate motion, and local topography.

NODULE CHARACTERISTICS

Occurrence, chemical and mineral compositions, and regional and local variability are characterized for the CPB nodules on the basis of 1,389 bottom samplings, 879 sea-bed photographs, 975 compositional data, and 803 X-ray diffraction analyses, along with published and unpublished seismic and sedimentological data.

External and Internal Morphology

The field description of nodule morphology during the ten cruises (Moritani et al., 1977) reveals two typical appearances of nodule surfaces in the CPB; type r (rough) and type s (smooth) as shown in the scanning electron micrographs of nodule surfaces (Fig. 3). The two types in the CPB are very similar to those in the northeastern Pacific nodule provinces (Meylan, 1974; Usui and Mita, 1987). Type r is believed to be formed within surface sediments through early diagenetic supply of metals from sediment interstitial water, while type s is believed to be formed at the interface between bottom waters and nodules/rocks

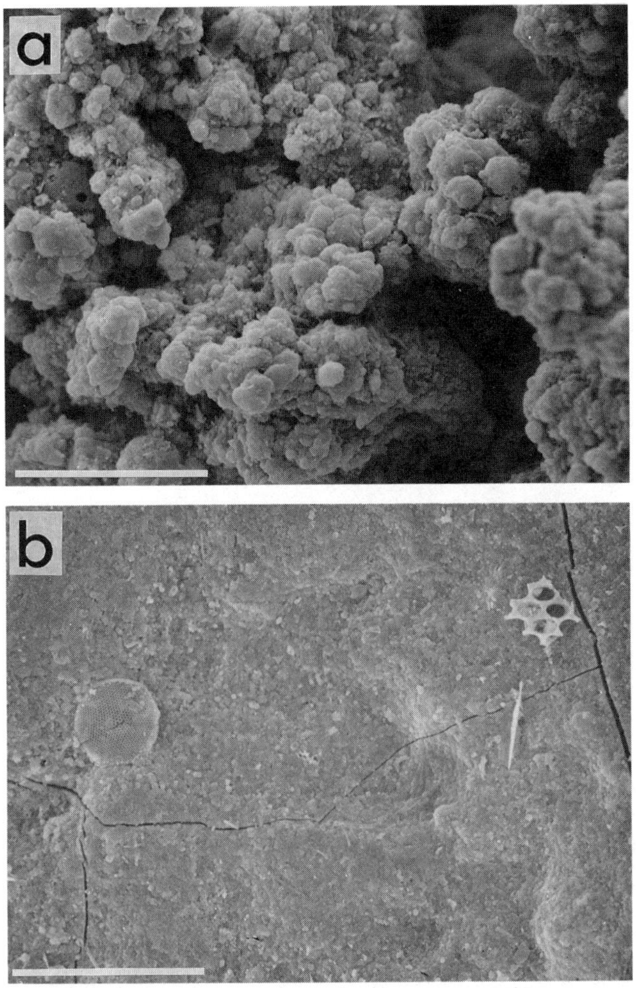

Figure 3. Scanning electron micrographs (SEM) of typical surface of (a) type r and (b) type s. Scale bar = 0.1 mm. Skeletons of radiolaria and diatom are attached on the smooth surface (b).

through accumulation of settling colloids from bottom waters (Usui et al., 1978; Halbach and Özkara, 1979). The rough structure consists of a number of growth cusps of several tens microns in diameter and length, and it usually forms the several-millimeter porous layers, which commonly include many detrital mineral particles, microfossils, and voids (Fig. 4a). The smooth surface consists of stratified and rather dense layers including finer clayey detrital particles (Fig. 4b). These two typical surface structures are also observed inside nodules as growth microstructures like stratigraphic records. An intermediate feature between the two typical surfaces is occasionally seen; it is composed of a very thin (less than 1 mm thick) coating of rough structure on smooth structure and vice versa.

It is notable that these two typical microstructures are well correlated to the two manganese minerals; 10Å manganate (similarly as buserite or todorokite) for rough structure, and δ-MnO_2 (similarly as vernadite) for smooth structure (Usui et al., 1978; Usui 1979a). Characteristic microstructures are shown in the micrographs of the cross section of the nodules (Fig. 5). The two manganese minerals represent particular chemical characteristics and origin (Tables 2 and 3) as mentioned later. Thus surface structure is the most important criteria of physical features in characterizing nodule morphology.

The nodule shape of the Central Pacific Basin samples is primarily spheroidal or discoidal. Other less common shapes have been described in detail by Meyer (1973), Meylan (1974), and Moritani et al. (1977), including the elongated, platy, fragmented, irregular, and polynucleate types. The external shape is principally determined by the shape of nuclei, e.g., rock fragments, fossils, and fragmented older nodules. Smaller nuclei are common in spheroidal type r nodules, while large rock/nodule fragments and shark teeth often serve as nuclei of type s nodules. It was noted that these nuclei are usually different in composition and in age from the associated surface sediments, suggesting that the nuclei were supplied in different sedimentary conditions from the present. For instance, Miocene shark teeth (Kuga and Usui, 1982) and consolidated zeolitic claystone often serve as nuclei of nodules resting on the Quaternary unconsolidated surface sediments.

Variation of nodule size has been described in the Eastern and Central Pacific Ocean (Piper et al., 1979; Heath, 1979; Sundkvist, 1983). Nodule size in the CPB varies with locality from a few millimeters to several centimeters in diameter; the size fraction 2-4 cm is dominant for type s while type r is generally smaller in the CPB as shown in Figure 6. The larger type s nodules are typically distributed in the northern part of the CPB and the Penrhyn Basin of the South Pacific, where pelagic/zeolitic clay surface sediments are dominant in flat-floor basins. Type r nodules are, in contrast, more variable in size with locality, ranging from micronodules (less than 1 millimeter) up to 8 cm, rarely within a single sampler (Usui, 1986).

Internal structure is of significance in that it may record the past depositional environment and growth history of nodules (Sorem, 1967; Sorem and Fewks, 1979). The CPB nodules have concentric and layered structure; occasionally each layer shows variable microstructures: laminated, dense, massive, network, columnar, and dendritic. Some nodules from the CPB show distinct growth stages inside with different mineralogy and microstructure, and the internal growth stages within nodules are in some cases correlated between stations at tens of kilometer intervals (Usui, 1979a;1984). However, the period of growth of each layer must be determined in

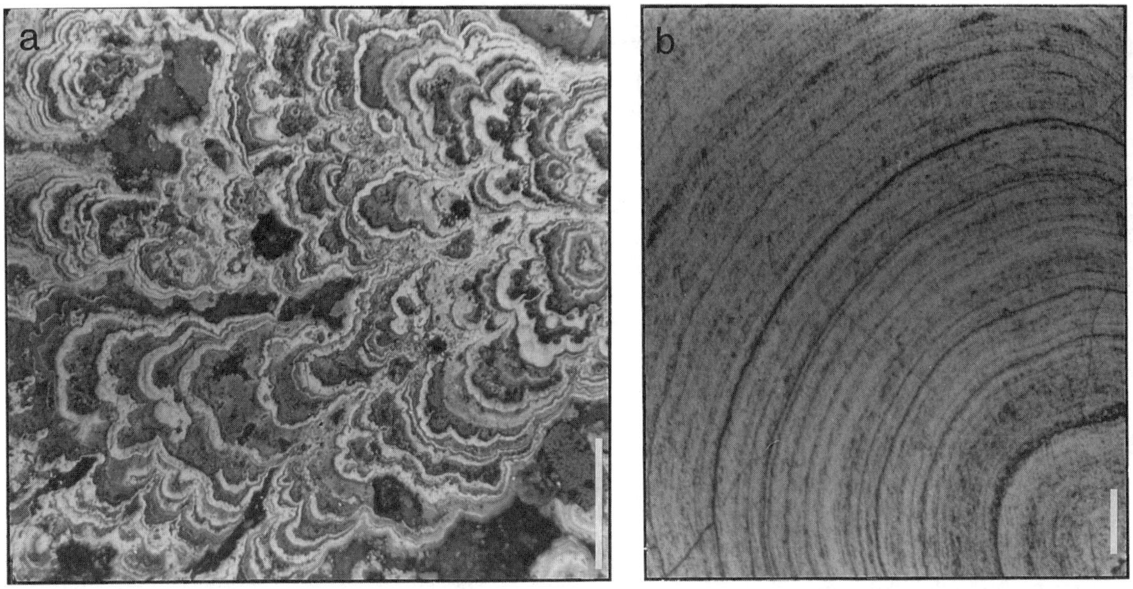

Figure 4. Micrographs of typical growth structures of nodules under reflecting lights Scale bar =0.1 mm. (a) Cuspate structure of 10Å manganate forming type r. (b) Layered structure of δ-MnO_2 forming type s.

further detail to correlate nodule internal structure to the change of marine environment.

Occurrence

The nodules of this area are clearly concentrated at or within surface sediments (several centimeters thick) rather than deep in the sediments, suggesting that nodules have preferentially grown at the sediment-water interface while being continuously uplifted during nodule growth and pelagic sedimentation. However, in rare cases small nodules of both types s and r occur buried at various depths in the sediment columns (found in 8 out of 80 cores; up to 8-meter penetration). No consistent strata-bound occurrence of buried nodules was observed even between cores located close together, but the nodules are diversely distributed within sediments of various ages from the middle Miocene to the Quaternary (Nishimura, 1986).

Detailed observation of nodule occurrence with box cores and sea-bed photographs revealed differences in position relative to the sediment-water interface for the two nodule types (Fig. 7). Type s is always exposed to sea water, occasionally forming a widely spread nodule pavement (coverage: 60-80%, abundance: 10-20 kg/m^2 or more) in the northern CPB and the Penrhyn Basin (GH83-3 in Fig. 8). The most extreme case of exposure is represented by manganese crusts covering hard rocks without surface sediments. In contrast, type r nodules are commonly buried within several centimeters by the sediments of high water contents. The abundance-coverage plots (Fig. 8) demonstrate the buried nature of type r nodules. For instance, the maximum abundance of 9.6 kg/m^2 was encountered for type r nodules in the siliceous clay area with surface coverage of zero percent (GH81-4 in Fig. 8). Figure 9 shows preferential occurrence of the two types in relation to surface sediment type. Type r nodules are most often associated with siliceous sediments. This occurrence

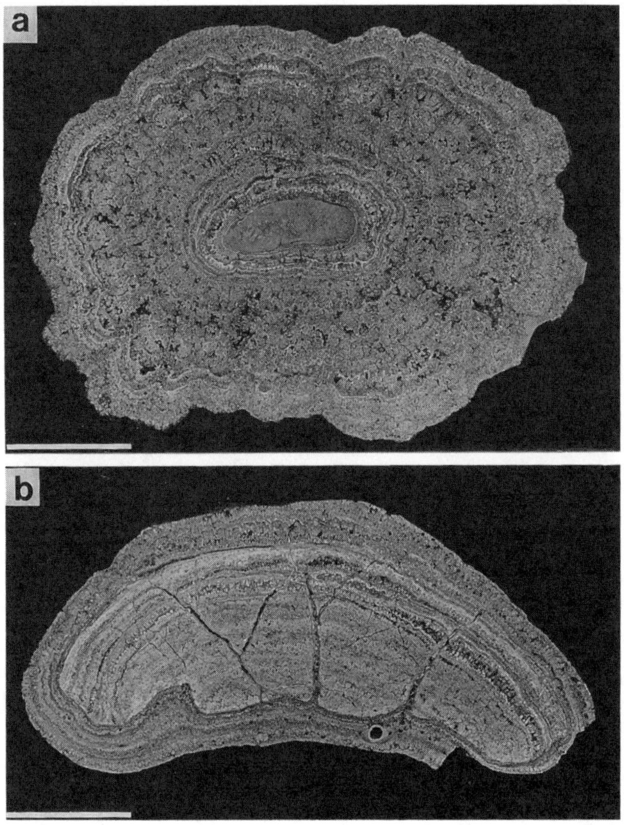

Figure 5. Polished cross sections under reflecting lights. Scale bar = 1 cm. (a) Type r nodule with concentric structure. (b) Type s nodule with two growth stages.

Table 2. Bulk chemical composition of manganese nodules in CPB (Phase II) in wt.% for 110°C dried basis.

Morphological type	rough				intermediate				smooth				total			
Origin	diagenetic				mixture				hydrogenous							
Number of analyses	86				35				314				435			
	ave.	s.d.	min.	max.	ave.	s.d.	min.	max.	ave.	s.d.	min.	max.	ave.	s.d.	min.	max.
Elements																
Mn	27.5	4.7	8.6	35.1	25.5	2.9	19.1	31.8	19.3	3.7	6.4	33.4	22.5	5.6	6.4	35.0
Cu	1.32	0.46	0.25	2.09	0.99	0.37	0.18	1.86	0.38	0.28	0.05	1.58	0.75	0.57	0.07	2.09
Ni	1.39	0.45	0.19	2.13	1.10	0.33	0.29	1.51	0.56	0.27	0.13	1.54	0.89	0.53	0.13	2.13
Zn	0.134	0.027	0.046	0.181	0.109	0.024	0.082	0.212	0.065	0.016	0.035	0.160	0.088	0.038	0.035	0.212
Fe	6.6	2.2	2.7	17.1	10.3	2.5	6.6	17.1	14.5	3.2	2.8	24.0	11.6	4.7	2.7	24.0
Co	0.14	0.04	0.07	0.37	0.20	0.05	0.14	0.33	0.32	0.13	0.06	0.62	0.25	0.13	9.95	0.62
Pb	0.026	0.012	0.001	0.073	0.036	0.019	0.001	0.068	0.080	0.028	0.002	0.208	0.055	0.031	0.001	0.208
Si	6.49	2.05	1.31	19.19	6.42	1.21	3.40	8.93	7.22	2.06	3.69	19.64	6.91	2.04	1.31	19.64
Al	2.54	0.80	1.06	6.72	2.29	0.55	1.11	3.75	2.66	0.76	1.11	5.63	2.59	0.77	1.06	6.72
H_2O	10.3	1.2	8.3	16.8	10.4	1.1	8.9	12.7	9.8	1.3	6.1	14.5	10.0	1.3	6.1	16.8
Cu + Ni + Zn	3.32	0.68	1.06	4.13	2.47	0.45	1.29	3.45	0.97	0.51	0.25	2.89	1.77	1.22	0.25	4.23
Mn/Fe	4.63	1.73	1.07	10.9	2.62	0.73	1.34	4.00	1.48	1.00	0.35	9.04	2.65	1.96	0.35	10.9

Table 3. Characteristics of CPB manganese nodules

Type	r	s
Surface feature	rough (gritty, granular)	smooth
Origin	diagenetic	hydrogenetic
Regional distribution	northern to central CPB	northern CPB, Penrhyn Basin
Associated sediment	siliceous ooze/clay	pelagic/zeolitic clay
Occurrence in sea bed	buried in surface sediments	exposed to sea water
Color	black	black, sometimes brownish
Dominant shape	spherical	discoidal, spherical, irregular
Internal structure	concentric, laminated	dense, massive, laminated, often with cracks
Abundance (approx.)	1 - 5 kg/m^2	10-20 kg/m^2
Size (mode of long axis)	1-2 cm	2-4 cm
Mineralogy	10Å manganate	δ-MnO_2
Chemistry	high Mn/Fe (>2)	low Mn/Fe (1-2)
	high Cu, Ni, Zn	low Cu, Ni, Zn

strongly suggests a genetic relationship of type r nodules to early diagenesis of surface siliceous sediments.

Mineral Components

Mineralogical characterization of nodules has been based mainly on X-ray powder diffraction analysis with microscopy and electron microprobe analysis (Burns and Brown, 1972; Schellman, 1977; Burns and Burns, 1977; Usui, 1979a; Sorem and Fewks, 1979). Despite disagreement regarding nomenclature of marine manganese minerals (Burns and Burns, 1977; 1979), normal deep-sea nodules consist of 10Å manganate (also called 10Å manganite, todorokite, or buserite) and δ-MnO_2 (also vernadite or

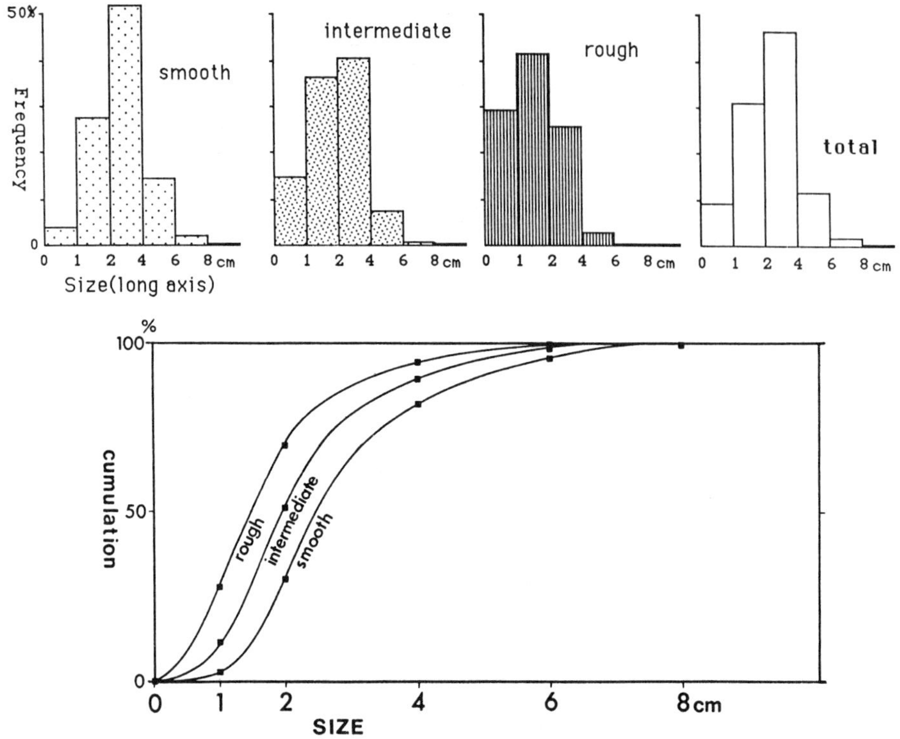

(left) **Figure 6.** Frequency distribution of nodule size (long axis) in the CPB, demonstrated as frequency of numbers of nodules normalized to the total number of each type. Numbers of stations (and total numbers of nodules) are 227 (31510) for type s, 62 (6824) for the intermediate type, 141 (6904) for type r, and 430 (45241) for the total.

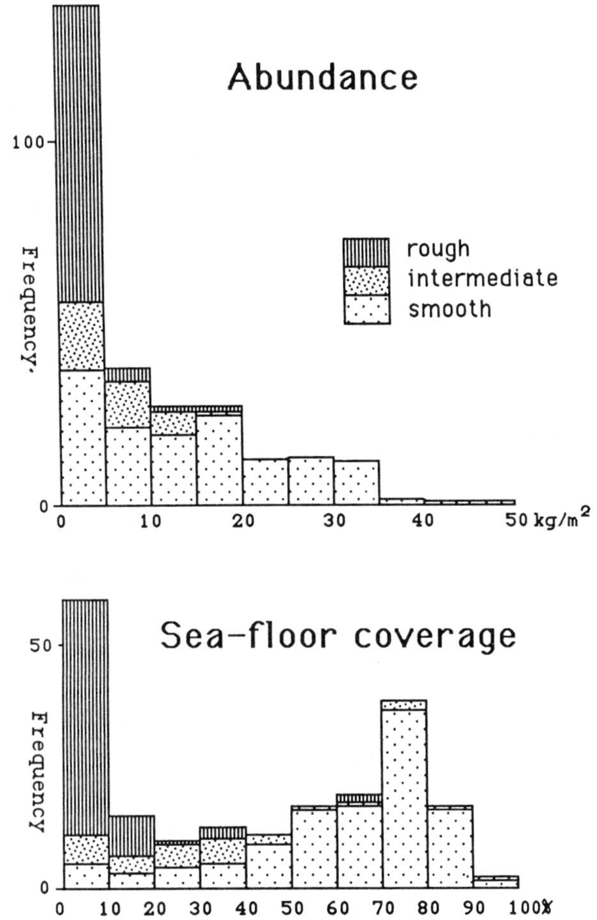

(below) **Figure 7.** Frequency distribution of nodule abundance and sea-floor coverage. Frequency is shown as numbers of stations of nodule occurrence. Total numbers of stations; 430 bottom-sampling for abundance and 243 photo stations for coverage.

disordered birnessite). Although the occurrence of 7Å manganate (also called birnessite) is reported to occur in hydrothermal deposits and low-metal diagenetic nodules (Glover, 1977; Corliss et al., 1978; Lonsdale et al., 1980; Stoffer et al., 1984), its presence in normal hydrogenous and high-grade diagenetic deposits appears to be misleading identification on dehydrated or heated samples (Usui et al., 1989).

The earlier studies have revealed that the two minerals (δ-MnO_2 and 10Å manganate) are principal ferromanganese phases of hydrogenous and diagenetic deep-sea nodules and crusts (Table 3), and that mineralogy strongly controls nodule surface structure, internal microstructure, and chemical composition (Halbach and Ozkara, 1979; Usui, 1979a,b; 1983). The characteristics of the two minerals are hereby summarized on the basis of earlier works:

(1) 10Å manganate: iron-free hydrous manganate of nearly constant chemical composition containing lattice-held Cu, Ni, Zn, and other specific transition elements. The atomic ratio (Cu + Ni + Zn)/Mn is about 0.16 (1:6) or slightly less. The mineral is precipitated from dissolved metal ions released through moderately reducing surface sediments during early diagenesis.

(2) δ-MnO_2: submicroscopically intergrowth of poorly crystallized δ-MnO_2 (with two X-ray reflections) and amorphous hydrous ferric oxides, containing similar amounts of Fe and Mn (Ostwald, 1984). The Cu + Ni + Zn content is much lower than in 10Å manganate but Co is more en-

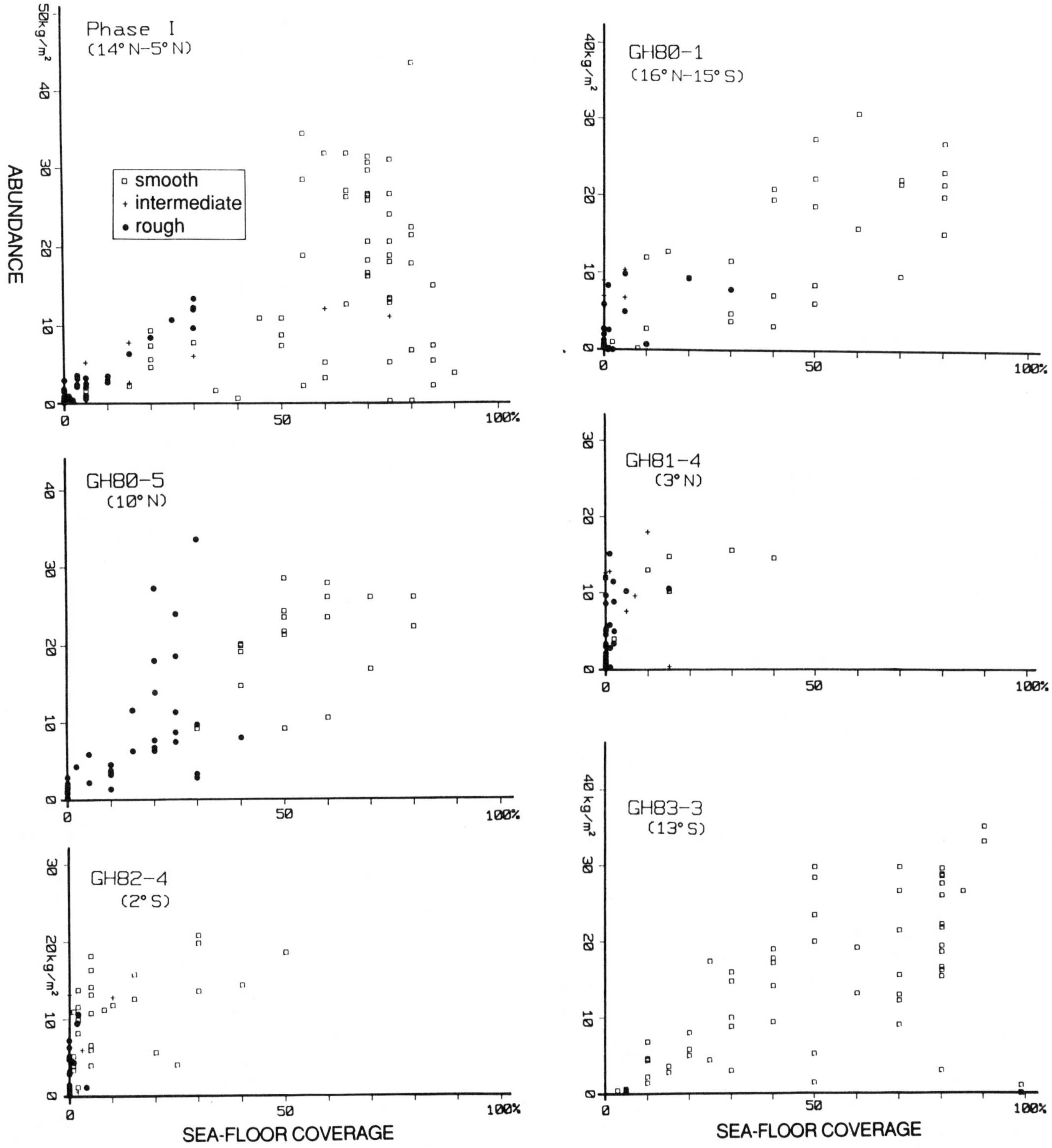

Figure 8. Plots of nodule abundance (recovered) versus sea-floor coverage (from photos).

riched, probably replacing Mn^{+4} in manganates (Burns and Burns, 1979). The mineral may be deposited from overlying normal sea waters as settling colloids under more oxidizing conditions.

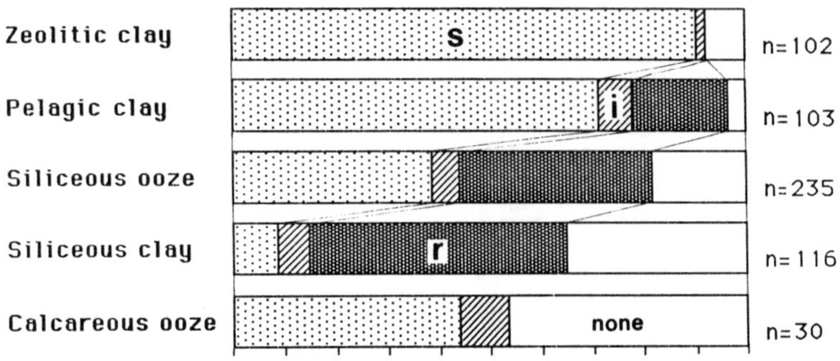

Figure 9. Relationship of nodule occurrence to associated surface sediments. Numbers on the right denote total bottom sampling stations for each sediment type. None: no nodules sampled.

To simplify the relationship between microstructure and mineral component in the CPB nodules, the cuspate growth structure is typical of 10Å manganate forming a rough and gritty nodule surface (type r) while the stratified structure is typical of δ-MnO_2, forming a smooth nodule surface (type s). Inside the nodules, layer-by-layer alternation of the two minerals (for instance, Sorem and Fewks, 1979) is not very common, but in rare cases nodules show two or more prominent growth stages composed of either mineral from nuclei to nodule surface (Fig. 4b). In the CPB, bulk nodule characteristics, in particular chemical composition, are quite well related to nodule surface features (smooth or rough), because of infrequent stratigraphic variation in manganese mineralogy inside nodules.

Accessory minerals are included as detrital, rock particles and nuclei. Quartz, plagioclase, smectite, and phillipsite most commonly occur within the oxide layers; these minerals are also common in normal pelagic surface sediments. Apatite, clinoptilolite, and potash feldspars are less common but present. Any unique iron oxide phases have not been discerned in this study, except for detrital hematite particles. According to microscopic observation, microfossils (radiolarians and foraminiferas) and very rarely cosmic metallic spherules are included in the oxide layers.

Chemistry

Major and minor elements of nodule samples were analyzed by atomic absorption to specify their regional variability, local small-scale variability, intra-nodule variability, and variation with mineralogy. Two distinct compositional features of the CPB nodules are (1) wide ranges of variation and (2) strong correlations between element concentrations.

Table 2, showing averages and ranges of ten components in relation to morphological type, illustrates the variability of element concentrations. The great variation is most prominently reflected by the ratio Mn/Fe (Table 3). Among nine metal elements, a 10-fold variation or more is observed mainly for minor elements, Cu, Ni, Zn, Co, Pb, and Fe, whereas the concentrations of Mn, Si, and Al are much less variable.

Ranges of compositional variation with locality are actually similar to that found inside nodules on a microscopic scale. For instance the ratio Mn/Fe ranges between 0.35 and 10.9 for bulk composition (Table 3) while between 0.8 to 80 on a microscopic scale (Usui, 1979b). This fact is ascribed, based on previous comparative chemical and mineral analysis, to preferential development of the two ferromanganese minerals in space and in time. Electron microprobe analyses and synthesis experiments (Giovanoli and Brutsch, 1979; Usui, 1979b) revealed that the chemical composition of each mineral is much more constant than nodule bulk composition on regional and local scales. However, variation between bulk compositions in a single site is relatively constant despite significant inside micro-scale variation,

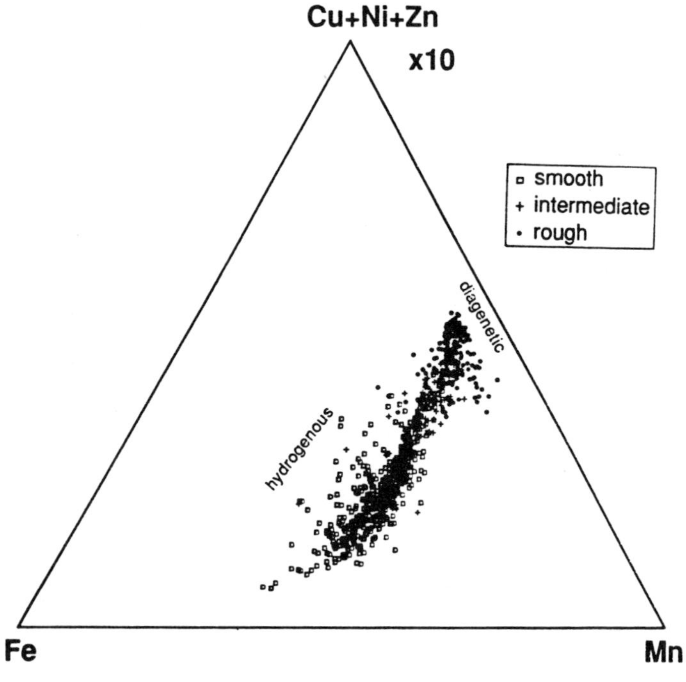

Figure 10. Ternary diagram for ratio Mn/Fe/(Cu + Ni + Zn) of nodules from Phases I and II (940 data sets), showing a linear dependence of Cu + Ni + Zn content on Mn and Fe contents.

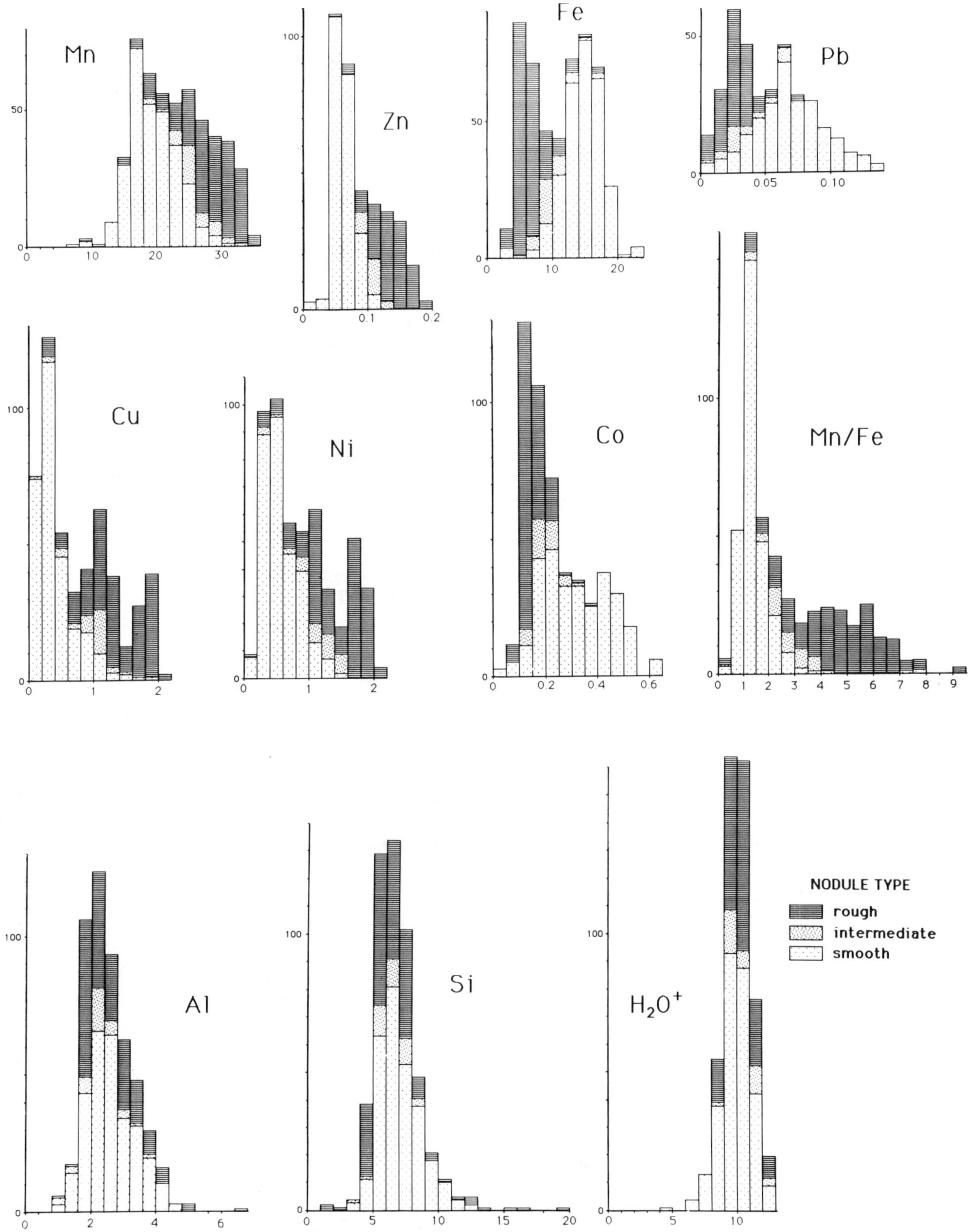

Figure 11. Frequency diagrams of element concentrations, H_2O^+ and Mn/Fe ratio of nodules (Phase II). Numerical data are presented in Table 2.

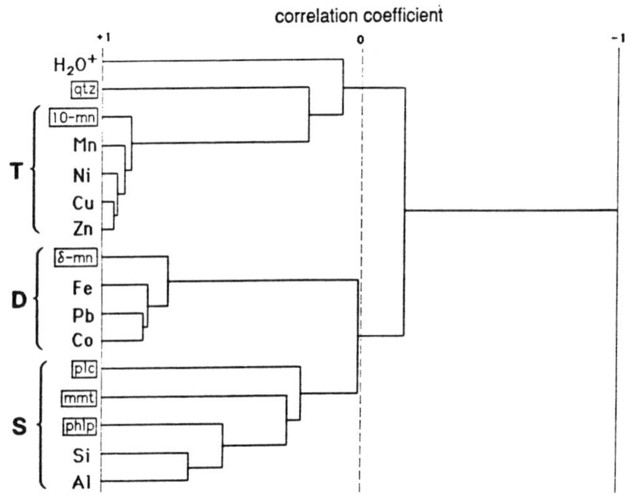

Figure 12. Results of cluster analysis for chemical and mineral compositions of nodules (Phase II), showing three groups; hosted by 10Å manganate (T), δ-MnO_2 (D), and silicates (S). Mineral contents were determined as relative values from the intensity on X-ray diffractograms. Mineral symbols are enclosed as qtz: quartz, 10-mn: 10Å manganate, δ-mn: δ−MnO_2, plc: plagioclase, mmt:montmorillonite, and phlp: phillipsite.

probably because of similar growth history within a single locality.

Using Bonatti's diagram (Fig. 10), the great variation of Mn/Fe and linear dependence of total (Cu + Ni + Zn) to Mn and Fe contents is remarkably illustrated. The two intersections of an extrapolated regression line and triangle axes represent the ideal composition of 10Å manganate and δ-MnO_2, respectively. Selective enrichment of Ni, Cu, and Zn into the 10Å manganate phase controls principally bulk composition and intra-nodule variation. For instance, type r nodules are always enriched in Mn, Ni, Cu, and Zn, and always depleted in Fe, Co, and Pb throughout the CPB. Diagrams of the frequency distribution of element concentrations (Fig. 11 and Table 2) clearly show different ranges between types s and r. This tendency is similar to that displayed by the Ni- and Cu-enriched nodules from the Northeast Pacific Manganese Nodule Belt (Sorem and Fewks, 1979; Piper and Blueford, 1982; Usui and Mita, 1987).

Another prominent chemical feature is a strong inter-element relationship within both bulk samples and intra-nodule subsamples. Negative correlation is marked between the two groups of Mn-Cu-Ni-Zn (10Å manganate hosted) and Fe-Co-Pb (δ-MnO_2 hosted), whereas strong positive correlations are between elements within each group (Table 4). These correlative features are quite consistent when the data set includes both nodule types s and r, i.e., 10Å manganate and δ -MnO_2. Cluster analysis for the ten components clearly suggests strong dependency of element concentration on mineral composition (Fig. 12). The silica plus alumina content is generally greater than about 13 wt.%, and the variation coefficients (standard deviation divided by average) of Si and Al are the smallest among the nine elements. The contents of silica and silicate minerals seem independent of morphological type (Table 4).

Rates of Growth

Several data on growth rates and ages for the CPB nodules can be used to postulate very slow rates of growth (several mm/10^6 yr), in contrast with sedimentation rates of several mm/10^3 yr for this area (Ku, 1977; Joshima, 1982). The nodule growth rates were estimated as 2.2 to 5.0 mm/10^6 yr (northern CPB, GH79-1) by the alpha-activity method

Table 4. Correlation matrix for chemical composition of manganese nodules (Phase II)

	Mn	Cu	Ni	Zn	Fe	Co	Pb	Si	Al	H_2O^+
Mn	1									
Cu	0.912	1								
Ni	0.903	0.953	1							
Zn	0.940	0.954	0.932	1						
Fe	-0.772	-0.866	-0.846	-0.875	1					
Co	-0.635	-0.753	-0.744	-0.740	0.819	1				
Pb	-0.662	-0.714	-0.765	-0.698	0.833	0.842	1			
Si	-0.493	-0.293	-0.231	-0.413	0.059	-0.134	-0.104	1		
Al	-0.485	-0.268	-0.275	-0.482	-0.050	0.021	0.141	0.665	1	
H_2O+	0.115	0.012	0.032	0.041	-0.080	-0.077	-0.060	-0.199	-0.191	1

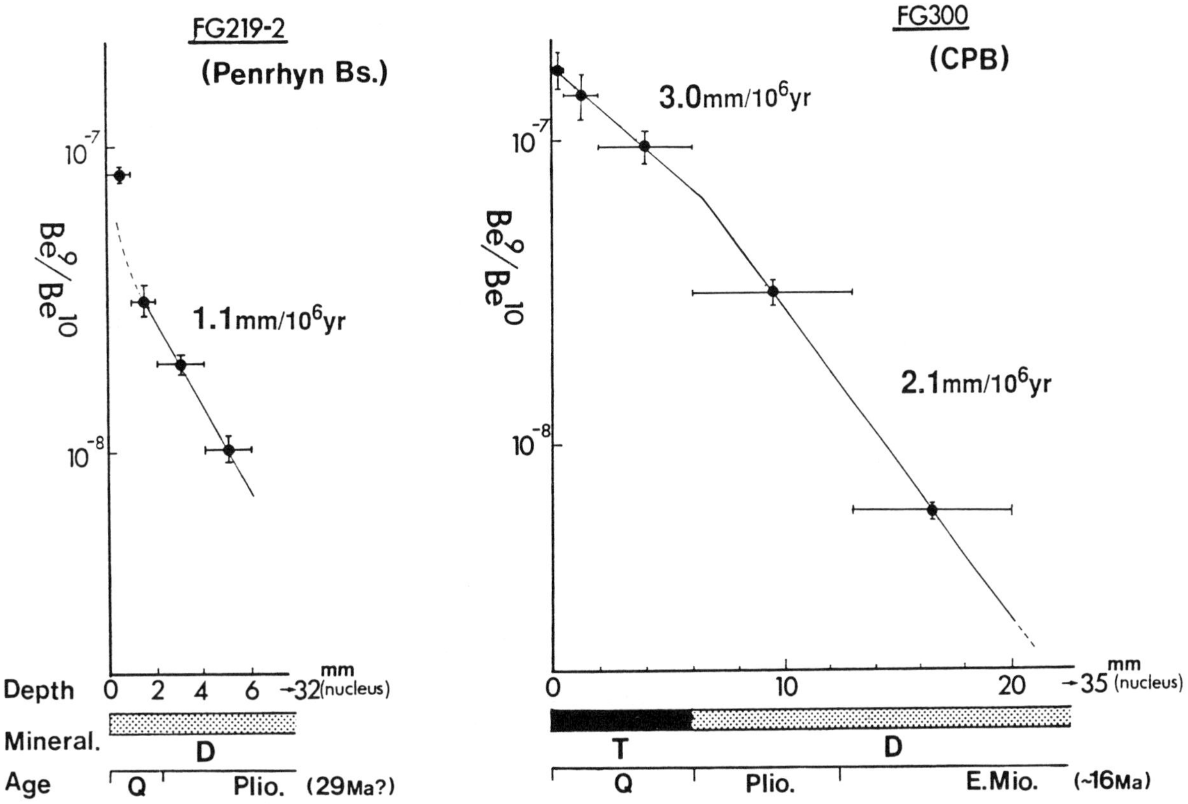

Figure 13. Be isotope dating for two nodules (Inoue, personal communication, 1983). Nodule FG219-2: 65 mm in diameter, 13° 47.2'S-159 ° 28.3'W, Penrhyn Basin (GH80-1), d = 5150 m. Nodule FG300: 70 mm in diameter, 9°03.5'N-174° 03.5'W, CPB (GH80-5), d = 5950 m. The time scales were estimated on the assumption that the age of nodule surface is Holocene and growth rate is constant. Mineralogy symbols with internal structure, T:10Å manganate, D:δ-MnO_2.

(Piper and Gibson, 1981), and 1.1 (Penrhyn Basin, GH83-3) and 2.1-3.0 mm/10^6 yr (central CPB, GH80-5) by the ^{10}Be method (Inoue, personal communication; method in Inoue et al., 1983). These rates for CPB nodules are within a range of earlier reported values for Central and South Pacific nodules (e.g., 1.8, 2.1, 10 mm/10^6 yr; by Ku, 1977). The growth rates inside nodules from cruise GH80-5 suggest a continuous growth from the Miocene despite an abrupt change of mineralogy within nodules (FG300 in Fig. 13), and a slightly higher rate during growth of type r. These rates are consistent with the occurrence of Miocene shark teeth surrounded by 20-mm-thick oxide layers of the nodules from the GH77-1 area, and with alpha-activity dating (Piper and Gibson, 1981).

DISTRIBUTION PATTERNS AND GROWTH ENVIRONMENTS

The reconnaissance survey of Phase I in the northern CPB (Usui et al., 1983; 1987) and detailed surveys on the Wake-Tahiti Transect of Phase II show that there are typical regional distribution patterns for the two morphological nodule types and composition of nodules (Figs. 2 and 14).

Type s nodules (hydrogenous origin) occupy the two areas of vast and continuous nodule pavements of high abundance and sea-floor coverage, typically in the northern CPB (8°N to 13°N) and in the north Penrhyn Basin (10°S to 17°S), where pelagic/zeolitic clay sediments are associated with the nodules. The former is an ENE-trending belt extending more than 1000 km with width varying from 100 to 300 km (Fig. 14). The ENE trend is oblique to the topographic trend WNW for abyssal hills and small valleys in the basin. The latter covers most of the north Penrhyn Basin, ending in the west at the foot of the Manihiki Plateau and in the south near the Aitutaki Passage. Northern and eastern extension for some distance beyond our survey area can be expected. Nodule abundance of these pavement zones often exceeds 20 kg/m^2, with the average of around 13. A patchy distribution of abundant type s nodules is found locally on isolated seamounts, abyssal hills, and ridges.

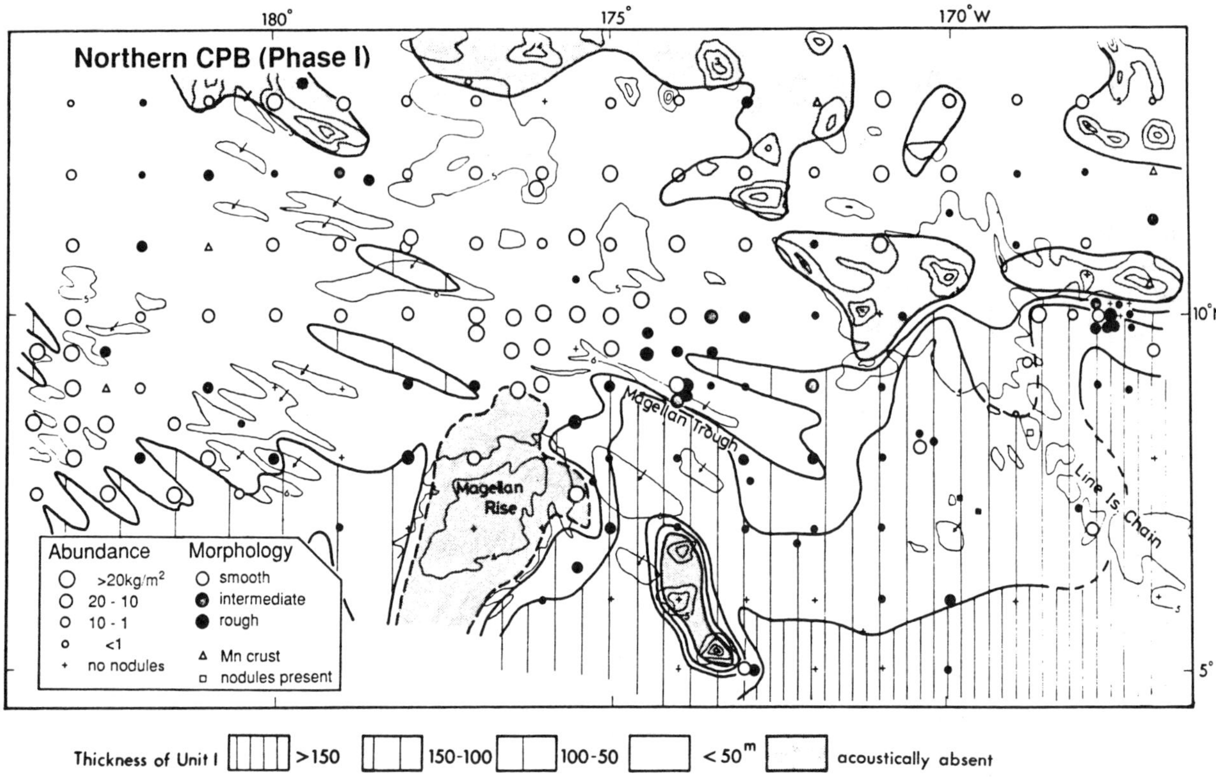

Figure 14. Regional variation of nodule abundance and type in the area of Phase I. The vast nodule province of type s is closely related to thin or scarce development of Unit I, while type r is related to moderate thickness of the unit. The isopach contours were compiled from GSJ cruise reports and unpublished data of M. Tanahashi, GSJ.

These deposits are occasionally associated with hydrogenous manganese crusts.

In contrast, type r nodules (diagenetic origin) are distributed widely in the central to southern CPB floor as lower-abundance deposits, where siliceous sediments are dominant. The abundance of type r nodules is lower than 5 kg/m^2 in 93 percent of the total sample locations where type r nodules occur.

The nodule abundance in this basin as a whole, when neglecting morphological type or composition, shows a reverse relationship to thickness of the acoustically transparent substrate, such as found within a small survey area by Mizuno et al. (1980). In general, the abundance-thickness plots in Figure 15 clearly demonstrate that abundance is typically low where Unit I is thicker than 40 m, irrespective of nodule type. This evidence supports the interpretation that nodule growth requires a slow sedimentation rate. Mizuno et al. (1980) estimated the average sedimentation rate of Unit I, and derived a rate of 5 mm/10^3 yr as an upper limit for abundant nodule occurrence. Relationships of nodule abundance and thickness of Unit I on regional (Fig. 14) and local small scales (Figs. 16 and 17) support the contention that rapid sedimentation prevents abundant nodule development.

On the basis of Figure 15, the following different processes of nodule growth are assumed. Type s nodules (hydrogenous) preferentially grow at zero or low sedimentation rate where continuous currents of oxygenated bottom waters prevent burial of slowly growing nodules and supply metals. Preferential conditions for deposition of hydrogenous manganese minerals may be encountered on topographic highs, e.g., as manganese crusts on rock outcrops. In contrast, the formation of type r nodules (diagenetic) requires moderate sedimentation and subsequent decomposition of organic materials. This process brings on a relatively reduced environment at the sediment-water interface. During early diagenesis of the surface sediments (Marchig and Gundlach, 1981; Halbach et al., 1982), in situ solid manganese phase is dissolved from the surface sediments into interstitial waters, and reprecipitated as 10Å manganate within sediments. If the accumulation of sediments is too rapid, reprecipitated oxides may be deeply buried in the sediment column and fixed as dispersed manganese micronodules. Under high primary productivity

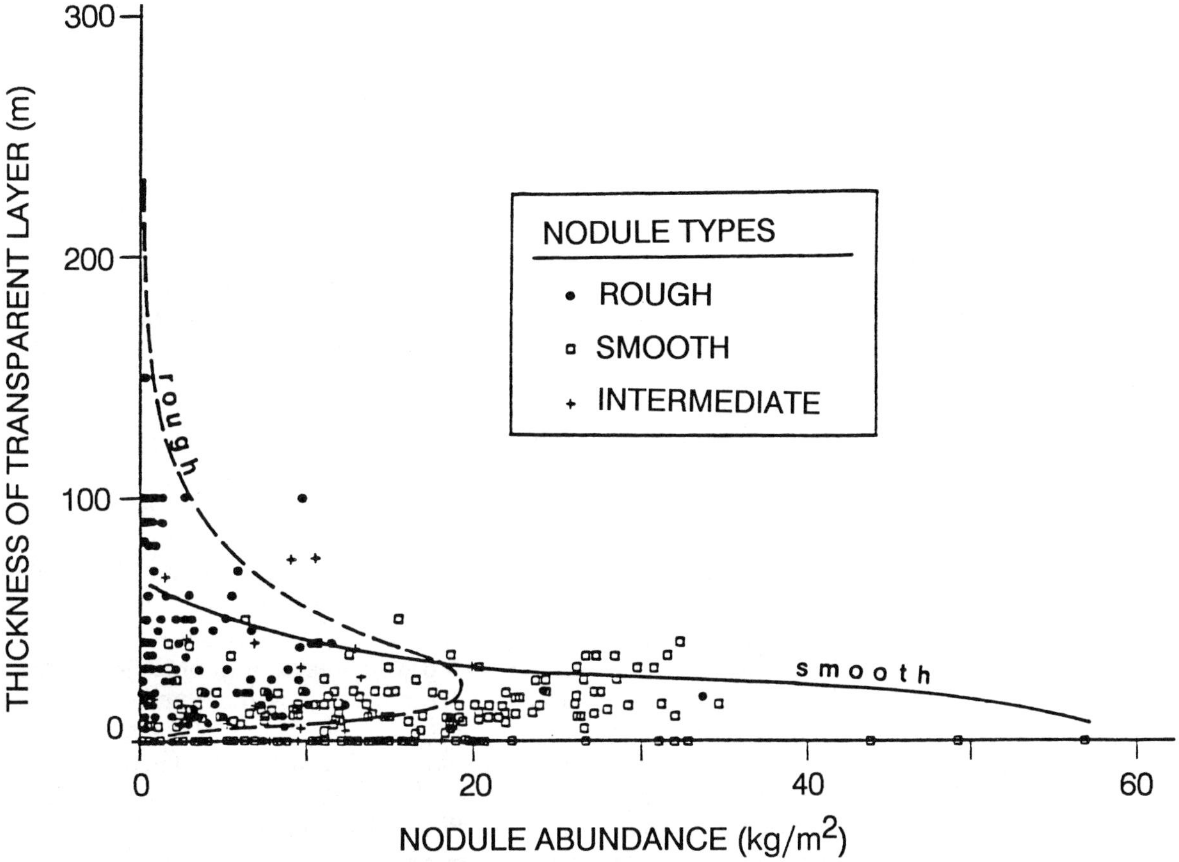

Figure 15. Relationship between nodule abundance and thickness of Unit I (on 3.5-kHz SBP), showing different trends for the two nodule types.

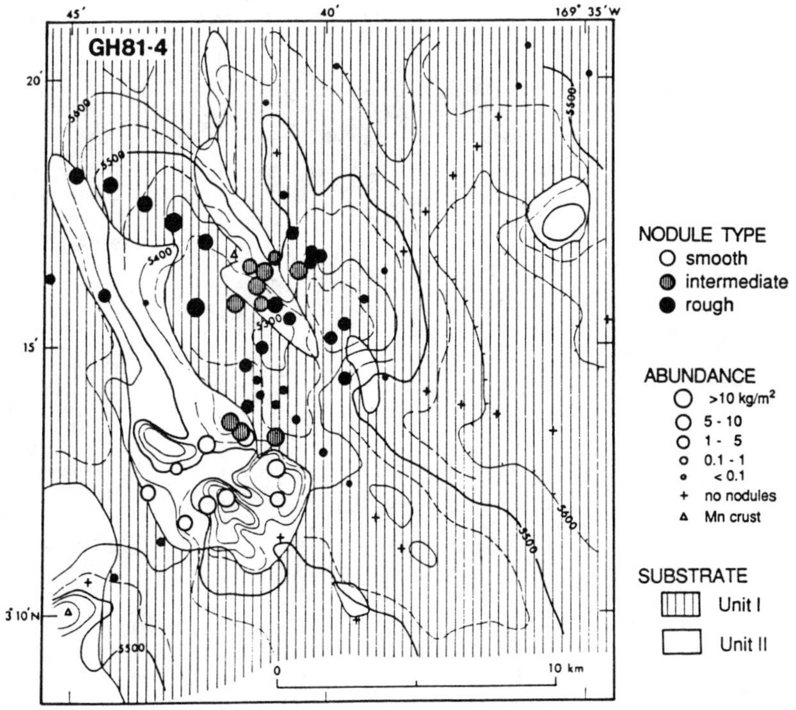

Figure 16. Local small-scale variation of nodule abundance and type (GH81-4, Phase II) modified from Usui et al.)(1983). Note a similar relationship to that on a regional scale in Fig. 14.

around the equatorial zone, the supply of organic matter yields a moderately reducing surface sediment through its decomposition. Thus the distribution of type r nodules is generally restricted in biosiliceous sediments area around the equator, although minor local variations are common. Figure 2 demonstrates the excellent correspondence of regional distribution of nodule type with the present sedimentary provinces: type r nodules on siliceous surface sediments and type s on pelagic/zeolitic clay.

The sedimentation rate may not be the unique factor controlling nodule type, hydrogenous or diagenetic, because the both manganese minerals occur in some areas where Unit I is of the same thickness (Fig. 15). Moderate and continuous sedimentation and a sufficient supply of organic matter, resulting in diagenesis of surface sediments, are requisites for development of diagenetic nodules (type r). Very rapid or very slow sedimentation prevents growth of type r nodules. However, the most optimal rate for the growth of type r nodules cannot be specified because of variable modes of sedimentation in space and time during growth.

In spite of only limited age data for the CPB nodules, a general growth history can be inferred from sedimentological and seismic data (Nishimura, this volume). Radiochemical datings by Be isotopes on the two nodules (Fig. 13) and alpha-track dating on ten nodules (Piper and Gibson, 1979) suggest that the initiation of nodule growth goes back to the middle-late Miocene at the earliest. The age of the large shark teeth which often serve as nuclei of type s nodules are Miocene (or Oligocene). The other available estimation is possible from Nishimura's work (1986 and this volume) on sediment cores; the claystone which is covered with manganese crusts is assigned Late Eocene to Oligocene from the assemblage of ichthyoliths. This claystone, buff in color, stiff, and often zeolitic, is probably correlated with the claystone fragments as nodule nuclei of type s of the CPB, and further with those found deep in DSDP cores of this region.

These data suggest that the abundant nodule deposits on the present sea floor of the basin started to grow from the Oligocene-Miocene before deposition of Unit I. The estimated commencement of growth is consistent with the age of the sediment hiatus determined by Nishimura (1986) between Units I and II.

The abundant nodule pavements of type s have grown in the northern CPB since the Oligocene-early Miocene, probably being promoted by intensified and fairly continuous bottom currents, as suggested by van Andel et al. (1975). Northward inflow of Antarctic Bottom Waters into the Central Pacific Basin has been considered to be the most probable factor controlling nodule development (Goodell et al., 1971; Watkins and Kennett, 1977; Pautot and Melguen, 1979; Mizuno et al., 1980). The type s nodules often have two growth stages as recorded in nodule internal structure; the older one is fragmented type s nodules, around which the youngest layers of rough or smooth structure surround entirely (Fig. 5b). On the other hand, type r nodules have less variable internal structure. The typical simple rhythmical lamination suggests (Fig. 5a) a uniform condition of growth. As type r nodules in the Phase I are generally small and the rough structure often covers older type s nodules (Fig. 13 and Usui, 1979a), the nodules may

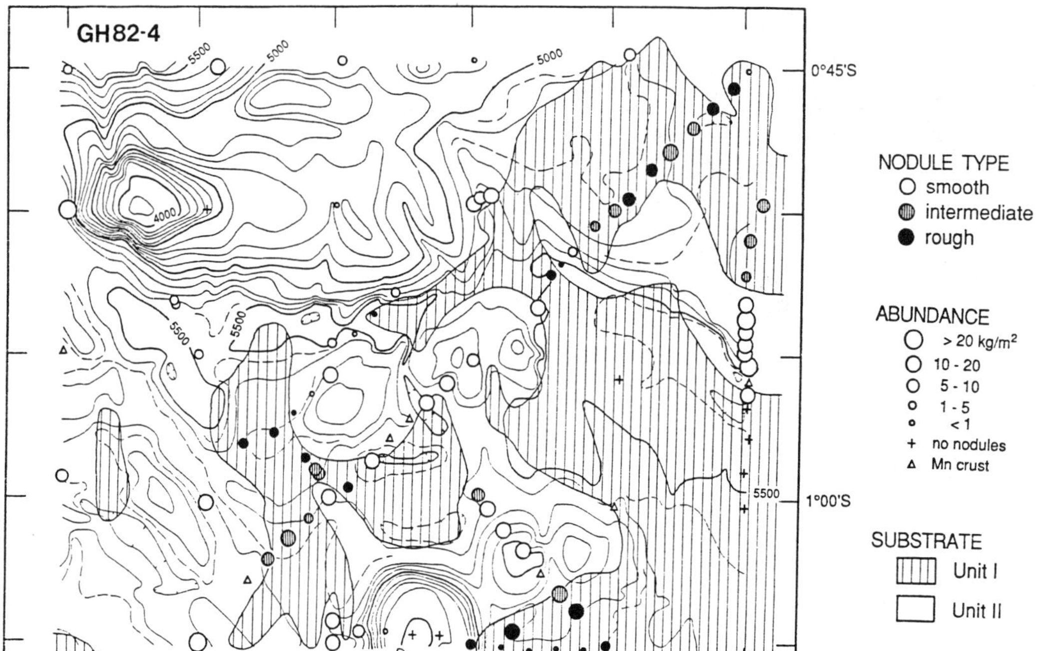

Figure 17. Local small-scale variation of nodule abundance and type (GH82-4, Phase II). Note a similar relationship to those in Figs. 14 and 16. Seismic profile data are from Usui and Tanahashi (unpublished).

be younger than type s, supposedly from the late Pliocene or the Quaternary according to the ^{10}Be age dating. This estimation is consistent with the proposed growth history of northern CPB nodules based on acoustic stratigraphy and sedimentological data.

ACKNOWLEDGMENTS

The authors are most grateful to the officers, crew, ship-board scientists during cruises of R/V *Hakurei Maru* and GSJ scientific staff for their support and collaboration. Drs. S. Nakao, A. Nishimura, M. Tanahashi, GSJ, and Professor Dr. A. Mizuno, Ehime University, are acknowledged for discussion and helpful comment. Special thanks are due to Dr. T. Inoue, Daiichi Radio Isotope Laboratory for providing us unpublished radiochemical data.

REFERENCES

Andrews, J.E., and H.W. Friedrich, 1979, Distribution patterns of manganese nodule deposits in the Northeast Equatorial Pacific: Marine Mining, v. 2, p. 1-44.

Berger, W.H., C.G. Adelseck, and L.A. Mayer, 1976, Distribution of carbonate in surface sediments of the Pacific Ocean: Journal of Geophyscial Research, v. 81, p. 2617-2627.

Burns, R.G., and B.A. Brown, 1972, Nucleation and mineralogical controls on the composition of manganese nodules, *in* D.R. Horn, ed., Ferromanganese Deposits on the Ocean Floor: Washington DC, NSF, p. 51-61.

Burns, R.G., and V.M. Burns, 1977, Mineralogy of Manganese Nodules, *in* G.P. Glasby, ed., Marine Manganese Deposits: Amsterdam, Elsevier Publ., p. 185-248.

Burns, R.G., and V.M. Burns, 1979, Manganese Oxides, *in* R.G. Burns, ed., Marine Minerals: Mineralogical Society of America Short Course Notes, v. 6, p. 1-46.

Calvert, S.E., N.B. Price, G.R. Heath, and T. Moore, 1978, Relationship between ferromanganese nodule composition and sedimentation in a small survey area of the equatorial Pacific: Journal of Marine Research, v. 36, p. 161-183.

Corliss, J.B., M. Lyle, J. Dymond, and K. Crane, 1978, The geochemistry of hydrothermal mounds near the Galapagos rift: Earth and Planetary Science Letters, v. 40, p. 12-24.

CPCEMR, 1984, Mineral-Resources Map of the Circum-Pacific Region: Northeast Quadrant, AAPG: Circum-Pacific Council for Energy and Mineral Resources, Houston, Texas, one-sheet map.

CPCEMR, 1985, Manganese Nodules, Seafloor Sediment, and Sedimentation Rates of the Circum-Pacific Region, AAPG: Circum-Pacific Council for Energy and Mineral Resources, Houston, Texas, one-sheet map.

Craig, J.D., 1979, The relationship between bathymetry and ferromanganese deposits in the north equatorial Pacific: Marine Geology, v. 29, p. 165-185.

Cronan, D.S., 1972, Regional geochemistry of ferromanganese nodules in the world ocean, *in* D.R. Horn, ed., Ferromanganese Deposits on the Ocean Floor: Washington DC, NSF, p. 19-30.

Cronan, D.S., 1980, Underwater Minerals: London, Academic Press, 362 pp.

Exon, N.F., 1983, Manganese nodule deposits in the Central Pacific Ocean and their variation with latitude: Marine Mining, v. 4, p. 79-107.

Fewks, R.H., W.D. McFarland, W.R. Reinhart, and R.K. Sorem, 1980, Evaluation of metal resources at and near proposed deep sea mining sites: U.S. Bureau of Mines, 242 pp.

Giovanoli, R., and R. Brütsch, 1979, Uber Oxidehydroxide des Mn(IV) mit Schichtengitter, 5: Mitteilung: stoichiometrie, Austauschverhalten und die Rolle bei der Bindung von Tiefsee-Mangankonkret: Chimia, v. 33, p. 372-376.

Glasby, G.P., 1974, Manganese Deposits in the South Pacific Ocean: N.ew Zealand. Oceanographic Institute, DSIR, Misc. Ser., no. 33-39., seven-sheet map.

Glover, E.D., 1977, Characterization of marine birnessite: American Mineralogist, v. 62, p. 278-285.

Goodell, H.G., M.A. Meylan, and B. Grant, 1971, Ferromanganese deposits of the South Pacific Ocean, Drake Passage, and Scotia Sea: American Geophysical Union, Antarctic Research Series, no. 15, p. 27-92.

Halbach, P., R. Giovanoli, and D. von Borstel, 1982, Geochemical processes controlling the relationship between Co, Mn, and Fe in early diagenetic deep-sea nodules: Earth and Planetary Science Letters, v. 60, p. 226-236.

Halbach, P., and M. Özkara, 1979, Morphological and geochemical classification of deep-sea ferromanganese nodules and its genetical interpretation, in C. Lalou, ed., La Genese des Nodules de Manganese: Colloq. Intl., Paris, C.N.R.S. no. 289, p. 77-89.

Heath, G.R., 1979, Burial rates, growth rates and size distribution of deep-sea manganese nodules: Science, v. 205, p. 903-904.

Horn, D.R., B.M. Horn, and M.N. Delach, 1972, Distribution of ferromanganese deposits in the world ocean, *in* D.R. Horn, ed., Ferromanganese Deposits on the Ocean Floor: Washington DC, NSF, p. 9-17.

Inoue, T., Z-Y. Huang, M. Imamura, S. Tanaka, and A. Usui, 1983, ^{10}Be and ^{10}Be/^{9}Be in manganese nodules: Geochemical Journal, v. 17, p. 307-312.

Joshima, M., 1982, Remanent magnetization of piston cores from the Wake-Tahiti Transect in the Central Pacific, *in* A. Mizuno and S. Nakao, eds., Geological Survey Of Japan Cruise Report, no. 18, p. 276-287.

Ku, T.L., 1977, Rates of accretion, *in* G.P. Glasby, ed., Marine Manganese Deposits: Amsterdam, Elsevier., p. 249-267.

Kuga, N., and A. Usui, 1982, Fossil teeth of Elasmobranchii associated with manganese nodules (abs.): Proceedings Annual Meeting. Geological Society of Japan, p. 337. (in Japanese).

Lonsdale, P., V.M. Burns, and M. Fisk, 1980, Nodules of hydrothermal birnessite in the caldera of a young seamount: Journal of Geology, v. 88, p. 611-618.

Marchig,V., and H. Gundlach, 1981, Separation of iron from manganese and. growth of manganese nodules as a consequence of diagenetic ageing of radiolarians: Marine Geology, v. 40, p. M35-M45.

Meyer, K., 1973, Surface sediments and manganese nodule facies, encountered on R/V *Valdivia* Cruise 1971/1973, in Origin and Distribution of Manganese Nodules in the Pacific and Prospects for Exploration: Honolulu, Hawaii Institute of Geophysics, p. 125-230.

Meylan, M.A., 1974, Field description and classification of manganese nodules, in Ferromanganese Deposits of the Ocean Floor: Hawaii Institute of Geophysics Technical Report, no. 74-9, p. 158-168.

Mizuno, A., ed., 1981, Deep sea mineral resources investigation in the northern part of Central Pacific Basin (GH79-l cruise): Geological Survey of Japan Cruise Report, no. 15, 309 pp.

Mizuno, A., and J. Chujo, eds., 1975, Deep sea mineral resources investigation in the eastern Central Pacific Basin (GH74-5 cruise): Geological Survey of Japan Cruise Report, no. 4, 103 pp.

Mizuno, A., T. Miyazaki, A. Nishimura, K. Tamaki, and M. Tanahashi, 1980, Central Pacific manganese nodules, and their relation to sedimentary history, *in* Proceedings 12th Offshore Technical Conference: Houston, v. 3, p. 331-340.

Mizuno, A., and T. Moritani, eds., 1977, Deep sea mineral resources investigation in the Central-Eastern Part of Central Pacific Basin (GH76-l cruise): Geological Survey of Japan Cruise Report, no. 8, 217 pp.

Mizuno, A., and S. Nakao, eds., 1982, Regional data of marine geology, geophysics, and manganese nodules: the Wake-Tahiti transect in the

Central Pacific (GH80-1 cruise): Geological Survey of Japan Cruise Report, no. 18, 399 pp.
Moore, T.C., and G.R. Heath, 1966, Manganese nodules, topography, and thickness of Quaternary sediments in the Central Pacific: Nature, v. 212, p. 983-985.
Morgan, W.J., 1972, Deep mantle convection plumes and plate motions: Bulletin, American Association of Petroleum Geologists, v. 56, p. 203-213.
Moritani, T., ed., 1979, Deep sea mineral resources investigation in the central-western part of Central Pacific Basin (GH77-1 cruise) : Geological Survey of Japan Cruise Report, no. 12, 256 pp.
Moritani, T., S. Maruyama, M. Nohara, K. Matsumoto, T. Ogitsu, and H. Moriwaki, 1977, Description, types, and distribution of manganese nodules, *in* T. Moritani, ed., Geological Survey of Japan Cruise Report, no. 8, p. 136-158.
Moritani, T., and S. Nakao, eds., 1981, Deep sea mineral resources investigation in the western part of Central Pacific Basin (GH78-1 cruise): Geological Survey of Japan Cruise Report, no. 17, 281 pp.
Nakao, S., ed., 1986, Marine geology, geophysics, and manganese nodules around deep-sea hills in the Central Pacific Basin (GH81-4 cruise): Geological Survey of Japan Cruise Report, no. 21, 257 pp.
Nakao, S., and Moritani, T., eds., 1984, Marine geology, geophysics, and manganese nodules in the northern vicinity of the Magellan Trough (GH80-5 cruise): Geological Survey of Japan Cruise Report, no. 20, 272 pp.
Nishimura, A., 1986, Deep-sea sediments in the central equatorial Pacific (GH81-4 area), *in* S. Nakao, ed., Geological Survey of Japan Cruise Report, no. 21, p. 56-83.
Nishimura, A., Sedimentation and hiatuses in the Central Pacific Basin — their relationship to manganese nodule formation. This volume.
Onodera, K., and Mizuno, A., 1981, Bathymetry of the GH79-1 Area, *in* A. Mizuno, ed., Geological Survey of Japan Cruise Report, no. 15, p. 45-56.
Orwig, T.L., 1981, Channeled turbidites in the eastern Central Pacific Basin: Marine Geology, v. 39, p. 33-57.
Orwig, T.L., and L.W. Kroenke, 1980, Tectonics of the eastern Central Pacific Basin: Marine Geology, v. 34, p. 29-43.
Ostwald, J., 1984, Ferruginous vernadite in an Indian Ocean ferromanganese nodule: Geol. Mag., v. 121, p. 483-488.
Pautot, G., and M. Melguen, 1979, Influence of deep water circulation and sea floor morphology on the abundance and grade of central South Pacific manganese nodules, *in* J.L. Bischoff and D.Z. Piper, eds., Marine Geology and Oceanography of the Pacific Manganese Nodule Provinces: New York, Plenum, p. 437-473.
Piper, D.Z., and J.R. Blueford, 1982, Distribution, mineralogy, and texture of manganese nodules and their relation to sedimentation at DOMES Site A in the equatorial North Pacific: Deep-Sea Research, v. 29, p. 927-952.
Piper, D.Z., and C.N. Gibson, 1981, Nodule growth rates in the GH79-1 area, *in* A. Mizuno, ed., Geological Survey of Japan Cruise Report, no. 15, p. 265-269.
Piper, D.Z., K. Leong, and W.F. Cannon, 1979, Manganese nodules and surface sediment compositions: DOMES Sites A, B, and C, *in* J.L. Bischoff and D.Z. Piper, eds., Marine Geology and Oceanography of the Pacific Manganese Nodule Provinces: New York, Plenum, p. 437-473.
Piper, D.Z., T.R. Swint-Iki, and F.W. McCoy, 1987, Distribution of ferromanganese nodules in the Pacific Ocean: Chem. Erde, v. 46, p. 171-184.
Rawson, M.D., and W.B.F. Ryan, 1978, Oceanfloor Sediments and Polymetallic Nodules: Lamont-Doherty Geological Observatory, Columbia University, one-sheet map.
Sager, W.W., and B.H. Keating, 1984, Paleomagnetism of Line Islands Seamounts: Evidence for Late Cretaceous and Early Tertiary volcanism: Journal of Geophysical Research, v. 89, p. 11135-11151.
Schellmann, W., 1977, Korrelationsanalysen mit der Mikrosonde am Beispiel einer Pazifischen Manganknolle: Geol. Jahrb., v. D23, p. 105-117.
Schlanger, S.O., M.O. Garcia, B.H. Keating, J.J. Naughton, W.W. Sager, J.A. Haggerty, and J.A. Philpotts, 1984, Geology and geochronology of the Line Islands: Journal of Geophysical Research, v. 89, p. 11,261-11,272.
Sorem, R.K., 1967, Manganese nodules: nature and significance of internal structures: Economic Geology, v. 62, p. 141-147.
Sorem, R.K., and R.H. Fewks, 1979, Manganese Nodules: Research Data and Methods of Investigation: New York, Plenum, 723 pp.
Stoffer, P., G.P. Glasby, and G. Frenzel, 1984, Comparison of the characteristics of manganese micronodules from the equatorial and southwest Pacific: Tschermaks Mineral Petrogr. Mitt., v. 33, p. 1-23.
Sundkvist, K.E., 1983, Size distribution of manganese nodules: Marine Mining, v. 4, p. 305-316.
Tamaki, K., 1977, Study on substrate stratigraphy and structure by continuous seismic reflection profiling survey, *in* A. Mizuno and T. Moritani, eds., Geological Survey of Japan Cruise Report, no. 8, p. 51-62.
Tamaki, K., 1984, Seismic reflection survey in the Central Pacific Basin during GH80-5 Cruise, *in* S. Nakao and T. Moritani, eds., Geological Survey of Japan Cruise Report, no. 20, p. 43-52.
Tamaki, K., M. Joshima, and R.L. Larson, 1979, Remanent Early Cretaceous spreading center in the Central Pacific Basin: Journal of Geophysical Research, v. 84, p. 4501-4510.
Tanahashi, M., 1986, Seismic reflection survey in eastern part of the Central Pacific Basin (GH81-4 Area), *in* S. Nakao, ed., Geological Survey of Japan Cruise Report, no. 21, p. 33-48.
Terashima, S., 1978, Atomic absorption analysis of Mn, Fe, Cu, Ni, Co, Pb, Zn, Si, Al, Ca, Mg, Na, K, Ti, and Sr in manganese nodules: Bulletin Geological Survey Japan, v. 29, p. 401-410 (in Japanese with English abstract).
Usui, A., 1979a, Minerals, metal contents, and mechanism of formation of manganese nodules from the Central Pacific Basin (GH76-1 and GH77-1) *in* J.L. Bischoff and D.Z. Piper, eds., Marine Geology and. Oceanography of the Pacific Manganese Nodule Provinces: New York, Plenum, p. 651-679.
Usui, A., 1979b, Nickel and copper accumulation as essential elements in 10Å manganate of deep-sea manganese nodules: Nature, v. 279. p. 411-413.
Usui, A., 1983, Regional variation of manganese nodule facies on the Wake-Tahiti Transect: morphological, chemical and mineralogical study: Marine Geology, v. 54, p. 27-51.
Usui, A., 1984, Mineralogy and internal structure of manganese nodules of the GH80-5 area, *in* S. Nakao and T. Moritani, eds., Geol. Surv. Japan Cruise Report, no. 20, p. 227-241.
Usui, A., 1986, Local variability of manganese nodule deposits around the small hills in the GH81-4 area, *in* S. Nakao, ed., Geological Survey of Japan Cruise Report, no. 21, p. 98-159.
Usui, A., T.A. Mellin, M. Nohara, and M. Yuasa, 1989, Structural stability of marine 10Å manganates from the Ogasawara (Bonin) arc: implication for low-temperature hydrothermal activity: Marine Geology, v. 86, p 41-56.
Usui, A., and N. Mita, 1987, Comparison of manganese nodules from the Northeast Equatorial Pacific (Cruise 50-25) with nodules from the Central Pacific Basin: Geol. Jahrb., v. D87, p. 287-313.
Usui, A., S. Nakao, and T. Moritani, 1983, Manganese Nodule Distribution in the Central Pacific Ocean: Geological Survey of Japan Marine Geology Map Series, no. 21, one-sheet map.
Usui, A., A. Nishimura, M. Tanahashi, and S. Terashima, 1987, Local variability of manganese nodule facies on small abyssal hills of the central Pacific Basin: Marine Geology, v. 74, p. 237-275.
Usui, A., S. Takenouchi, and T. Shoji, 1978, Mineralogy of deep sea manganese nodules and synthesis of manganese oxides: implication to genesis and geochemistry: Mining Geology, v. 28, p. 405-420 (in Japanese with English abstract).
Usui, A., and M. Tanahashi, unpublished, The relationship of local variation of manganese nodule deposits to acoustic stratigraphy in the GH82-4 area, southern Central Pacific Basin: prepared for Geological Survey of Japan Cruise Report.

Van Andel, T.H., G.R. Heath, and T.C. Moore, 1975, Cenozoic History and Paleooceanography of the Central Equatorial Pacific Ocean: Geological Society America Memoir, no. 143, 134 pp.

Watkins, N.D., and J.P. Kennett, 1977, Erosion of deep-sea sediments in the Southern Ocean between longitudes 70°E and 190°E and contrasts in manganese nodule development: Marine Geology, v. 23, p. 103-111.

Winterer, E.L., 1976, Anomalies in the tectonic evolution of the Pacific, *in* G.H. Sutton, M.H. Manghnani, and R. Moberly, eds., The Geophysics of the Pacific Ocean Basin and its Margin, Amer. Geophys. Union, Geophysical Monograph Series, v. 19, p. 269-278.

Winterer, E.L., J.I. Ewing, et al., eds., 1973, Initial Reports of the Deep Sea Drilling Project, v. 17, 930 pp.

Keating, B.H., and Bolton, B.R., editors, 1992,
Geology and offshore mineral resources of the
central Pacific basin, Circum-Pacific Council for
Energy and Mineral Resources Earth Science
Series, V. 14. New York, Springer-Verlag.

CHEMISTRY AND MINERALOGY OF FERROMANGANESE DEPOSITS FROM THE EQUATORIAL PACIFIC OCEAN

Eric H. De Carlo and Charles M. Fraley
Hawaii Institute of Geophysics, Honolulu, Hawaii 96822 U.S.A.

ABSTRACT

Ferromanganese nodules, crusts, and phosphatic substrates were recovered during two cruises of the R/V *Moana Wave* to the Cook Islands, Kiribati, and Tuvalu. We report the geochemistry and mineralogy of ferromanganese crusts and seamount nodules only. Samples were collected from flanks, gentle slopes, and nearly flat summits; nodules were absent from the steeper slopes. Cobalt, the metal of highest commercial interest, ranges between 0.19% and 1.90% by weight in deposits consisting of stains of less than one millimeter to a maximum of 12 cm thick. The highest Co dredge averages of 1.72% and 1.30% were found in crusts recovered near Vaitupu and Enderbury islands respectively. Most crusts from the Manihiki Plateau were less than one centimeter thick and exhibited low metal values in contrast to those from a region of mud volcanoes on the plateau in which Co concentrations averaged 1.0%. Copper and Ni dredge averages ranged from 0.05% to 0.14% and from 0.19% to 0.89% respectively.

The results of this study are generally in agreement with observations in other areas of the Pacific, and confirm the following criteria for seeking high metal grades in vernadite-rich ferromanganese deposits: 1) water depth of 1000-2000 m, 2) latitude 2-7°S, and 3) low sedimentation rates. Crusts of potential commercial interest occur primarily in the vicinity of Vaitupu and Enderbury Islands, but other gently sloped ridges and seamount summits in the Line Islands (near 1°S, 156°W) have abundant nodules or encrusted cobbles averaging 0.78% Co and 0.63% Ni.

The mineralogy of crusts is dominated by vernadite with few samples enriched in todorokite. Seamount nodules are chemically and mineralogically distinct from their abyssal counterparts; like crusts they are primarily composed of vernadite. No crystalline iron phases were identified within Fe-Mn deposits. Phosphate bands are generally absent within crusts but commonly occur in substrates of samples collected above 2500 m. The phosphorite material in crusts and substrates is primarily carbonate fluorapatite replacing calcium carbonate.

INTRODUCTION

In early 1986 and 1987 the R/V *Moana Wave* carried out two one-month mineral assessment cruises within territorial waters of the Cook Islands, Kiribati, Tuvalu and Western Samoa, in the south equatorial Pacific. This paper summarizes results of chemical and mineralogical studies on ferromanganese crusts and seamount nodules; deep-sea nodules and phosphorites from the region are discussed elsewhere in this volume (Bolton et al., 1990; Rao and Burnett, 1990; Usui and Moritani, 1990).

Studies of deep-sea nodules in the past two decades have led to numerous papers on their distribution, abundance, geochemistry and mineralogy (Glasby, 1977; Meylan et al., 1981; and references therein). Although the existence of deep-sea nodules and of Fe-Mn crusts, recently termed Co-rich crusts, has been known since the HMS Challenger expedition of the 1870's (Glasby, 1977), crusts have not until recently received the attention given to nodules. The shift in interest has been attributed to the high content of the strategic metal Co in crusts (three to five times that of abyssal nodules), their potential as a source of

Pt and their occurrence in areas under the jurisdiction of sovereign nations. Although present day economics of Co recovery from crusts are marginal, the Co in crusts is significant because major land deposits are concentrated in nations where disruptions of supplies are likely to occur (Johnson and Clark, personal communication, 1988).

Crusts, a common occurrence on seamounts of the world's oceans, were first reported in the Mid-Pacific Mountains by Hamilton (1956), later found in the Hawaiian Archipelago by Moore (1966), and are nearly ubiquitous within the subtropical climatic belt of the Pacific. Recent crust studies in the Pacific Basin include: Craig et al. (1982), Commeau et al. (1984), Halbach et al. (1982, 1983), Halbach and Manheim (1984), Halbach and Puteanus (1984), Aplin and Cronan (1985a), Hein et al. (1985a,b; 1988), Johnson et al. (1985), De Carlo et al. (1987a,b), and Pichoki and Hoffert (1987). Crusts are generally found on seamounts and submerged slopes of islands and atolls and form primarily where Mn and Fe oxides accrete on sediment-free outcrops. Slopes exposed to strong currents are particularly suitable for crust accumulation. Highly sedimented or rubble-covered slopes of limestone-capped edifices are generally devoid of crusts (Frank et al., 1976; Craig et al., 1982; De Carlo et al., 1987a; Hein et al., 1988).

STUDY AREA AND ANALYTICAL METHODS

Regional Setting

The sparsely distributed northern atolls of the Cook Islands are located on the Manihiki Plateau (Fig. 1) between 8°S and 14°S and 158°W and 166°W. The three volcanic chains of the Gilbert, Phoenix, and Line islands lie on oceanic crust believed to be of Jurassic and late Cretaceous age (Larson et al., 1974; Keating et al., 1984) and may reflect the movement of crust over several hot spots in the underlying mantle. The old inactive or extinct volcanoes are buried deep beneath coralline atolls or form deep-water seamounts. The Ellice Islands of Tuvalu lie directly to the southeast of the Gilbert Islands but may be geologically distinct from the former. The study area ranges from 5°N to 12°S and 170°E to 155°W. Summits of seamounts studied range from 300 m to near 2000 m below sea-level.

Sampling Strategy

Dredging was carried out in water depths of 1000-5000 m and concentrated on east-facing slopes of the surveyed seamounts and on escarpments of the eastern Manihiki Plateau. Results of previous expeditions in the area (Exon, 1982; Cronan, 1983, 1984; Aplin and Cronan, 1985a,b) suggested that eastern slopes of seamounts would be less sedimented than western slopes due to local currents. In order to accurately determine the depth dependence of metal enrichments in crusts we attempted to dredge narrow intervals. This approach was successful in the Ellice, Gilbert and Phoenix islands, but difficult to realize in the central Line Islands due to currents and local topography.

Chain bag rock dredges or pipe dredges (RD or PD designation in sample list) were deployed with guidance from 3.5-kHz echo sounding and SeaMARC II side-scan acoustic imaging. Dredge recoveries included manganese crust fragments with or without underlying substrate, manganese coated basalt, highly brecciated volcanoclastics, unaltered as well as phosphatized carbonates or lithified car-

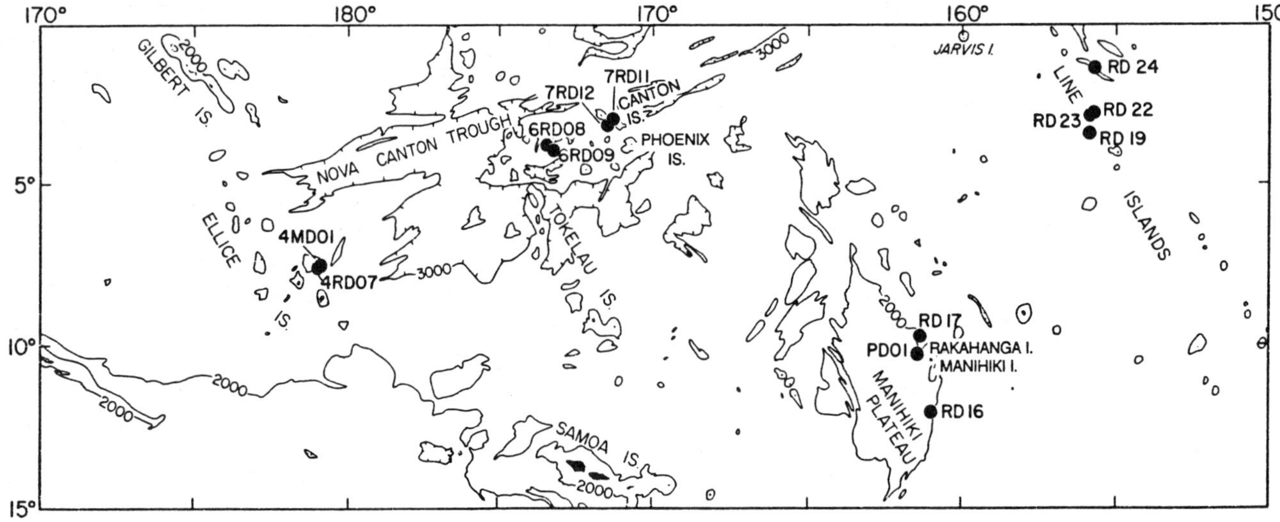

Figure 1. Site location map of dredge sites during MW86-02 and MW87-02 cruises.

bonate ooze, sediments and uncoated material. Samples were described and catalogued upon recovery. Representative specimens were selected from each dredge, slabbed, examined under a binocular microscope, sealed in plastic bags (to minimize changes in mineralogy) and subsequently refrigerated. Larger samples were wrapped in seawater-soaked paper towels, placed in large plastic bags and refrigerated. We present here results of bulk chemical and mineralogical analyses carried out on a subset of the sample collection from the two cruises.

Analytical Procedures

Portions of samples were air-dried, separated from substrate material, manually ground to pass through SPEX standard 100 mesh nylon sieves and stored in glass vials. Substrates were ground manually or in a SPEX ball and mill but not sieved.

Mineralogy was studied with a Philips XRG 3100 automatic X-ray diffractometer equipped with long-fine-focus CuK_α radiation, graphite curved-crystal monochromator, and scintillation detector, or with a computer controlled Scintag PAD-V powder diffractometer equipped with a solid-state Ge-Li detector. Samples were analyzed as bulk powders in aluminum or plastic holders or as dried slurries on glass slides. Samples were scanned at a rate of 2° 2θ/min over the range of 2-68°.

Chemical analyses were performed on a 110°C dry basis. Triplicate splits were dissolved with mineral acids in Parr bombs using a modification of a previously described method (Bernas, 1968; De Carlo et al., 1982, 1987a) or in a microwave oven. The microwave technique (method adapted after Nadkarni, 1984; Kingston and Jassie, 1986) is much more efficient and requires less than one minute of digestion whereas two hours are needed in an induction oven at 110°C. Solutions were analyzed for selected major and minor constituents by atomic absorption (AA) and inductively coupled plasma (ICP) emission spectroscopies. A Perkin-Elmer 603 AA spectrometer was used for K and Zn. All other elements as well as Zn were determined with a Leeman Labs Plasma Spec I ICP system. The relative precision of analyses was better than 7% of the reported value; accuracy of determinations was ascertained by analysis of U.S. Geological Survey (USGS) standard Fe-Mn nodules A-1 and P-1; results were generally within 5% relative to recommended values.

RESULTS AND DISCUSSION

Sample Recoveries and Distribution

The sampling sites (Fig. 1 and Table 1) are distributed throughout the major island groups of the CCOP/SOPAC member nations except in the Cook Islands where only areas near atolls on the Manihiki Plateau were studied. The material recovered during the two cruises is characterized by a large degree of heterogeneity. Thick crusts (30-120 mm) were encountered below 2000 m near the Phoenix Islands and in the central Line Islands. Thin crusts (0.1-35 mm) occurred on limestone and/or phosphatized substrates in shallow waters (< 1500 m) of the Ellice and Gilbert Island areas as well as on semi-indurated sediments of the Manihiki Plateau. A relationship between crust thickness and water depth is noted in this study and agrees with observations by Johnson and Clark (personal communication, 1988) for selected areas of the Pacific Ocean, but contrasts findings of Halbach and Puteanus in the Johnston Island EEZ (Halbach and Puteanus, unpublished results). The abundant thick crusts from the Phoenix Islands (6RD08, 6RD09) and seamount nodules from the central Line Islands (RD23 and RD24) do not typically display very high valuable metal concentrations (see below). However the mass of thick material recovered in each of these dredges represents the largest potential tonnage of Co and Ni sampled in the area. De Carlo et al. (1987b) recently suggested that such areas may yield greater resources than where higher concentrations of Co and Ni are found in thin deposits.

Photographic surveys

A towed camera sled equipped with a down-looking Benthos camera and strobe programmed to take photographs every 10 seconds was deployed at one station (STN23-BC05) in the central Line Islands. The paucity of bottom photographs obtained suggests the camera spent a large portion of the time gliding in canyons and reflects the very rugged topography of many Line Island seamounts. Photographs (Fig. 2) reveal Fe-Mn encrusted volcanic outcrops and calcareous ooze filled depressions (possibly covering additional Fe-Mn accumulations). Evidence of the coexistence of nodules and crusts as observed in other areas (De Carlo et al., 1987a; Hein et al., 1988, 1989) was also obtained. Both gentle slopes and very rugged topography were encountered.

Mineralogy

Crusts

A compilation of crust and substrate mineralogy is given in Appendix I. Identifications of mineral phases are based on peak positions and relative intensities as compared to standard reference patterns (JCPDS). A typical pattern for the principal Mn phase, vernadite (δ-MnO2), is shown in Figure 3. It is identified by two broad peaks at

Table 1. Dredge Locations and Average Chemical Composition

Dredge	Latitude/ and longitude	Depth (m)	Samples (no.)	Crust thick-ness (mm)	Mn/Fe	Mn (%)	Fe (%)	Co (%)	Ni (%)	Cu (%)	Zn (%)	Si (%)	Al (%)	Ti (%)	V (%)	P (%)	Ca (%)	Mg (%)	Ba (%)	K (%)
RD16	12:04 S 160:57 W	2275	11	16.0	0.990	18.42	18.58	0.478	0.336	0.131	0.066	5.31	1.31	0.95	0.059	0.43	2.70	0.99	0.129	0.561
RD17	09:47 S 161:25 W	3990	1	6.0	0.791	14.45	18.27	0.347	0.194	0.091	0.055	8.70	2.48	1.12	0.052	0.35	1.93	2.05	0.104	0.862
RD01	10:19 S 161:28 W	1688	7	10.0	1.557	23.99	15.41	1.139	0.525	0.081	0.074	2.59	0.53	1.26	0.060	0.47	3.30	1.13	0.139	0.520
RD19	03:22 S 155:50 W	1330	3	27.0	3.636	25.72	7.11	0.739	0.676	0.049	0.098	1.06	0.13	0.56	0.060	3.57	10.25	1.08	0.291	0.628
RD22	02:49 S 155:47 W	3050	2	15.0	1.297	23.21	17.90	0.664	0.435	0.095	0.077	5.48	0.64	1.15	0.066	0.39	2.35	1.03	0.158	0.510
RD23	02:51 S 155:48 W	2095	6	22.0	1.604	19.62	12.33	0.584	0.567	0.100	0.077	5.53	1.36	1.03	0.051	1.30	4.39	1.17	0.155	0.843
RD24	01:19 S 155:48 W	2163	34	22.0	1.745	25.32	14.66	0.761	0.639	0.119	0.083	3.35	0.59	0.99	0.061	0.49	2.77	1.13	0.164	0.610
4MD01	07:27 S 179:02 E	1230	2	12.5	2.833	32.47	11.46	1.720	0.889	0.065	0.094	0.93	0.24	1.05	0.077	0.49	2.94	1.47	0.184	n.d.
4RD07	07:25 S 179:03 E	1690	16	39.6	1.481	23.55	16.07	0.781	0.578	0.099	0.077	2.76	0.82	1.22	0.077	0.82	3.33	1.17	0.179	n.d.
6RD08	03:50 S 173:20 W	2265	8	49.4	1.591	24.97	15.82	0.798	0.626	0.105	0.084	2.81	0.64	1.17	0.075	0.86	3.57	1.15	0.188	n.d.
6RD09	03:48 S 173:30 W	4800	7	65.7	1.238	22.95	18.66	0.587	0.331	0.111	0.056	3.54	1.00	1.28	0.082	0.45	2.16	1.08	0.166	n.d.
7RD11	03:00 S 171:23 W	1465	15	13.2	2.778	28.44	10.80	1.294	0.873	0.076	0.090	1.00	0.36	0.77	0.079	1.46	5.16	1.43	0.218	n.d.
7RD12	03:01 S 171:27 W	3130	2	52.0	1.501	26.39	17.58	0.678	0.537	0.136	0.079	2.66	0.51	1.21	0.089	0.46	2.48	1.07	0.209	n.d.

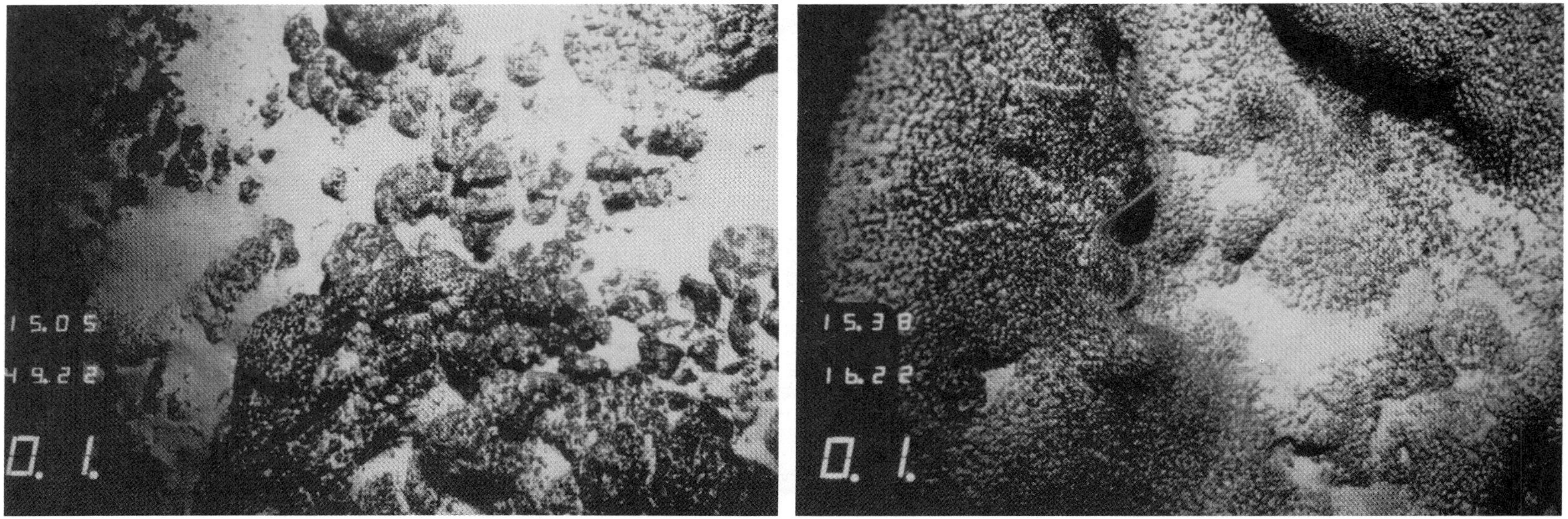

Figure 2. Bottom photographs of Fe-Mn encrusted outcrops showing (left) sediment-filled depressions and (right) nearly continuous pavement coverage with reduced roughness.

2.43Å and 1.41Å (Burns and Burns, 1977) and represents hydrogenetic manganese oxide precipitated under highly oxidizing conditions. The majority of crusts also contain small amounts of detrital volcanic material (feldspars) and/or calcite and apatite depending on site and depth of recovery. A greater proportion of detrital matter is found in thin deposits than in crusts thicker than 2 cm; the latter often contain fine-grained apatite filling voids in the Fe-Mn oxides. Samples from RD19, recovered from a depth of 1350 m in the south central Line Islands, contain significant amounts of phosphatic material admixed with the ferromanganese oxides and are characterized by an apatite layer between crust and substrate. Crusts from water depths less than 2500 m are usually more enriched in apatite (both in the Fe-Mn oxide itself and in substrate) than those recovered at greater depths.

Highly lustrous vernadite with very little admixed todorokite and phosphatic material characterizes crusts from RD24. These samples are highly reflective under the light microscope and reveal a regular hexagonal or pentagonal columnar structure extending over several hundred micrometers. Sample 6RD09-1 collected near McKean Island at 4800 m is particularly noteworthy as its thick (~10 cm) smooth crust contains nearly 5 cm of the columnar

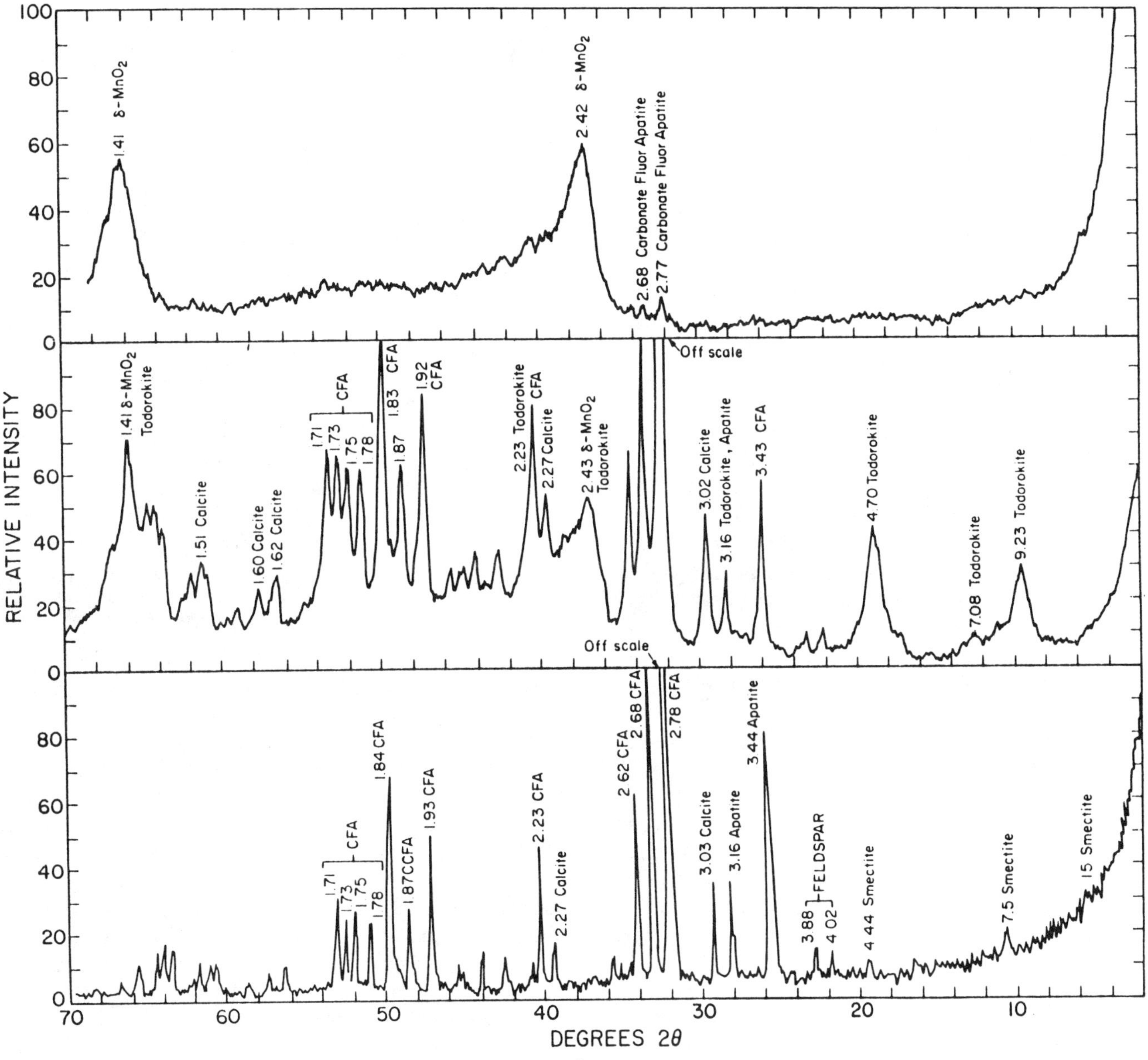

Figure 3. X-ray diffractograms: (top) vernadite-rich crust, (center) apatite- and todorokite-rich crust, (bottom) apatite-rich substrate material. CFA: carbonate fluorapatite.

structure. To our knowledge the large-scale morphological regularity shown in Figure 4 has not been previously reported in the literature for vernadite-rich crusts; the micromorphological characteristics of this columnar structure will be reported elsewhere (Alvarez et al., 1990). Many samples display the columnar structure on a smaller scale as well as irregular cuspate growth patterns typical of Fe-Mn nodules. Recent work by Hein (personal communication) indicates similar structures exist in crusts from other areas. Smooth vernadite-rich abyssal nodules collected northeast of Canton Island also display a columnar structure similar to that observed in the above crusts whereas rough todorokite-rich nodules only exhibit cuspate growth patterns (Bolton et al., 1990).

Only a few samples contained significant amounts of todorokite, identified by reflections at 9.6Å, 4.7Å and 3.2Å (Figure 3), along with vernadite. Three samples, 7RD11-03, RD19-S3 and RD23-SF, which are todorokite-enriched, are also chemically anomalous. The first is a complex crust with discrete areas rich in apatite, vernadite, and todorokite as well as areas where dendritic todorokite is admixed in the phosphatic phase. Rao and Burnett (1990) have suggested that this type of crust formed during early diagenesis of carbonate sediments when the supporting seamount was near the ocean surface. Mn and Fe oxide replacements of carbonates released by reductive processes may have reprecipitated and led to the contemporaneous deposition of apatite and todorokite. The other todorokite enriched samples, RD19-S3 and RD23-SF, do not display an unusual morphology relative to vernadite-rich crusts. Sample RD3-SF is similar to other encrusted cobbles recovered at the same station; its todorokite content is significantly less than that of 7RD11-03 (decreased intensities of characteristic X-ray diffraction peaks). Since this sample was collected on gentle slopes near the top of a ridge it is possible that the todorokite results from diagenetic interactions with underlying sediment. A mixture of vernadite and todorokite is common in abyssal nodules found at the sediment/seawater interface; the sea-water exposed top is enriched in vernadite whereas the bottom, buried in the sediment, is enriched in todorokite (Bolton et al., 1990 and references therein). The only other samples with discernible amounts of todorokite were recovered from RD19. Each of the samples displays a high Mn/Fe ratio but only RD19-S3 contains enough todorokite to significantly affect its chemistry; it contains only approximately 0.42% Co whereas other crusts from this depth typically contain approximately 1% Co by weight (De Carlo et al., 1987a,b).

A 7.2Å reflection is evident in several samples from the Manihiki Plateau and the southernmost station in the Line Islands. Since major birnessite reflections other than at 7.2Å were not identified in the X-ray diffractograms, and this mineral is only known to occur in marine diagenetic and

Figure 4. Scanning electron photomicrograph of columnar structure in sample 6RD09-01. Scale bar is 1 μm. Photo credit J.P. Cowen and R. Alvarez.

hydrothermal ferromanganese deposits (Burns and Burns, 1977), we suggest that the 7.2Å reflection observed in this study may be attributed to vernadite. Aged synthetic vernadite prepared in our laboratory exhibits a low intensity reflection at 7.2Å which is not present in fresh vernadite. This observation combined with recent work in our laboratory (Alvarez et al., unpublished information) suggests that vernadite crystallinity changes with time and that this mineral may be more crystalline than commonly believed. Although unlikely, it must be noted that the 7.2Å could also result from the presence of bobierrite, a magnesium phosphate mineral which has been identified in a few substrate samples (see Appendix I).

Substrates

The diverse mineralogy of substrates is evident from the individual sample date presented in Appendix 1. Dominant phases include phosphatized carbonates for dredges 4RD07, 6RD08, 7RD11 (Ellice and Phoenix islands), consolidated Fe-rich foraminiferal ooze for PD01 (mud-volcano area on Manihiki Plateau), phosphatized volcanic breccia for RD19 (south central Line Islands) and RD24 (central Line Islands), basaltic/volcanoclastic matter for RD22, RD23, and RD26 (central Line Islands) (substrates from RD23 are also enriched in apatite). The rocks underlying crusts from the eastern Manihiki Plateau (RD16 and RD17) consist of mixtures of volcanic matter, its weathering products, and phosphatized carbonaceous material. The most common substrate material in this area consists of a mixture of detrital aluminosilicates, calcite, and carbonate fluorapatite. As noted above, phosphatized substrates are more commonly encountered in dredge hauls from shallow areas than from deep water. A representative diffractogram of an apatite-rich substrate is presented in Figure 3. Commonly phosphatic substrates of encrusted cobbles contain large (0.5-1.5 cm) fragments of Fe-Mn oxides as well as dendritic Mn oxide probably of diagenetic origin (Halbach and Puteanus, 1984).

Chemical composition

Major and minor elements

The average bulk composition of crusts from each dredge is presented in Table 1; elemental abundances for each individual sample are given in Appendix II. Although the data presented here are representative of the samples recovered during this work, use of other sample collection methods in this area could result in the collection of somewhat different material. This results from the fact that ship-based dredging preferentially selects loose Fe-Mn encrusted cobbles over firmly attached crusts (Hein et al., 1985a) and underestimates crust pavement thickness by a factor of two or more.

Inter-element correlations

An inter-element correlation matrix based on 16 variables including depth and Mn/Fe ratio for the 114 crusts is presented in Table 2. Potassium has not been included since it was only determined in samples from the MW87-02 cruise. For the sample population (n = 114) correlations greater than r = 0.22 are statistically significant at the 99% confidence level. Selected positive correlations in decreasing order include: (r > 0.9) Ca with P; (r > 0.8) Al with Si; Zn and Mn/Fe with Ni; (r > 0.7) Co and Ni with Mn; Zn with Mn/Fe; (r > 0.6) Mg with Ni; Zn with Mn; Fe with Ti; (r > 0.5) Ba, Ca, P and Mg with Mn/Fe; Fe with depth; V with Mn; Fe with Si. Significant negative correlations include: (r < -0.7) Al and Si with Mn; (r < -0.6) Si and Ti with Mn/Fe; Ni, Ca, P and Zn with Fe; Co and Ni with Si. In general crust inter-element correlations are not as strong as found by Bolton et al. (1989) for abyssal nodules recovered within the northeastern region of the study area, and probably reflect the much wider geographical area sampled for crusts.

The strong covariance of Ca and P (r = 0.985) is evident in the scatter plot shown in Figure 5 and follows from the presence of apatite identified by XRD; it indicates that little Ca in the crusts is present in calcite, aluminosilicates or their weathering products and agrees with a similar correlation found by Bolton et al. (1990) in abyssal nodules. These authors suggest that Ca and P enrichment in nodules results from the incorporation of biogenic matter, whereas we favor the diagenetic mechanism proposed by Rao and Burnett (1990). Direct precipitation of apatite from seawater could also lead to presence of this phase in crusts.

The incorporation of traces of detrital aluminosilicates in crusts is evident from the low concentrations of these elements in the individual samples (Appendix II) and the strong correlation between Al and Si (r = 0.875) (Figure 5B). Based on the extraction of Al from Fe-Mn oxides by

Table 2. Interelement correlation matrix.

	Z	Mn/Fe	Mn	Fe	Co	Ni	Cu	Zn	Si	Al	Ti	V	P	Ca	Mg	Ba
Z	1.000															
Mn/Fe	-.464	1.000														
Mn	-.271	.615	1.000													
Fe	.531	-.878	-.315	1.000												
Co	-.466	.553	.760	-.388	1.000											
Ni	-.510	.815	.733	-.680	.597	1.000										
Cu	.112	-.029	.020	.047	-.083	.111	1.000									
Zn	-.504	.744	.634	-.607	.435	.844	.069	1.000								
Si	.375	-.696	-.718	.504	-.655	-.626	.102	-.485	1.000							
Al	.335	-.576	-.722	.395	-.589	-.515	.093	-.461	.875	1.000						
Ti	.361	-.604	-.054	.632	-.049	-.387	.025	-.315	.237	.234	1.000					
V	.095	.133	.559	.161	.427	.141	-.093	.077	-.458	-.333	.292	1.000				
P	-.259	.544	-.140	-.628	-.106	.230	-.077	.278	-.253	-.083	-.485	-.189	1.000			
Ca	-.304	.564	-.113	-.638	-.065	.269	-.061	.317	-.274	-.116	-.483	-.223	.985	1.00		
Mg	-.191	.518	.423	-.393	.476	.656	.070	.495	-.203	.026	-.151	.153	.120	.144	1.000	
Ba	-.196	.565	.462	-.407	.277	.407	.003	.498	-.470	-.324	-.160	.393	.396	.368	.283	1.000

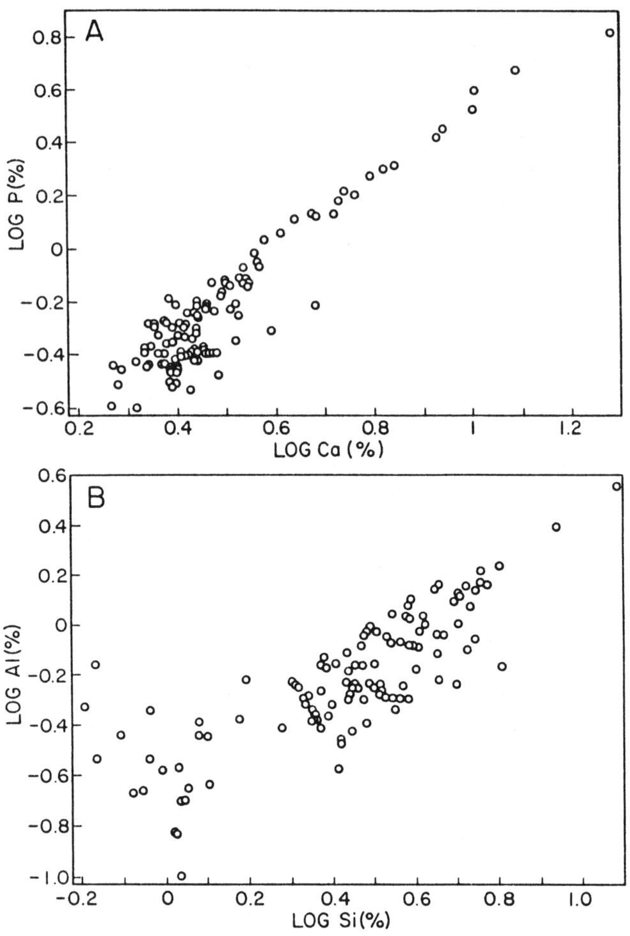

Figure 5. Scatter plots of (A) Ca vs P, and (B) Al vs Si.

weakly acidic reagents, Robbins et al. (1984) suggest that this metal may also be present as poorly crystallized oxyhydroxides, a hypothesis they support by a weak correlation between Al and Fe in amorphous oxyhydroxides. However, detrital Fe in volcanic weathering products (De Carlo et al., 1987a) as well as the presence of amorphous Fe-aluminosilicates in crusts (Halbach and Puteanus, 1984) can also lead to such a correlation. Bolton et al. (1989) find a correlation between Ti and Fe in abyssal nodules (r = 0.99) which is much stronger than observed here (r = 0.632). This may result from an enrichment of TiO_2 in the interlayer FeOOH of vernadite in smooth abyssal nodules (Halbach and Ozkara, 1979) whereas Ti in crusts is present both in vernadite and in Ti-containing Fe-rich clay minerals.

The associations of Co, Ni, Zn with Mn and Mn/Fe (Fig. 6) reflect the predominantly hydrogenetic origin of crusts in this work and are consistent with previous studies of seamount crusts (Aplin and Cronan, 1985a; Halbach et al., 1982, Halbach and Puteanus, 1984; Hein et al., 1985b, 1988; De Carlo et al., 1987a). Removal of todorokite-rich samples (identified in Figs. 6A, 6B, and 6E) from the data set enhances these correlations further supporting the proposed hydrogenetic origin of transition metals in vernadite-rich crusts. It is interesting to compare plots of Co, Ni, and Zn versus Mn (Figs. 6A, 6C, and 6E) with those versus Mn/Fe (Figs. 6B, 6D, and 6F). Differences in the scatter after normalizing to Fe content reflect the dominant enrichment processes for these elements in crusts. The direct association of Co with vernadite through oxidative scavenging, and its relationship with Ni reflect hydrogenetic processes. The enhanced incorporation of Zn in todorokite-rich rather than vernadite-rich crusts reflects its preferred diagenetic enrichment mechanism in crusts exhibiting increased Mn/Fe ratios.

The positive correlation between Ni and Mg in this study (r = 0.656) is stronger than reported by Hein et al. (1988) in the Marshall Islands but contrasts our findings in the Hawaiian Archipelago (De Carlo et al., 1987a) where these elements display no interdependence. This finding suggests a variability in the processes which influence the supply of Ni among the three areas. There appears to be a significant detrital input of Ni in the crusts from this study and from the Marshall Islands, but the Ni in Hawaiian crusts is primarily of hydrogenetic origin.

The diversity in chemical constituents within individual dredge hauls, compounded with the error associated with the sampling depth range reported, leads to considerable scatter in depth/chemistry plots and results in a lack of strong depth correlations in Table 2. Some correlations are nonetheless evident. That between depth and Fe (r = 0.531) is attributed to a greater supply of detrital volcanic material at depth (De Carlo et al., 1987a) but may also partially result from increased dissolution of Fe oxide-coated calcareous tests falling through the water column (Halbach and Puteanus, 1984). Correlations of Al, Si and Ti with depth, although very weak, suggest a greater availability of detrital matter with depth or slower growth rates.

Inverse relationships between Co, Ni, Zn, and Mn/Fe with depth are typical of hydrogenetic ferromanganese crusts (Halbach et al., 1982, 1983; Halbach and Puteanus, 1984; Aplin and Cronan, 1985a; De Carlo et al., 1987a; Hein et al., 1988). Advective transport, within the O_2 minimum zone, of Mn^{2+} remobilized from near-shore reducing sediments (Martin and Knauer, 1984) and its subsequent oxidation to MnO_2 below the O_2 minimum zone has been suggested to lead to an inverse relation between Mn and depth in crusts (Halbach and Puteanus, 1984; De Carlo et al., 1987a). Recent metal profiles from the northwest Indian Ocean measured during periods of extended oxygen depletions associated with the monsoon season support this mechanism (Saager et al., 1988, German and Elderfield, 1988). The lack of a strong inverse relationship between Mn

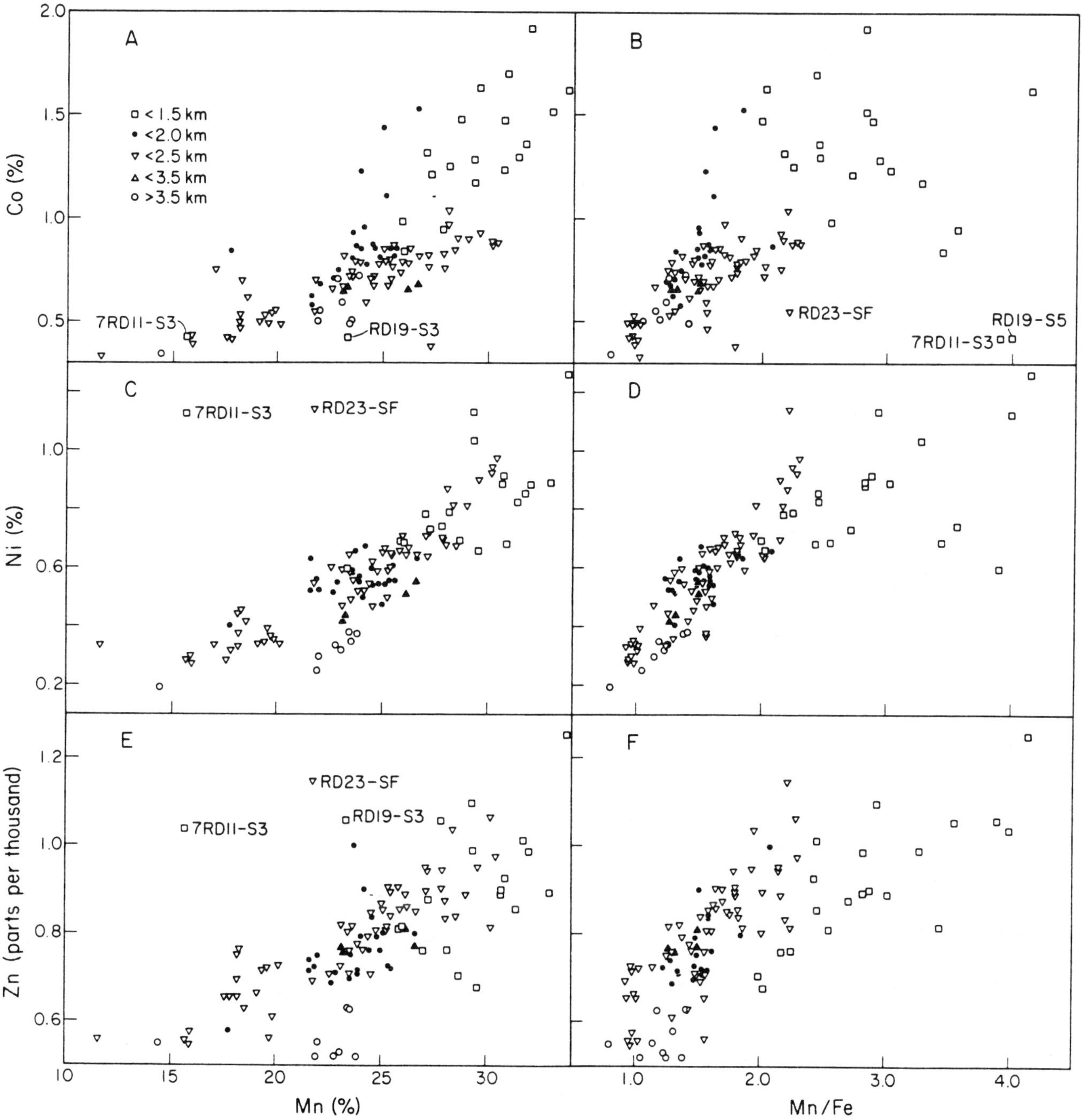

Figure 6. Scatter plots of (A) Co vs Mn, (B) Ni vs Mn, (C) Zn vs Mn, (D) Co vs Mn/Fe, (E) Ni vs Mn/Fe, and (F) Zn vs Mn/Fe.

and water depth indicates that lateral advective transport of Mn does not play as important a role here or in the Marshall Islands (Hein et al., 1988) as it does in the Hawaiian Archipelago (De Carlo et al., 1987a). We further suggest that in the equatorial zone of high productivity, downward particulate fluxes may obliterate evidence of any lateral dissolved metal transport processes.

We reported above a correlation between crust thickness and water depth. This covariance and the inverse relationship between Co concentration and depth then lead to the inverse relation between Co content and crust thickness previously suggested by Halbach (personal communication, 1987).

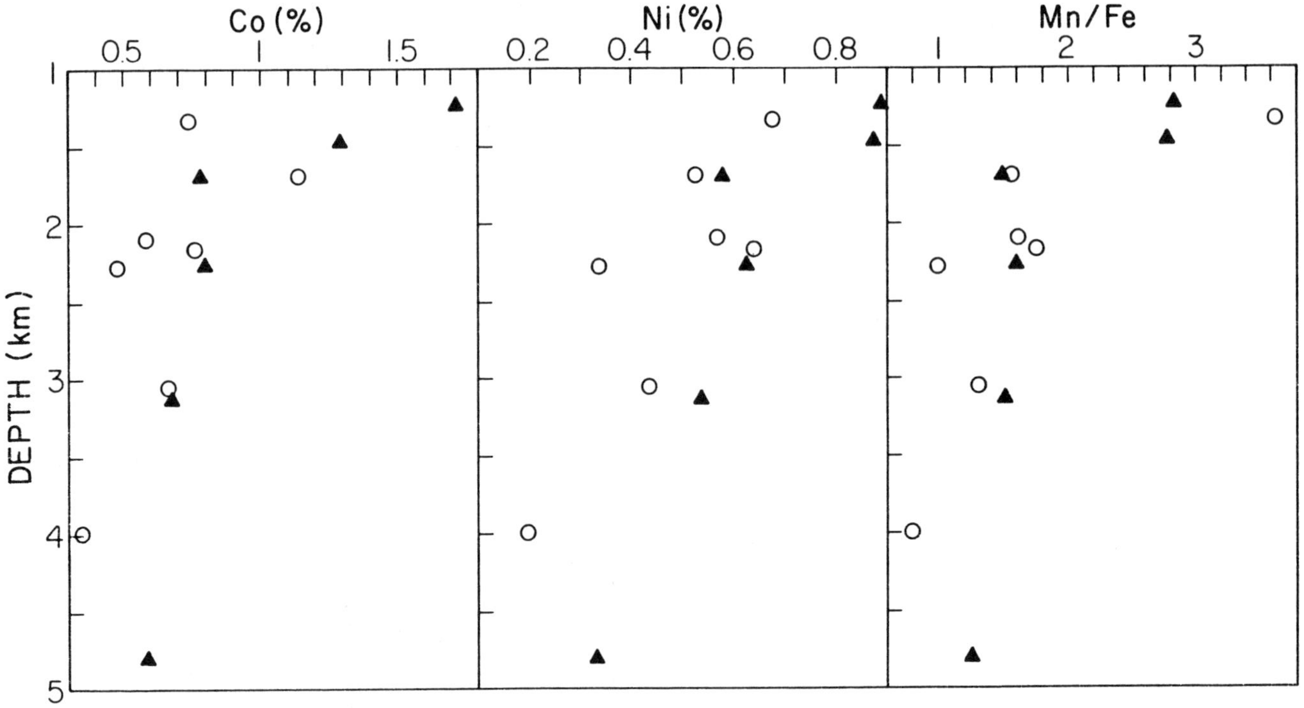

Figure 7. Dredge average (left) Co, (center) Ni, and (right) Mn/Fe vs depth. Open symbols are averages from cruise MW86-02, closed symbols from cruise MW87-02.

In order to minimize the scatter in the depth relations we use dredge average concentrations of each element to evaluate these relationships. Observation of Figure 7 reveals the utility of this approach. Significant trends between the economically interesting metals and depth are now more evident. The criterion that high metal values will occur in water depths between 1000 and 1600 m suggested by Hein et al. (1985a) is supported by our data although we further suggest potential economic deposits may be found as deep as 2000 m. The inverse relation between Mn/Fe and depth in Table 2 correlates with the abundance of Co and Ni in crusts and reflects the dilutionary effect of Fe. It must be pointed out that since most samples in this study are enriched in vernadite, variations in minor element concentrations are not principally governed by the relative abundances of todorokite and vernadite but rather by the detrital Fe content of crusts (Halbach and Puteanus, 1984; De Carlo et al., 1987a).

Rare earth element concentrations

Rare earth element (REE) concentrations were determined on a subset of crusts (26) from this study. The data were normalized to the North American Shale Composite for comparison with the abundances in other materials from the marine environment. Figure 8 summarizes the results in a comparison of shale-normalized plots of crusts from the Hawaiian Archipelago (De Carlo and McMurtry, 1990), apatite isolated from a substrate (this work) and seawater (De Baar et al., 1985). The results indicate that crusts from

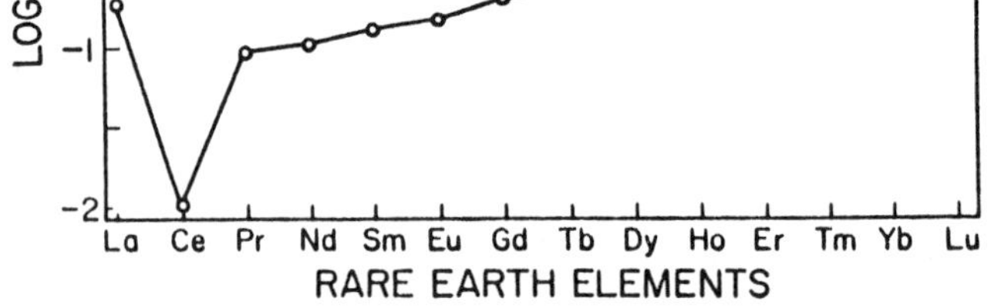

Figure 8. Rare earth elements in the marine environment: (from top to bottom) crusts from the Hawaiian Archipelago (De Carlo et al., 1987a), crusts from this work, apatite isolated from a substrate, and seawater ($\times 10^6$), DeBaar et al. (1985).

Table 3. Comparison of Pacific seamount ferromanganese crust compositions.

	MW-86-02[1]	MW87-02[2]	Hawaiian EEZ[3]	Line Islands[4]	Central Pacific[5]	Marshall Islands[6]	Pacific Ocean average[7]
No. of samples	50	64	32	59	26 to 46	12	251 to 803
Element (wt %)							
Mn	25.69	23.17	23.30	20.40	24.60	20.40	23.1
Fe	14.73	15.08	15.60	17.00	14.50	12.30	16.1
Ti	1.08	1.00	0.95	1.20	0.96	0.77	--
Co	0.945	0.736	0.900	0.550	0.790	0.843	0.73
Cu	0.095	0.098	0.060	0.154	0.065	0.038	0.16
Ni	0.650	0.557	0.440	0.390	0.490	0.388	0.47
Mn/Fe	1.74	1.54	1.49	1.20	1.70	1.66	1.43
Average depth (m)	2189	2126	1546	---	2179	---	---

[1]De Carlo et al., 1987b.
[2]De Carlo and Fraley, this work.
[3]De Carlo et al., 1987a.
[4]Aplin and Cronan, 1985a.
[5]Halbach and Manheim, 1984.
[6]Hein et al., 1988.
[7]Manheim, 1986.

the study area contain slightly lower concentrations of REE than in the Hawaiian Archipelago but display the large positive Ce anomaly typical of hydrogenous Fe-Mn deposits (De Carlo and McMurtry, 1989, and references therein). Apatite is considerably less enriched in REE than crusts with near-shale concentrations, although it displays a negative Ce anomaly and REE fractionation which parallels that of seawater.

CONCLUSIONS

Results of exploratory cruises to the study area have demonstrated that the island nations of Kiribati and Tuvalu have potential Fe-Mn resources on the slopes of seamounts in the Ellice, Gilbert, Phoenix, and Line islands. Limited surveys of Cook Island waters on the Manihiki Plateau have not yielded evidence of extended mineral deposits. Crusts were recovered from most areas dredged including seamounts, submerged ridges, slopes of islands and atolls, and escarpments on the Manihiki plateau. Slopes of islands and atolls consist of carbonates which are rarely covered with thick Fe-Mn crust accumulations. Escarpments on the Manihiki are generally not within the optimum depth range for high metal values and mass wasting may prevent significant accumulations of Fe-Mn crusts.

Concentrations of the economically more valuable metals Co and Ni reach 1.92% and 1.26% in the most enriched crusts, and average 0.83% and 0.58%, respectively, for all samples analyzed in this study. Average metal values are slightly lower than observed in the Hawaiian Archipelago (De Carlo et al., 1987a) reflecting the greater average depth of recovery in this study. Ore grades compare well with the Mid-Pacific Mountains (Halbach and Manheim, 1984) and the Marshall Islands (Hein et al., 1988). The highest grades are found in the Phoenix and Ellice island areas (sampled during the MW86-02 cruise) whereas crusts from the Manihiki Plateau and the central Line Islands (MW87-02 cruise) are generally not as enriched in Co and Ni (Table 3). Deposits of potentially higher tonnage per square meter but lower grade exist in deeper waters (~2000 m) of the Phoenix Islands and on gently sloped ridges of the central Line Islands.

The optimum range of water depth for crusts of high metal values is between 1000 and 1600 m where Co and Ni contents are typically 1.5% and 0.9% respectively, but the potential high total metal content of thick and abundant lower grade deposits near 2000 m water depth may yield a greater resource than found in areas displaying high Co and Ni concentrations in thin crusts. Abundant Fe-Mn encrusted cobbles and nodules appear in the central Line Islands. Although easier to collect than firmly attached crusts on volcanic outcrops they are often relatively thin (1-3 cm).

Phosphatic material is ubiquitous in substrates recovered from water less than 2500 m depth, and may represent an additional mineral which could be recovered as a byproduct of future Fe-Mn crust mining operations.

Manganese mineralogy is dominated by vernadite (δ-MnO_2) but a few chemically distinct samples also contain significant amounts of todorokite. No crystalline Fe minerals were identified in crusts. Phosphatic material is predominantly composed of carbonate fluorapatite occurring as massive deposits or as cements in highly brecciated volcanic material. Chemical and mineralogical characteristics place these deposits within the hydrogenetic field of Fe-Mn deposits.

ACKNOWLEDGMENTS

We are grateful to the captain and crew of the R/V *Moana Wave* for their professionalism and kind assistance at sea. We express our appreciation to the co-chief scientists of the two cruises: B. Bolton and B. Keating (MW86-02), and W. Coulbourn and P. Hill (MW87-02) as well as to the other shipboard colleagues. We gratefully acknowledge the comments of two anonymous reviewers which helped improve this manuscript. The technical assistance of J. Lin, A. Murakami, P. Pennywell and G. Taogoshi contributed to the success of this project. Particular thanks are due to W. Coulbourn for numerous refreshing morale-boosting episodes during marathon SeaMARC II and dredging sessions during the MW87-02. One of us (EDC) gratefully acknowledges the research support provided through an institutional grant to the Hawaii Institute of Geophysics by the U.S. AID. This is Hawaii Institute of Geophysics contribution No. 2223.

REFERENCES

Alvarez, R., E.H. De Carlo, J.P. Cowen, and G. Andermann, 1990, Micromorphological characteristics of a marine ferromanganese crust: Marine Geology, v. 94, p. 239-249.

Aplin, A.C. and D.S. Cronan, 1985a, Ferromanganese oxide deposits from Central Pacific Ocean. I: Encrustations from the Line Island Archipelago: Geochimica et Cosmochimica Acta, v. 49, p. 427-436.

Aplin, A.C., and D.S. Cronan, 1985b, Ferromanganese oxide deposits from the Central Pacific Ocean II: Nodules and associated sediments: Geochimica et Cosmochimica Acta, v. 49, p. 437-451.

Bernas, B., 1968, A method for decomposition and comprehensive analysis of silicates by atomic absorption spectrophotometry: Analytical Chemistry, v. 40, p. 479-482.

Bolton, B., J. Bogi, and D.S. Cronan, Geochemistry and mineralogy of ferromanganese nodules from the Kiribati region of the eastern Central Pacific, this volume.

Burns, R.G. and V.M. Burns, 1977, Mineralogy, *in* G. P. Glasby (ed.), Marine Manganese Deposits: Elsevier Oceanography Series, 15, Elsevier, Amsterdam, p. 185-248.

Commeau, R.F., A. Clarke, C. Johnson, F.T. Manheim, P.J. Muscavage, and M.C. Lane, 1984, Ferromanganese crust resources in the Pacific and Atlantic Oceans: Proceedings of Oceans, IEEE, Washington, DC, p. 421-430.

Craig, J.D., J.E. Andrews, and M.A. Meylan, 1982, Ferromanganese deposits in the Hawaiian Archipelago: Marine Geology, v. 45, p. 127-157.

Cronan, D.S., 1983, Metalliferous Sediments in the CCOP/SOPAC region of the Southwest Pacific, with particular reference to geochemical exploration for the deposits. CCOP/SOPAC Technical Bulletin No. 4, Technical Secretariat of CCOP/SOPAC, Mineral Resources Department, Private Bag, Suva, Fiji, 55 pp.

Cronan, D.S., 1984, Criteria for the recognition of potentially economic Mn nodules and encrustations in the CCOP/SOPAC region: South Pacific Marine Geological Notes, 3, 17 pp.

de Baar, H.J.W., M.P. Bacon, P.G. Brewer and K.W. Bruland, 1985, Rare earth elements in the Pacific and Atlantic Oceans: Geochimica et Cosmochimica Acta, v. 49, p. 1943-1959.

De Carlo, E.H. and McMurtry, G.M., Rare earth element geochemistry of ferromanganese deposits from the Hawaiian Archipelago, Chem. Geol., in press.

De Carlo, E.H., H. Zeitlin and Q. Fernando, 1982, Separation of Cu, Co, Ni, and Mn from deep-sea ferromanganese nodules by adsorptive colloid flotation: Analytical Chemistry, v. 54, p. 898-902.

De Carlo, E.H., G.M. McMurtry and K.H. Kim, 1987a, Geochemistry of ferromanganese crusts from the Hawaiian Archipelago-I. Northern survey areas: Deep-Sea Research, v. 34, p. 441-467.

De Carlo, E.H., P.A. Pennywell and C.M. Fraley, 1987b, Geochemistry of Ferromanganese Deposits from the Kiribati and Tuvalu Region of the West Central Pacific Ocean: Marine Mining, v. 6, p. 301-321.

Exon, N.F., 1982, Manganese nodules in the Kiribati Region, Equatorial Western Pacific: South Pacific Marine Geological Notes, v. 2, p. 77-102.

Frank, D.J., M.A. Meylan, J.D. Craig and G.P. Glasby, 1976, Ferromanganese deposits of the Hawaiian Archipelago, Hawaii Institute of Geophysics Technical Report no. HIG 76-14, University of Hawaii, Honolulu, 71 pp.

German, C.R. and H. Elderfield, 1988, REE in the NW Indian Ocean: Biogeochemical cycling in a sub-oxic marine environment [abs.]: Eos, Transactions of the American Geophysical Union, v. 68, p. 1755.

Glasby, G.P., editor, 1977, Marine Manganese Deposits: Elsevier Oceanography Series, 15, Elsevier, Amsterdam, 523 pp.

Halbach, P. and Ozkara, M., 1979, Morphological and geochemical classification of deep-sea ferromanganese nodules and its genetical interpretation, *in* C. Lalou (Editor, La Genese des Nodules de Manganese, Colloques Internationaux du C.N.R.S. No. 289, C.N.R.S., Paris, p. 77-88.

Halbach, P. and F.T. Manheim, 1984, Potential of cobalt and other metals in ferro-manganese crusts on seamounts of the Central Pacific Basin: Marine Mining, v. 4, p. 319-336.

Halbach, P. and D. Puteanus, 1984, Influence of the carbonate dissolution rate on the growth and composition of Co-rich ferromanganese crusts from Central Pacific seamount areas: Earth and Planetary Science Letters, v. 68, p. 73-87.

Halbach, P., F.T. Manheim and P. Otten, 1982, Co-rich ferromanganese deposits in the central seamount regions of the Central Pacific Basin-results of the Midpac '81: Erzmetall, v. 35, p. 447-453.

Halbach, P., M. Segl, D. Puteanus and A. Mangini, 1983, Co-fluxes and growth rates in ferromanganese deposits from Central Pacific seamount areas: Nature, v. 304, p. 716-719.

Hamilton, E.L., 1956, Sunken Islands of the Mid-Pacific Mountains: Geological Society of America Memoir 64, 97 pp.

Hein, J.R., F.T. Manheim, W.C. Schwab and A.S. Davis, 1985a, Ferromanganese from Necker Ridge, Horizon Guyot and S. P. Lee Guyot: Geological considerations: Marine Geology, v. 69, p. 25-54.

Hein, J.R., F.T. Manheim, W.C. Schwab, A.S. Davis, C.L. Daniel, R.M. Bouse, L.A. Morgenson, R.E. Sliney, D. Clague, G.B. Tate and D.A. Cacchione, 1985b, Geological and geochemical data for seamounts and associated ferromanganese crusts in and near the Hawaiian, Johnston Island, and Palmyra Island Exclusive Economic Zones: U.S. Geological Survey, Open-File Report, 85-292, 129 pp.

Hein, J., W.C. Schwab and A.S. Davis, 1988, Co and Pt-rich Ferromanganese Crusts and associated substrate rocks from the Marshall Islands: Marine Geology, v. 78, p. 255-283.
Hein, J.R., M.S. Morrison and L.A. Gein, Central Pacific Co-rich ferromanganese crusts: Historical perspective and regional variability, this volume.
Johnson, C.J., A.L. Clark, J.M. Otto, D.K. Pak, K.T.M.T.M. Johnson, and C.L. Morgan, 1985, Resource Assessment of Cobalt-Rich Ferromanganese Crusts in the Hawaiian Archipelago. Final report to Department of Interior, Minerals Management Service, Cooperative Agreement No. 14-13-0001-30177, 135 pp.
Keating, B., D.P. Mattey, C.E. Helsley, J.J. Naughton, D. Epp, A. Lazerewicz and D. Schwank, 1984, Evidence for a hot spot origin of the Caroline Islands: Journal of Geophysical Research, v. 89, p. 9937-9948.
Kingston, H.M. and L.B. Jassie, 1986, Microwave energy for acid decomposition at elevated temperatures and pressures using biological and botanical samples: Analytical Chemistry, v. 58, p. 2534- 2541.
Larson, R. L., S. M. Smith and R.G. Chase, 1974, Magnetic lineations of early Cretaceous age in the Western Equatorial Pacific Ocean: Earth and Planetary Science Letters, v. 15, p. 315-319.
Manheim, F.T., 1986, Marine cobalt resources: Science, v. 232, p. 600-608.
Martin, J.H., and G.M. Knauer, 1984, VERTEX: Manganese transport through oxygen minima: Earth and Planetary Science Letters, v. 67, p. 35-47.
Meylan, M.A., G.P. Glasby and L. Fortin, 1981, Bibliography and index to literature on manganese nodules, 1861-1979: State of Hawaii, Dept. of Planning and Economic Development, Honolulu, 530 pp.
Moore, J. G., 1966, Rate of palagonitization of submarine basalt adjacent to Hawaii: U.S. Geological Survey Professional Paper, 550D, 163.
Nadkarni, R.A., 1984, Applications of microwave oven sample dissolution in analysis: Analytical Chemistry, v. 56, p. 2233-2237.
Pichoki, C. and M. Hoffert, 1987, Characteristics of Co-rich ferromanganese nodules and crusts sampled in French Polynesia: Marine Geology, v. 77, p. 109-119.
Rao, V.P, and W.C. Burnett, Phosphatic rocks from seamounts in the EEZ of Kiribati and Tuvalu, Central Pacific Ocean, this volume.
Robbins, J.M., Lyle, M. and Heath, G.R., 1984, A sequential extraction procedure for partitioning elements among co-existing phases in marine sediments: Oregon State University, College of Oceanography Technical Report 84-3, 54 pp.
Saager, P.M., H.J.W. de Baar and H. Elderfield, 1988, Manganese in the Indian Ocean [abs.]: Eos, Transactions of the American Geophysical Union, v. 68, p. 1754.
Usui, A., and Moritani, T., Manganese nodule deposits in the central Pacific Basin: Distribution, geochemistry, mineralogy, and genesis, this volume.

APPENDIX I

MINERALOGY OF CRUSTS AND SUBSTRATES*

Sample	Crust	Substrate
RD16-S1	V/P (B?)	CFA/C, Ct
RD16-S2A	V/C, P (T?	Q, S/C (P,V)
RD16-S2B	V/P, B (C)	S/P, C, Mj (V)
RD16-S3	V/C, P	C/P (CFA, S, V)
RD16-S5A	V/C, P, Q	S/P, C (W)
RD16-S6	V/C, P	CFA/C (Ct)
RD16-S8	V/C, P (B?)	C/V (P, CFA)
RD16-S9	V/C, P (B?)	CFA/P, C (Ct, B?)
RD16-S10	V/C, P, Q	Q, S/C, Ct (CFA)
RD16-S11	V/C, P, (B, T)	An, S/Ct, C (B, K?)
RD16-S12	V/C, P	V/C (P)
RD17-S1	V/A (B, T)	S/P (C)
PD01-S1	V/C, P	C, CFA (V)
PD01-S2	V/C (B7)	C (CFA)
PD01-S3	V/C, CFA	C (CFA)
PD01-S4	V/C	0
PD01-S6	V/C, CFA	CFA/C
PD01-S7		C, CFA
PD01-S10	V/C (B?)	CFA (C)
PD01-S12	V/C, P, CFA	C/CFA (P)
RD19-S1	V/CFA C	CFA/So, P, C (B, Ct, An?)
RD19-S2	V/CFA, C, P, (B?, T)	CFA/P (C, K, B, Ct)
RD19-S3	CFA/C, P, T	CFA/P (C, B, K, Ct)
RD22-S1A	V/C, Q, CFA, P (T, B?)	P, C/Ct, Q (CFA)
RD23-S1	V/P, Q (T?, B?, C)	0
RD23-S2	V/P, CFA (B?, C, T?)	C/CFA (B)
RD23-S3	V, CFA/C, P, (B?)	CFA/C, P (S)
RD23-S4	P, CFA, V (B?, C)	D, P/B, K, Ct, S (CFA)
RD23-SA	V/T, CFA (B?)	CFA, P/S (V)
RD23-SF	V, T/CFA, C, P (B?)	P/CFA, L, K (B, Ct, C)
RD24-S	V/C, CFA	0
RD24-Frag	P, B, CFA, C	0
RD24-S1	V (B?)	/P, S (B, Ct, V)
RD24-S2	V/CFA (B?, C	CFA/B, K, An (C)
RD24-S4	V/CFA	0
RD24-S5	V/C, CFA, P (B?)	CFA, P/S, Ct, B, K (C, Q, An)
RD24-S6	V/CFA, C, P	P (B, C?)
RD24-S7	V/B, P	P/CFA, B, K, Ct (C)
RD24-S8	V/CFA, P	P/S, Ch

(table continued)

Sample	Crust	Substrate
	(table continued)	
RD24-S9	V/C, Q (B? , CFA)	CFA, P/Ct, B, K (C, An)
RD24-S10	V/B, CFA, C	S/P (B)
RD24-S11	V/CFA, P, Q, T (B?)	CFA/C, P (Ct, B, K)
RD24-S12	V/B, C, P	P/CFA, S, B, (Ct, C)
RD24-S14	V/CFA, P, C, Q (B? T)	CFA/C, P (B, Ct)
RD24-S13	V/T, P	0
RD24-S15	V/P, CFA	0
RD24-S15G	V/C	0
RD24-S16	V/CFA, P	CFA, P/S, B, K, C (Ct)
RD24-S17	V/CFA, P, C, T	CFA/S, P, C (B, K, Ct)
RD24-S18	V/CFA, C, P (B?)	P/CFA, B (S, C, Ct? , An?)
RD24-S19	V/B, CFA, P	CFA/C (Ct)
RD24-S20	V/CFA, B, C, P	CFA/C, P (Ct, B, Bo, S?)
RD24-S21	V/C, B, P, CFA	0
RD24-S22	V/C, P, CFA (B?)	CFA/C, P (Ct)
RD24-S23	V/Q, CFA?	V/Q (Ch?)
RD24-S24	V/P, CFA (B?, T)	P/CFA, S, K, Ct (C)
RD24-S25	V/Q, P, C, CFA, T (B?)	P/CFA, Ct, K, S (C)
RD24-S26	V/C, P, CFA	CFA/P (B, Ct, C?)
RD24-S27	V/T, B, C, Q, P	P/Ct, CFA, V, B, K
RD24-SNSI	V/T, P, C, CFA	CFA/C, P (Ct, B, K)
RD24-SB	V/CFA, Q, T? , C	CFA/C (B)
RD24-SC	V/CFA, P, C, T (B?)	0
RD24-SG	V/CFA, P, C	CFA/C (P, Ct, B, K)
RD24-SD	V/CFA, Q, B, T, C, P	CFA
RD24-SE	V/CFA, Q, B, P, T	CFA/P, C, De (Ct, B, K)
RD24-SF	V/T, CFA, P	CFA (C, P)
4MD01-01	V	0
4MD01-02	V (CFA)	0
4RD07-01	V	0
4RD07-02	V (CFA, C)	CFA, C
4RD07-03	V/CFA (C)	CFA, C (S)
4RD07-04	V (CFA, C)	CFA, C (S)
4RD07-05	V (CFA, C, P)	C, CFA
4RD07-07	V (CFA, C)	CFA, P (S, Mag)
4RD07-08	V (CFA, C)	C, CFA (S)
4RD07-09	V (CFA, C)	CFA, C
4RD07-10	V (CFA, C)	CFA, C (Mag, S, V)
4RD07-11	V/CFA (P, S)	CFA, C (S)
4RD07-12	V (C, P, S)	CFA, C
4RD07-13	V (P)	P, C
4RD07-14	V (P)	0
4RD07-15	V (CFA, C)	CFA, C
4RD07-16	V/CFA	CFA, P (S)
4RD07-17	V	P, C, CFA
4RD07-18	V	0

(table continued)

Sample	Crust	Substrate
		(table continued)
6RD08-01	V	P, CFA, (S, Il, Mag)
6RD08-03	V	CFA, C (Mag, S)
6RD08-04	V	P, CFA, Mag, V (S)
6RD08-05	V	CFA, P (C, Mag)
6RD08-06	V	P, C (I)
6RD08-07	V	CFA (C, P, Mag)
6RD08-08	V CFA	CFA (C, Mag)
6RD09-1A	V (Q)	S, CFA, P (V, Mag)
6RD09-02	Q (S)	S, P, Mag (V, I)
6RD09-03	V	0
6RD09-04	V (CFA)	0
6RD09-05	V (Q)	S, Mag
6RD09-06	V (Q)	S, Mag
7RD11-X	V	CFA, C
7RD11-01	V (CFA, S)	CFA, C (V)
7RD11-02	V	CFA, C
7RD11-03	T, V/CFA (C)	C, CFA
7RD11-04	V (CFA)	CFA, C
7RD11-05	V	CFA, C
7RD11-06	V	CFA, C
7RD11-07	V (CFA)	CFA, C
7RD11-08	V/CFA (S)	CFA, C
7RD11-10	V (CFA)	CFA, C
7RD11-11	V	CFA (C, Mag)
7RD11-12	V (CFA, S)	CFA, C
7RD11-13	V/CFA (C, S)	CFA, C
7RD11-15	V (CFA)	CFA, C
7RD12-1A	V	0
7RD13-1B	V	0

*Order of abundance: Major/Minor (trace)

1) Mn minerals

B = Birnessite
Ch = Chalcophanite
Mj = Manjiroite
T = Todorokite
V = Vernadite

2) Fe Ti minerals

Mag = Maghemite/Magnetite
Il = Ilmenite

3) Ca Mg minerals

Bo = Bobierrite
C = Calcite
CFA = Carbonate-fluorapatite

4) Detrital minerals

An = Analcime
Ct = Cristobalite
De = Delafossite
D = Diopside
K = Kaoliite
L = Lizardite
P = Plagioclase
S = Sodalite
Sm = Smectite
W = Wairakite
0 = no substrate recovered

APPENDIX II. Bulk Elemental Analysis of Crusts and Seamount Nodules

Sample	Depth (m)	Mn/Fe	Mn (%)	Fe (%)	Co (%)	Ni (%)	Cu (%)	Zn (ppm)	Si (%)	Al (%)	Ti (%)	V (%)	P (%)	Ca (%)	Mg (%)	Ba (%)	K (%)
RD16-S1	2275	0.952	15.72	16.51	0.421	0.287	0.156	556	5.75	1.65	0.982	0.055	0.66	2.41	0.89	0.115	0.519
RD16-S2A	2275	0.925	18.18	19.66	0.494	0.332	0.116	694	5.60	1.39	1.072	0.063	0.41	2.88	0.94	0.134	0.592
RD16-S2B	2275	0.936	17.60	18.81	0.424	0.285	0.112	655	6.38	1.72	0.912	0.057	0.38	2.48	1.04	0.130	0.771
RD16-S3	2275	0.970	20.15	20.78	0.487	0.343	0.118	728	5.08	1.02	0.985	0.061	0.39	2.56	1.05	0.122	0.541
RD16-S5A	2275	1.013	17.83	17.60	0.415	0.320	0.101	655	5.96	1.46	0.879	0.056	0.41	3.02	1.00	0.128	0.607
RD16-S6	2275	1.034	19.58	18.93	0.493	0.394	0.167	722	5.30	1.43	0.959	0.059	0.42	2.84	1.05	0.142	0.574
RD16-S8	2275	1.145	23.13	20.20	0.671	0.472	0.129	726	4.50	0.92	0.993	0.068	0.46	3.29	1.19	0.135	0.355
RD16-S9	2275	0.983	15.95	16.23	0.395	0.274	0.111	575	4.19	1.01	0.805	0.049	0.43	2.22	0.83	0.119	0.462
RD16-S10	2275	0.980	19.41	19.81	0.529	0.349	0.118	718	5.42	1.19	0.969	0.066	0.41	2.93	1.00	0.141	0.653
RD16-S11	2275	0.962	15.91	16.53	0.433	0.301	0.155	547	5.09	1.34	0.854	0.050	0.37	2.51	0.92	0.116	0.519
RD16-S12	2275	0.993	19.14	19.27	0.499	0.342	0.156	664	5.13	1.30	1.000	0.062	0.40	2.62	1.00	0.138	0.577
RD17-S1	3990	0.791	14.45	18.27	0.347	0.194	0.091	550	8.70	2.48	1.115	0.052	0.35	1.93	2.05	0.104	0.862
PD01-S1	1688	1.616	25.00	15.47	1.443	0.476	0.080	762	2.73	0.51	1.426	0.060	0.40	2.66	1.10	0.128	0.546
PD01-S2	1688	1.849	26.63	14.40	1.533	0.635	0.073	799	1.90	0.39	1.366	0.056	0.34	3.05	1.29	0.146	0.573
PD01-S3	1688	1.311	17.78	13.56	0.842	0.403	0.076	579	2.79	0.56	0.957	0.047	0.50	3.89	0.94	0.102	0.431
PD01-S4	1688	1.487	24.09	16.20	0.958	0.502	0.083	794	2.90	0.56	1.242	0.064	0.41	2.68	1.13	0.138	0.524
PD01-S6	1688	1.609	25.14	15.62	1.112	0.546	0.094	802	2.77	0.53	1.320	0.063	0.60	3.22	1.09	0.163	0.529
PD01-S10	1688	1.480	25.36	17.13	0.854	0.559	0.092	727	2.86	0.69	1.124	0.067	0.43	2.84	1.18	0.154	0.530
PD01-S12	1688	1.544	23.94	15.51	1.233	0.554	0.072	710	2.17	0.48	1.379	0.059	0.62	4.78	1.20	0.143	0.506
RD19-S1	1330	3.562	27.82	7.81	0.950	0.744	0.052	1056	1.08	0.10	0.677	0.065	2.62	8.47	1.20	0.170	0.590
RD19-S2	1330	3.445	26.01	7.55	0.842	0.687	0.055	816	1.05	0.15	0.626	0.059	3.34	10.06	1.05	0.312	0.673
RD19-S3	1330	3.900	23.32	5.98	0.425	0.598	0.040	1058	1.06	0.15	0.379	0.057	4.75	12.22	0.98	0.390	0.621
RD22-S1	3050	1.274	23.14	18.17	0.662	0.425	0.094	770	6.43	0.68	1.156	0.067	0.40	2.37	1.02	0.151	0.435
RD22-S1A	3050	1.320	23.28	17.63	0.665	0.444	0.095	761	4.53	0.60	1.134	0.065	0.37	2.34	1.05	0.164	0.584

(table continued)

Sample	Depth (m)	Mn/Fe	Mn (%)	Fe (%)	Co (%)	Ni (%)	Cu (%)	Zn (ppm)	Si (%)	Al (%)	Ti (%)	V (%)	P (%)	Ca (%)	Mg (%)	Ba (%)	K (%)
(table continued)																	
RD23-S1	2095	1.459	18.28	12.53	0.697	0.455	0.079	763	5.56	0.88	1.136	0.051	0.26	1.84	0.95	0.127	0.812
RD23-S2	2095	1.499	22.59	15.07	0.659	0.601	0.065	708	4.03	0.82	0.988	0.062	0.60	2.86	1.11	0.136	0.645
RD23-S3	2095	1.560	18.20	11.67	0.470	0.377	0.096	655	4.15	1.09	0.880	0.052	3.96	10.14	0.84	0.199	0.631
RD23-S4	2095	1.023	11.60	11.34	0.327	0.335	0.066	557	12.23	3.57	1.197	0.036	1.36	4.72	1.53	0.092	1.697
RD23-SA	2095	1.867	25.28	13.54	0.799	0.595	0.080	815	2.65	0.34	1.044	0.067	0.36	2.43	0.99	0.173	0.561
RD23-SF	2095	2.214	21.76	9.83	0.551	1.141	0.211	1146	4.54	1.46	0.938	0.041	1.29	4.35	1.58	0.206	0.710
RD24-S1	2163	1.700	27.84	16.38	0.830	0.702	0.088	902	3.30	0.55	1.078	0.064	0.41	2.89	1.19	0.195	0.909
RD24-S2	2163	1.700	28.01	16.48	0.976	0.679	0.111	875	2.84	0.59	1.075	0.069	0.41	2.76	1.26	0.212	0.669
RD24-S4	2163	2.241	30.19	13.47	0.877	0.945	0.120	815	2.65	0.35	1.044	0.067	0.36	2.50	1.35	0.192	0.056
RD24-S5	2163	1.825	25.91	14.20	0.790	0.704	0.095	856	3.96	0.66	0.990	0.060	0.50	2.73	1.15	0.161	0.645
RD24-S6	2163	1.563	24.55	15.71	0.675	0.470	0.059	708	3.27	0.58	0.825	0.058	1.15	4.08	1.04	0.118	0.523
RD24-S7	2163	2.300	30.40	13.22	0.885	0.974	0.116	976	2.31	0.43	0.925	0.059	0.31	2.51	1.39	0.186	0.670
RD24-S8	2163	2.015	23.44	11.63	0.721	0.645	0.096	804	2.36	0.39	0.824	0.052	0.25	2.07	1.03	0.159	0.511
RD24-S9	2163	1.805	26.15	14.49	0.784	0.645	0.080	888	3.25	0.53	1.026	0.062	0.42	2.73	1.19	0.150	0.570
RD24-S10	2163	2.282	30.15	13.21	0.889	0.925	0.120	1065	3.64	0.51	0.962	0.067	0.31	2.50	1.41	0.141	0.694
RD24-S11	2163	1.559	19.75	12.67	0.546	0.370	0.034	561	2.60	0.27	0.658	0.058	0.53	2.38	0.81	0.102	0.381
RD24-S12	2163	2.146	29.57	13.78	0.930	0.901	0.105	952	2.23	0.41	0.926	0.060	0.30	2.66	1.39	0.181	0.668
RD24-S13	2163	1.954	28.34	14.50	0.849	0.815	0.188	1037	2.51	0.48	1.206	0.065	0.30	2.45	1.31	0.204	0.623
RD24-S14A	2163	1.305	19.88	15.23	0.553	0.359	0.083	610	3.81	0.51	0.915	0.057	0.31	1.90	0.84	0.137	0.472
RD24-S14B	2163	1.260	18.20	14.44	0.533	0.447	0.090	752	5.74	1.48	1.061	0.050	1.89	6.20	1.05	0.139	0.960
RD24-S15	2163	2.166	29.00	13.39	0.901	0.813	0.103	889	1.50	0.42	0.989	0.058	0.38	2.77	1.28	0.178	0.604
RD24-S16	2163	1.787	27.21	15.23	0.385	0.719	0.111	944	3.38	0.90	1.158	0.063	0.59	2.99	1.20	0.191	0.737
RD24-S17	2163	1.418	18.56	13.09	0.617	0.419	0.030	629	3.56	0.46	0.727	0.050	0.38	2.07	0.90	0.084	0.518
RD24-S17A	2163	1.441	23.90	16.59	0.787	0.522	0.043	776	5.01	0.58	0.928	0.063	0.46	2.69	1.12	0.110	0.533
RD24-S18	2163	2.201	28.06	12.75	1.040	0.870	0.895	835	3.36	0.51	0.881	0.056	0.38	2.72	1.30	0.145	0.651
RD24-S19	2163	2.145	27.87	12.99	0.759	0.697	0.130	945	2.98	0.50	0.885	0.069	0.86	3.68	1.20	0.215	0.640
RD24-S20	2163	2.022	27.13	13.42	0.769	0.640	0.118	896	2.28	0.42	0.938	0.069	0.63	3.29	1.14	0.199	0.589

(table continued)

Sample	Depth (m)	Mn/Fe	Mn (%)	Fe (%)	Co (%)	Ni (%)	Cu (%)	Zn (ppm)	Si (%)	Al (%)	Ti (%)	V (%)	P (%)	Ca (%)	Mg (%)	Ba (%)	K (%)
RD24-S21C	2163	1.382	24.41	17.66	0.711	0.546	0.073	794	5.31	0.79	1.145	0.059	0.37	2.42	1.12	0.145	0.645
RD24-S22	2163	1.529	21.80	14.26	0.699	0.551	0.080	693	2.71	0.60	0.903	0.048	0.47	2.59	1.00	0.143	0.573
RD24-S23	2163	1.487	23.53	15.82	0.718	0.491	0.087	709	3.69	0.57	1.082	0.065	0.37	2.19	1.02	0.212	0.610
RD24-S24	2163	1.832	28.52	15.57	0.905	0.677	0.074	839	2.80	0.38	0.892	0.061	0.32	2.42	1.15	0.151	0.573
RD24-S25	2163	1.277	23.65	18.52	0.791	0.559	0.092	817	3.82	0.84	1.090	0.052	0.36	2.40	1.09	0.182	0.649
RD24-S26	2163	1.628	25.03	15.37	0.853	0.656	0.092	867	4.66	0.91	1.049	0.057	0.35	2.42	1.12	0.159	0.725
RD24-S27	2163	1.591	25.10	15.78	0.794	0.664	0.087	855	3.08	0.59	0.970	0.059	0.36	2.17	1.14	0.149	0.570
RD24-NS1	2163	1.245	17.03	13.68	0.752	0.338	0.077	766	4.48	0.77	1.017	0.057	0.67	3.08	0.86	0.095	0.564
RD24-SB	2163	1.605	25.23	15.72	0.676	0.497	0.112	807	3.03	0.41	1.043	0.066	0.73	3.49	0.98	0.191	0.548
RD24-SD	2163	1.550	24.17	15.59	0.594	0.520	0.101	762	3.16	0.56	1.029	0.065	0.50	2.73	0.98	0.198	0.614
RD24-SF	2163	1.936	27.10	14.00	0.824	0.711	0.104	949	3.51	0.51	1.074	0.067	0.34	2.49	1.21	0.171	0.626
RD24-SG	2163	1.802	25.81	14.32	0.742	0.658	0.121	905	3.16	0.70	0.965	0.066	0.37	2.37	1.20	0.191	0.628
RD24-FRAG	2163	1.648	25.35	15.38	0.708	0.600	0.132	905	3.88	0.83	1.233	0.068	0.49	2.74	1.06	0.200	0.778
4MD-01-S01	1230	2.834	31.97	11.28	1.921	0.885	0.067	987	0.98	0.26	1.140	0.072	0.41	2.54	1.56	0.182	
4MD-01-S02	1230	2.832	32.96	11.64	1.518	0.892	0.063	893	0.88	0.22	0.964	0.081	0.57	3.34	1.38	0.186	
4RD-07-01	1690	1.303	22.68	17.41	0.709	0.517	0.069	687	3.21	0.94	1.110	0.074	0.53	2.53	1.12	0.159	
4RD-07-02	1690	1.263	22.03	17.44	0.683	0.525	0.100	750	3.77	1.08	1.200	0.077	0.76	3.14	1.17	0.174	
4RD-07-03	1690	1.590	24.60	15.47	0.856	0.543	0.126	837	2.08	0.56	1.350	0.082	1.35	5.22	1.12	0.207	
4RD-07-04	1690	1.589	25.59	16.10	0.857	0.558	0.100	842	2.20	0.52	1.270	0.088	0.77	3.47	1.18	0.202	
4RD-07-05	1690	1.289	21.62	16.77	0.624	0.523	0.109	740	3.88	1.28	1.410	0.076	0.64	2.75	1.22	0.191	
4RD-07-07	1690	1.516	24.22	15.98	0.780	0.675	0.130	901	3.10	0.99	1.160	0.085	0.97	3.60	1.26	0.192	
4RD-07-09	1690	1.585	23.93	15.10	0.854	0.571	0.107	717	2.05	0.58	1.260	0.074	0.75	3.16	1.13	0.177	
4RD-07-10	1690	1.485	24.83	16.72	0.812	0.547	0.121	793	2.72	0.77	1.300	0.081	0.78	3.38	1.16	0.202	
4RD-07-11	1690	1.349	21.63	16.03	0.579	0.632	0.115	715	3.50	1.12	1.040	0.070	1.08	3.77	1.23	0.190	
4RD-07-12	1690	1.537	25.51	16.60	0.822	0.606	0.096	720	2.43	0.67	1.260	0.085	0.58	2.71	1.22	0.174	
4RD-07-13	1690	1.351	22.88	16.94	0.751	0.553	0.068	710	2.94	0.82	1.100	0.074	0.51	2.44	1.10	0.150	
4RD-07-14	1690	1.233	21.89	17.76	0.691	0.562	0.093	724	3.81	1.21	1.250	0.074	0.62	2.48	1.23	0.167	
4RD-07-15	1690	1.495	23.57	15.77	0.934	0.586	0.087	753	2.40	0.74	1.270	0.069	0.85	3.42	1.17	0.160	
4RD-07-16	1690	2.079	23.74	11.42	0.868	0.659	0.106	1000	1.20	0.41	1.280	0.068	1.98	6.58	1.08	0.194	
4RD-07-17	1690	1.470	23.57	16.03	0.804	0.588	0.087	697	2.56	0.70	1.090	0.073	0.54	2.36	1.11	0.152	
4RD-07-18	1690	1.570	24.50	15.61	0.875	0.601	0.075	762	2.37	0.69	1.120	0.077	0.48	2.29	1.17	0.175	

(table continued)

Sample	Depth (m)	Mn/Fe	Mn (%)	Fe (%)	Co (%)	Ni (%)	Cu (%)	Zn (ppm)	Si (%)	Al (%)	Ti (%)	V (%)	P (%)	Ca (%)	Mg (%)	Ba (%)	K (%)
(table continued)																	
6RD-08-01	2265	1.304	23.51	18.03	0.743	0.586	0.099	760	4.07	0.95	1.110	0.074	0.58	2.63	1.04	0.154	
6RD-08-02	2265	1.363	23.13	16.97	0.820	0.596	0.092	820	3.65	0.86	1.120	0.068	0.73	3.21	1.20	0.163	
6RD-08-03	2265	1.650	26.25	15.91	0.855	0.672	0.120	860	2.74	0.66	1.160	0.077	0.51	2.58	1.21	0.187	
6RD-08-04	2265	1.527	25.46	16.67	0.871	0.648	0.113	840	2.97	0.69	1.200	0.076	0.56	2.75	1.19	0.195	
6RD-08-05	2265	1.586	24.79	15.63	0.783	0.586	0.097	807	2.36	0.55	1.220	0.075	0.62	2.88	1.09	0.199	
6RD-08-06	2265	1.741	26.65	15.31	0.818	0.646	0.096	850	2.29	0.44	1.150	0.080	0.78	3.40	1.13	0.196	
6RD-08-08	2265	1.803	25.42	14.10	0.770	0.651	0.122	897	2.14	0.51	1.270	0.073	1.59	5.77	1.23	0.211	
6RD-08-S8	2265	1.755	24.52	13.97	0.723	0.620	0.103	847	2.25	0.46	1.160	0.076	1.51	5.34	1.10	0.200	
6RD-09-01A	4800	1.414	23.45	16.58	0.495	0.382	0.129	630	4.96	1.25	1.090	0.074	0.40	2.29	1.05	0.213	
6RD-09-1AF	4800	1.050	21.92	20.88	0.504	0.251	0.087	520	3.48	0.85	1.250	0.091	0.52	2.24	1.05	0.160	
6RD-09-02	4800	1.184	23.54	19.88	0.514	0.350	0.153	627	3.85	1.06	1.260	0.089	0.53	2.20	1.04	0.159	
6RD-09-03	4800	1.147	22.01	19.19	0.556	0.298	0.116	553	4.44	1.40	1.250	0.079	0.43	2.15	1.02	0.158	
6RD-09-04	4800	1.236	23.08	18.68	0.599	0.321	0.094	530	3.04	0.94	1.340	0.084	0.51	2.25	1.08	0.158	
6RD-09-05	4800	1.385	23.84	17.21	0.728	0.377	0.085	520	2.02	0.60	1.330	0.078	0.41	2.15	1.26	0.150	
6RD-09-06	4800	1.252	22.80	18.21	0.711	0.338	0.115	520	2.99	0.91	1.420	0.077	0.37	1.86	1.08	0.164	
7RD-11-X	1465	2.457	31.72	12.91	1.365	0.855	0.057	1013	1.13	0.22	0.900	0.087	0.55	2.77	1.30	0.175	
7RD-11-01	1465	2.433	30.87	12.69	1.701	0.682	0.113	927	1.56	0.60	1.020	0.093	0.75	2.95	1.62	0.501	
7RD-11-02	1465	2.027	29.54	14.57	1.632	0.660	0.042	678	1.11	0.20	0.792	0.083	0.63	2.88	1.32	0.139	
7RD-11-03	1465	3.997	15.67	3.92	0.427	1.127	0.127	1037	0.68	0.69	0.225	0.032	6.57	19.10	1.72	0.207	
7RD-11-04	1465	3.277	29.33	8.95	1.176	1.035	0.098	990	0.68	0.29	0.675	0.064	1.32	4.82	1.44	0.201	
7RD-11-05	1465	2.249	28.13	12.51	1.256	0.790	0.053	763	1.07	0.27	0.827	0.081	0.52	2.60	1.37	0.152	
7RD-11-06	1465	1.997	28.66	14.35	1.478	0.694	0.035	705	1.27	0.23	0.898	0.092	0.61	2.75	1.26	0.132	
7RD-11-07	1465	2.460	31.37	12.75	1.301	0.828	0.057	856	1.09	0.20	0.865	0.084	0.75	3.51	1.33	0.228	
7RD-11-08	1465	2.935	29.29	9.98	1.288	1.134	0.082	1097	0.78	0.36	0.681	0.065	1.64	5.48	1.56	0.193	
7RD-11-10	1465	3.026	30.68	10.14	1.238	0.890	0.089	890	0.92	0.46	0.852	0.100	0.75	3.41	1.41	0.266	

(table continued)

Sample	Depth (m)	Mn/Fe	Mn (%)	Fe (%)	Co (%)	Ni (%)	Cu (%)	Zn (ppm)	Si (%)	Al (%)	Ti (%)	V (%)	P (%)	Ca (%)	Mg (%)	Ba (%)	K (%)
(table continued)																	
7RD-11-11	1465	2.883	30.70	10.65	1.476	0.915	0.058	900	0.83	0.22	0.836	0.072	0.44	2.38	1.43	0.167	
7RD-11-12	1465	4.150	33.78	8.14	1.623	1.263	0.117	1253	0.64	0.47	0.534	0.094	0.69	3.10	1.84	0.269	
7RD-11-13	1465	2.559	25.87	10.11	0.987	0.688	0.083	810	0.91	0.29	0.852	0.085	2.82	8.73	1.15	0.276	
7RD-11-14	1465	2.722	27.25	10.01	1.216	0.733	0.049	877	1.26	0.36	0.903	0.084	2.05	6.92	1.24	0.180	
7RD-11-15	1465	2.174	27.02	12.43	1.321	0.784	0.067	760	1.20	0.36	0.867	0.073	0.89	3.65	1.31	0.147	
7RD-12-1A	3130	1.502	26.15	17.41	0.664	0.515	0.125	813	2.46	0.43	1.210	0.090	0.47	2.51	1.11	0.212	
7RD-12-1B	3130	1.500	26.62	17.75	0.691	0.558	0.146	773	2.86	0.58	1.200	0.088	0.45	2.45	1.03	0.206	
average	2162	1.756	24.31	14.88	0.822	0.598	0.104	795	3.16	0.72	1.036	0.068	0.80	3.45	1.17	0.173	0.616
std. dev.	794	0.668	4.21	3.10	0.315	0.207	0.081	143	1.72	0.47	0.209	0.013	0.90	2.27	0.20	0.053	0.189
maximum	4800	4.150	33.78	20.88	1.921	1.263	0.895	1253	12.23	3.57	1.426	0.100	6.57	19.10	2.05	0.501	1.697
minimum	1230	0.791	11.60	3.92	0.327	0.194	0.030	520	0.64	0.10	0.225	0.032	0.25	1.84	0.81	0.084	0.056
A-1	1.634	18.42	11.24	0.326	0.636	0.107	629	1.80	1.98	0.291	0.072	0.54	10.40	2.95	0.151	0.524	
P-1	4.781	30.07	6.29	0.240	1.359	1.222	1625	6.40	2.29	0.275	0.059	0.21	2.09	2.04	0.358	1.023	
Recommended																	
A-1	1.696	18.54	10.93	0.311	0.636	0.110	587	1.78	2.05	0.320	(0.080)	0.60	11.03	2.87	0.167	0.500	
P-1	4.476	29.14	6.51	0.231	1.340	1.150	1595	6.51	2.55	0.300	(0.055)	0.20	2.19	1.99	0.335	1.050	

Keating, B.H., and Bolton, B.R., editors, 1992, Geology and offshore mineral resources of the central Pacific basin, Circum-Pacific Council for Energy and Mineral Resources Earth Science Series, V. 14. New York, Springer-Verlag.

GEOCHEMISTRY AND MINERALOGY OF FERROMANGANESE NODULES FROM THE KIRIBATI REGION OF THE EASTERN CENTRAL PACIFIC BASIN

Barrie R. Bolton*
Department of Geology and Geophysics, The University of Adelaide, P.O. Box 498, Adelaide, South Australia 5001

Julius Bogi
Department of Chemistry and Earth Science, New South Wales Institute of Technology, P.O. Box 123, Broadway, N.S.W., 2007

David S. Cronan
Department of Geology, Imperial College of Science and Technology, London, SW7, England

ABSTRACT

The bulk chemistry and mineralogy of 16 ferromanganese nodules collected from the eastern Central Pacific Basin are presented. The nodules were collected in water depths ranging from 5400 to 5525 m from a rough seabed consisting dominantly of brown biosiliceous clay.

The eight successful bottom samplings used free-fall grabs and revealed two characteristic nodule types. The dominant nodules in the survey area were small to medium, intergrown, and spheroidal to ellipsoidal with smooth to slightly gritty surface textures (s-type). Typically these nodules show a weakly developed, concentrically laminated internal structure and a large nucleus (often comprising nodule fragments or clays). Concentric layers consist dominantly of vernadite and, to a lesser degree, todorokite. Nodules of this type typically yield low Mn/Fe ratios (ave. 1.37) and Cu + Ni + Zn (ave. 1.02%) values. The second nodule type (r-type) consists of spheroidal to ellipsoidal nodules with rough, gritty, surface textures and little or no nucleus. Typically, these nodules exhibit a complex, finely dendritic to globular structure and consist mainly of todorokite; they usually show high Mn/Fe ratios (ave. 3.37) and high contents of Cu + Ni + Zn (ave. 2.25%).

The two principal nodule morphologies (smooth and rough) and the observed variations in internal structure and chemistry result mainly from preferential deposition of hydrogenetic (vernadite) or diagenetic (todorokite) manganese minerals. Moderate to high metal grades (max. % Cu + Ni + Co = 3.0%) and high nodule concentrations (max. 34 kg/m^2) indicate the area has potential for commercially significant nodule deposits, and that further detailed surveys are warranted, although high grades and concentrations are yet to be found together at one station.

INTRODUCTION

This paper reports the results of a geochemical and mineralogical study of 16 ferromanganese oxide nodules collected by free-fall grabs from a single station near the Phoenix Islands (Kiribati) in the eastern Central Pacific Basin (Fig. 1). The samples were recovered from water depths of 5400 to 5525 m during a 1986 cruise of the University of Hawaii research vessel *Moana Wave*. Details of the ship track and the geological and geophysical sampling results are given in Keating, Bolton and Shipboard Party (1986). The chemistry and mineralogy of extensive ferromanganese crusts dredged from nearby elevated parts of the seafloor during this cruise are presented by De Carlo et al. (1987) and De Carlo and Fraley (this volume).

There has been a steady increase in our knowledge of the manganese nodule distribution, character (size, shape, and texture), geochemistry, and mineralogy in the Central

*Present address: BHP-Utah International Inc, 200 Fairbrook Drive, Suite 101, Herndon, VA 22070-5200.

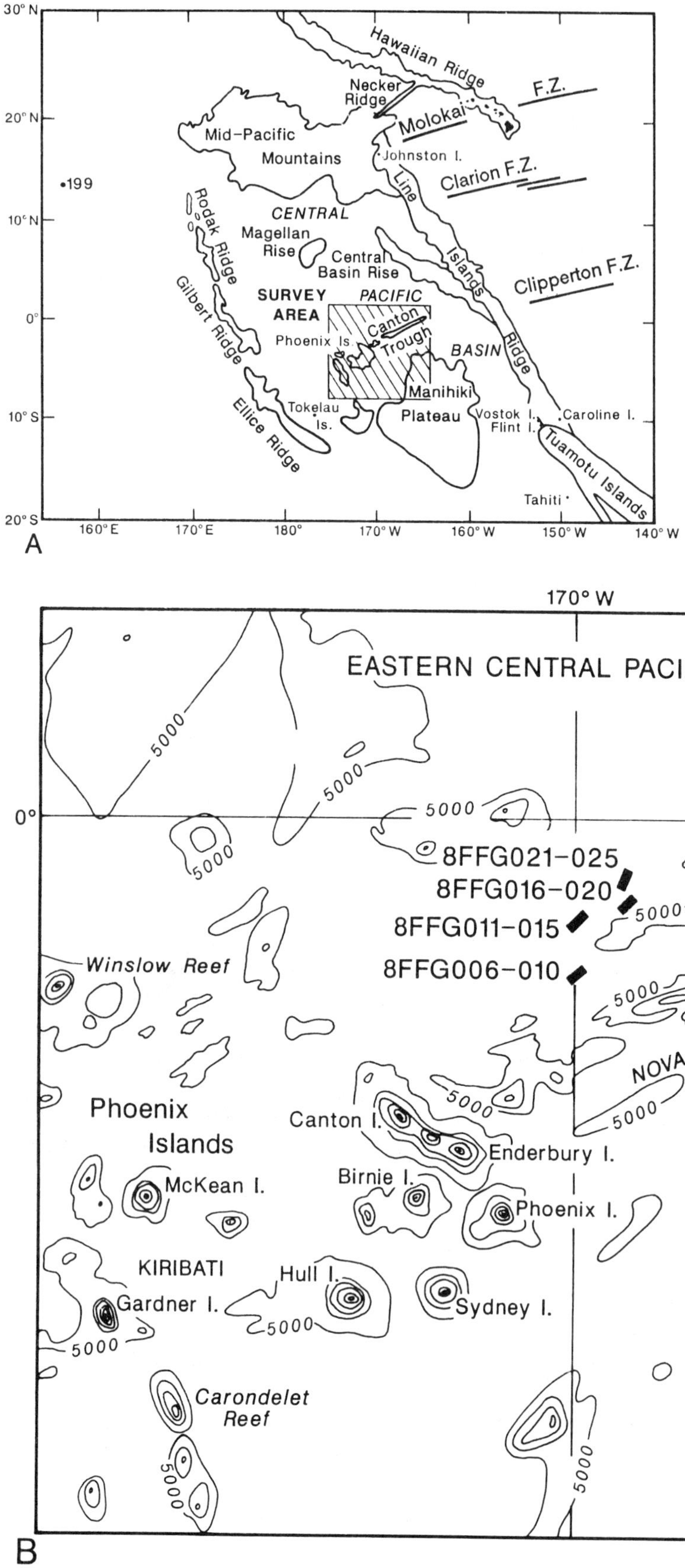

Figure 1. A. Map of the Central Pacific Basin showing the location of major seamount and island groups. B. Locality map showing positions of samples examined. Bathymetric contours in meters.

Pacific Basin during the 1970's and 1980's. The most extensive investigations have been carried out by the Geological Survey of Japan over the ten year period from 1974 to 1983; their results are presented in Mizuno and Moritani (1977), Moritani (1979), Mizuno and Nakao (1982), Usui (1983), Usui et al. (1987), and several papers in this volume.

Specifically, these workers showed that the most common shapes of nodules are spheroidal, ellipsoidal, discoidal, polylobate, or intergrown aggregates; that nodules vary greatly in size; and that surface textures range from rough (r-type—where nodule surface is buried) to smooth (s-type—where nodule surface is exposed)(see Usui, this volume). Nodules which are dominantly smooth occur widely and in high abundances, but are fairly low grade. Nodules which are dominantly rough presumably formed below the sediment surface; they are less widespread and of lower abundance than the smooth nodules, but are of fairly high grade.

Additional information on the distribution and chemistry of nodules in the Central Pacific Basin comes from numerous surveys carried out by CCOP/SOPAC on the UNDP charter vessel *Machias* (Gauss, 1980; Exon, 1981; McDougall and Eade, 1981). Comprehensive summaries of manganese nodule data for the eastern Central Pacific Basin were given by Exon (1983) and Cronan (1984) and used to plan the work described here.

Exon (1982, 1983) and Exon (1984) concluded that the prime areas for potentially economic nodules within the study region are likely to be those meeting the following criteria:

(a) Water depth of 5500–5600 m.
(b) Presence of Antarctic Bottom Water
(c) Elevated biological productivity in surface waters.

Of these (b) is generally thought to be met throughout the area, whereas (a) and (c) have a patchy distribution. The minimum prospective depth of 5000 m is thought to coincide with the approximate level of the CCD, although this is likely to be slightly deeper where productivity is higher (Aplin and Cronan, 1985). The maximum depth of such nodules was estimated to be 5500 m by Cronan (1984) and 5600 m by Exon (1982).

In this paper we present data on the geochemistry and mineralogy of polymetallic nodules in the Phoenix Island region. We also discuss the origin of the deposits, compare them with ferromanganese deposits described elsewhere, and attempt assessment of their economic potential.

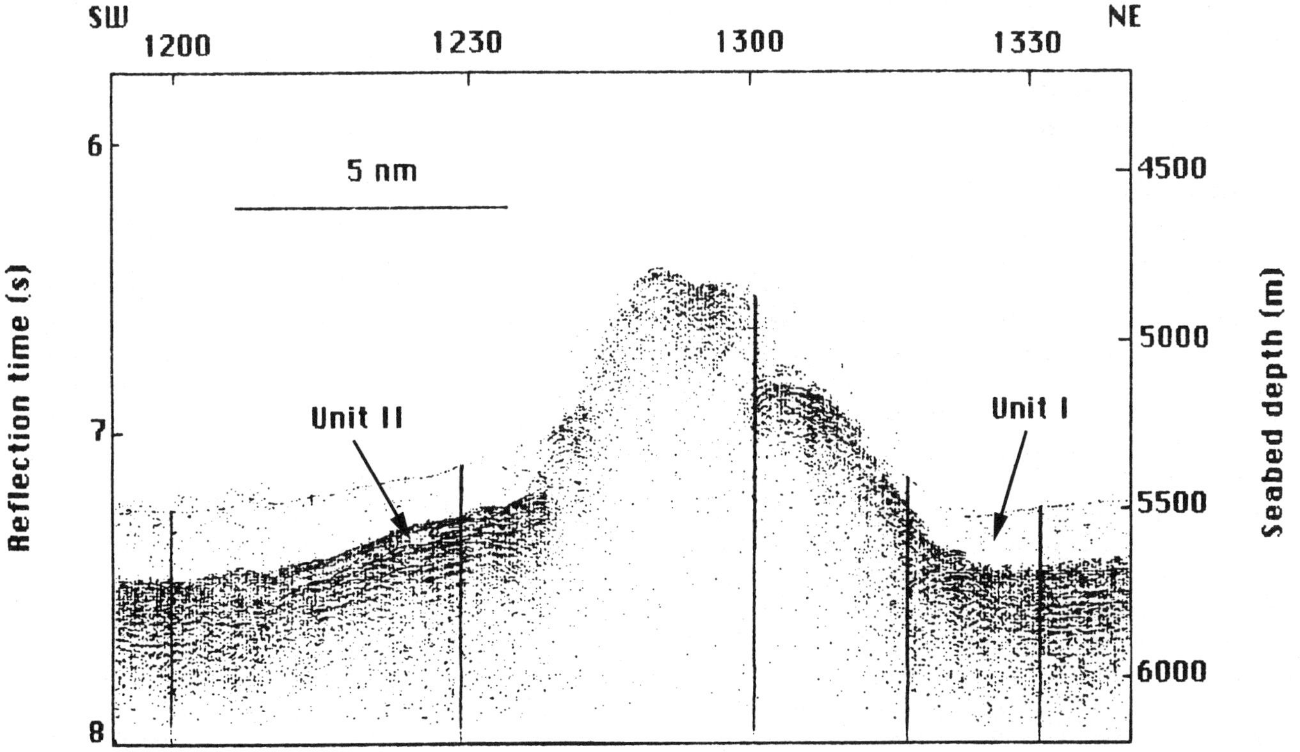

Figure 2. Single-channel air-gun seismic reflection profile typical of those in survey area, showing two seismic patterns, the transparent Unit I and the layered Unit II.

DESCRIPTION OF THE AREA

The East Central Pacific Basin is bounded by the Howland-Phoenix-Tokelau island chain to the west, the Manihiki Plateau and the North Penrhyn Basin to the south, and the Line Islands to the east. Water depths are less than 5000 m in the east and 5000–6000 m elsewhere. According to Exon (1982), the southern part of the basin is floored by calcareous clay, which suggests that it must be just above the CCD. By contrast, the northern part of the basin is floored with siliceous ooze, indicating that it is below the CCD. Aplin and Cronan (1985) have shown siliceous remains to comprise 10-20% of the sediment north of about 5°S, and 2-10% from 5-8°S.

The samples examined in this study were all collected from the southern part of the eastern Central Pacific Basin in waters greater than 5,000 m deep (Fig. 1). Single-channel seismic and 3.5-kHz acoustic profiles show a rugged seafloor characterized by abyssal hills or knolls with a relief of 100 to 700 m (Fig. 2). Similar topography has been recognized elsewhere in the Central Pacific Basin by Japanese researchers (Murakami and Moritani, 1979; Tamaki and Tanahashi, 1981; Mizuno and Okuda, 1982).

Seismic reflection records show that the survey area is largely underlain by two acoustic units, which in turn overlie acoustic basement. At the surface, beneath a very weak reflection, is a unit some 0-0.25 sec thick and consisting of nearly acoustically transparent material that shows no internal structure (Unit I). Underlying this unit is one characterized by an upper part of semi-opaque to opaque material, up to 0.05 sec thick, and a lower part of semi-opaque to transparent material up to 0.1 sec thick, which rarely shows stratification (Unit II). Both units typically pinch out against the flanks of abyssal hills, where acoustic basement appears to crop out (Fig. 2).

The seismic stratigraphy described above is tentatively correlated with that proposed by Winterer et al. (1973), Tamaki (1977), Murakami and Moritani (1979), Tamaki and Tanahashi (1981), and Mizuno and Okuda (1982). Correlation with DSDP Site 166 (Winterer et al., 1973) suggests that Unit I consists of a thick (up to 200 m) accumulation of radiolarian ooze or siliceous clay with nannos of Quaternary to middle Miocene age. Unit II is possibly predominately middle Eocene to Early Cretaceous chert radiolarian ooze.

Sediment cores, taken close to the free-fall grab samples examined in this study, are dominantly yellowish to reddish brown biosiliceous ooze and mud. The fossils are dominantly radiolarians and diatoms, with less sponge spicules and silicoflagellates. All cores almost totally lack anisotropic grains, and contain only minor amounts of volcaniclastic material (usually less than 10%). The sediments of all cores are structureless except for mottles caused by bioturbation. We have no micropaleontologic data for determination of sedimentation rates. The core lithologies are correlated to Unit I on seismic records.

MANGANESE NODULES

Occurrence

Northeast of the Phoenix Islands group, twenty free-fall grabs were deployed over an area of approximately 1400 square miles (Fig. 1; Table 1). Manganese nodules were recovered from eight stations in the southwestern sector of the area surveyed.The nodules were found to vary significantly both in size and shape. They range from small, spheroidal types to large, discoidal forms. Small to medium nodules (2 to 4 cm in diameter) of spheroidal to ellipsoidal shape, and frequently intergrown, are dominant (Fig. 3). Most nodules have a smooth to slightly gritty surface texture and are polynucleate (Fig. 3). There is seldom much difference between the top and bottom surfaces, in contrast with nodules observed elsewhere (Moritani et al., 1977). Nodule abundance ranges from 7 to 34 kg m^{-2}.

The grabs deployed elsewhere in the survey area recovered a coarse fraction dominated by manganese stained pebbles of pumice and rock fragments, with rare sharks' teeth, in abyssal muds, but no manganese nodules.

Mineralogy and Internal Structure

The samples were analyzed by X-ray diffractrometry (XRD)(Cu radiation, 50 kV, 30 mA), optical microscopy, scanning electron microscopy (SEM), and electron probe microscopy (EPMA), in order to determine the mineralogy, internal structure, and microtexture of the nodules.

Analysis of the nodules shows that the principal manganese minerals are todorokite (also called 10Å manganate or buserite) and vernadite (also called δ-MnO_2). No birnessite (or 7Å manganate) was observed. Treatment of the samples with ammonium oxalate/oxalic acid mixture (Dymond and Heath, 1977), failed to detect any crystalline iron phases. In the absence of crystalline Fe-bearing minerals the dominant Fe phase in the nodules is assumed to be a fine-grained disordered Fe oxyhydroxide ($FeOOH_x$-H_2O)(Burns and Burns, 1977).

A number of silicate phases were identified in all nodules, though they were generally not abundant. These include fine particles of clay minerals (montmorillonite), zeolites (mainly phillipsite), phosphates (mainly apatite), and plagioclase and quartz of probable aeolian origin. Fossils, often wholly or partly replaced by manganese oxides, were also common (Fig. 4).

Table 1. Location and desciption of ferromanganese nodules from the eastern Central Pacific Basin.

Sample number	Latitude(S) Longitude(W)	Water depth (m)	Nodule size, shape, and surface texture	Mineral composition*	Internal structure
8FFG-006-1	1°25.99 170°00.14	5525	small size, spheroidal and rough texture	D:todorokite P:vernadite,feldspar quartz, phillipsite	laminated concentric layers; no nucleus apparent
8FFG-006-2	1°25.99 170°00.14	5525	medium size,spheroidal and rough texture	D:todorokite C:vernadite P:plagioclase, quartz phillipsite	laminated concentric layers; no nucleus apparent
8FFG-006-3	1°25.99 170°00.14	5525	medium size,spheroidal intergrown,rough	D:todorokite P: vernadite, quartz, feldspar phillipsite	laminated concentric layers, no nucleus apparent
8FFG-007-1	1°26.65 170°00.92	5525	small size,spheroidal ellpsoidal/intergrown, smooth texture	D:vernadite,todorokite C:feldspar P: quartz, phillipsite	contains fractures, concentrically laminated, fragmented nucleus, altered
8FFG-007-2	1°26.65 170°00.92	5525	medium size,flattened/ ellipsoidal/intergrown, smooth texture	D:vernadite,todorokite C:feldspar,quartz, P: phillipsite	contains fractures, concentrically laminated, fragmented nucleus, altered
8FFG-007-3	1°26.65 170°00.92	5525	medium size, flattened/ ellipsoidal/spheroidal/ intergrown, smooth	D:vernadite,todorokite C:phillipsite P: quartz	contains fractures, concentrically laminated, fragmented nucleus, altered
8FFG-007-4	1°26.65 170°00.92	5525	large size, flattened/ ellipsoidal/spheroidal/ intergrown, smooth	D:vernadite,todorokite C:phillipsite P: quartz, feldspar	contains fractures, concentrically laminated fragmented nucleus, altered
8FFG-008-1	1°27.29 170°01.59	5455	small size,flattened/ ellipsoidal/intergrown, smooth	Not determined	contains fractures, concentrically laminated, nucleus-nodule frag.
8FFG-008-2	1°27.29 170°01.59	5455	medium size, flattened/ ellipsoidal/intergrown, smooth	D:vernadite P:todorokite,feldspar quartz	contains fractures, concentrically laminated, large nucleus.
8FFG-008-3	1°27.29 170°01.59	5455	medium size,flattened/ discoidal/intergrown, smooth	D:vernadite P:todorokite,feldspar, quartz	contains fractures massive, large nucleus-nodule frag.
8FFG-008-4	1°27.29 170°01.59	5455	large size,flattened/ discoidal, intergrown smooth	D:vernadite P:feldspar quartz	contains fractures, massive, large nucleus-nodule frag.
8FFG-010-1	1°29.21 170°03.54	5585	small size,spheroidal, rough texture	D:vernadite,todorokite P:feldspar,quartz, phillipsite	concentrically laminated, no nucleus apparent
8FFG-014-1	0°57.63 170°03.53	5490	medium size,ellipsoidal/ irregular, rough texture	D:vernadite P:todorokite,quartz, feldspar, phillipsite	concentrically laminated, no nucleus apparent
8FFG-016-1	0°42.32 169°29.94	5500	small size, flattened/ discoidal, rough texture	D: todorokite C: feldspar, quartz, P: vernadite, phillipsite	concentrically laminated, no nucleus apparaent
8FFG-017-1	0°43.19 169°30/78	5415	small size, irregular/ ellipsoidal, rough texture	D: vernadite C: todorokite, quartz P: feldspar	concentrically laminated, no nucleus apparent
8FFG-018-1	0°44.04 169°31.81	5390	small size, irregular/ ellipsoidal, rough texture	D: vernadite C: quartz, feldspar P: todorokite	concentrically laminated no nucleus apparent

*D = dominant; C = common; P = present

Figure 3. Typical nodules from the survey area. Those nodules in the top line are spheroidal to ellipsoidal, intergrown and polynucleate, and have smooth surface textures. Those in the bottom line are spheroidal and have rough surface textures.

Under the microscope (including SEM), todorokite occurs mainly as complex finely dendritic to globular growth structures. Vernadite occurs typically in cuspate or columnar growth structures which are radially oriented (with respect to the nucleus) and which commonly lie entirely within parallel to slightly undulating layers or, less frequently, forms large columnar growth structures radiating outward from the nucleus without obvious layering. The columnar structures are further characterized by abundant, lighter colored detrital material filling the spaces between columns (Fig. 5).

Significantly, where todorokite is the dominant manganese phase, the nodules have gritty to rough surface textures (r-type). In contrast, those dominated by vernadite show generally smooth surface textures (s-type). Silicate minerals and fossils are scattered in the oxide layers or form nucleii (Figs. 4,5).

In detail, most nodules exhibit a crude to well-defined concentric layering that may be continuous (representing sequential growth) or discontinuous (representing intermittent growth) about the central nucleus (Figs. 4,5). In most cases the boundary layers separate the two Mn phases. In some samples, however, layering reflects varying proportions of detrital material . Under high magnification, most concentric layers are themselves laminated on a micron scale (Figs. 4,5). In most cases the alternating light and dark layers mirror the shape of the nucleus.

Nodules from the same grab show little variation in internal structure. Appreciable differences are, however, apparent when samples from different grabs are compared, even those only one to two nautical miles apart (Figs. 4,5). We have insufficient data to assess these variations in terms of factors such as topography, substrate lithology and thickness, and water depth.

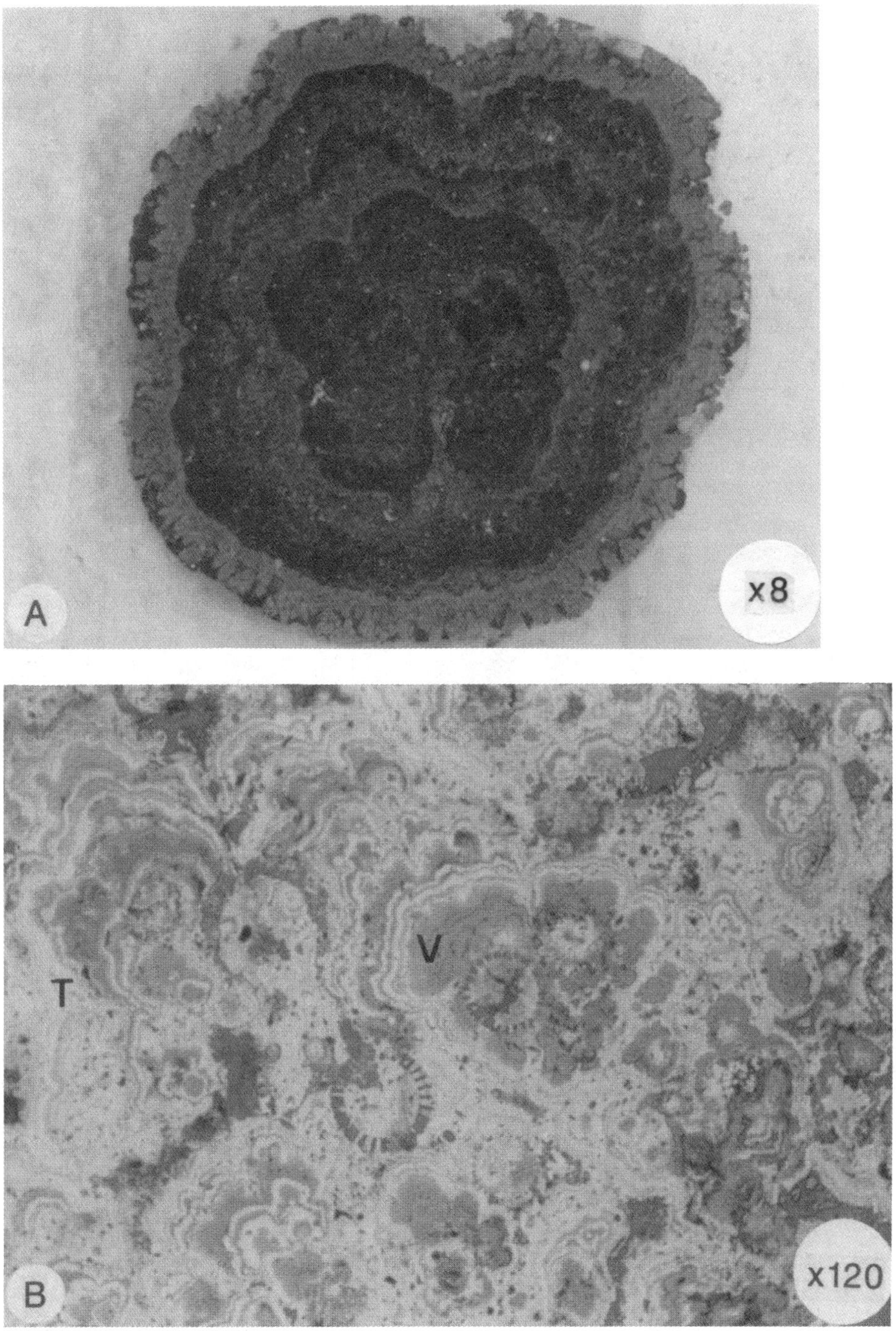

Figure 4. A. Polished section of a rough spheroidal nodule (8FFG6-1) showing concentrically laminated internal structure. Note also the absence of a recognizable nucleus and the prominent dendritic growth structures. B. Polished section showing detail of complex dendritic and globular growth structure in A. Note also the entrapped microfossils (diatoms). V - vernadite; T - todorokite.

Chemistry

The bulk composition of nodules is listed in Table 2; the most significant feature is the wide variation in composition between samples from different grabs separated by only one or two nautical miles. In contrast, the bulk compositions of whole nodules are similar within individual grabs (Table 2).

The data in Table 3 show that the highest values of Mn/Fe, Mn, Cu, Ni, and Zn are encountered in nodules with rough surface texture (also todorokite-rich) while those nodules richest in Fe, Co, and Pb typically have a smooth surface texture (vernadite-rich).

When the proportions of Ni, Cu, and Co (where Ni + Cu + Co = 100%) are plotted against one another (Fig. 6), the results suggest that the relative proportions of nickel and copper remain fairly constant at about 1:1, but the proportion of cobalt varies. Nodules with smooth surface textures (s-type) contain proportionally more Co than nodules with rough surface textures (r-type).

Variation in chemical composition with surface texture is also indicated in a plot showing the proportions of Fe, Mn, and (Cu + Ni + Zn) (Fig. 7). In r-type nodules Ni, Cu, and Zn are preferentially concentrated and Fe is depleted . In contrast, s-type nodules show relative depletion in Cu, Ni, and Zn and enrichment in Fe. The marked difference between types s and r is similar to that found in nodules previously analyzed from the Ccntral Pacific Basin and the Penhryn Basin. Thus, taken together, Figures 6 and 7 confirm that, in general, the composition of nodules can be correlated with differences in surface texture, which reflects mineralogical variations. However, in three samples (014, 017, 018), bulk composition is not in accordance with surface texture. A similar mismatch has been

Table 2. Bulk chemistry of ferromanganese nodules from

Sample	Nod.type*	Mn*	Fe*	Ni	Cu	Co	Zn	Pb	Si*	Al*	Mg*
8FFG-006-1	r	26.06	5.01	13080	13360	1148	1187	230	5.74	2.08	1.87
8FFG-006-2	r	26.45	5.67	13460	14380	1135	1127	230	5.37	2.22	1.98
8FFG-006-3	r	20.21	4.72	11090	11250	972	794	210	4.84	1.97	1.62
8FFG-007-1	s	15.81	10.55	5460	4882	1597	535	480	13.69	3.35	1.35
8FFG-007-2	s	18.98	10.65	7140	6640	1686	662	505	12.00	2.66	1.54
8FFG-007-3	s	15.70	10.04	5440	5060	1500	520	445	11.24	3.21	1.33
8FFG-007-4	s	17.27	10.49	6250	5450	1576	601	470	9.60	2.93	1.31
8FFG-008-1	s	14.17	15.83	3116	2838	1897	520	670	12.01	2.42	1.09
8FFG-008-2	s	16.96	14.89	3796	3348	2314	486	650	6.33	1.64	1.09
8FFG-008-3	s	15.72	16.04	3322	3035	1944	487	630	6.43	1.78	1.07
8FFG-008-4	s	19.82	13.76	5930	5180	2507	652	655	6.19	1.67	1.26
8FFG-010-1	r	26.57	5.11	15180	13860	994	1124	245	14.77	2.03	1.70
8FFG-014-1	r	17.96	10.10	9810	7205	1221	802	341	11.74	4.32	1.53
8FFG-016-1	r	22.28	10.05	12156	9024	1419	939	410	9.20	3.11	1.70
8FFG-017-1	r	17.47	10.35	8896	6667	1368	739	416	--	--	--
8FFG-018-1	r	18.07	9.27	5552	7640	1384	801	1267	--	--	--
Mean		19.34	10.16	8105	7489	1541	749	491	9.23	2.53	1.46
Std. Dev.		1.00	0.92	976	960	111	59	65	0.90	0.21	0.08
Minimum		14.17	4.72	3116	2838	972	486	210	4.84	1.64	1.07
Maximum		26.57	16.04	15180	14380	2507	1187	1267	14.77	0.32	1.98

* analyses expressed in wt%; all other values in ppm.
* r=rough surface texture; s=smooth surface texture
Analyses by ICP; analyst, D.S. Cronan

found in some Pacific nodules by Sorem and Fewkes (1977, 1979), Marchig and Halbach (1982), and Usui et al. (1987). Usui et al. (1987) ascribe this to variations in mineral composition within the nodules; a similar explanation is suggested for at least one of our samples (017), which has both vernadite and todorokite in approximately equal proportions.

Inter-element relationships

A correlation matrix of the nodule elemental compositions is presented in Table 4. The patterns of inter-element relationships for all the nodules studied are in general consistent with those reported for abyssal nodules elsewhere (Calvert and Price, 1977; Piper and Leong, 1979; Frazer and Fisk, 1981). The most significant positive correlations are those between Mn, Cu, Ni, and Zn and those between Fe, Co, and Pb. There is a negative correlation between the two groups, Mn-Cu-Ni-Zn and Fe-Co-Pb (Table 4)(c.f. Usui, 1984). Manganese also shows a strong positive correlation with Mo, Li, and Cd; Fe is strongly positively correlated with V, Sr, and Be.

Other significant positive correlations include Al with K and Na ($r = 0.95$ and 0.29, respectively) as would be expected from their volcanic derivation (primarily from sodium- and potassium-rich aluminosilicates). Calcium and P correlate directly and are probably associated with detrital material of probable biogenic origin. The high degree of correlation between the Fe and Ti concentrations ($r = 0.99$) has also been observed by Halbach and Özkara (1979) in nodules from the northeast Pacific; they ascribe it essentially to the enrichment of TiO_2 in the iron-rich vernadite of s-type nodules (their type B nodules). Anenrichment of TiO_2 in nodules dominated by vernadite is similar to that found by Halbach and Özkara (1979).

DISCUSSION

Origin of deposits

All the nodules examined show evidence of both hydrogenetic and diagenetic processes. Those in which the latter is dominant have rough surface textures, an abundance of todorokite relative to vernadite, higher Mn/Fe ratios, and enrichment in the trace metals, Ni, Cu, and Zn. All these are features which workers elsewhere have suggested to be diagnostic of nodule growth by precipitation of dissolved metals from the interstitial water of unconsolidated surface sediments during early diagenesis (Usui et al., 1978; Halbach and Özkara, 1979; Marchig and the eastern Central Pacific Basin near the Phoenix Islands.

Ca*	Na*	Ti	K	P	Sr	Mo	V	Li	Cd	Be	Mn/Fe
2.04	1.97	2862	6550	2932	534	410	322	109	20	1.40	5.20
2.03	1.93	3090	7220	2936	552	411	335	84	20	1.60	4.66
7.90	1.82	2220	5990	25650	681	334	263	71	15	1.70	4.28
2.06	1.83	6280	11600	3484	765	212	386	35	6	2.90	1.50
1.80	1.66	5970	7640	2480	795	323	428	49	9	3.00	1.78
2.07	1.76	5860	12280	3844	741	243	381	34	6	2.90	1.56
2.00	1.76	5710	11770	3456	777	301	418	38	8	2.90	1.65
1.83	1.52	9630	6530	2760	966	154	447	14	5	4.10	0.90
1.85	1.35	8690	4740	2720	1035	255	489	15	4	4.30	1.14
1.86	1.35	8380	4680	2984	1026	245	506	13	4	4.30	0.98
2.03	1.46	8080	5420	3072	1005	319	504	26	8	4.10	1.44
1.98	2.20	2964	6200	1868	543	470	383	84	17	1.30	5.20
2.36	1.43	7973	11373	1833	--	--	--	--	--	--	1.78
2.07	1.34	4916	7638	1658	--	--	--	--	--	--	2.22
--	--	--	--	--	--	--	--	--	--	--	1.69
--	--	--	--	--	--	--	--	--	--	--	1.95
2.42	1.67	5902	7831	4405	785	306	405	48	10	2.90	2.37
0.42	0.07	652	731	1643	54	26	22	9	2	0.30	1.50
1.80	1.34	2220	4680	1658	534	154	263	13	4	1.30	0.90
7.90	2.20	9630	12280	25650	1035	470	506	109	20	4.30	5.20

Gundlach, 1981; Halbach et al., 1982, Aplin and Cronan, 1985). On the other hand, those nodules with smooth surface textures, an abundance of vernadite relative to todorokite, relative depletion in Ni,Cu, and Zn, and lower Mn/Fe ratios, have probably developed by slow hydrogenetic precipitation of dissolved metals either from the overlying normal seawater within sediments but without significant diagenesis (Halbach and Özkara, 1979; Usui, 1979; Usui et al., 1987; Aplin and Cronan, 1985).

Comparison with ferromanganese deposits elsewhere

In shape, size, and surface texture, the nodules described in this study are very similar to nodules described by Craig (1979), Mizuno et al. (1980) and, more recently, by Usui (1982), Aplin and Cronan (1985), and Usui et al. (1987) from elsewhere in the central and eastern Pacific Basin. They similarly show wide local variability in these parameters and a direct relationship between surface texture, internal structure, and mineralogy.

Average bulk compositions are similar to those previously reported for abyssal nodules from the eastern Central Pacific Basin, the Penrhyn Basin, and the Central Pacific Basin as a whole (Table 5). Manganese values are significantly lower than those reported for the Northeast Pacific Basin and are considerably higher than those from the western Central Pacific Basin.

Iron percentages (Table 5) are similar to previous values for the Central Pacific Basin and for the North Penrhyn Basin, but significantly lower than those for the Northeast Pacific, the South Penrhyn Basin, and eastern Central Pacific Basin.

The Mn/Fe ratios vary considerably from basin to basin (Table 5). The mean value of 2.37 is similar to that for the entire Central Pacific Basin, and is significantly higher than those for the eastern and western Central Pacific Basin, the South and North Penrhyn basins, and the entire Pacific Basin. The ratio is, however, lower than in the Northeast Pacific Basin.

Economic considerations

Deep sea ferromanganese deposits have long been considered a potential economic source of the metals Ni, Cu, Co, and to a lesser extent, Mn. It is generally considered that for ferromanganese deposits to be considered of potential economic interest, they must occur in concentrations of

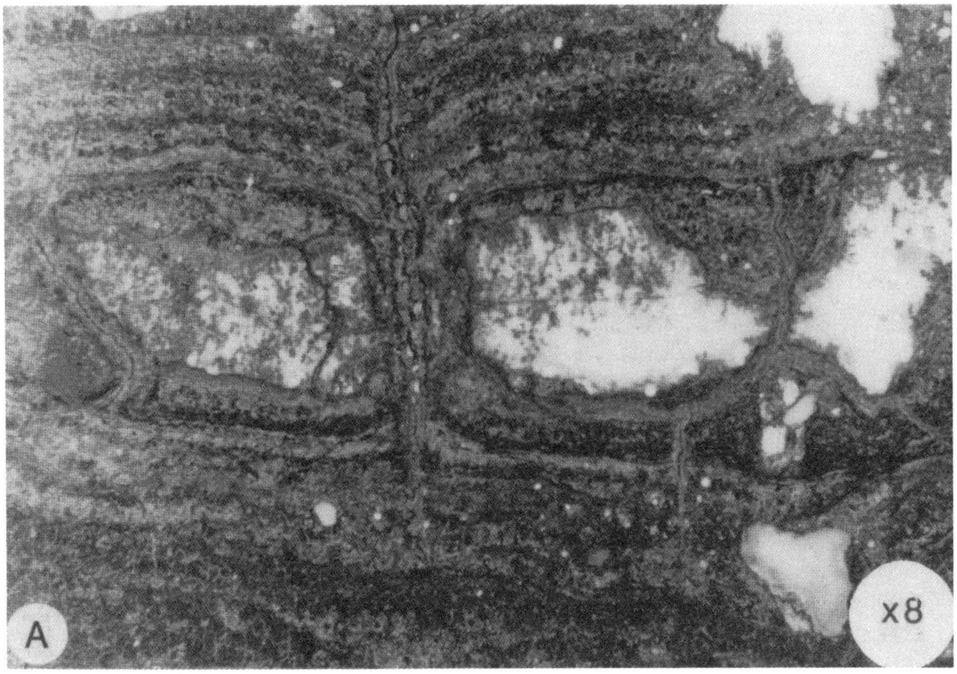

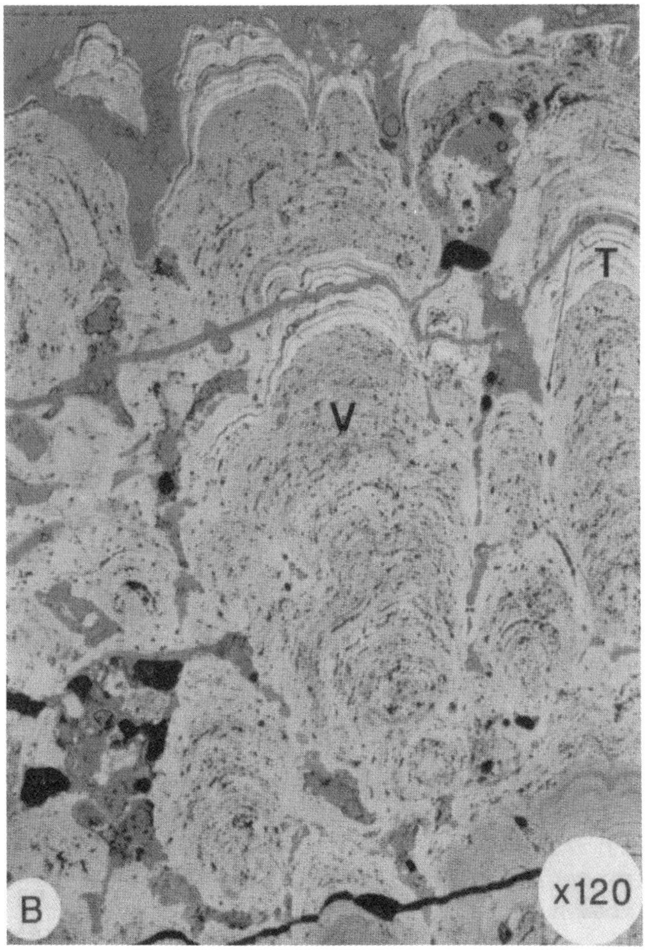

Figure 5. A. Polished section showing internal structure of nodule (8FFG7-2) with a smooth surface texture. Note the large nucleus and well laiminated structure. B. Polished section showing detail of large columnar segregation growth structures in A. Note also the thin, highly reflective laminae of todorokite in the middle and upper part of the central column. V = vernadite; T = todorokite.

more than 10 kg "wet" nodules/m^2 and have combined contents of Ni + Cu + Co of more than 2%.

Nodule concentration and grade data from the eastern Central Pacific Basin have been analyzed by Exon (1983). Nodules were present in 68% of 140 deep-water stations. The average concentration from 30 stations was 7.83 kg/m^2 and the maximum recorded was 31.6 kg/m^2. All concentrations greater than 10 kg/m^2 lay in water depths of 5250–5720. Chemical analyses from 57 stations gave an average Mn/Fe ratio of 1.89, and an average grade of 1.73% Ni + Cu + Co. The maximum Mn/Fe ratio was 6.00 and the maximum

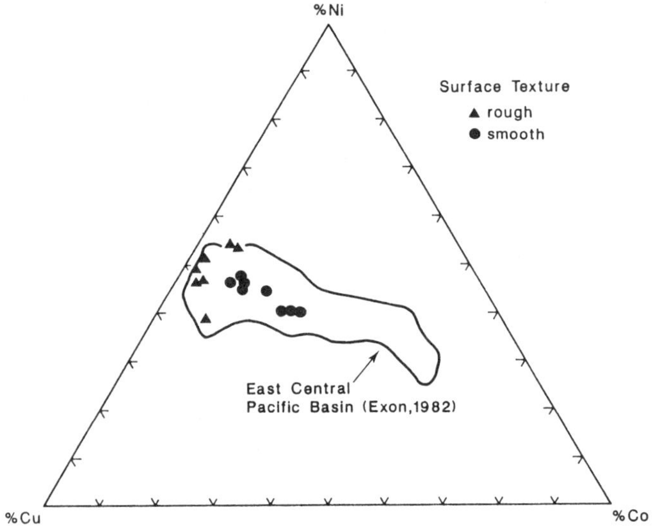

Figure 6. Ternary diagram of ratios Ni/Co/Cu. Values are percentages where Ni + Co + Cu = 100%. Note the distinct compositional differences with nodule type. Compositional field for eastern Central Pacific Basin is from Exon (1982).

Table 3. Composition of ferromanganese nodules on the basis of surface texture.

Textural type		Rough (8)*	Smooth (8)
Mn	Mean:	21.88	16.81
	Minimum:	17.47	14.17
	Maximum:	26.57	19.82
Fe	Mean:	7.53	12.78
	Minimum:	4.72	10.04
	Maximum:	10.35	16.04
Cu	Mean:	1.04	0.46
	Minimum:	0.67	0.28
	Maximum:	1.44	0.66
Ni	Mean:	1.12	0.51
	Minimum:	0.56	0.31
	Maximum:	1.52	0.71
Co	Mean:	0.12	0.19
	Minimum:	0.09	0.15
	Maximum:	0.14	0.25
Zn	Mean:	0.09	0.06
	Minimum:	0.07	0.05
	Maximum:	0.12	0.07
Pb	Mean:	0.04	0.06
	Minimum:	0.02	0.04
	Maximum:	0.13	0.07
Cu+Ni+Zn	Mean:	2.25	1.02
	Minimum:	1.41	0.65
	Maximum:	3.02	1.44
Mn:Fe	Mean:	3.37	1.37
	Minimum:	1.69	0.91
	Maximum:	5.21	1.78

* Number of samples

grade was 3.55%, both very high values. All grades of Ni + Cu + Co higher than 1.3% lay in water depths of 4900–5600 m.

Table 6 presents a summary of available grade and concentration information for the current survey. As can be seen, the concentration of nodules varies considerably, ranging from zero to a maximum concentration of 34 kg/m^2 at station 8 (8FFC-008) northwest of Canton Island. Of the 22 grabs deployed, 32% obtained nodules. This is about half the average recovery rate for the eastern Central Pacific Basin, as given by Exon (1983). Thus, nodules are relatively scarce in the present survey (Fig. 1). However, for those successful grabs, 8FFG 006-10, the average abundance is 11.85 kg/m^2, considerably higher than the 7.83 previously determined by Exon (1983). In essence the present survey indicates a patchy distribution of nodules but, where present, they seem to occur in moderate to high concentrations. However, as was shown by Exon (1983) for the eastern Central Pacific Basin as a whole, high grades do not coincide with high concentrations.

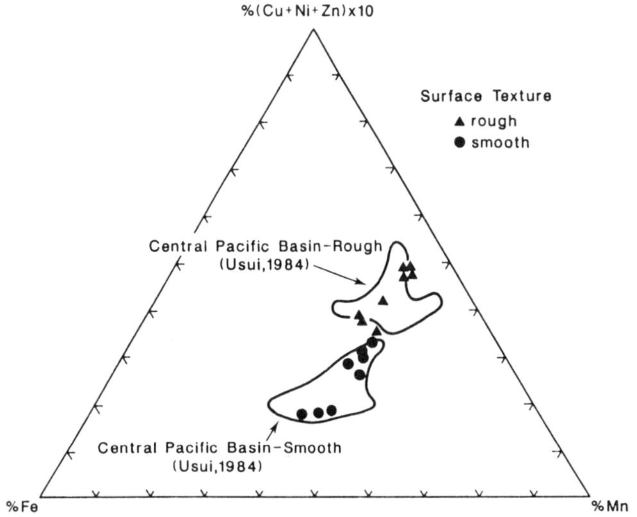

Figure 7. Ternary diagram of ratio Mn/Fe/(Cu + Ni + Zn) x 10. Note the differences of chemical characteristics with nodule type. Compositional fields for nodule facies are from Usui (1984).

The combined contents of Ni + Cu + Co, average 1.71% for the area surveyed (Table 6). This value is very close to the 1.73% determined for the eastern Central Pacific Basin as a whole by Exon (1983). The maximum grade of 3.0% was recorded from grab sample, FFG-010 (Table 6) and is comparable to the 3.55% presented by Exon (1982), for an eastern Central Pacific Basin sample.

The data indicate significant variation in grade and concentration along a four nautical mile traverse line. For samples FFG-006 to FFG-10, the grades range from 0.79% to 3.0% (ave. 1.68%) and the concentrations range from 1.73 to 34 kg/m^2 (ave. 11.85 kg/m^2).

In summary, although the sample numbers are low, the present survey indicates that there exist nodules of both potentially economic metal grades and concentrations in the study area, although high grades and high concentrations are yet to be found in the same sample. Nevertheless, further, closely spaced sampling in selected areas is probably warranted to fully assess the region.

Table 4. Correlation matrix for chemical compositions of ferromanganese nodules (n = 16).

	Mn	Fe	Ca	Al	Na	Mg	Ti	Cu	Ni	Pb	Zn	Co	Mo	V	Sr	K	P	Li	Cd
Fe	**-0.79**	1.00																	
Ca	0.07	-0.43	1.00																
Al	-0.35	-0.08	-0.17	1.00															
Na	**0.72**	**-0.90**	0.15	0.29	1.00														
Mg	**0.89**	**-0.92**	0.21	-0.01	**0.82**	1.00													
Ti	**-0.78**	**0.99**	-0.47	-0.05	**-0.86**	**-0.90**	1.00												
Cu	**0.95**	**-0.92**	0.29	-0.18	**0.83**	**0.95**	**-0.92**	1.00											
Ni	**0.95**	**-0.92**	0.26	-0.17	**0.86**	**0.93**	**-0.91**	**0.99**	1.00										
Pb	**-0.78**	**0.98**	-0.44	-0.05	**-0.88**	**-0.91**	**0.98**	**-0.93**	**-0.92**	1.00									
Zn	**0.97**	**-0.83**	0.10	-0.25	**0.78**	**0.92**	**-0.81**	**0.96**	**0.96**	**-0.83**	1.00								
Co	-0.60	**0.90**	-0.42	-0.23	**-0.87**	**-0.79**	**0.90**	**-0.81**	**-0.80**	**0.93**	**-0.70**	1.00							
Mo	**0.95**	**-0.78**	0.11	-0.33	**0.70**	**0.83**	**-0.79**	**0.90**	**0.92**	**-0.76**	**0.90**	-0.58	1.00						
V	-0.52	**0.89**	-0.61	-0.23	**-0.76**	**-0.77**	**0.88**	**-0.75**	**-0.72**	**0.91**	-0.62	**0.90**	-0.46	1.00					
Sr	**-0.77**	**0.95**	-0.20	-0.24	**-0.95**	**-0.92**	**0.93**	**-0.89**	**-0.89**	**0.95**	**-0.84**	**0.92**	**-0.73**	**0.85**	1.00				
K	-0.28	-0.14	-0.14	**0.95**	0.31	0.01	-0.11	-0.14	-0.12	-0.11	-0.21	-0.22	-0.23	-0.24	-0.27	1.00			
P	0.01	-0.38	**0.99**	-0.14	0.10	0.17	-0.42	0.23	0.21	-0.40	0.04	-0.38	0.05	-0.59	-0.16	-0.12	1.00		
Li	**0.90**	**-0.93**	0.25	-0.07	**0.84**	**0.95**	**-0.92**	**0.96**	**0.95**	**-0.93**	**0.94**	**-0.82**	**0.85**	**-0.79**	**-0.92**	-0.04	0.20	1.00	
Cd	**0.94**	**-0.89**	0.27	-0.21	**0.79**	**0.95**	**-0.88**	**0.98**	**0.96**	**-0.90**	**0.97**	**-0.76**	**0.87**	**-0.74**	**-0.87**	-0.16	0.22	**0.96**	1.00
Be	**-0.79**	**0.98**	-0.34	-0.12	**-0.94**	**-0.92**	**0.97**	**-0.93**	**-0.93**	**0.98**	**-0.85**	**0.93**	**-0.78**	**0.88**	**0.98**	-0.16	-0.30	**-0.94**	**-0.90**

Significant values are >0.62 at the 99% confidence level (bold values)

Table 5. Metal contents and ratios of ferromanganese nodules from this study and comparison with previous chemical data. All elements in wt%.

	This study	East Central Pacific Basin Exon (1982)	West Central Pacific Basin Exon (1982)	Central Pacific Basin Usui (1984)	Northeast Pacific Manganese belt Hein (1977)	Southwest Pacific South Penrhyn Basin Exon (1982)	Southwest Pacific North Penrhyn Basin Exon (1982)	Pacific Basin Price and Calvert (1980)
number of analyses	16	57	14	112	1261	48	18	202
	Mean	Mean	Mean	Mean	Mean	Mean	Mean	Mean
Mn	19.34	20.30	14.70	19.40	26.70	15.20	18.60	18.00
Fe	10.16	10.70	14.60	10.80	6.40	15.40	11.50	12.10
Mn:Fe	2.37	1.89	1.08	2.25	4.18	0.99	1.62	1.48
Ni	0.81	0.80	0.37	0.73	1.25	0.41	0.60	0.72
Cu	0.75	0.74	0.32	0.61	1.05	0.22	0.47	0.53
Co	0.15	0.21	0.20	0.26	0.24	0.36	0.25	0.36
Ni+Cu+Co	1.71	1.73	0.89	1.60	2.54	0.99	1.32	1.61
Zn	0.07	--	--	0.08	0.13	--	--	--
Pb	0.05	--	--	0.08	0.05	--	--	0.09

ACKNOWLEDGMENTS

The authors gratefully acknowledge the assistance of shipboard colleagues and the crew of the *Moana Wave*. We are also deeply indebted to the scientists and technicians of the Hawaii Institute of Geophysics for their support. In this respect special mention must be made of Patricia Pennywell and Bruce Tsutsui, curator of the HIG core analysis laboratory. This study was carried out on behalf of CCOP/SOPAC under a Tripartite agreement between Australia, New Zealand, and the United States of America. BRB and JB thank the Australian Development Assistance Bureau (ADAB) for financial support. N. Exon and K. Chave are thanked for their constructive reviews.

Table 6. Summary of nodule grades and concentrations.

Sample	%Ni+Cu+Co	Concentration (kg/m^2)
8FFG-006-1	2.76	
8FFG-006-2	2.89	
8FFG-006-3	2.33	4.33
8FFG-007-1	1.19	
8FFG-007-2	1.55	
8FFG-007-3	1.20	
8FFG-007-4	1.33	7.35
8FFG-008-1	0.79	
8FFG-008-2	0.95	
8FFG-008-3	0.83	
8FFG-008-4	1.36	34.00
8FFG-010-1	3.00	1.73
8FFG-014-1	1.82	Tr
8FFG-016-1	2.26	Tr
8FFG-017-1	1.69	Tr
8FFG-018-1	1.46	Tr
Minimum	0.79	Tr
Maximum	3.00	34.00
Mean	1.71	

Tr=Trace

REFERENCES

Aplin, A.C. and D.S. Cronan, 1985, Ferromanganese oxide deposits from the Central Pacific Ocean, II. Nodules and associated sediments: Geochimica et Cosmochimica Acta, v. 49, p. 437–451.

Burns, R.G. and V.M. Burns, 1977, Mineralogy, *in* G.P. Glasby, ed., Marine Manganese Deposits: Elsevier, Amsterdam, p. 185–248.

Calvert, S.E., and N.B. Price, 1977, Geochemical variation in ferromanganese nodules and associated sediments from the Pacific Ocean: Marine Chemistry, v. 5, p. 43–74.

Craig, J.D., 1979, The relationship between bathymetry and ferromanganese deposits in the Northern Pacific: Marine Geology, v. 29, p. 165–185.

Cronan, D.S. 1984, Criteria for the recognition of areas of potentially economic manganese nodules and encrustations in the CCOP(SOPAC) region of the central and southwestern tropical Pacific: South Pacific Marine Geological Notes, v. 3, p. 1–17.

De Carlo, E.H., and C.M. Fraley, Chemistry and mineralogy of ferromanganese deposoits from the equatorial Pacific Ocean, this volume.

De Carlo, E.H., P.A. Pennywell, C.M. and Fraley, 1987, Geochemistry of ferromanganese deposits from the Kiribati and Tuvalu region of the west Central Pacific Ocean: Marine Mining, v. 6, p. 301–321.

Dymond, J. and G.R. Heath, 1977, Copper-Nickel enriched ferromanganese nodules and associated crusts from the Bauer Basin, northwest Nazca Plate: Earth and Planetary Science Letters, v. 35, p. 55–64.

Exon, N.F., 1981, Kiribati manganese nodules survey cruise, KI-81(1), 4-15 April, 1981: CCOP/SOPAC Cruise Report No. 56 (unpublished), 3 pp.

Exon, N.F., 1982, Manganese nodules in the Kiribati region, equatorial Western Pacific: South Pacific Marine Geological Notes, v. 2, p. 77–102.

Exon, N.F., 1983, Manganese nodule deposits in the Central Pacific Ocean and their variation with latitude: Marine Mining, v. 4, p. 79–107.

Frazer, J.L. and M.B. Fisk, 1981, Geological factors related to characteristics of sea-floor manganese nodule deposits: Deep-Sea Research, v. 28A, p. 1533–1551.

Gauss, G. A., 1980, Kiribati offshore survey, 9-25 February, 1980: KI-80(1): CCOP/SOPAC Cruise Report, no. 37 (unpublished), 5 pp.

Halbach, P. and M. Özkara, 1979, Morphological and geochemical classification of deep-sea ferromanganese nodules and its genetical interpretation, *in* C. Lalou, Ed., Colloques Internationaux du C.N.R.S. No. 289, La Genese desNodulesdeManganese. C.N.R.S., Paris, pp.77–88.

Halbach, P., F.T. Manheim, and P. Otten, 1982, Co-rich ferromanganese deposits in the marginal seamount regions of the Central Pacific Basin – results of the Midpac '81: Ermetall., v. 35, p. 447–453.

Keating, B., B.R. Bolton, and Shipboard Party, 1986, Initial Report of 1986 R/V *Moana Wave* Criuse, MW86002 in the Kiribati/Tuvalu Region, Central Pacific Ocean. CCOP/SOPAC Cruise Report No. 121, Southpac, 1986, 39 pp.

Marchig, V. and H. Gundlach, 1981, Separation of iron from manganese and growth of manganese nodules as a consequence of diagenetic aging of radiolarians: Marine Geology, v. 40, p. M35–M45.

Marchig, V. and P. Halbach, 1982, Internal structures of manganese nodules related to conditions of sedimentation: Tschermaks Mineral. Petrogr. Mitt., v. 30, p. 81–110.

McDougall, J.C. and J.V. Eade, 1981, Manganese nodules in western Kiribati (Gilbert Islands): South Pacific Marine Geological Notes, v. 2, p. 67–75.

Mizuno, A., and T. Moritani, 1977, Deep sea mineral resources investigation in the central-eastern part of the Central Pacific Basin, January-March, 1976 (GH76-1): Geological Survey of Japan Cruise Report, no. 8, 217 pp.

Mizuno, A., T. Miyazaki, A. Nishimura, K. Tamaki, and M. Tanahashi, 1980, Central Pacific manganese nodules, and their relation to sedimentary history: Proceedings 12th Annual Offshore Technical Conference, Houston, v. 3, p. 331–340.

Mizuno, A. and S. Nakao, 1982, Regional data of marine geology, geophysics and manganese nodules: the Wake-Tahiti transect in the central Pacific, January-March 1980 (GH80-1): Geological Survey of Japan Cruise Report, no. 18, 399 pp.

Mizuno, A., and Y. Okuda, 1982, Seismic reflection profiles of the Wake-Tahiti transect in the Central Pacific Ocean, *in* A. Mizuno, ed., Geological Survey of Japan Cruise Report, no. 18, p. 46–89.

Moritani, T., 1979, Deep sea mineral resources investigation in the central-western part of Central Pacific Basin, January-March, 1977 (GH77-1 Cruise): Geological Survey of Japan Cruise Report, no. 12, 256 pp.

Moritani, T., S. Maruyama, M. Nohara, et al., 1977, Description, classification, and distribution of manganese nodules, *in* A. Mizuno, and T. Moritani, T., eds., Geological Survey of Japan Cruise Report, no. 8, p. 136-158.

Murakami, F. and T. Moritani, 1979, Continuous seismic reflection profiling survey, *in* T. Moritani, ed., Geological Survey of Japan Cruise Report, no. 12, p. 74–102.

Piper, D.Z. and K. Leong, 1979, Manganese nodule and surface sediment composition, Domes sites A, B, and C, *in* J.L. Bischoff, and D.Z. Piper, eds., Marine Geology and Oceanography of the Pacific Manganese Nodule Province: Plenum, New York, N.Y., pp. 437–473.

Sorem, R.K. and R.H. Fewkes, 1977, Internal characteristics of manganese nodules, *in* G.P. Glasby, ed., Marine Manganese Deposits: Elsevier, Amsterdam, p. 147–184.

Sorem, R. and R.H. Fewkes, 1979, Manganese Nodules: Research Data and Methods of Investigation: Plenum, New York, N.Y.,. 723. p

Tamaki, K., 1977, Study on substrate stratigraphy and structure by continuous seismic reflection profiling survey, *in* A. Mizuno and T. Moritani, eds., Deep Sea Mineral Resources Investigation in the Central Pacific Basin (GH79-1 Cruise), Geological Survey of Japan Cruise Report, no. 15, p. 77–99.

Tamaki, K. and M. Tanahashi, 1981, Seismic reflection survey in the northeastern margin of the Central Pacific Basin, *in* A. Mizuno, ed., Geological Survey of Japan Cruise Report, no. 15, p. 77–99.

Usui, A. and T. Moritani, Manganese nodule deposits in the Central Pacific Basin: distribution, geochemistry, mineralogy and genesis: this volume.

Usui, A., S. Takenouchi, and T. Shoji, 1978, Mineralogy of deep-sea manganese nodules and synthesis of manganese oxides: implication to genesis and geochemistry: Mining Geology, v. 28, p. 405–420.

Usui, A., 1979. Minerals, metal contents, and mechanism of formation of manganese nodules from the Central Pacific Basin (GH76-1 and GH77-1 areas), *in* J.L. Bischoff and D.Z. Piper, eds., Marine Geology and Oceanography of the Pacific Manganese Provinces: New York, Plenum, pp. 651–679.

Usui, A., 1982, Variability of manganese nodule deposits: The Wake to Tahiti Transect: Geological Survey of Japan Cruise Report, no. 18, p. 138–223.

Usui, A., 1983, Regional variation of manganese nodule facies on the Wake-Tahiti transect: Morphological, chemical and mineralogical study: Marine Geology, v. 54, p. 27–51.

Usui, A., 1984, Mineralogy and internal structure of manganese nodules of the GH8O-5 area, *in* S. Nakao and T. Moritani, eds., Marine Geology, Geophysics, and Manganese Nodules in the northern vicinity of the Magellan Trough (GH8O-5 Cruise): Geological Survey of Japan Cruise Report, no. 20, p. 215–226.

Usui, A., A. Nishimura, M. Tanahashi, and S. Terashima, 1987, Local variability of manganese nodule facies on small abyssal hills of the Central Pacific Basin: Marine Geology, v. 74, p. 237–275.

Winterer, E.L., J.I. Ewing, R.G. Douglas, Y. Lancet, R.M. Moberly, T.C. Moore, Jr., P.H. Roth, P.H., and S.O. Schlanger, 1973, Site report 166, *in* P.H. Roth, J.R. Herring et al., Initial Reports of the Deep Sea Drilling Project, v. 17, p. 103–143.

Keating, B.H., and Bolton, B.R., editors, 1992,
Geology and offshore mineral resources of the
central Pacific basin, Circum-Pacific Council for
Energy and Mineral Resources Earth Science
Series, V. 14. New York, Springer-Verlag.

CENTRAL PACIFIC COBALT-RICH FERROMANGANESE CRUSTS: HISTORICAL PERSPECTIVE AND REGIONAL VARIABILITY

James R. Hein, Marjorie S. Schulz, and Lisa M. Gein
U.S. Geological Survey, 345 Middlefield Road, Mail Stop 999, Menlo Park, California 94025, USA

ABSTRACT

Interelement correlations coupled with X-ray mineralogy, chemical analyses, and Q-mode factor analysis define five phases that compose ferromanganese crusts: δ-MnO_2, Fe-phases including Fe oxyhydroxide and Fe silicate, detrital aluminosilicate, biogenic phosphate, and biogenic-nonphosphate. This last phase is derived primarily from the dissolution of biogenic carbonate and silica. These five phases are characterized by the following elements respectively: Co, Mn, Ni, Pb; Fe, Si, As; Al, Cr, Si, Ti, K; P, Ca; and Cu, Ni, Ba, Zn, Cd. The distribution of the δ-MnO_2 phase is related to latitude and is controlled by the equatorial zone of higher productivity, which produces a strong and extensive oxygen-minimum zone. The iron phases are similarly distributed, but the overall variability is not as great as for the δ-MnO_2 phase. The detrital phase has the inverse distribution to the δ-MnO_2 phase and is greatest at higher latitudes, especially in areas beneath the trade-wind belt. The detrital phase is composed of eolian debris and volcanogenic debris eroded from submarine outcrops by bottom currents. The phosphate phase is not clearly distributed with latitude, longitude, or the equatorial zone of high biological productivity. The biogenic-nonphosphate phase is most prominent in the equatorial zone of high biological productivity.

The Co content of bulk crusts in the central Pacific averages 0.79% (308 analyses), Ni averages 0.47% (311 analyses), and Pt averages 0.24 ppm (29 analyses). Cobalt and Ni are richest in crusts from the EEZ of Tuvalu, followed in decreasing order by Kingman-Palmyra Islands, Howland-Baker Islands, Johnston Island, Marshall Islands, Hawaiian Islands, and Kiribati Islands. However, based on 11 geologic, oceanographic, and geochemical criteria, we rank the resource potential of Pacific nations and islands in decreasing order Marshall, Federated States of Micronesia, Johnston, French Polynesia, Kiribati, Hawaiian, Tuvalu, Kingman-Palmyra, and Howland-Baker islands.

The knowledge gained from ten cruises devoted to the study of cobalt-rich ferromanganese crusts in the central Pacific has contributed to the understanding of the origin and evolution of crusts and their localized distributions. These ten cruises were undertaken by the Technical University of Clausthal (F.R. Germany), U.S. Geological Survey, and University of Hawaii. The scientific results presented by each group are outlined and put into a historical perspective.

INTRODUCTION

Ferromanganese deposits are ubiquitous in the ocean basins and occur in nearly all tectonic and geomorphological environments (Table 1). Four general types of ferromanganese deposits occur: nodules and micronodules; crusts and pavements; sediment-hosted stratabound layers and lenses; and cements and fracture and vein infillings. Ferromanganese oxides are also dispersed in sediments and are somewhat more concentrated in metalliferous sediments, but these occurrences commonly have bulk manganese contents of only a few percent (Böstrom and Peterson, 1966, 1969); if more, the presence of micronodules is likely to be the reason.

Table 1. Form, processes of formation, and environment of formation of marine ferromanganese oxide deposits[1].

	Hydrogenetic	Hydrothermal	Diagenetic	Hydrogenetic and Hydrothermal	Hydrogenetic and Diagenetic
Nodules	Abyssal plains, oceanic plateaus, seamounts[2]	Rare in submerged calderas and fracture zones	Abyssal plains, oceanic plateaus	Rare in submerged calderas	Abyssal plains, oceanic plateaus[2]
Crusts	Midplate volcanic edifices[3]	Spreading centers, fracture zones	---	Active volcanic arcs, spreading centers off axis seamounts, fracture zones	Rare on abyssal hills
Sediment-hosted, strata-bound layers and lenses	---	Active volcanic arcs, large mid-plate volcanic edifices, sediment-covered spreading axes	Continental margins[4]	---	---
Cements	Midplate volcanic edifices[5]	Active volcanic arcs,[6] large midplate volcanic edifices[6]	Midplate volcanic edifices[5]	---	Midplate volcanic edifices[5]

[1] This table is based primarily on published and unpublished data of Hein (Hein et al. 1979, 1985a, 1987a,b, 1988a,b) but also on previous compilations (Mero, 1965; Bonatti et al., 1972; Rona, 1978; Toth, 1980; Heath, 1981; Cronan, 1984; Baturin, 1988).
[2] Less common on ridges, continental slope, and shelf
[3] Includes seamounts, guyots, ridges, plateaus
[4] Ferromanganese carbonate lenses and concretions
[5] Mostly fracture and vein fill, cement for volcanic breccia
[6] Mostly cement for sandstone and siltstone

Oceanic ferromanganese deposits are formed from hydrothermal solutions via leaching of sediments or oceanic crust, hydrogenetic[1] precipitation from ambient seawater, and diagenetic processes including precipitation from pore waters and reconstitution of deposits (Table 1). Various combinations of these three processes produce a variety of deposits. Different types of ferromanganese deposits can be distinguished by their mineralogy, Fe/Mn ratios, trace metal contents, rare-earth element patterns, and growth rates. Seven combinations of deposit types, depositional mechanisms, and environments of formation are common and widespread in the oceans.

1. Abyssal ferromanganese-oxide nodules form by a combination of hydrogenetic and diagenetic processes (Heath, 1981),
2. Ferromanganese-oxide crusts form by purely hydrogenetic precipitation on hard-rock substrates of midplate volcanic edifices (Hein et al., 1987a),
3. Manganese-oxide or iron-oxide crusts form by predominantly hydrothermal precipitation at oceanic spreading axes (Toth, 1980),
4. Ferromanganese-oxide crusts form by a combination of hydrogenetic and hydrothermal precipitation in active volcanic arcs (Hein et al., 1987b),
5. Stratabound layers and lenses of manganese-oxide precipitate from hydrothermal fluids that leach volcaniclastic sediments in active volcanic arc environments (Hein et al., 1987b),
6. Manganiferous sandstone forms from hydrothermal fluids that cement volcaniclastic sandstone produced in active volcanic arcs (Hein et al., 1987b), and
7. Ferromanganese carbonate lenses form by diagenetic processes within continental-margin sedimentary sequences (Hein et al., 1979). Other, apparently relatively minor deposits, occur in various ways and in various places within the ocean basins (Table 1).

The part of the central Pacific considered here contains twelve EEZs that compose four major linear island chains (Fig. 1) and contains many large submarine volcanic edifices, some as large as continental mountain ranges (Fig. 2). Hydrogenetic cobalt-rich ferromanganese crusts and abyssal ferromanganese nodules comprise more than 99% of all the ferromanganese deposits collected in the Central

1 We use the term hydrogenetic to refer to precipitation from ambient seawater and hydrogenous as a general term to indicate precipitation from seawater and pore waters.

Pacific Basin (Glasby and Andrews, 1977; Craig et al., 1982; Halbach et al., 1982; Nakao and Moritani, 1984; Hein et al., 1985a,b, 1988a; Aplin and Cronan, 1985; Nakao, 1986; Chave et al., 1986; De Carlo et al., 1987a). Relatively minor stratabound manganese oxides and manganiferous sandstones recovered off the island of Hawaii (Hein and Clague, unpublished data) and from Gardner Pinnacles, central Hawaiian Ridge (De Carlo et al., 1987a) were produced by hydrothermal activity during or following submarine flank fissure eruptions.

This paper provides a historical perspective of studies of central Pacific ferromanganese crusts, discusses the distribution and compositional variability of ferromanganese crusts in the central Pacific (Fig. 1) through statistical analyses of chemical data, and ranks the island groups with regards to possible resource potential. This work is an extension of the compilation and discussion by Hein et al. (1987a).

METHODS

We compiled all available published and unpublished chemical data on crusts collected from all water depths in the central Pacific (Figs. 1, 3). Element distribution maps for the entire Pacific are presented by Hein et al. (1987a). All chemical data were closely scrutinized and only internally consistent analyses were used (see Hein et al., 1987a for a complete discussion of the selection of samples and methods used for chemical analyses). Our data set for the central Pacific consists of 311 samples of bulk crusts (Table 2) and 102 analyses for the outer layers (10 ± 5 mm) of crusts (Table 3). Samples are grouped according to the 370 km (200 nm) Exclusive Economic Zones (EEZs) of central Pacific nations and territories (Fig. 1). A 370 km boundary was drawn around each island or island group (Fig. 1), which is not necessarily the EEZ boundary officially accepted by each nation. Analyses of crusts that were col-

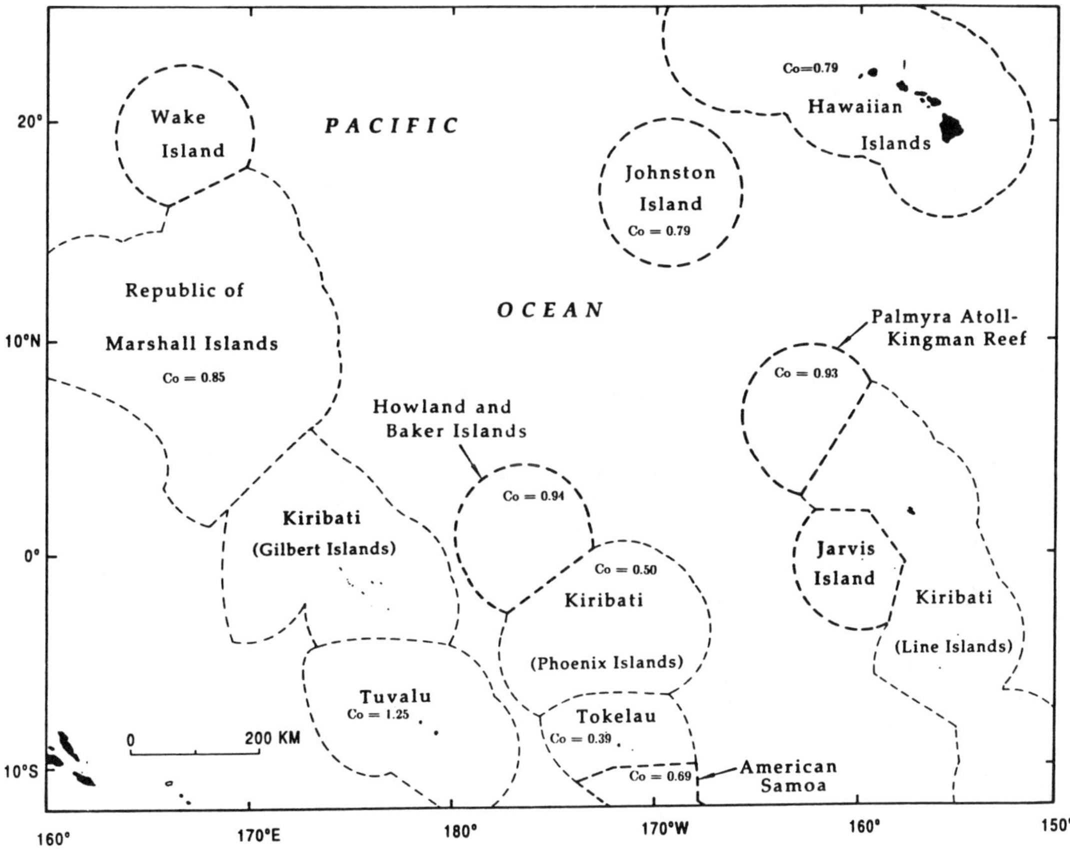

Figure 1. Location of the central Pacific region considered in this paper. A 370 km (200 nm) boundary is drawn around each island that is a U.S. Possession, the State of Hawaii, and the central Pacific island nations. This boundary is not necessarily the officially recognized Exclusive Economic Zone (EEZ) of each nation, but will be referred to here as the EEZ. Areas that fall outside these boundaries are collectively called international waters. The average Co content of crusts is indicated for each EEZ and is taken from Table 2. The number for Kiribati is for the Phoenix and Line Islands; no data are available for the Gilbert Islands. No data are available for the EEZ of Wake Island, and only one sample analysis is available for the EEZ of Jarvis Island and one for American Samoa.

Table 2. Number of samples, average chemical composition, and standard deviations of bulk ferromanganese crusts from the Exclusive Economic Zone of central Pacific nations and from international waters. Data are in weight percent, except Pt is in ppm.

	Fe/Mn	Mn	Fe	Co	Ni	Cu	Pb	Ti	Si	P	Pt
International Waters											
n	63	63	63	62	63	53	36	49	35	42	—
Mean	0.72	22.7	15.4	0.77	0.47	0.17	0.18	1.12	3.57	1.00	—
S.D.	0.24	5.36	3.06	0.36	0.20	0.21	0.04	0.34	2.17	1.09	—
Howland-Baker Islands											
n	7	7	7	7	7	4	3	6	6	6	—
Mean	0.66	26.7	17.3	0.94	0.56	0.09	0.14	1.09	2.73	0.62	—
S.D.	0.14	2.47	2.68	0.31	0.17	0.02	0.006	0.34	0.69	0.23	—
Johnston Island											
n	86	86	86	86	86	56	53	76	50	57	7
Mean	0.68	22.7	14.87	0.79	0.48	0.11	0.17	1.07	4.54	1.06	0.18
S.D.	0.19	4.55	2.89	0.29	0.16	0.11	0.11	0.27	5.46	0.89	0.12
Kingman-Palmyra Islands											
n	29	29	29	29	29	29	28	27	27	27	3
Mean	0.58	26.6	15.0	0.93	0.59	0.13	0.16	1.04	2.39	1.05	0.22
S.D.	0.19	2.63	3.63	0.28	0.23	0.19	0.03	0.21	1.30	0.88	0.11
Kiribati Islands											
n	13	13	13	13	13	13	8	9	9	9	—
Mean	0.71	23.2	15.2	0.50	0.47	0.18	0.08	0.89	3.40	0.99	—
S.D.	0.29	6.52	4.97	0.35	0.26	0.17	0.06	0.43	1.71	1.57	—
Marshall Islands											
n	13	13	13	13	13	13	13	13	13	13	8
Mean	0.62	20.3	12.5	0.85	0.39	0.04	0.14	0.83	2.51	0.43	0.32
S.D.	0.10	1.21	1.30	0.18	0.06	0.02	0.02	0.22	1.38	0.21	0.13
Tokelau Islands											
n	2	2	2	2	2	2	2	—	—	—	—
Mean	1.11	17.1	18.7	0.39	0.25	0.19	0.09	—	—	—	—
S.D.	0.28	2.47	2.12	0.03	0.16	0.08	0.01	—	—	—	—
Tuvalu Islands											
n	2	2	2	2	2	2	—	2	2	2	—
Mean	0.52	28.0	13.8	1.25	0.73	0.08	—	1.14	1.84	0.66	—
S.D.	0.23	6.36	3.25	0.66	0.22	0.02	—	0.12	1.29	0.24	—
Hawaii (on axis)											
n	49	49	49	47	49	47	16	35	18	—	—
Mean	0.72	23.69	16.35	0.96	0.45	0.08	0.22	1.00	4.84	—	—
S.D.	0.24	4.05	3.79	0.29	0.18	0.15	0.07	0.25	3.04	—	—
Hawaii (off axis)											
n	45	45	45	45	45	44	43	39	39	25	11
Mean	0.91	21.1	18.3	0.62	0.38	0.10	0.18	1.25	6.47	0.58	0.24
S.D.	0.28	3.81	3.23	0.25	0.19	0.05	0.06	0.31	4.69	0.43	0.23
Average Central Pacific Crusts (this compilation)											
n	311	311	311	308	311	265	204	256	199	181	29
Mean	0.72	23.0	15.7	0.79	0.47	0.12	0.17	1.08	4.22	0.91	0.24
S.D.	0.24	4.67	3.51	0.33	0.19	0.15	0.07	0.30	3.95	0.90	0.17
Average Central Pacific Crusts (Hein et al., 1987a)											
Mean	0.71	22	15	0.78	0.44	0.08	0.16	1.1	2.9	0.65	—
Average Central Pacific Crusts (Manheim, 1986)											
Mean	—	22.6	—	0.76	—	—	—	—	—	—	—
Average Pacific Crusts (Hein et al., 1987a)											
Mean	0.81	22	15	0.63	0.44	0.08	0.16	0.98	5.1	—	—
Average Pacific Crusts (Manheim, 1986)											
Mean	0.70	23.1	16.1	0.73	0.47	0.16	0.16	—	3.7	—	—

— means no data available.
Data sources for compilation listed in separate reference section.

lected outside these EEZs are grouped under the heading international waters. Jarvis, Howland-Baker, Wake, Johnston, and Kingman-Palmyra islands are U.S. possessions. Analysis of only one crust is available from the Jarvis Island EEZ and one from the American Samoa EEZ. These two analyses are included in the central Pacific average (Table 2), but are not listed separately.

A Q-mode factor analysis was performed on samples from several EEZs where data sets with the largest number of well constrained analyses for many elements (24 elements in each set) are available. Twenty-four elements were included in several of the correlation coefficient matrices and in all of the Q-mode factor analyses, but only ten elements are presented in the EEZ averages in Table 2. For Q-mode factor analysis, each variable percentage was scaled to the percent of the maximum value before the samples were row normalized and the cosine theta coefficients were calculated. The factors presented here were derived from orthogonal rotations of the principal component eigenvectors using the Varimax method (Klovan and Imbrie, 1971).

HISTORICAL PERSPECTIVE

The central Pacific is the most intensively studied area of the world's oceans for ferromanganese crusts. Ten cruises have been devoted to crust studies (Table 4) and many cruises dedicated to other studies collected crusts that were the subject of a number of publications.

Table 3. Number of samples, average chemical composition, and standard deviations of outer ferromanganese crusts from the Exclusive Economic Zones of central Pacific nations, and from international waters. Data are in weight percent.

	Fe/Mn	Mn	Fe	Co	Ni	Cu	Pb	Ti	Si	P
International Waters										
n	22	22	22	22	22	22	19	21	18	18
Mean	0.61	26.2	14.9	1.00	0.52	0.06	0.16	0.98	2.87	0.44
S.D.	0.26	5.06	2.96	0.37	0.16	0.04	0.05	0.28	2.48	0.27
Johnston Island										
n	35	35	35	35	35	14	12	35	35	35
Mean	0.54	26.7	13.9	1.12	0.58	0.07	0.17	0.92	2.54	0.53
S.D.	0.15	3.82	2.78	0.31	0.20	0.05	0.04	0.14	1.27	0.53
Kingman-Palmyra Islands										
n	11	11	11	11	11	11	11	10	10	10
Mean	0.49	29.4	13.8	1.37	0.61	0.05	0.17	0.99	2.17	0.59
S.D.	0.16	4.98	2.84	0.59	0.18	0.02	0.03	0.17	1.80	0.55
Marshall Islands										
n	22	22	22	22	22	22	22	22	22	22
Mean	0.61	20.6	12.4	0.89	0.40	0.04	0.14	0.81	2.05	0.36
S.D.	0.09	1.26	1.25	0.25	0.09	0.024	0.015	0.13	0.56	0.11
Hawaii										
n	12	12	12	10	12	10	10	12	12	10
Mean	0.83	22.5	17.2	0.59	0.38	0.10	0.15	1.16	4.76	0.49
S.D.	0.34	5.36	6.11	0.11	0.23	0.05	0.03	0.38	2.32	0.15
Average										
n	102	102	102	100	102	79	74	100	97	95
Mean	0.60	25.05	14.2	1.02	0.51	0.06	0.16	0.94	2.73	0.48
S.D.	0.22	4.93	3.40	0.40	0.19	0.04	0.04	0.24	1.82	0.39

Data sources for compilation listed in separate reference section.

The earliest description of crusts from the central Pacific that we are aware of was by Moore (1965) who reported crusts composed of todorokite up to 100-mm thick from the Hawaiian chain. Moore (1966) attempted to use the crust thicknesses to determine the age of lava flows. Nodules in the central Pacific were first described in detail by Mero (1965), although earlier studies first described nodules and their general distribution (Murray and Renard, 1891; Agassiz, 1902, 1906; see Glasby and Read, 1976 for a historical perspective on nodule studies in the Pacific). Mero used nodules as a general term and some of the seamount samples he described are actually crusts. In the early 1970's, a series of descriptive papers on crusts was presented by Morgenstein and coworkers (e.g., Morgenstein, 1972) that correctly interpreted the source of metals as hydrogenetic, but incorrectly suggested (as pointed out by Glasby and Andrews, 1977) that the presence of iron oxides was necessary to catalyze the precipitation of the manganese oxides. Follow-up work on those Hawaiian crusts provided detailed chemical, mineralogical, and descriptive data (Frank et al., 1976; Glasby and Andrews, 1977; Craig et al., 1982). These

Table 4. Cruises dedicated to the study of ferromanganese crusts in the Central Pacific Basin.

Ship/Cruise	Date	Institution	Study Areas
Sonne SO 18	June-July, 1981	Technical University of Clausthal, F.R. Germany	Mid-Pacific Mts.: U.S. Johnston I. EEZ; Line Is.-Kiribati EEZ; U.S. Kingman-Palmyra EEZ
Sonne SO 33	July-August, 1984	Technical University of Clausthal, F.R. Germany	Hawaiian, Palmyra-Kingman, and Johnston Is. EEZs.
Sonne SO 37	May-June, 1985	Technical University of Clausthal, F.R. Germany	U.S. Howland-Baker and Johnston Is. EEZs
Sonne SO 46	November-December, 1986	Technical University of Clausthal, F.R. Germany	Marshall Is. EEZ; Johnston I. EEZ.
S.P. Lee L5-83-HW	October-November, 1983	U.S. Geological Survey	Hawaiian, Johnston, Palmyra-Kingman EEZs
S.P. Lee L9-84-CP	August, 1984	U.S. Geological Survey	Marshall Is. EEZ
Farnella F7-86-HW	November-December, 1986	U.S. Geological Survey, U.S. Bureau of Mines	Johnston I. EEZ
Kana Keoki KK84-04-28-05 and KK84-08-24-02	June and August, 1984	University of Hawaii	Hawaiian Is. EEZ
Moana Wave MW-86-02	February-March, 1986	University of Hawaii	Phoenix Is.-Kiribati EEZ; Tuvalu EEZ
Moana Wave MW-87-02	February, 1987	University of Hawaii	Cook Is. EEZ; Line Is.-Kiribati EEZ

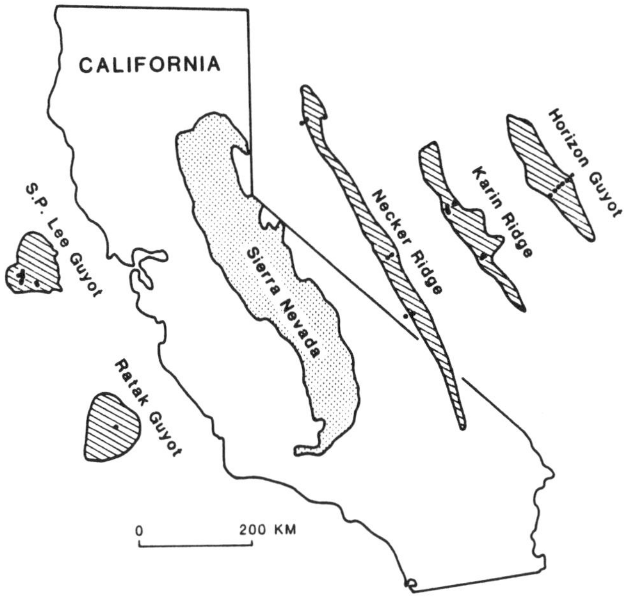

Figure 2. Comparison of the size of several volcanic edifices in the central Pacific with the Sierra Nevada in California and the location of dredge stations (solid circles). The dredge sites are not to scale and overemphasize the actual area sampled. Horizon Guyot and Karin Ridge are in the Johnston Island EEZ, Necker Ridge is in the Hawaiian EEZ, Ratak Guyot is in the Marshall Islands EEZ, and S.P. Lee Guyot is in the northern Line Islands in the Palmyra-Kingman EEZ.

authors noted that crusts up to 50-mm thick are thicker on volcanogenic substrates than on carbonate substrates. Frank et al. (1976) suggested that Mn, Ni, Cu, and Co in crusts increase and Fe decreases with increasing latitude along the Hawaiian chain. Craig et al. (1982) revised that conclusion and suggested that Ni, Co, and crust thickness increase and Ti and Fe decrease to the northwest; δ-MnO_2 (vernadite) was identified as the predominant X-ray crystalline phase. Frank et al. (1976) recognized the potential economic value of the cobalt in the crusts and stated that nickel and manganese are of secondary economic interest. Also, Craig et al. (1982) concluded that crust chemistry is not influenced by substrate type, that seamount nodule fields are covered and exhumed by migrating sediment ripples, and that Ni and Co contents are strongly positively correlated with Mn in the δ-MnO_2 phase. However, they did not find a correlation between cobalt content and water depth, a correlation firmly established in all subsequent studies, and originally proposed by Menard (1964) and Mero (1965). Glasby and Andrews (1977) also suggested that diagenetic migration of elements may occur within the crusts.

Following these early studies, a large number of papers appeared in the 1980's based on the work derived from the ten crust-dedicated research cruises (Table 4). The ten cruises were conducted by the Technical University of Clausthal (F.R. Germany), the U.S. Geological Survey, and the University of Hawaii.

F.R. Germany Studies

Chemical data and corresponding sample locations have been published for only one of the four R.V. Sonne cruises, SO 18. Six papers were generated from analysis of nine crusts collected during SO 18 (Halbach et al., 1982, 1983, 1984; Halbach and Manheim, 1984; Halbach and Puteanus, 1984; Mangini et al., 1987). Two of these papers also presented average chemical compositions for crusts from areas sampled on R.V. Sonne cruise SO 33, but no specific sample locations were provided (Halbach, 1986; Mangini et al., 1987). Results from these studies confirmed that cobalt contents of crusts increase with decreasing water depth and showed that (1) platinum is enriched in crusts; (2) cobalt content is greater in the outer layers of crusts than in the inner layers, whereas platinum content is greater in the inner layers; (3) two characteristic growth generations separated by a phosphorite parting occur in crusts thicker than 4 cm; (4) the degree of cobalt enrichment is inversely proportional to the ferromanganese oxide accumulation rate; (5) the water depths where cobalt is enriched in crusts correspond to the mid-water oxygen-minimum zone; and (6) growth rates of crusts vary between 1 and 5 mm/m.y.

Halbach et al. (1983) also suggested that there is a constant flux of cobalt to crusts with time at all water depths; however, this idea needs further support (Hein et al., 1988a). In another study, Halbach and Puteanus (1984) suggested that the increased supply of iron to deeper-water crusts is related to the dissolution of iron-bearing biogenic calcite below the calcite compensation depth, but Aplin and Cronan (1985) and De Carlo et al. (1987a) rejected calcite dissolution as the source of iron below 2000 m water depth.

U.S. Geological Survey Studies

The first of three USGS central Pacific cruises (Table 4) addressed the geology, sedimentology, geochemistry, and physical oceanography of Necker Ridge (Hawaiian Islands EEZ), Horizon Guyot (Johnston Island EEZ), and the informally named S.P. Lee Guyot (Kingman-Palmyra Islands EEZ) (Hein et al., 1985a,b; Schwab, 1986; Davis, 1987; Schwab et al., 1988). Results showed that reworking and mass-wasting on the slopes of volcanic edifices partly control the thickness of crusts; sediment mobility on guyots

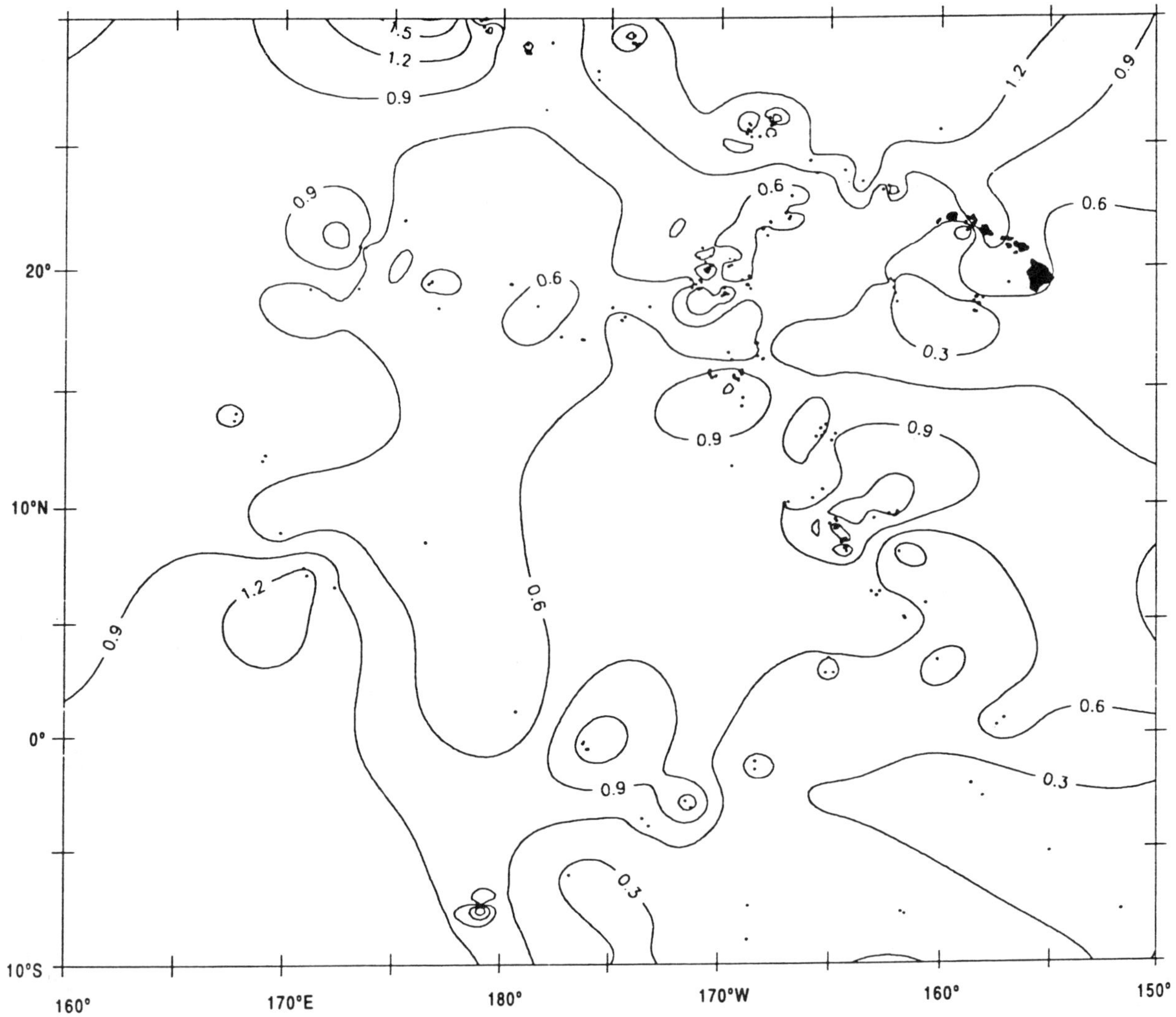

Figure 3. Distribution of cobalt in ferromanganese crusts in the central Pacific, based on 308 data points (dots). Contour interval is 0.3 weight percent.. Map was generated by U.S. Geological Survey NAVPLT programand includes an area that extends about 5° latitude north of that included in Figure 1.

buries and exhumes both nodule fields and crust-coated outcrops; eolian input to crusts (quartz and some feldspar) and biological constituents in crusts (phosphorites) vary with latitude and are related to zonal wind patterns and to regional and local upwelling; high current velocities (up to 20 cm/s) occur as much as 180 m above Horizon Guyot, which can account for sediment migration; the oxygen-minimum zone can vary markedly from the summit to the flanks of a seamount; Co, Ni, and Mn contents increase and Fe decreases with decreasing latitude, opposite to the relationship noted by Frank et al. (1976) and Craig et al. (1982) for crusts along the Hawaiian Ridge; metals positively correlated with Mn include Ni, Mo, Co, and Cd; Fe correlates with Cu; and water depth correlates with Si, Al, Ti, Fe, and K. Ratios of strontium isotopes of crusts and phosporites compared with the same ratios that define the Tertiary seawater curve can be used to determine ages of crusts and their phosphatic substrates (Futa et al., 1987, 1988). Growth rates determined by the strontium isotope method are consistent with those determined by Halbach et al. (1983) and Mangini et al. (1987) using ^{10}Be dating techniques.

The second USGS cruise (Table 4) was to the eastern chain of the Marshall Islands (Schwab et al., 1986; Davis et al., 1986; Hein et al., 1988a). Results generally agreed with those from areas farther to the east in the central Pacific but, in addition, demonstrated that rhodium is strongly enriched in crusts and covaries with platinum, whereas palladium concentrations are relatively constant at about the earth's crustal average. Rare-earth elements (REE) are enriched in crusts and chondrite-normalized patterns are heavy REE depleted, show a strong positive cerium anomaly, and a small positive gadolinium anomaly. Two of these REE pattern characteristics are the same as the REE pattern of seawater and the other two are opposite to seawater, indicating that REE are fixed in several different phases including amorphous iron oxyhydroxide, δ-MnO_2, minor apatite and barite, and amorphous iron silicate. The concentrations of REE and other elements fluctuate significantly through the crusts on a fine-scale (millimeters). The Ti, Co, Ni, Zn, and Pb are all fixed by lattice substitution for Mn^{4+} in the δ-MnO_2 phase. Iron oxyhydroxides crystallize to goethite in many samples and Fe and Cu contents do not correlate as found for crusts from other areas. Large atolls inhibit the formation of thick crusts to water depths as great as 2,000 m because mass wasting continuously sheds reef debris down the volcano flanks. The first K-Ar ages of volcanic rocks dredged from the Marshall Islands are 82 to 85 Ma.

Analyses are still underway on samples from the third U.S. Geological Survey cruise in the central Pacific, which surveyed Karin Ridge and an unnamed ridge south of Johnston Island, both within the Johnston Island EEZ (Hein et al., 1987c). The crusts from these ridges are the thickest crusts recovered to date that are enriched in Co and Pt (average 0.70% Co; 0.36 ppm Pt). The thickest crust recovered is 160 mm and one dredge haul had an average crust thickness of 65 mm and another dredge haul an average thickness of 100 mm. Thick crusts are composed of up to eight distinct ferromanganese-oxide layers in contrast to the two layers (growth generations) noted in earlier studies (Halbach et al., 1982).

University of Hawaii Studies

Two cruises in 1984 by the University of Hawaii (Table 4) sampled crusts from the Hawaiian EEZ and subdivided the sampling sites into on-axis volcanoes (Cenozoic Hawaiian Ridge) formed along the main hot-spot trace, and off-axis volcanoes (Cretaceous seamounts), which occur away from the main hot-spot trace, but within the 370 km EEZ (Chave et al., 1986; De Carlo et al., 1987a). These studies mainly supported previous conclusions based on data from other areas and the earlier crust studies of the Hawaiian Ridge. However, Chave et al. (1986) also concluded that the chemistry of on-axis and off-axis crusts are essentially the same. Shallow-water crusts (<1500 m) are compositionally and fundamentally different from deeper-water crusts. Deeper-water crusts may have had a more complex origin that involved several processes of accretion. Based on the work of Martin and Knauer (unpublished and 1985), Chave et al. (1986) concluded that cobalt enrichment is the result of both large fluxes of leachable cobalt at water depths of 500 to 1,500 m and the westward advection of manganese along equal-density surfaces within the oxygen-minimum zone (see also Aplin and Cronan, 1985).

De Carlo et al. (1987a) confirmed the presence of manjiroite in crusts, as reported earlier by Hein et al. (1985b), and applied multivariate statistical techniques to chemical data to better understand end-member compositions. De Carlo et al. (1987a) determined the following:

1. Multivariate statistical analysis defined three groups of elements that they interpreted as detrital (Si, Al, K, Ti, Fe, Ca, Cu, Pb, Na), hydrogenetic (Pb, V, Mn, Co, Ni, Ca, Fe), and hydrothermal (Cd, Mg, Zn, Ba, Mn, Al, K, Na, Si) end members. Most crusts are simple mixtures of the hydrogenetic and detrital end members.
2. REE patterns for most samples show typical patterns for hydrogenetic precipitation of ferromanganese oxides with a large positive Ce anomaly. Two crusts of possible hydrothermal origin have low total REE and no Ce anomaly, which is consistent with the REE data of Hein et al. (1987b).
3. Element distributions that vary with water depth in other areas, for example cobalt, also occur in the Hawaiian Archipelago, in contrast to the earlier results of Frank et al. (1976) and Craig et al. (1982).
4. The relative enrichment of iron in deeper-water crusts resulted from a combination of increased detrital input with increasing water depth and increased manganese flux at shallower-water depths.

The next cruise by the University of Hawaii (Table 4) was in 1986 to Kiribati (Phoenix Islands) and Tuvalu (Ellice Islands) where six dredge hauls were taken for crust studies (Keating et al., 1986; De Carlo et al., 1987b). Data on samples from these dredges support previous work. In addition, a positive correlation was found between average crust thickness and water depth (De Carlo et al., 1987b). Because cobalt contents decrease with increasing water depth, average cobalt contents decrease with increasing average crust thickness. This relationship has not been found in other areas and it may be fortuitous, resulting from the small sample size (six dredges). However, if valid, it indicates that cobalt grade decreases with increasing tonnage of the mineral deposit for the Tuvalu and Phoenix Islands areas.

The final cruise by the University of Hawaii was in 1987 to the northern Cook Islands on the Manihiki Plateau and to the central Line Islands (Kiribati) (De Carlo and Fraley, this volume). They found generally thin crusts on the Manihiki Plateau volcanic edifices. Cobalt and nickel contents are higher in crusts collected near a field of submarine mud volcanoes than at other places on Manihiki Plateau. Crusts and nodules were recovered from a 2,000-m deep ridge in the central Line Islands. The cobalt contents of these Line Island crusts are not as high as those in the Phoenix Island group of Kiribati (De Carlo, personal communication, 1988; De Carlo and Fraley, this volume).

Other Central Pacific Studies

Ferromanganese crusts from the Line Islands (Kiribati) were collected from 16 sites between 15°S and 17°N latitudes in 1979 by the University of Hawaii. Samples were studied chemically and mineralogically by scientists at Imperial College, London (Aplin, 1984; Aplin and Cronan, 1985). Their main conclusions include: (1) Most compositional variability is related to water depth; (2) manganophile elements are enriched in crusts from water depths of 1-2 km relative to deeper-water crusts, and are enriched to a greater degree than is manganese; (3) crust compositions are mainly determined by Mn/Fe ratios and scavenging of dissolved trace metals by oxide flocs suspended in seawater; (4) cobalt and lead are oxidized at the MnO_2 surface; (5) manganese is supplied via advective-diffusion in the oxygen-minimum zone from continental borderlands; (6) shallower-water crusts (1-2 km) are 50% depleted in total REE, are more fractionated, and are heavy REE enriched compared to deeper-water crusts; and (7) Fe oxides are REE enriched and exhibit flat shale-normalized patterns compared to Mn oxides, which exhibit heavy REE enriched patterns. The internal structure of a crust from the northern Line Islands was mapped by Kang (1987). Because of the broad boundaries between growth layers, he suggested that remobilization of elements occurred within the crust, and that diagenesis increased with increasing crust thickness.

In addition to the area-specific studies cited above, several data compilations and syntheses have been presented (Cronan, 1984; Hein et al., 1986, 1987a; Manheim and Hein, 1986; Manheim, 1986; Manheim and Lane, eds., 1987). Most of the ideas on crust formation and chemical characteristics cited above (compositions and interelement associations) published prior to 1987 are summarized by Hein et al. (1987a).

REGIONAL VARIABILITY

As an introduction to the following discussions of compositional variability of crusts in the central Pacific, we will summarize the broad- scale, pan-Pacific distribution of Mn, Fe, Co, Ni, Cu, Pb, Ti, Si, and P in ferromanganese deposits (crusts and stratabound oxide deposits; exclusive of abyssal nodules) presented by Hein et al. (1987a). Manheim (1986) provided average Co and Mn values of crusts from each ocean basin and for subregions of the Pacific. Considering crusts from the entire Pacific, Hein et al. (1987a) showed that Co and Pt are highest in crusts from the central Pacific and from French Polynesia. Si, Al, and other elements related to detrital minerals increase in crusts with proximity to continents and active island arcs. The Fe/Mn ratio is greatest for island arc and borderland areas. Crusts are lowest in Co, Ni, Pb, and other minor and trace metals at oceanic spreading axes, where either Mn oxide or Fe oxyhydroxide is the dominant phase. Concentrations of Mo are highest in island arc stratabound manganese deposits and in some spreading center deposits (Hein et al., 1988b)

The chemical variability of Co- and Pt-rich crusts from different island groups of the central Pacific is minor when compared to Pacific ocean ferromanganese deposits as a whole, yet important differences in central Pacific crusts are noted (Tables 2, 3) and can be related to oceanographic and geologic conditions.

Distribution of Elements

The average cobalt content of central Pacific crusts is 0.79%, somewhat higher than previous estimates (Table 2). This concentration of cobalt is near the grade considered as the cut-off average grade (0.80%) for crusts to be of potential economic interest (Manheim, 1986; Hein et al., 1987a). Other strategically important metals in central Pacific bulk crusts include Mn, Ni, and Pt which average 23%, 0.47%, and 0.24 ppm, respectively.

Cobalt is higher than the regional average (0.79%) in bulk crusts from the EEZs of Howland-Baker, Kingman-Palmyra, Marshall, and Tuvalu islands, (Table 2; Figs. 1,3). Cobalt is lower than the regional average in bulk crusts from the EEZs of Kiribati and Tokelau islands. And, cobalt is at the regional average for the EEZs of Johnston Island (28% of the reported values), Hawaiian Islands (30% of the reported values) and crusts from international waters, which is expected because crusts from international waters are distributed throughout the central Pacific (Table 2). Conclusions based on the few samples from Tuvalu and Tokelau should be viewed with caution. The Tuvalu average is based on the average of 18 analyses from two dredge hauls (De Carlo et al., 1987b), whereas the Tokelau average is based on the average of only two analyses from one dredge haul.

Relative to the regional average (0.24 ppm), platinum is enriched only in crusts from the Marshall Islands (0.32 ppm; 28% of the reported values). More data are needed to provide a clearer picture of the distribution of platinum.

Analysis of the outer layers of a crust provides a better picture of the recent hydrogenetic precipitation. The outer 10 mm of a crust may represent 2 to 5 m.y. of accretion, whereas the bulk crusts may represent 5 to 30 m.y. of accretion depending on the thickness and changes in growth rates with time. Data are available on a third as many outer layers of crusts (102) as on bulk crusts (311) (Table 3). Cobalt averages 1.02% for the outer layers of crusts from the central Pacific. Cobalt and nickel are enriched over the central Pacific average in outer crusts from the EEZ of Johnston Island and Kingman- Palmyra Islands, and cobalt and nickel are depleted in outer crusts from the EEZ of the

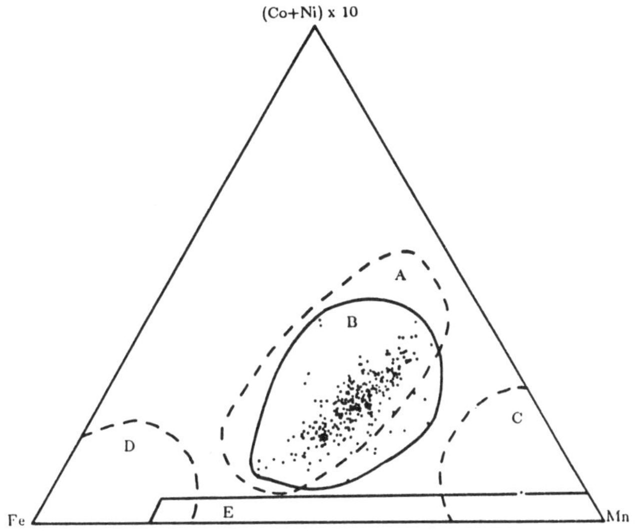

Figure 4. Ternary diagram of (Co + Ni) × 10:Fe:Mn for 308 samples from the central Pacific. Field A is defined by 161 Pacific hydrogenous ferromanganese nodules (based on Co + Ni + Cu × 10) from Mero (1965); Field B is defined by 308 hydrogenetic crusts from the central Pacific; Field C is defined by a mixture of 18 diagenetic and 3 hydrothermal deposits (located at the Mn apex); Field E is defined by 30 samples of hydrothermal manganese from the Mariana Islands arc (Hein et al., 1987b); Field D is defined by 10 hydrothermal samples from the East Pacific Rise and 19 hydrothermal deposits from volcanic edifices. Fields A, C, and D are as presented by Bonatti et al. (1972).

Marshall Islands and Hawaiian Islands (Table 3). The outer crust layers, relative to bulk crusts, are enriched in Mn, Co, and Ni; depleted in Fe, Cu, Si, Ti, P; and are similar in Pb content. Platinum is also depleted in outer crust layers relative to inner layers (Halbach et al., 1984; Hein et al., 1988a).

Bonatti et al. (1972) classified marine ferromanganese deposits by their (Cu + Co + Ni) × 10:Fe:Mn contents plotted on a ternary diagram. Data in this report from the central Pacific define a hydrogenetic crust field that is enriched in Mn and depleted in Co + Ni + Cu relative to the hydrogenous nodule field (Bonatti et al., 1972) (Fig. 4). This is partly due to the fact that we use (Co + Ni) × 10, leaving out Cu (8.7% of the Cu + Ni + Co average) in our plot because, in crusts, Cu follows a different geochemical pathway than do Ni and Co (see next section); we therefore exclude it from the association with Co and Ni in the ternary plot (Figs. 4,5). The Mn-rich point on Figure 4 is for a crust from the Nova Trough (located in international waters between the EEZs of Jarvis Island and the Phoenix Islands), which probably has a hydrothermal or diagenetic origin. On the average, one crust sample per seamount dredged in the central Pacific is not purely hydrogenetic in origin, but rather has a hydrothermal or diagenetic origin as indicated by its minor element chemistry and todorokite and birnessite mineralogy (Hein et al., 1985b, 1987a; De Carlo et al., 1987a). The δ-MnO_2 mineralogy is characteristic of hydrogenetic crusts.

The average (Co + Ni) × 10:Fe:Mn values for each island group show a relatively narrow range of average Mn values but broad ranges for average Fe and average (Co + Ni) × 10 values (Fig. 5). Tokelau has the most Co + Ni depleted and Fe enriched crusts, whereas Tuvalu shows the opposite trend. Although these marked chemical differences from geographically close sampling sites may prove to be valid, the few analyses available for crusts from these two areas make us cautious about drawing conclusions. Crusts from international waters and the average for bulk crusts plot together, which is expected. Crusts from the Kingman-Palmyra EEZ and the average outer crust layers also plot together. This is expected because most Kingman-Palmyra crusts are thin owing to extensive reworking of deposits on the slopes of S.P. Lee Guyot, the main seamount sampled in that area (Hein et al., 1985b). The thin crusts represent the same growth generation as the outer layers of thicker crusts. Analyses of crusts from the Johnston Island and the Howland-Baker Islands EEZs plot together (Fig. 5) even though those islands are separated by 17° of latitude (~1850 km).

Interelement Associations

We calculated correlation coefficients for chemical analyses of bulk crusts from ten areas of the central Pacific, and for chemical analyses of the outer layers of crusts from six areas[2]. In addition, correlation matrices have been published for four areas of the central Pacific[2]. The following

---

[2]correlation coefficients were calculated for chemical analyses of bulk crusts from all areas combined (based on 10 chemical elements), international waters (9 elements), Howland-Baker (9 elements), Kingman-Palmyra (24 elements), Marshall Islands (25 elements), Johnston Island (25 elements), Kiribati (9 elements), Hawaii on-axis (8 elements), Hawaii off-axis (10 elements), Hawaii off- and on-axis combined (10 elements); for outer layers of crusts from all areas combined (10 elements), international waters (10 elements), Kingman- Palmyra (24 elements), Marshall Islands (24 elements), Johnston Island (24 elements), and Hawaii (9 elements). Correlation matrices have been published for Hawaii (17 elements; De Carlo et al., 1987a), Tuvalu and Phoenix Islands (Kiribati 11 elements; De Carlo et al., 1987b), the Line Islands (Kiribati Mn, Fe, and Al against 14 elements; Aplin and Cronan, 1985), and the Marshall Islands (24 elements, Hein et al., 1988a).

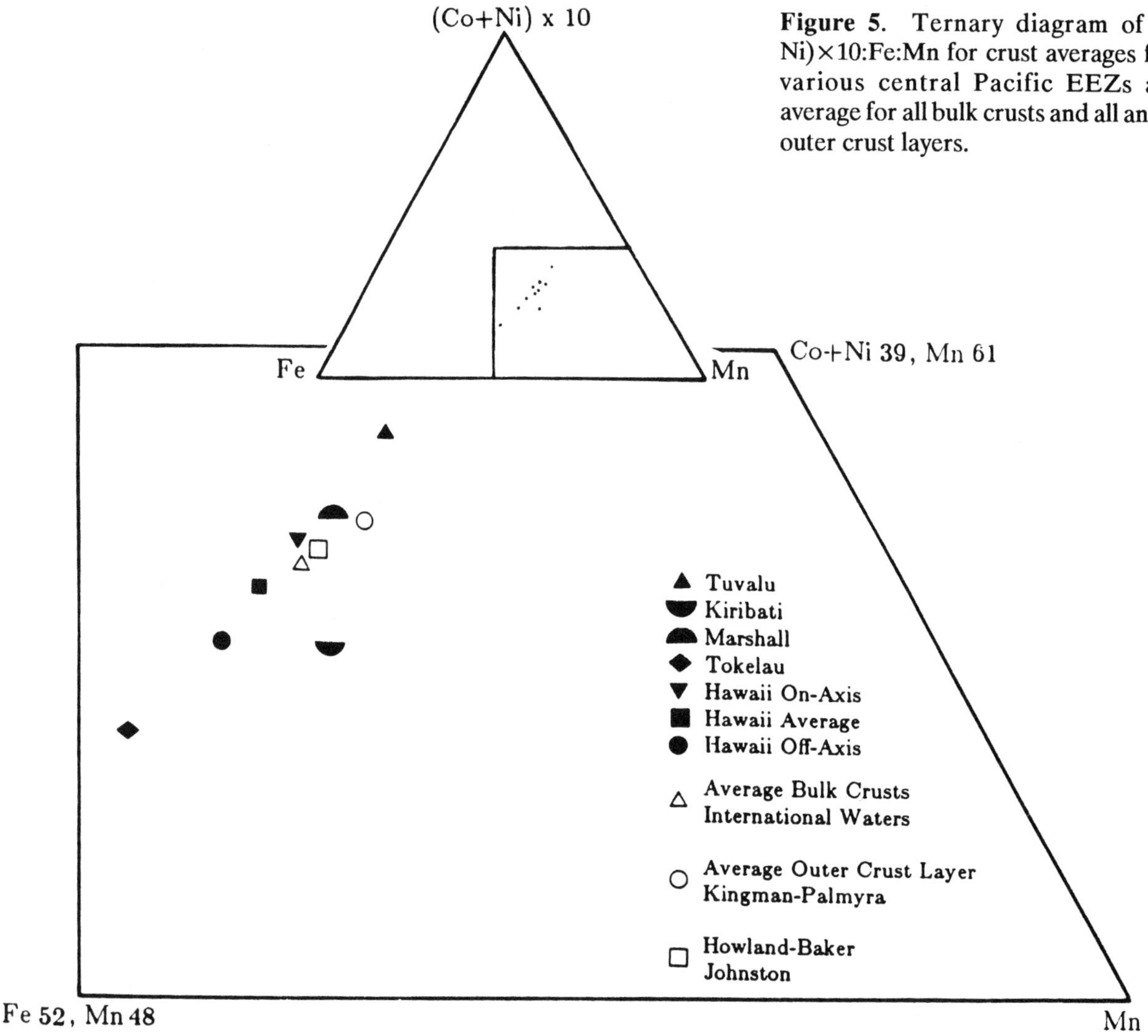

Figure 5. Ternary diagram of (Co + Ni)×10:Fe:Mn for crust averages from the various central Pacific EEZs and the average for all bulk crusts and all analyses of outer crust layers.

discussion is based on analysis of all sixteen of the correlation matrices constructed for this study. However, because of space limitations, only four matrices are presented: bulk crusts from all areas combined (Table 5), bulk crusts from the Kingman-Palmyra Islands EEZ (Table 6), outer layers of crusts from all areas combined (Table 7), and outer crusts layers from the Marshall Islands EEZ (Table 8). These four matrices were chosen because they show most of the relations discussed in the succeeding sections. Other matrices can be provided on request to those interested.

Interelement correlations vary from area to area, but for bulk crusts three correlations occur in all ten matrices: Mn with Co, Si with Al, and Ca with P. For bulk crusts, common positive correlations among the major elements include Mn with Co, Ni, and Pb; Fe with Ti, Si, As, and water depth; Si with Al, Ti, Fe, Cr and water depth; Al with Cr, Ti, Si, K, and water depth; P with Ca, and commonly also with Ni, Cu, Sr, and Y; Mg with Cd, Zn, and Ni; Ba with Ce and Cu; and Zn with Ce, Mg, Cd, and Ni. Pt is most commonly correlated with Mn or manganophile elements. There is also a consistent negative correlation of Mn and manganophile elements with water depth.

All six correlation matrices for the outer layers of crusts show positive correlations of Mn with Ni and Co; Fe with Sr; Al with Si and K; P with Ca; Mg with Ni and Zn; and Zn with Cu (for example see Tables 7 and 8). Other common positive correlations include Mn with Pb and Pt; Fe with Si and Ti; Si with Fe, Ti, Al, and water depth; Al with Na and Cr; P with Cr and Cu; Mg with Pb, Na, Mn, and Cd; Ba with Ce, Zn, Sr, and Cu; Zn with Ba, K, Mg, and Ce; and Pt with Ni and Co. A negative correlation commonly exists between Mn and water depth. Other correlations occur among different elements in different island groups (see Tables 5–8 for examples).

Interelement correlations are generally stronger and more abundant in the correlation matrices for outer crust layers compared to the matrices of bulk crusts (Tables 5–8). This is expected because the outer layers represent

Table 5. Correlation coefficient matrix for bulk crusts, all data. Correlations at greater than 95% confidence level are boldfaced and depend on the number of samples. Fewer Pt analyses are available than analyses for the other elements, which vary little in number of available analyses.

	Fe/Mn	Mn	Fe	Co	Ni	Cu	Pb	Ti
Mn	-0.730							
Fe	**0.770**	-0.210						
Co	-0.555	**0.650**	-0.256					
Ni	-0.671	**0.561**	-0.524	**0.410**				
Cu	0.039	-0.121	-0.052	-0.208	0.116			
Pb	-0.035	**0.184**	0.075	**0.322**	0.072	-0.068		
Ti	**0.600**	-0.305	**0.600**	-0.200	-0.351	0.067	-0.040	
Si	**0.489**	-0.449	**0.363**	-0.531	-0.279	-0.001	-0.164	**0.220**
P	-0.244	-0.004	-0.357	-0.089	**0.158**	**0.371**	0.074	-0.301
Pt	-0.445	0.184	-0.365	0.333	-0.017	-0.342	-0.368	-0.307
Depth	**0.463**	-0.450	**0.261**	-0.527	-0.308	**0.219**	-0.285	**0.349**
Lat.	0.099	-0.084	**0.110**	0.082	-0.154	-0.068	**0.367**	0.055
Long.	-0.059	-0.123	-0.150	0.005	-0.022	**0.127**	-0.002	-0.191

	Si	P	Pt	Depth	Lat.
P	-0.207				
Pt	-0.210	0.137			
Depth	**0.164**	-0.248	-0.520		
Lat.	**0.284**	0.007	-0.159	-0.404	
Long.	-0.093	-0.062	0.414	-0.014	-0.044

the integration of hydrogenetic precipitation over shorter time intervals, and therefore incorporate fewer compositional variations caused by changing oceanographic, atmospheric, and geologic conditions (Halbach et al., 1982; Segl et al., 1984; Hein et al., 1985a,b). Some specific differences among interelement correlations of bulk crusts and outer layers of crusts include a weaker association of Fe, Si, and Al with water depth in the outer crust layers; and a stronger association of Mn with Ni, Pb, Pt, Mg, Mo, Sr, Ba, and water depth in outer crust layers.

Interelement associations can be related to the mineralogical composition of the crusts by noting the distribution of major elements among the various correlations and by knowing the X-ray mineralogy (XRD) of the samples. It is clear from analysis of all the matrices that Mn occurs predominantly in one phase, $\delta-MnO_2$ as determined by XRD, only rarely is it associated with a biogenic phase. In contrast, the associations of Fe are complex and at least three phases are indicated by the correlations, including Fe in the aluminosilicate detrital phase (eolian and volcanic debris), an X-ray amorphous Fe oxyhydroxide phase, and an X-ray amorphous Fe silicate phase. From the correlation coefficients and the X-ray mineralogy, we interpret the major phases and their associated elements to be: (1) δ-MnO_2, Mn, Co, Ni, Pb, Zn, Mg, Ce, Mo, Cd, ±Sr, Ba, Ti, Na, V, K. This phase decreases with increasing water depth. (2) Detrital-aluminosilicate phase, Si, Al, Ti, Fe, Cr, K, V, ±Na, Mg, Ca, Cu, As. This phase increases with increasing water depth. (3) Amorphous $FeOOH \cdot xH_2O$, Fe, Sr, As, Y, ±Ce, Cu, Cd. This phase increases with increasing water depth. (4) Amorphous Fe silicate, Fe, Si, ±Pb, V, Cd, Al, Ti. This phase probably increases with increasing water depth. (5) Phosphorite-biogenic carbonate fluorapatite, P, Ca, ±Cr, Cu, Ni, Zn, Pb, Ce, Mg, Mo, Ba, Sr, Na, Y, Cd, Mn. This phase decreases with increasing water depth. (6) Biogenic-barite, calcite, amorphous phases, Ba, Mg, Zn, Sr, Ca, Ce, ±Pb, Cd, Ni, Na, Mo, Cu, Y, Ti, As, Pt. This phase probably decreases with increasing water depth.

Some elements, for example, Fe, Ti, and V, occur in several phases, whereas others, for example Mn, Co, Al, and P are essentially restricted to one phase. The abundance of

Table 6. Correlation coefficient matrix for bulk crusts, Kingman-Palmyra EEZ. Correlations at greater than 95% confidence level are boldfaced.

	Fe/Mn	Si	Na	Ca	Fe	K	Mg	P	As	Ba	Cd	Co	Cr	Cu	Mn	Mo
Si	**0.849**															
Na	-0.249	-0.042														
Ca	**-0.603**	**-0.544**	-0.203													
Fe	**0.954**	**0.766**	-0.162	**-0.686**												
K	-0.200	0.205	0.158	0.020	-0.299											
Mg	**-0.638**	**-0.416**	0.148	**0.462**	**-0.687**	**0.463**										
P	**-0.605**	**-0.539**	-0.223	**0.996**	**-0.698**	0.062	**0.509**									
As	0.259	0.010	-0.037	**-0.511**	**0.432**	-0.338	**-0.599**	**-0.533**								
Ba	-0.319	**-0.508**	-0.310	**0.482**	-0.330	-0.027	0.346	**0.491**	-0.140							
Cd	**-0.759**	**-0.506**	0.212	**0.526**	**-0.780**	**0.395**	**0.874**	**0.553**	**-0.552**	0.203						
Co	**-0.622**	**-0.396**	**0.553**	0.047	**-0.506**	**0.412**	**0.445**	0.060	0.100	0.040	**0.484**					
Cr	-0.026	0.225	0.251	-0.050	-0.093	**0.576**	0.381	-0.023	-0.461	-0.258	0.269	0.084				
Cu	-0.147	-0.113	-0.185	**0.461**	-0.176	0.020	0.278	**0.453**	-0.349	**0.515**	0.197	-0.075	-0.297			
Mn	**-0.699**	**-0.761**	**0.455**	0.195	**-0.474**	-0.059	**0.422**	0.178	0.106	0.248	**0.511**	**0.652**	-0.062	0.100		
Mo	**-0.547**	**-0.817**	0.273	0.278	**-0.430**	-0.326	0.114	0.250	0.256	0.360	0.221	0.320	-0.321	-0.059	**0.603**	
Ni	**-0.752**	**-0.540**	0.129	**0.489**	**-0.773**	0.361	**0.915**	**0.518**	**-0.567**	0.263	**0.900**	**0.397**	0.374	0.294	**0.490**	0.171
Pb	**-0.663**	**-0.704**	**0.469**	0.287	**-0.532**	-0.107	0.284	0.272	0.122	0.121	**0.410**	**0.656**	-0.018	-0.115	**0.678**	**0.719**
Sr	0.186	-0.285	-0.215	-0.076	0.314	**-0.677**	**-0.610**	-0.112	**0.558**	0.267	**-0.615**	-0.314	**-0.612**	-0.037	0.093	**0.470**
Ti	**0.650**	**0.616**	0.100	**-0.762**	**0.738**	-0.025	**-0.534**	**-0.771**	**0.517**	**-0.392**	**-0.564**	-0.009	-0.021	-0.345	-0.314	**-0.384**
V	0.241	-0.161	-0.052	-0.373	**0.400**	**-0.477**	**-0.539**	**-0.395**	**0.601**	0.064	**-0.454**	-0.210	**-0.514**	-0.232	0.182	**0.459**
Zn	-0.205	-0.263	-0.240	0.306	-0.220	0.127	**0.678**	0.341	**-0.500**	**0.479**	**0.557**	-0.072	0.053	0.358	0.206	-0.146
Al	0.347	**0.704**	0.149	-0.192	0.210	**0.606**	0.248	-0.157	**-0.447**	-0.320	0.060	-0.020	**0.687**	0.052	**-0.533**	**-0.779**
Y	**-0.463**	**-0.443**	-0.158	**0.668**	**-0.560**	0.076	**0.491**	**0.692**	**-0.428**	0.352	0.348	0.150	0.079	0.236	0.077	0.117
Ce	-0.052	-0.366	0.207	-0.039	0.072	**-0.389**	-0.372	-0.084	0.322	0.193	-0.349	0.147	-0.191	-0.104	0.330	**0.519**
Depth	**0.830**	**0.849**	-0.239	**-0.464**	**0.719**	0.000	-0.375	**-0.456**	-0.141	-0.295	**-0.463**	**-0.635**	0.091	0.044	**-0.653**	**-0.714**
Lat.	-0.295	-0.121	-0.250	0.173	-0.273	0.236	0.208	0.187	-0.071	0.163	0.215	0.316	-0.252	0.148	0.226	-0.144
Long.	0.003	-0.001	**0.438**	0.131	-0.048	-0.362	-0.069	0.092	-0.328	-0.141	0.094	-0.203	-0.065	-0.129	-0.044	0.232

	Ni	Pb	Sr	Ti	V	Zn	Al	Y	Ce	Depth	Lat.
Pb	0.288										
Sr	**-0.580**	0.112									
Ti	**-0.630**	-0.140	0.094								
V	**-0.510**	0.062	**0.797**	0.228							
Zn	**0.592**	-0.120	-0.305	-0.327	-0.151						
Al	0.084	**-0.410**	**-0.719**	0.308	**-0.606**	0.099					
Y	**0.478**	0.088	-0.105	**-0.611**	-0.316	0.286	-0.075				
Ce	-0.350	**0.481**	**0.644**	0.179	**0.525**	-0.264	**-0.512**	-0.020			
Depth	**-0.440**	**-0.811**	-0.136	0.327	-0.078	0.022	**0.492**	-0.364	-0.350		
Lat.	0.243	-0.001	-0.261	-0.059	-0.307	0.274	0.026	0.212	-0.189	-0.118	
Long.	-0.001	0.088	-0.040	-0.186	-0.012	-0.025	-0.114	0.105	0.148	0.060	-0.189

Table 7. Correlation coefficient matrix for outer crust layers, all data. Correlations at greater than 95% confidence level are boldfaced.

	Fe/Mn	Mn	Fe	Co	Ni	Cu	Pb	Ti
Mn	**-0.746**							
Fe	**0.820**	**-0.335**						
Co	**-0.701**	**0.797**	**-0.439**					
Ni	**-0.824**	**0.827**	**-0.626**	**0.666**				
Cu	0.181	-0.072	0.159	**-0.334**	-0.009			
Pb	-0.182	**0.493**	0.115	**0.440**	**0.278**	-0.183		
Ti	**0.615**	**-0.223**	**0.690**	**-0.249**	**-0.360**	**0.460**	-0.040	
Si	**0.850**	**-0.548**	**0.703**	**-0.567**	**-0.588**	**0.327**	-0.165	**0.669**
P	-0.149	0.018	**-0.211**	-0.094	0.190	0.140	0.148	-0.105
Pt	-0.733	0.679	-0.854	0.611	**0.954**	-0.219	-0.024	**-0.628**
Depth	**0.477**	**-0.528**	0.193	**-0.463**	**-0.442**	0.130	**-0.518**	**0.301**
Lat	**0.438**	**-0.224**	**0.411**	**-0.350**	**-0.251**	**0.358**	0.119	**0.275**
Long	0.053	**-0.495**	**-0.241**	**-0.222**	**-0.324**	**-0.288**	**-0.244**	**-0.258**

	Si	P	Pt	Depth	Lat
P	0.008				
Pt	**-0.719**	0.321			
Depth	**0.433**	-0.135	-0.413		
Lat	**0.511**	0.047	-0.614	-0.157	
Long	**-0.202**	-0.165	**0.505**	0.191	**-0.350**

each phase varies among the island groups, so we used Q-mode factor analysis to discriminate these variations.

Grouping of Elements: Q-mode Factor Analysis

A Q-mode factor analysis was computed for chemical analyses of bulk crusts collected by the USGS from the EEZ of the Marshall Islands, Kingman-Palmyra Islands, and Johnston Island (Figs. 6, 7). Four factors for the Johnston and Marshall Islands and five factors for Kingman-Palmyra account for 99% of the variance in each original data set. The commonality for each sample in each data set is greater than 0.9. We interpret the five factors to represent $\delta-MnO_2$ phase, Fe phases, detrital (aluminosilicate phase), biogenic phosphate phase, and biogenic-nonphosphate phases (predominantly from dissolution of biogenic debris-residual). These are the same phases determined from analysis of the correlation coefficients, except that only one iron phase is identified by Q-mode analysis. De Carlo et al. (1987a) interpreted a factor containing a similar set of elements as our biogenic-nonphosphate factor as a hydrothermal factor. However, the elements associated with this factor are not characteristic of known hydrothermal ferromanganese deposits (Hein et al., 1987b), but rather are characteristic of elements (Cu, Zn, Ba, Cd, Mg) associated with biological removal from the water column (Bruland, 1983; Martin and Knauer, 1985).

P and Ca, characteristic of our biogenic phosphate factor, have high factor scores for the detrital factor of the Marshall Islands crusts (Fig. 6C). The average P content is very low for Marshall Islands crusts (Fig. 8) and a biogenic phosphate factor was not interpreted from the Q-mode analysis. However, phosphatized volcanogenic rocks are common crust-substrate rocks in the Marshall Islands (Hein et al., 1988a), and P should covary with Si, Al, etc., as a detrital component. The Q-mode results show that detrital phosphorite (reworked) is more important than primary biogenic phosphorite in Marshall Islands crusts. Biogenic phosphate and biogenic-nonphosphate factors occur together only for crusts from Kingman-Palmyra Islands, which is the only area of the three analyzed that occurs solely within the equatorial zone of high biological productivity.

Table 8. Correlation coefficient matrix for outer crust layers, Marshall Islands EEZ. Correlations at greater than 95% confidence level are boldfaced.

	Fe/Mn	Si	Na	Ca	Fe	K	Mg	P	As	Ba	Cd	Co	Cr	Cu	Mn	Mo
Si	**0.743**															
Na	-0.158	-0.103														
Ca	-0.045	0.099	-0.017													
Fe	**0.953**	**0.619**	-0.165	-0.093												
K	-0.239	0.114	-0.049	0.154	-0.244											
Mg	**-0.517**	-0.295	0.377	0.316	**-0.545**	0.242										
P	0.346	0.415	0.004	**0.857**	0.245	-0.066	0.123									
As	**0.762**	**0.494**	-0.022	-0.025	**0.806**	-0.325	**-0.637**	0.311								
Ba	-0.134	-0.168	-0.185	0.056	-0.138	0.016	**0.524**	-0.098	**-0.441**							
Cd	**-0.557**	-0.352	-0.005	-0.070	**-0.565**	**0.406**	**0.537**	-0.282	**-0.688**	0.390						
Co	**-0.726**	**-0.539**	0.069	-0.090	**-0.768**	0.155	0.128	-0.324	**-0.685**	-0.205	0.341					
Cr	0.268	0.280	-0.301	**0.567**	0.166	-0.114	0.146	**0.657**	-0.097	0.340	-0.058	-0.163				
Cu	-0.230	-0.219	0.251	0.007	-0.212	0.090	**0.801**	-0.074	-0.370	**0.764**	**0.472**	-0.258	0.090			
Mn	**-0.865**	**-0.782**	0.170	-0.063	**-0.679**	0.133	0.355	**-0.450**	**-0.520**	0.061	0.374	**0.534**	-0.400	0.178		
Mo	-0.059	-0.230	0.142	-0.192	0.121	-0.296	-0.046	-0.145	0.411	0.070	-0.125	-0.394	-0.362	0.248	0.340	
Ni	**-0.729**	**-0.492**	0.117	0.079	**-0.700**	0.291	**0.872**	-0.235	**-0.775**	**0.598**	**0.662**	0.264	-0.024	**0.749**	**0.601**	0.052
Pb	0.097	0.030	-0.046	-0.062	0.020	-0.172	-0.255	0.179	0.235	-0.181	-0.152	0.166	0.047	-0.195	-0.200	0.199
Sr	**0.522**	0.372	-0.229	0.042	**0.644**	-0.197	-0.371	0.214	**0.598**	0.050	-0.365	**-0.567**	0.199	-0.127	-0.206	**0.478**
Ti	**-0.541**	**-0.436**	-0.202	-0.125	**-0.595**	0.211	0.373	-0.399	**-0.885**	**0.474**	**0.553**	**0.679**	0.125	0.220	0.332	**-0.539**
V	**0.599**	0.383	0.015	0.014	**0.739**	-0.283	-0.409	0.269	**0.842**	-0.203	**-0.479**	**-0.762**	-0.013	-0.111	-0.254	**0.649**
Zn	**-0.629**	-0.395	0.066	0.143	**-0.595**	0.376	**0.855**	-0.172	**-0.781**	**0.609**	**0.660**	0.216	0.052	**0.717**	**0.512**	-0.094
Al	0.316	**0.676**	0.060	0.158	0.196	**0.559**	0.322	0.278	-0.071	0.195	0.145	-0.322	0.316	0.326	**-0.474**	-0.379
Y	**0.523**	0.260	**-0.497**	0.021	**0.608**	-0.151	**-0.765**	0.143	**0.646**	**-0.472**	**-0.511**	-0.254	-0.014	**-0.712**	-0.269	0.047
Ce	-0.376	-0.349	-0.229	-0.083	**-0.468**	0.181	0.202	-0.288	**-0.712**	0.286	0.407	**0.684**	0.044	0.003	0.141	**-0.613**
Depth	0.082	-0.014	-0.109	-0.104	0.233	-0.105	-0.122	-0.305	-0.001	0.282	0.015	-0.288	-0.061	0.068	0.197	0.137
Lat.	0.243	0.162	-0.033	0.076	0.077	-0.219	-0.052	0.391	0.201	-0.056	-0.165	0.030	0.250	-0.041	**-0.467**	-0.014
Long.	**-0.481**	-0.334	0.063	-0.141	-0.316	0.268	0.132	**-0.477**	-0.345	-0.019	0.342	0.234	-0.355	0.045	**0.649**	0.013

	Ni	Pb	Sr	Ti	V	Zn	Al	Y	Ce	Pt	Depth
Pb	-0.343										
Sr	-0.404	0.050									
Ti	**0.541**	-0.123	-0.497								
V	**-0.497**	0.048	**0.804**	**-0.851**							
Zn	**0.946**	**-0.457**	-0.388	**0.569**	**-0.484**						
Al	0.138	-0.149	-0.015	-0.013	-0.037	0.257					
Y	**-0.710**	-0.002	**0.548**	**-0.422**	**0.507**	**-0.647**	-0.308				
Ce	0.305	0.071	**-0.515**	**0.905**	**-0.826**	0.345	-0.073	-0.230			
Depth	0.058	**-0.674**	0.293	0.042	0.221	0.087	-0.160	0.206	-0.111		
Lat.	-0.276	**0.872**	-0.035	-0.096	-0.037	-0.312	0.056	-0.083	0.120	-0.799	
Long.	0.382	**-0.735**	-0.077	0.223	-0.112	**0.430**	-0.157	-0.029	-0.019	**0.554**	-0.900

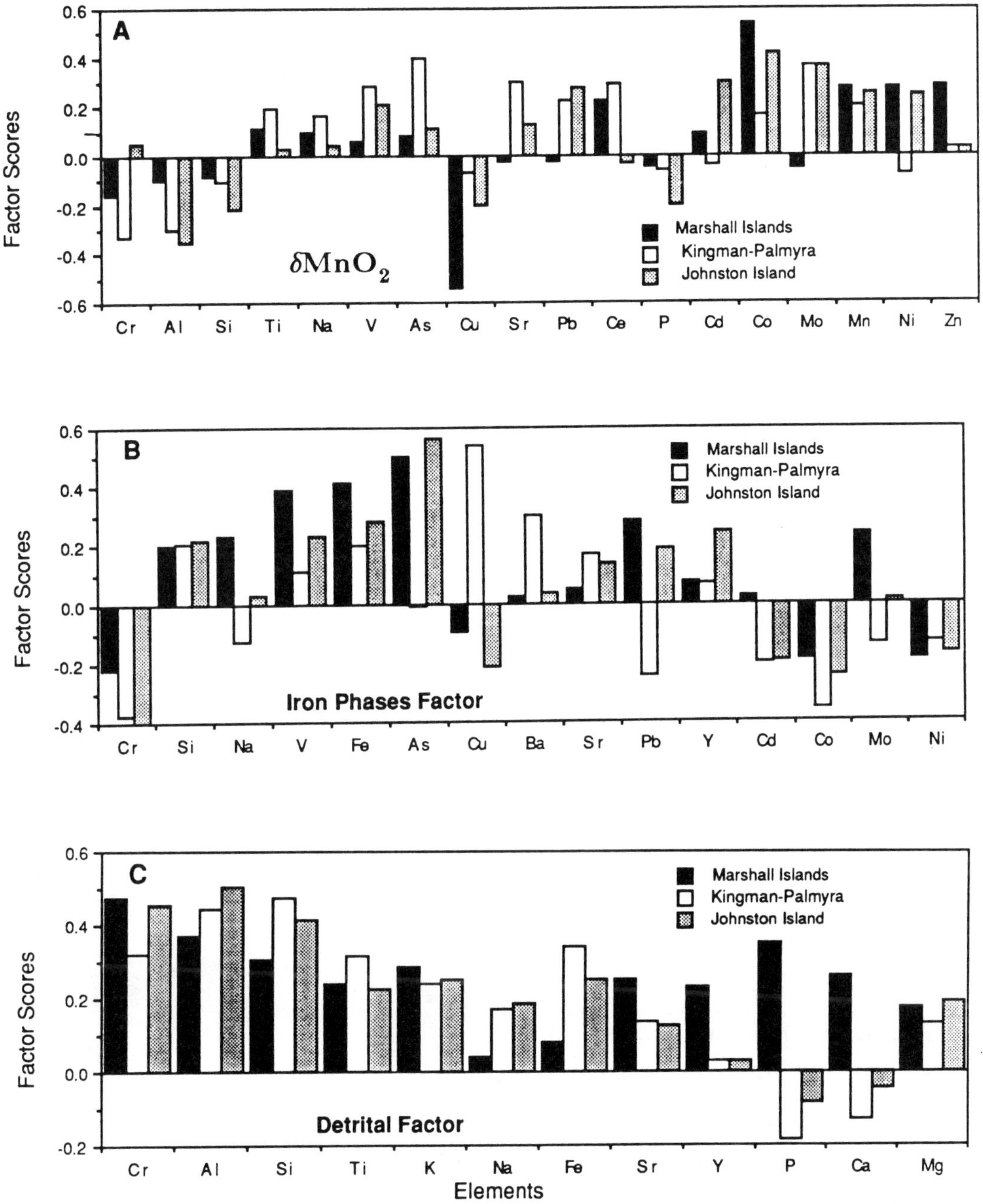

Figure 6. Q-mode factor scores for (A) δ-MnO_2 factor, (B) iron phases factor, and (C) detrital factor for crusts from the EEZs of the Marshall, Kingman-Palmyra, and Johnston Islands. If all three areas had factor scores for any element between 0 and | 0.1600 |, those elements were excluded from the graph; because of random noise it is difficult to resolve the orientation of the factor to within 10° of an absolute direction in variable space.

The dominant positive factor scores that define each factor include: δ-MnO_2, Co and Mn for crusts from all three EEZs, Ni, Ce, Mo, Pb, and V for two EEZs, and Zn, Cd, As, Sr, and Ti for one of the three EEZs (Fig. 6A); Fe-phases, Fe and Si for crusts from all three EEZs, As, V, and Pb for two EEZs, and Mo, Na, Y, Cu, Ba, and Sr for one of the three EEZs (Fig. 6B); the detrital factor is the most consistent of the five factors from area to area and includes Al, Cr, Si, Ti, and K for crusts from all three EEZs, Fe, Mg, and Na for two EEZs, and P, Ca, Sr, and Y for only the Marshall Islands, which has a detrital phosphate component as discussed above (Fig. 6C). Consequently, phos-

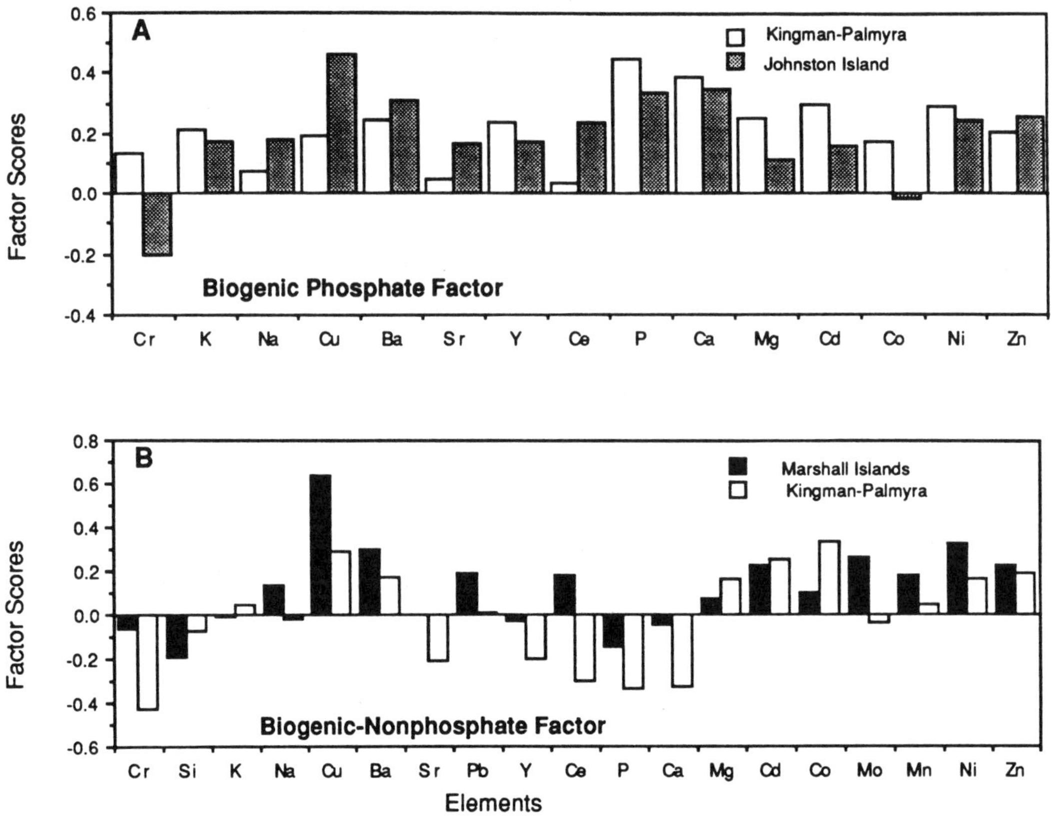

Figure 7. Q-mode factor scores for (A) biogenic phosphate factor and (B) biogenic nonphosphate factor. Factor A does not occur in the Marshall Islands crusts, and factor B is absent from the Johnston Island crusts. See Figure 5 caption for additional information.

phate in the Marshall Islands is part of the detrital factor and is represented by P, Ca, Sr, and Y, which are elements present in the biogenic phosphate factors for the other two areas. In addition, the biogenic phosphate factor for Johnston and Kingman-Palmyra EEZs includes Ba, Cu, Zn, Ni, and K, and Ce, Cd, and Mg occur in the phosphate factor from one or the other of these two EEZs (Fig. 7A). Biogenic-nonphosphate factors were derived from data from the Marshall Islands and Kingman-Palmyra, where Cu, Ni, Ba, Zn, and Cd occur for both areas (Fig. 7B).

Generally, the elements associated with the various crust phases interpreted from the interelement correlations agree well with those grouped in the Q-mode factors. Correlation coefficients commonly allowed placement of more elements with each crust phase and Q-mode did not separate Fe oxyhydroxide and Fe silicate phases. The detrital phase elements were the same for both methods, including the detrital phosphate in the Marshall Islands. The most notable differences include a strong As association with δ-MnO_2, strong Ba association with Fe phases, and strong Co association with biogenic-nonphosphate in Kingman-

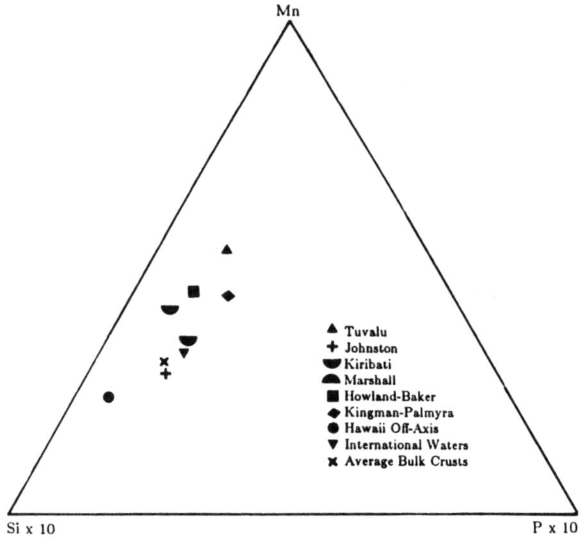

Figure 8. Ternary plot of Mn:Si × 10:P × 10 for crust averages from the various central Pacific EEZs and the average for all bulk crusts.

Palmyra crusts as determined by Q-mode; Cr is strongly positively correlated with P and Ca in many areas, but does not show up in the biogenic phosphate factor by Q-mode analysis.

Summary and Discussion of Regional Variability

Based on correlation coefficients and Q-mode factor analyses, elements are assigned to five major phases, δ-MnO_2, Fe, detrital, phosphate, and biogenic. The relative importance of these phases to the make-up of the crusts varies in the central Pacific.

The importance of the δ-MnO_2 phase relative to the other phases in crusts from the various EEZs is, in decreasing order: Tuvalu, Howland-Baker, Kingman-Palmyra, Marshall, Kiribati, Johnston, and Hawaiian Islands. This distribution is strongly related to latitude and is controlled by the equatorial zone of high biological productivity, which produces a strong and extensive oxygen-minimum zone. Mn and manganophile elements are enriched in the oxygen minimum zone via advection-diffusion from eastern borderlands (see discussion in Historical Perspective section). The Fe phase follows the same distribution as Mn for the areas studied, but the overall variability of the Fe phase is not as great as the Mn phase. Even though Fe and Mn are negatively correlated and Fe increases with water depth whereas Mn decreases with water depth, when bulk crusts from all depths of water are considered, the Fe and Mn phases follow the same primary regional trend in the central Pacific.

Typically, the detrital phase is greatest in Hawaiian crusts and decreases in relative importance in the order from Johnston, to Kiribati, Marshall, Howland-Baker, Kingman-Palmyra, and finally Tuvalu Islands. This distribution is in reverse order to the δ-MnO_2 distribution (Fig. 8). The highest detrital content is for the area beneath the trade wind belt (northernmost area) and thus, the detrital component is probably partly eolian (see Hein et al., 1985b). Equatorial zone crusts ($0\pm8°$ latitude) have less of a detrital component. Kiribati covers a large range of latitudes and longitudes, but most samples analyzed are within the $0\pm10°$ latitude zone. The eastern island groups (Hawaii, Johnston, Line Islands; Hawaii and Johnston also being the northernmost) have a high detrital component, possibly due to input to the crusts of eroded substrate material resuspended by strong bottom currents in the eastern part of the study area. Detrital input is apparently related to both eolian and erosional (bottom current) processes and affects the northernmost and easternmost island groups to the greatest degree.

The phosphate phase is greatest in crusts from the Kingman-Palmyra EEZ and decreases in relative importance in the order Kiribati, Johnston, Tuvalu, Howland-Baker, Marshall Islands, and Hawaii. The phosphate phase is not as variable as the δ-MnO_2 and detrital phases on a regional scale. The phosphate phase of bulk crusts is not clearly distributed by latitude, longitude, or proximity to the equatorial zone of high productivity. Some equatorial areas do not have strong phosphate signatures, although as mentioned, the overall variability is not great. This lack of a clear distribution may be the result of biogenic phosphate being concentrated in the inner layers of crusts that were precipitated during the Miocene or earlier times (Halbach et al., 1983; Segl et al., 1984; Futa et al., 1987, 1988), when the crusts were at different geographic locations (due to plate motion) and at different water depths (due to subsidence of the volcanic edifices). The phosphate phase for the outer layers of the crusts has an importance in the various EEZs in the reverse order as that of the bulk crusts, with the greatest contribution in Hawaii followed by Marshall, Johnston, and Kingman-Palmyra Islands. This inverse distribution demonstrates that the phosphate phase (and also the P content) in crusts has a complex distribution that is not directly related to biological productivity in surface waters.

Co plus Ni, the main elements associated with the δ-MnO_2 phase and the elements of main economic importance, are distributed in crusts in decreasing average abundance, in the EEZs of Tuvalu, Kingman-Palmyra, Howland-Baker, Johnston, Marshall, Hawaiian, Kiribati, and Tokelau islands. With the exception of Tokelau, higher abundances roughly correspond to the equatorial zone of high biological productivity with the highest values occurring between about 8°N and 8°S latitudes, the next highest group of values at 8-17° latitude, and a third group at $>17°$ of latitude. Platinum data are insufficient to evaluate fully its distribution, but platinum appears to be in the highest concentrations in crusts from a zone between about 10 to 20° latitude.

RESOURCE CONSIDERATIONS

The central Pacific is considered the prime area for the economic exploitation of metals in ferromanganese crusts. Crusts would be mined primarily for Co, with Mn, Ni, and Pt as byproducts and Ti, Ce, Rh, P, Cd, and other metals enriched in the crusts as other possible byproducts.

Hein et al. (1988a) outlined eight criteria that can be used for the exploration for crusts of economic potential and three criteria for their exploitation. Exploration criteria include (1) large volcanic edifices shallower than 1,500–1,000 m; (2) substrates older than 20 Ma; (3) areas of strong current activity; (4) volcanic structures not capped by large atolls or reefs; (5) a shallow and well-developed oxygen-minimum zone; (6) slope stability; (7) absence of local volcanism; and (8) areas isolated from input of abundant fluvial and eolian debris. Exploitation criteria include average cobalt contents equal to or greater than 0.8%,

Table 9. Estimated total resource potential of crusts within the EEZ of selected Pacific nations, Hawaii, and U.S. possessions.

Pacific Area Johnson, Clark, and Otto (1985)	Relative Ranking	Potential	Pacific Area This Work	Relative Ranking	Potential
Micronesia	1	High	Marshall Is.	1	High
Marshall Is.	2	High	Micronesia	2	High
Northern Mariana Is.	3	High	Johnston Is.	3	High
Kingman-Palmyra	4	High	French Polynesia	4	High
Johnston Is.	5	High			
			Kiribati Is.	5	Medium
Hawaii-Midway	6	Medium	Hawaii Is.	6	Medium
Wake	7	Medium	Tuvalu Is.	7	Medium
			Kingman-Palmyra	8	Medium
			Howland-Baker	9	Medium
Jarvis	8	Low	Northern Mariana Is.	10	Low
Belau (Palau)	9	Low	Wake	11	Low
Guam	10	Low	Jarvis	12	Low
Howland-Baker	11	Low	Samoa	13	Low
Samoa	12	Low	Tokelau	14	Low

average crust thicknesses equal to or greater than 4.0 cm, and subdued small-scale topography. Hein et al. (1988a) ranked the resource potential of the Pacific nations on the basis of these 11 criteria and on the area of seafloor for each nation that lies in less than 2,400 m water depth (Johnson et al., 1985). New insights gained by the statistical analyses and geochemical data presented here allow us to modify our ranking of Pacific nations (Table 9). For comparison, we include nations that occur outside the central Pacific area considered in this paper for which we have data in addition to that presented by Hein et al. (1988a). Data are still scarce for all of these areas (Table 9) except for Johnston, Hawaiian, Kingman-Palmyra, and the Northern Mariana Islands. If the large submerged volcanic complexes in Micronesia and the Marshall Islands are capped by coral reefs, then the resource potential for these two nations will significantly decrease. The few samples that have been analyzed from French Polynesia (Pichochi and Hoffert, 1987) and Tuvalu show high grades of cobalt and thick crusts, but additional samples are needed to verify the extent of the high grades. The analyses available for French Polynesia are for crusts collected only from very shallow-water depths (800 – 1500 m), thus producing higher average cobalt values than would be produced by analyses from crusts from a larger range of water depths. At the present state of knowledge, any of the areas listed in the high and medium categories could, with additional data, emerge at the top of the list.

The few published studies on grade and tonnage estimates (Halbach and Manheim, 1984; Johnson et al. 1985; Ritchey, 1987; Mangini et al., 1987) are based on the chemical composition of only the ferromanganese material. However, this material plus the rock substrate, which may also be sampled in the mining process, constitutes the ore. The only published chemical analyses of ore (bulk dredges, crust plus substrate) show cobalt contents of 0.1 to 0.5% averaging 0.2% for 24 dredges from the Johnston Island EEZ, similar to the grade of cobalt mined on land (Hein et al., 1987c). Only one report provides quantitative evaluation of the amount of crust-covered rock that is exposed on the upper flanks of seamounts (Ritchey, 1987). Ritchey showed that the average extent of exposed crusts is 52% for Horizon Guyot and 24% for S.P. Lee Guyot; estimates used in the other resource analyses used 40 to 60%, but 25 to 45% might be more realistic. We have some knowledge of the grade, little knowledge of the tonnage, and no knowledge of the small-scale topography (criteria 9, 10, and 11 listed above) of seamounts and ridges in the central Pacific. In order to develop mining equipment, it is necessary to know what the topography is like on the scale of meters to tens of meters. To obtain this information, deep-towed side-scan sonar, deep-towed wide-angle photographic surveys, and high-resolution bathymetry are required.

Finally, development of equipment to measure the in situ thickness of crusts is needed to determine the variations in crust thickness throughout the area of a potential mine site. Development of this equipment will be difficult because some substrate types have acoustic signatures similar to the crusts. In addition, the substrate types can vary over short distances; five substrate lithologies were recovered from a small area of central Karin Ridge in the Johnston Island EEZ (Hein et al., 1987c).

Clearly, we are making steady progress in understanding how crusts form and grow and what controls their chemical diversity and variability. Nonetheless, it is also apparent that we are many years away from being able to make intelligent decisions concerning the location of mine sites and the best types of equipment needed to economically mine crusts from submerged volcanic edifices.

ACKNOWLEDGMENTS

We thank Marlene Noble and James Gardner (both at the U.S. Geological Survey) for advice and help with the Q-mode analysis. Carla Horst and Phyllis Swenson provided technical support. Marlene Noble, James Gardner, Randolph Koski (U.S. Geological Survey), Jung-Keuk Kang (Korean Ocean Research and Development Institute), David Cronan (Imperial College, London), Barrie Bolton (University of Adelaide), and Geoff Glasby (New Zealand Oceanographic Institute, DSIR) kindly reviewed this paper.

REFERENCES CITED

Aggasiz, A., 1902, Reports on the scientific results of the expedition to the tropical Pacific, in charge of Alexander Agassiz, by the U.S. Fish Commission Steamer "Albatross," from August 1899, to March 1900, Commander Jefferson F. Moser, U.S.N., Commanding. I. Preliminary report and list of stations: Mem. Mus. Comp. Zool. Harvard, v. 26, 108 p.

Aggasiz, A., 1906, Reports on the scientific results of the expedition to the tropical Pacific, in charge of Alexander Agassiz, by the U.S. Fish Commission Steamer "Albatross," from October 1904, to March 1905, Commander L.M. Garrett, U.S.N., Commanding. V. General report of the expedition: Mem. Mus. Comp. Zool. Harvard, v. 33, 75 p.

Aplin, A.C., 1984, Rare earth element geochemistry of central Pacific ferromanganese encrustations: Earth and Planetary Science Letters, v. 71, p. 13-22.

Aplin, A.C., and D.S. Cronan, 1985, Ferromanganese oxide deposits from the central Pacific Ocean, I. Encrustations from the Line Islands Archipelago: Geochimica et Cosmochimica Acta, v. 49, n. 2, p. 427-436.

Baturin, G.N., 1988, The geochemistry of manganese and manganese nodules in the ocean: D. Reidel, Dordrecht, 342 p.

Bonatti, E., T. Kraemer, and H. Rydell, 1972, Classification and genesis of submarine iron-manganese deposits, in D.R. Horn, ed., Ferromanganese Deposits on the Ocean Floor: Washington, DC, National Science Foundation, p. 149-166.

Boström, K., and M.N.A. Peterson, 1966, Precipitates from hydrothermal exhalations on the East Pacific Rise: Economic Geology, v. 61, p. 1258-1265.

Boström, K., and M.N.A. Peterson, 1969, The origin of aluminum-poor ferromanganoan sediments in areas of high heat flow on the East Pacific Rise: Marine Geology, v. 7, p. 427-447.

Bruland, K.W., 1983, Trace elements in seawater, in J.P. Riley and R. Chester, eds., Chemical Oceanography, v. 8: London, Academic Press, p. 157-220.

Chave, K.E. , C.L. Morgan, and W.J. Green, 1986, A geochemical comparison of manganese oxide deposits of the Hawaiian Archipelago and the deep sea: Applied Geochemistry, v. 1., p. 233-240.

Craig, J.D., J.E. Andrews, and M.A. Meylan, 1982, Ferromanganese deposits in the Hawaiian Archipelago: Marine Geology, v. 45, p. 127-157.

Cronan, D.S., 1984, Criteria for the recognition of areas of potentially economic manganese nodules and encrustations in the CCOP/SOPAC region of the central and southwestern tropical Pacific: South Pacific Marine Geological Notes, v. 3, n. 1, 17 p.

Davis, A.S., 1987, Geochemistry and petrology of basalt dredged on cruise L5-83-HW from Necker Ridge, Horizon and S.P. Lee Guyot, central Pacific Ocean: U.S. Geological Survey Open-File Report 87-513, 26 p.

Davis, A.S., W.C. Schwab, and J.A. Haggerty, 1986, Geochemistry and petrology of basaltic rocks from the Marshall Islands: U.S. Geological Survey Open-File Report 86-273, 23 p.

De Carlo, E.H., G.M. McMurtry, and K.H. Kim, 1987a, Geochemistry of ferromanganese crusts from the Hawaiian Archipelago-I. Northern survey areas: Deep Sea Research, v. 34, p. 441-467.

De Carlo, E.H., P.A. Pennywell, and C.M. Fraley, 1987b, Geochemistry of ferromanganese deposits from the Kiribati and Tuvalu Region of the west central Pacific Ocean: Marine Mining, v. 6, p. 301-321.

De Carlo, E.H., and C.M. Fraley, Chemistry and mineralogy of ferromanganese deposits from the equatorial Pacific Ocean, this volume.

Futa, K., Z.E. Peterman, and J.R. Hein, 1987, Variation of Sr and Nd isotope ratios in Fe-Mn crusts in the central Pacific [abs.] : EOS, Transactions of the American Geophysical Union, v. 68, p. 449.

Futa, K., Z.E. Peterman, and J.R. Hein, 1988, Sr and Nd isotopic variations in ferromanganese crusts from the central Pacific: Implications for age and source provenance: Geochimica et Cosmochimica Acta, v.. 52, p. 2229–2233..

Frank D.J., M.A. Meylan, J.D. Craig, and G.P. Glasby, 1976, Ferromanganese deposits of the Hawaiian Archipelago: Hawaii Institute of Geophysics Report HIG-76-14, 71 p.

Glasby, G.P., and A.J. Read, 1976, Deep Sea Manganese Nodules, in K.H. Wolf, ed., Handbook of Strata-bound and Stratiform Ore Deposits. Elsevier, Amsterdam, p. 295-340.

Glasby, G.P., and J.E. Andrews, 1977, Manganese crusts and nodules from the Hawaiian Ridge: Pacific Science, v. 31, p. 363-379.

Halbach, P., and F.T. Manheim, 1984, Potential of cobalt and other metals in ferromanganese crusts on seamounts of the central Pacific Basin: Marine Mining, v. 4, n. 4, p. 319-336.

Halbach, P., and D. Puteanus, 1984, The influence of the carbonate dissolution rate on the growth and composition of Co-rich ferromanganese crusts from central Pacific seamount areas: Earth and Planetary Science Letters, v. 68, p. 73-87.

Halbach, P., F.T. Manheim, and P. Otten, 1982, Co-rich ferromanganese deposits in the marginal seamount regions of the central Pacific Basin-results of the Midpac '81: Erzmetall, v. 35, n. 9, p. 447-453.

Halbach, P., D. Puteanus, and F.T. Manheim, 1984, Platinum concentrations in ferromanganese seamount crusts from the central Pacific: Naturwissenschaften, v. 71, p. 577-579.

Halbach, P., M. Segl, D. Puteanus, and A. Mangini, 1983, Co-fluxes and growth rates in ferromanganese deposits from central Pacific seamount areas: Nature, v. 304, p. 716-719.

Halbach, P., 1986, Cobalt-rich and platinum-bearing manganese crusts-nature, occurrence, and formation, *in* C.J. Johnson and A.L. Clark, eds., Pacific Mineral Resources Physical, Economic, and Legal Issues: Proceedings of the Pacific Marine Mineral Resources Training Course, East-West Center, Honolulu, Hawaii, p. 137-160.

Heath, G.R., 1981, Ferromanganese nodules of the deep sea: Economic Geology, 75th anniversary volume, p. 736-765.

Hein, J.R., J.R. O'Neil, and M.G. Jones, 1979, Origin of authigenic carbonates in sediments from the deep Bering Sea: Sedimentology, v. 26, p. 681-705.

Hein, J.R., F.T. Manheim, W.C. Schwab, A.S. Davis, C.L. Daniel, R.M. Bouse, L.A. Morgenson, R.E. Sliney, D. Clague, G.B. Tate, and D.A. Cacchione, 1985a, Geological and geochemical data for seamounts and associated ferromanganese crusts in and near the Hawaiian, Johnston Island, and Palmyra Island Exclusive Economic Zones: U.S. Geological Survey Open-File Report 85-292, 129 p.

Hein, J.R., F.T. Manheim, W.C. Schwab, and A.S. Davis, 1985b, Ferromanganese crusts from Necker Ridge, Horizon Guyot and S.P. Lee Guyot : Geological considerations: Marine Geology, v. 69, p. 25-54.

Hein, J.R., F.T. Manheim, W.C. Schwab, and D.A. Clague, 1986, Cobalt-rich ferromanganese crusts from the central Pacific: Proceedings of the Offshore Technology Conference, Richardson, TX, OTC 5234, p. 119-126.

Hein, J.R., L.A. Morgenson, D.A. Clague, and R.A. Koski, 1987a, Cobalt-rich ferromanganese crusts from the Exclusive Economic Zone of the United States and nodules from the oceanic Pacific, *in* D.W. Scholl, A. Grantz, and J.G. Vedder, eds., Geology and Resource Potential of the Continental Margin of Western North America and Adjacent Ocean Basins-Beaufort Sea to Baja California, Circum-Pacific Council for Energy and Mineral Resources, Earth Science Series, v. 6: Houston, TX, Circum- Pacific Council for Energy and Mineral Resources, p. 753-771.

Hein, J.R., C.L. Fleishman, L.A. Morgenson, S.H. Bloomer, and R.J. Stern, 1987b, Submarine ferromanganese deposits from the Mariana and Volcano Volcanic arcs, west Pacific: U.S. Geological Survey Open-File Report 87-281, 67 p.

Hein, J.R., W.C. Schwab, D.G. Foot, Y. Masuda, A. Usui, A.S. Davis, C.L. Fleishman, D.L. Barna, L.B. Pickthorn D.A. Larson, P. Ruzzi, L.M. Benninger, and L.M. Gein, 1987c, Farnella Cruise F7-86-HW, Cobalt-rich ferromanganese crust data report for Karin Ridge and Johnston Island, central Pacific: U.S. Geological Survey Open- File Report 87-663, 34 p.

Hein, J.R., W.C. Schwab, and A.S. Davis, 1988a, Cobalt- and platinum-rich ferromanganese crusts and associated substrate rocks from the Marshall Islands: Marine Geology, v. 78, p. 255-283.

Hein, J.R., L.M. Gein, and M.S. Morrison, 1988b, Submarine ferromanganese mineralization in active volcanic arc systems, *in* Proceedings of PACON 88, Pacific Congress on Marine Science and Technology, Honolulu, Hawaii, May 1988, p. MRM3/83–MRM3/88.

Johnson, C.J., A.L. Clark, and J.M. Otto, 1985, Pacific Ocean minerals: The next twenty years: East-West Center, Resource Systems Institute, Honolulu, HI, (unpublished).

Kang, J.-K., 1987, Mineralogy and internal structures of a ferromanganese crust from a seamount, central Pacific: Journal Oceanological Society of Korea, v. 22, p. 168-178.

Keating, B., B. Bolton, and shipboard party, 1986, Initial report of the 1986 R.V. Moana Wave Cruise MW-86-02 in the Kiribati/Tuvalu Region, central Pacific Ocean, CCOP/SOPAC Cruise Report No. SouthPAC 1986: Honolulu, HI, Hawaii Institute of Geophysics, 68 p.

Klovan, J.E., and J. Imbrie, 1971, An algorithm and FORTRAN-IV program for large-scale Q-mode factor analysis and calculation of factor scores: Mathematical Geology, v. 3, p. 61-77.

Mangini, A., P. Halbach, D. Puteanus, and M. Segl., 1987, Chemistry and growth history of central Pacific Mn-crusts and their economic importance, *in* P.G. Teleki, M.R. Dobson, J.R. Moore, and U. von Stackelberg, eds., Marine Minerals, Advances in Research and Resource Assessment: Dordrecht, Holland, D. Reidel, p. 205-220.

Manheim, F.T., 1986, Marine cobalt resources: Science, v. 232, p. 600-608.

Manheim, F.T.,and J.R. Hein, 1986, Ferromanganese crusts, *in* M. Lockwood, ed., Proceedings of the Exclusive Economic Zone Symposium, Exploring the New Ocean Frontier: Rockville, Maryland, U.S. Dept. Comm., NOAA, May 1986, p. 79-95.

Manheim, F.T. , and C.M. Lane, eds., 1987, World ocean database on the chemical composition of ferromanganese crusts: Final administrative report to Minerals Management Service, U.S. Geological Survey, Office of Energy and Marine Geology, Atlantic Gulf Branch, 500 p.

Mantyla, A.W., and J.L. Reid, 1983, Abyssal characteristics of the World Ocean waters: Deep-Sea Research, v. 30, n. 8A, p. 805-833.

Martin, J.H., and G.A. Knauer, 1985, Lateral transport of Mn in the north-east Pacific gyre oxygen minimum: Nature, v. 314, p. 524-526.

Menard, H.W., 1964, Marine geology of the Pacific: New York, McGraw Hill, 271 p.

Mero, J.L., 1965, The mineral resources of the sea, Oceanography Series 1: Amsterdam, Elsevier, 312 p.

Moore, J.G., 1965, Petrology of deep-sea basalt near Hawaii: American Journal of Science, v. 263, p. 40-52.

Moore, J.G., 1966, Rate of palagonitization of submarine basalt adjacent to Hawaii: U.S. Geological Survey Professional Paper 550D, p. 163-171.

Morgenstein, M., 1972, Manganese accretion at the sediment-water interface at 400 to 2400 meters depth, Hawaiian Archipelago, *in* D.R. Horn, ed., Ferromanganese Deposits on the Ocean Floor: Washington, DC, National Science Foundation, p. 131-138.

Murray, J., and A.F. Renard, 1891, Deep Sea Deposits. Report of Scientific Results, Exploration Voyage of the Challenger, 525 p.

Nakao, S., ed., 1986, Marine geology, geophysics, and manganese nodules around deep-sea hills in the central Pacific basin, August-October 1981 (GH81-4 Cruise): Cruise Report No. 21, Komiyama, Tokyo, 257 p.

Nakao, S., and T. Moritani, eds., 1984, Marine geology, geophysics, and manganese nodules in the northern vicinity of the Magellan Trough, August-October 1980 (GH80-5 Cruise): Cruise Report No. 20, Sumitomo, Tokyo, 272 p.

Pichocki, C., and M. Hoffert, 1987, Characteristics of Co-rich ferromanganese nodules and crusts sampled in French Polynesia: Marine Geology, v. 77, p. 109–119.

Reid, J.L., 1986, On the total geostrophic circulation of the South Pacific Ocean: Flow patterns, tracers and transports, *in* M.V. Angel and R.L. Smith, eds., Progress in Oceanography, v. 16, p. 1-61.

Ritchey, J.L., 1987, Assessment of cobalt-rich manganese crust resources on Horizon and S.P. Lee Guyots, U.S. Exclusive Economic Zone: Marine Mining, v. 6, p. 231-243.

Rona, P.A., 1978, Criteria for recognition of hydrothermal mineral deposits in oceanic crust: Economic Geology, v. 73, p. 135-160.

Schwab, W.C., ed., 1986, Sedimentological study of Horizon Guyot, mid-Pacific Mountains (Johnston I. EEZ) : U.S. Geological Survey Open-File Report 86-433, 137 p.

Schwab, W.C. , H.J. Lee, R.E. Kayen, P.J. Quinterno, and G.B. Tate, 1988, Erosion and slope instability on Horizon Guyot, Mid-Pacific Mountains: Geo-Marine Letters, v. 8, p. 1–10..

Schwab, W.C., J.R. Hein, A.S. Davis, L.A. Morgenson, C.L., Daniel, and J.A. Haggerty, 1986, Geological and geochemical data for seamounts and associated ferromanganese crusts in the Ratak Chain, Marshall Islands: U.S. Geological Survey Open-File Report 86-338, 36 p.

Segl, M., A. Mangini, G. Bonani, H.J. Hofmann, M. Nessi, M. Suter, W. Wölfli, G. Friedrich, W.L. Plüger, A. Wiechowski, and J. Beer, 1984, ^{10}Be-dating of a manganese crust from the central

North Pacific and implications for ocean palaeocirculation: Nature, v. 309, p. 540-543.

Toth, J.R., 1980, Deposition of submarine crusts rich in manganese and iron: Geological Society of America Bulletin, Part I, v. 91, p. 44-54.

REFERENCES TO CHEMICAL DATA USED IN THE COMPILATION OF TABLES 2–3

Anderson, M.E., 1978, Accumulation rates of manganese nodules and sediments : An alpha track method: M.A. thesis, University of California, San Diego, Scripps Institution of Oceanography Dissertations, 113 p.

Andrews, J.E., M. Morgenstein, C.D. Fein, M.A. Meylan, S.V. Margolis, G. Andermann, and G.P. Glasby, 1972, Investigations of ferromanganese deposits from the central Pacific: Hawaii Institute of Geophysics Report HIG-72-23, 133 p.

Aplin, A.C., 1984, Rare earth element geochemistry of central Pacific ferromanganese encrustations: Earth and Planetary Science Letters, v. 71, p. 13-22.

Aplin, A.C., and D.S. Cronan, 1985, Ferromanganese oxide deposits from the central Pacific Ocean, I. Encrustations from the Line Islands Archipelago: Geochimica et Cosmochimica Acta, v. 49, n. 2, p. 427-436.

Bacon, J.I., 1967, A geochemical study of some manganese nodules: M.A. thesis, University of Tulsa, Tulsa, Oklahoma, 53 p.

Barnes, S.S., 1967, The formation of oceanic ferromanganese nodules: Ph.D. thesis, University of California, San Diego, California, 59 p.

Bezrukov, P.L., ed., 1976, Ferromanganese nodules of the Pacific Ocean [in Russian] : Transactions of the P.P. Shirshov Institute of Oceanology, v. 109, 301 p.

Chave, K.E., C.L. Morgan, and W.J. Green, 1986, A geochemical comparison of manganese oxide deposits of the Hawaiian Archipelago and the deep sea: Applied Geochemistry, v. 1., p. 233-240.

Craig, J.D., J.E. Andrews, and M.A. Meylan, 1982, Ferromanganese deposits in the Hawaiian Archipelago: Marine Geology, v. 45, p. 127-157.

Cronan, D.S., and J.S. Tooms, 1969, The geochemistry of manganese nodules and associated pelagic deposits from the Pacific and Indian Oceans: Deep-Sea Research, v. 16, p. 335-359.

De Carlo, E.H., G.M. McMurtry, and K.H. Kim, 1987a, Geochemistry of ferromanganese crusts from the Hawaiian Archipelago-I. Northern survey areas: Deep Sea Research, v. 34, p. 441-467.

De Carlo, E.H., P.A. Pennywell, and C.M. Fraley, 1987b, Geochemistry of ferromanganese deposits from the Kiribati and Tuvalu Region of the west central Pacific Ocean: Marine Mining, v. 6, p. 301-321.

Fisk, M.B., J.Z. Frazer, J.S. Elliott, and L.L. Wilson, 1979, Availability of copper, nickel, cobalt and manganese from ocean ferromanganese nodules (II) : Scripps Institute of Oceanography Reference 79-17, 46 p.

Frazer, J.Z., M.B. Fisk, J. Elliott, M. White, and L. Wilson, 1978, Availability of copper, nickel, cobalt, and manganese from ocean ferromanganese nodules: Scripps Institute of Oceanography Reference 78-25, 45 p.

Goldberg, E.D., 1954, Marine geochemistry 1: Chemical scavengers of the sea: Journal of Geology, v. 62, n. 3, p. 249-265.

Halbach, P., and D. Puteanus, 1984, The influence of the carbonate dissolution rate on the growth and composition of Co-rich ferromanganese crusts from central Pacific seamount areas: Earth and Planetary Science Letters, v. 68, p. 73-87.

Halbach, P., F.T. Manheim, and P. Otten, 1982, Co-rich ferromanganese deposits in the marginal seamount regions of the central Pacific Basin-results of the Midpac ' 81: Erzmetall, v. 35, n. 9, p. 447- 453.

Halbach, P., D. Puteanus, and F.T. Manheim, 1984, Platinum concentrations in ferromanganese seamount crusts from the central Pacific: Naturwissenschaften, v. 71, p. 577-579.

Halbach, P., M. Segl, D. Puteanus, and A. Mangini, 1983, Co-fluxes and growth rates in ferromanganese deposits from central Pacific seamount areas: Nature, v. 304, p. 716-719.

Hein, J.R., F. T. Manheim, W.C. Schwab, A.S. Davis, C.L. Daniel, R.M. Bouse, L.A. Morgenson, R.E. Sliney, D. Clague, G.B. Tate, and D.A. Cacchione, 1985, Geological and geochemical data for seamounts and associated ferromanganese crusts in and near the Hawaiian, Johnston Island, and Palmyra Island Exclusive Economic Zones: U.S. Geological Survey Open-File Report 85-292, 129 p.

Hein, J.R., L.A. Morgenson, D.A. Clague, and R.A. Koski, 1987, Cobalt-rich ferromanganese crusts from the Exclusive Economic Zone of the United States and nodules from the oceanic Pacific, *in* D.W. Scholl, A. Grantz, and J.G. Vedder, eds., Geology and Resource Potential of the Continental Margin of Western North America and Adjacent Ocean Basins-Beaufort Sea to Baja California, Circum-Pacific Council for Energy and Mineral Resources, Earth Science Series, v. 6: Houston, TX, Circum- Pacific Council for Energy and Mineral Resources, p. 753-771.

Hein, J.R., W.C. Schwab, and A.S. Davis, 1988, Cobalt- and platinum- rich ferromanganese crusts and associated substrate rocks from the Marshall Islands: Marine Geology, v. 78, p. 255-283.

Hein, J.R., 1988, unpublished chemical data of R.V. Farnella Cruise F7-86-HW to Karin Ridge and south Johnston Island Ridge.

Hewett, D.F., M. Fleischer, and N. Conklin, 1963, Deposits of the manganese oxides: Supplement: Economic Geology, v. 58, n. 1, p. 1-51.

Manheim, F.T., and C.M. Lane, eds., 1987, World ocean database on the chemical composition of ferromanganese crusts: Final administrative report to Minerals Management Service, U.S. Geological Survey, Office of Energy and Marine Geology, Atlantic Gulf Branch, 500 p.

Mero, J.L., 1965, The mineral resources of the sea, Oceanography Series 1: Elsevier, Amsterdam, p. 182-219.

Morgan, C.L., and J.R. Moore, 1975, Role of the nucleus in formation of ferromanganese nodules: Processing guidelines for the marine miner: Proceedings of the Offshore Technology Conference, v. 1, paper n. OTC 2243, p. 943-953.

Okada, A., and M. Shima, 1970, Study on the manganese nodule [in Japanese] : Journal of the Oceanographical Society of Japan, v. 26, n. 3, p. 151-158.

Piper, D.Z., and M.E. Williamson, 1977, Composition of Pacific Ocean ferromanganese nodules: Marine Geology, v. 23, p. 285-303.

Toth, J.R., 1980, Deposition of submarine crusts rich in manganese and iron: Geological Society of America Bulletin, Part I, v. 91, p. 44-54.

Keating, B.H., and Bolton, B.R., editors, 1992,
Geology and offshore mineral resources of the
central Pacific basin, Circum-Pacific Council for
Energy and Mineral Resources Earth Science
Series, V. 14. New York, Springer-Verlag.

PHOSPHATIC ROCKS AND MANGANESE CRUSTS FROM SEAMOUNTS IN THE EEZ OF KIRIBATI AND TUVALU, CENTRAL PACIFIC OCEAN

V. Purnachandra Rao,* W.C. Burnett
Department of Oceanography, Florida State University, Tallahassee, Florida, 32306

ABSTRACT

Phosphorite-manganese crusts were dredged from the flanks of seamounts of the Kiribati and Tuvalu EEZ (Exclusive Economic Zone) from water depths between 1450 and 2355 m. Samples from Kiribati consist of phosphatized algal laminations, coated grains, filamentous and unicellular cyanobacteria. Manganese oxide occurs in the phosphate matrix in the form of dendritic growths, as a coating on intraclasts, and as infillings in foraminifers and algal tissues. Weathered clasts derived from mafic igneous rocks are the dominant constituents in the samples from Tuvalu seamounts and phosphate occurs mainly as a cementing material for these clasts. Calcite and carbonate fluorapatite are the dominant minerals present in the matrix material of both areas.

It appears that the Kiribati seamounts were at very shallow depths during the phosphatization process, and microbial activity may have taken an active role during the diagenetic transformation of algal sediments into phosphorites. The Tuvalu samples, however, showed no evidence of shallow water conditions although some microbial influence during phosphogenesis was still apparent. Thin manganese crusts in Kiribati and thick crusts in the Tuvalu samples suggest that depth and/or length of exposure determined the relative crust thicknesses observed.

INTRODUCTION

During the *Moana Wave* cruise MW 86-02 in 1986, phosphorite-manganese crusts were dredged from the flanks of seamounts in the EEZ of Kiribati and Tuvalu, Central Pacific Basin, from water depths between 1450 and 2355 m. The objectives of the cruise were to investigate the origin, growth and distribution of cobalt-rich manganese and phosphate crusts and assess these materials as resources for possible future exploitation. An evaluation of the general geology of these islands was a secondary objective of the expedition. The purpose of this paper is to describe the phosphatic crusts and discuss their origin and relation to associated ferromanganese materials.

Phosphorites, phosphorite-manganese crusts, and ferromanganese crusts occur either together or separately on many seamounts in the world's oceans (Hamilton, 1956; Baturin, 1982; Hein et al., 1985). Most seamount phosphorites are thought to have formed either by replacement of carbonate or other precursor materials by phosphate during diagenesis, or by the mobilization of guano-derived phosphate initially deposited on emergent islands (Braithwaite, 1980; Cullen and Burnett, 1986; and Burnett et al., 1987). Since it was recognized that ferromanganese crusts are usually enriched in cobalt, manganese crusts on seamounts in the Central Pacific have been the object of study of several investigators (Halbach et al., 1982; Halbach and Manheim, 1984; Hein et al., 1985; Manheim, 1986; and DeCarlo et al., 1987).

Phosphorite-manganese crusts on oceanic plateau areas have also been studied by many investigators (Heezen et al., 1966; Summerhayes, 1967 and 1973; Manheim et al., 1980; and Von Rad and Kudrass, 1984) with the general conclusion that the phosphatic component is formed in variable cycles of erosion, reworking, partial dissolution and reprecipitation. The phosphatic component of the phos-

*On leave from the National Institute of Oceanography, Dona Paula, Goa 403 004, India phorite-manganese crusts found on seamounts are the least studied, but nevertheless are of considerable interest for several reasons, including: (1) the Mn component of this type of crustal material is also enriched with cobalt (Halbach and Manheim, 1984); (2) the genesis of these crusts, and the relationship between phosphate and manganese is not well understood; and (3) the alternating layers of manganese-phosphorite may reflect past changes in oceanographic conditions.

STUDY AREA AND METHODS

The Kiribati and Tuvalu Islands are in the Central Pacific Basin, lying roughly within the area between 170°E and 170°W and 5°N and 10°S (Fig. 1). The Gilbert and Ellice islands are on the western margin of the area and the Phoenix Islands are on the eastern margin. These islands are aligned in a NW-SE direction parallel to the Line Islands. Seismic reflection data of the seamounts of Kiribati have indicated that they are cone-like features similar to subaerial cinder cones. The water depths at the crests of the seamounts range from 300 to 2000 m. The absolute ages of the seamounts are not available, but the sea-floor age of both island groups are believed to be Cretaceous (Keating et al., 1986; De Carlo et al., 1987).

Polished and thin sections of phosphatic rocks from seamounts in the Kiribati and Tuvalu regions (Table 1) were prepared for all samples. Thin sections were studied using a standard petrographic microscope and SEM studies were carried out with a JEOL model 840 scanning electron microscope equipped with an energy dispersive X-ray analyzer. Powdered bulk samples were analyzed for mineralogical composition from 2° to 70° 2θ at 1° 2θ/min. On a Philips X-ray diffractometer using nickel-filtered CuK_α radiation. Slow scan speeds (1/2° 2θ/min) were used for XRD estimation of the structural carbonate concentration of carbonate fluorapatite (CFA) by the method of Gulbrandsen (1970).

Chemical analyses of bulk samples for P_2O_5 and F were obtained by taking 100 mg of the dried and powdered rock sample, leaching with 2N HCl, and separating the insoluble

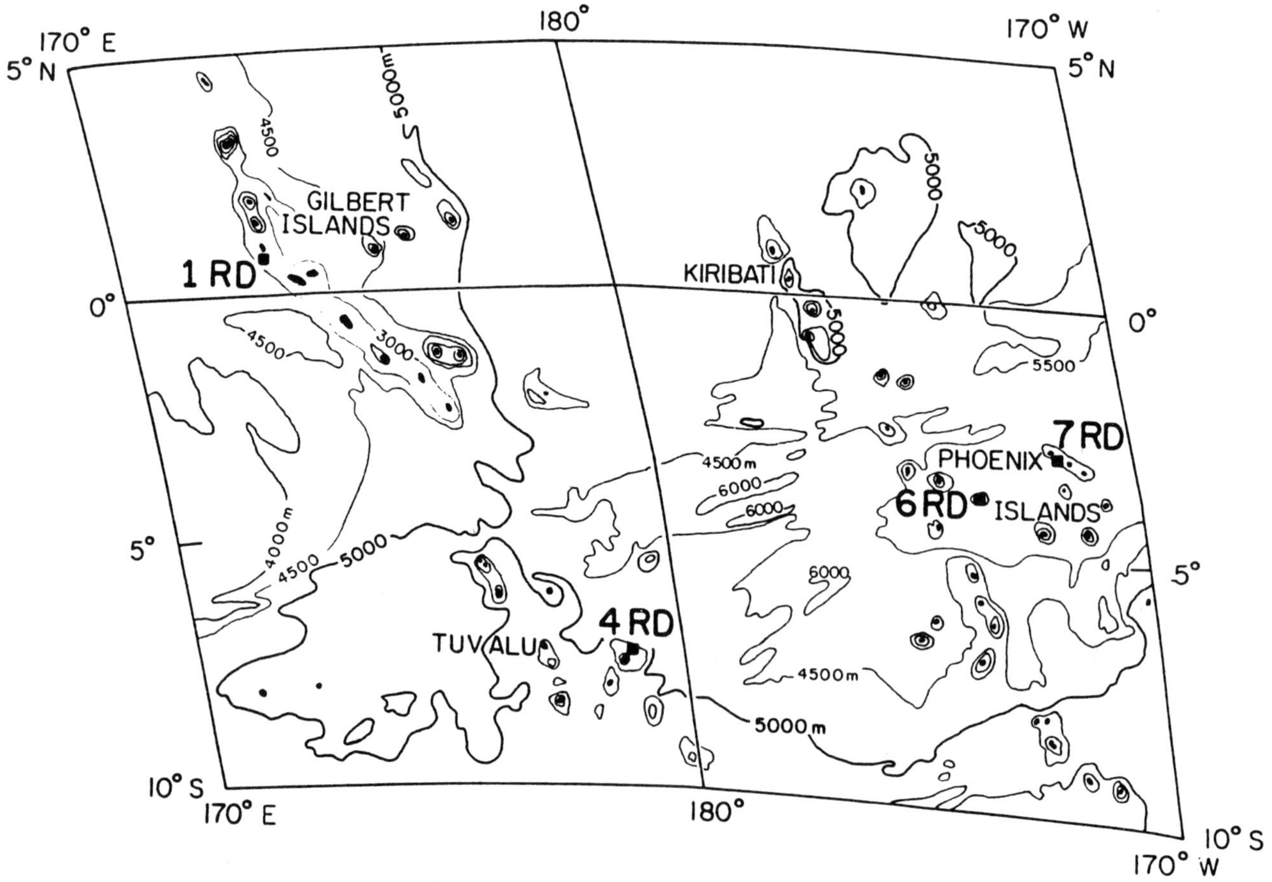

Figure 1. Map of Central Pacific Basin showing sampling locations.

residue by filtration. These leachates were diluted to the appropriate concentrations and used for both analyses. The P_2O_5 content was determined by colorimetry following the standard method of Murphy and Riley (1962). The P_2O_5 contents determined for NBS standard rocks 120b and 694 were within 2% of reported values. Fluoride was determined potentiometrically using an Orion F^- ion specific electrode and citrate buffer (Edmond, 1969). The accuracy of these values was within 7%.

RESULTS

Phosphatic rocks from Kiribati (stations 1 and 7)

a) Petrology

Polished sections of the rock sample from station 1 showed a phosphatized pink substrate in the basal parts

Table 1. Location and description of phosphorite-manganese crusts from Cruise MW-86-02.

Sample No	Latitude	Longitude	Water Depth (m)	Description
			Kiribati	
1RD001-1	00° 46.58'N	173° 01.33'E	1450	Algal limestones with stains of manganese oxide on the the surface: dominantly algal fragments and calcite.
1RD001-1	"	"	"	Substrate for algal limestones: ironstained, phosphatized and no other skeletal components can be identified except the benthic foram *Lepidocyclina*.
7RD011-03	02° 59.52'S	171°22.82'W	1730	Mn crust (15 mm thick) on algal limestones: phosphate replaced bioclasts.
7RD011	"	"	"	Mn crust (4 mm thick) on algal foram limestones: phosphate as well as Mn oxide replaced forams and coated grains. Botroiydal and dendritic growths of Mn oxide.
7RD011-P	"	"	"	Mn crust (3 cm thick) on algal lime stones: fine laminations and phosphate grain between algal filaments.
7RD011-R	"	"	"	2-4 mm thick Mn crust on algal foram limestones: 4-6 mm thick Mn oxide layer between phosphate layers.
7RD011-07	"	"	"	Mn crust (6 mm thick) on algal limestones: rounded chert grains and coated grains.
			Tuvalu	
4RD007-02	07° 24.47'S	179 °02.16'E	1620	Dominantly Mn crust (15-20 mm thick) on weathered clasts and carbonate: planktonics in the matrix.
4RD007-21A	"	"	"	Dominantly weathered clasts and Mn crust (15-20 mm thick): planktonics in carbonate-phosphate matrix.
4RD007-21B	"	"	"	Thin (3 mm) Mn crust. Dominantly weathered clasts and phosphatized carbonate.
6RD008-02	03° 50.23'S	173°19.74'W	2355	Thick (60 mm) Mn crust on weathered clasts and carbonate. Mn infilled forams and oxide growth around bioclasts.
6RD008-03	"	"	"	10-15 mm thick Mn crust: weathered rock clasts, micromanganese nodules and planktonics in the matrix.

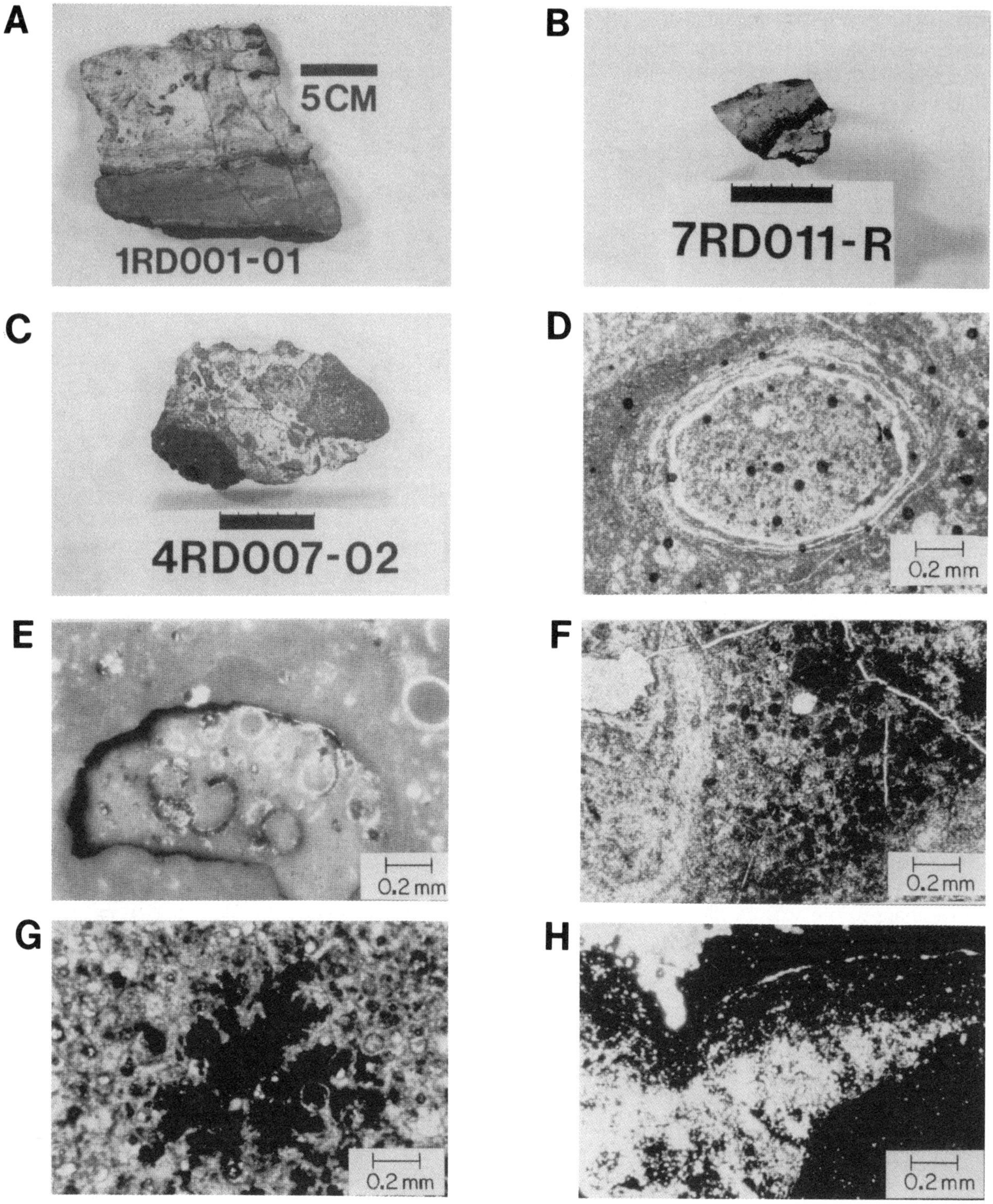

Figure 2. (A) and (B) are the polished sections of rocks sampled from Kiribati: (A) rock showing pink phosphatized carbonate as a substrate overlain by white carbonate limestone; (B) a sample characterized by alternate layers of phosphorite and manganese oxide; (C) polished section of a rock sampled from Tuvalu, showing weathered clasts embedded within a phosphate matrix; (D), (E), (F), and (G) are thin section photographs of rocks sampled from Kiribati : (D) coated grain (7RD011-07); (E) intraclasts coated by manganese oxide (7RD011-07); (F) manganese oxide distribution within the algal tissue material (7RD011); (G) dendritic growths of manganese oxide in a phosphatized carbonate matrix (7RD011-03); and (H) fine-grained phosphate separating metal oxide layers (7RD011-07).

overlain by white coralline limestone (Fig. 2A). In thin section, the pink coloring of the substrate proved to be caused by iron staining. It is difficult to identify the other skeletal components in this material except the benthic foraminifera *Lepidocyclina*, an index fossil of Eocene age. The white substrate of the rock consists of coralline algal fragments with occasional gastropods. Manganese oxide stains on the surface were also observed. Polished sections of the rocks from station 7 show distinct manganese oxide crusts on the outer edges and phosphatized, finely laminated structures on the interior. Discontinuous phosphate layers from 5 to 10 mm thick are present between the metal oxide layers (Fig. 2B). Thin sections show that these rocks consist of phosphatic grains and planktonic foraminifers that are bound by cryptalgal laminations. This laminated fabric interspersed with sediment particles is well-known in cyanobacterial mat activity (Monty, 1976) and is the result of concurrent finely laminated mat growth and sediment deposition. Among the phosphatic components, grains coated with micritic phosphate (Fig. 2D), similar in appearance to the coated grains reported by Peryt (1983), and phosphatic intraclasts with manganese oxide rims (Fig. 2E) are commonly observed. Coralline algal fragments, planktonic forams and fine-grained phosphate are found in the phosphatic layers present between metal oxide layers (Fig. 2H). Rounded, reddish brown chert grains are occasionally embedded in the algal layers.

Our SEM observations on freshly broken surfaces of the rocks from station 7 confirmed and expanded the previous petrologic observations. Rock cavities exhibit irregularly encrusted endolithic algal filaments (Fig. 3A) similar to those shown by Schroeder (1972). These filaments are coated by well-crystallized CFA. The filaments are empty, as shown by the cavities at the end, and the diameter of each filament is about 10 mm, with a hollow diameter of about 3 mm size (Fig. 3B). Phosphatic peloids formed by the accretion of cryptocrystalline diagenetic CFA are also present in the cavities of the rock. Tubular filaments (Fig. 3C) and straight tubes of 5 to 7 mm length enveloped by phosphatized sheaths (Fig. 3D) are also seen in the matrix. Similar structures, thought to be of cyanobacterial origin, have been reported by Krumbein (1977) in calcareous deposits and by Soudry and Champetier (1983) in Cretaceous phosphorites from the Negev Desert, Israel. These rocks also contain phosphatized colonies of unicells of possible cyanobacterial origin. These cells range from 3 to 4 mm size, and aggregation of cells into pellets at interfaces between mats and colonies of cyanobacterial cells was observed (Fig. 4A and 4B). Phosphatic "lumps" within the phosphatized carbonate shells, marginal grain corrosion, and phosphatic borings within shells were found within the matrix (Fig. 4C). Under high magnification, these lumps appear to be an amalgamation of cyanobacterial cells (Fig. 4D). These cells are larger than those observed in other areas (Fig. 4B), confirming the observations of Zanin et al. (1985) that cavities and voids are often sites for larger organisms.

Examination of freshly fractured surfaces of these samples show that crystals morphologically resembling gypsum, with well-developed prism faces (Fig. 4E), are occasional components of these phosphorites. Gypsum crystals formed by authigenesis in the presence of organic matter are stubby and tend to lack the prismatic forms of those observed here (Sonnenfeld, 1984). Since gypsum crystals also occur on cyanobacterial filaments (Fig. 4F), they apparently have not formed after sample recovery by dehydration. Rather, the presence of these crystals indicates that shallow evaporative conditions were prevalent at some stage in the history of these phosphorites.

Manganese oxide in these samples (station 7) occurs in two modes. First, it occurs as crusts on the surface of the phosphatic rocks. The thickness of the crusts varies from surficial rinds to crusts about 15 mm thick. Secondly, manganese oxide occurs within the phosphate matrix, infilling foraminiferal chambers, replacing carbonate in algal tissue material (Fig. 2F), and accreting in the matrix in dendritic patterns (Fig. 2G). Cuspate growths of iron oxides were observed within these dendritic patterns. The manganese oxides are often distributed along fine laminations. It is clear from the internal structure (Fig. 2G) that manganese oxide is associated with the phosphatic rather than with the calcareous material, suggesting that phosphate is a preferred host for manganese. This may be due to the relative insolubility of CFA relative to calcite. Manganese oxides distributed along microlaminations in a dendritic pattern are also present (Fig. 4F). The X-ray energy dispersive analyses indicated that magnesium, manganese, calcium, and phosphorus are all abundant in these laminations, suggesting that the microlaminations are filamentous algal structures in which manganese oxide replacement has taken place.

b) Mineralogy and Geochemistry

Analysis by X-ray diffractograms of the phosphatized material shows that calcite and CFA are the major minerals present. The structural CO_2 content of CFA in all samples estimated by the peak-pair X-ray method (Gulbrandsen, 1970), is relatively high, approximately 6%. The 'a' cell dimensions are about 9.3Å indicating that the CO_3:PO_4 mole ratio of these apatites is approximately 0.22 (Table 2). The replacement of PO_4 by CO_3 changes the cell parameters of the apatite lattice in such a way that there is a decrease in the 'a' unit cell parameter with increasing carbonate substitution (McClellan and Lehr, 1969). The XRD studies of samples from the manganese oxide layer, however, showed no discernible manganese mineral peaks, probably due to the poor crystalline nature and/or the ex-

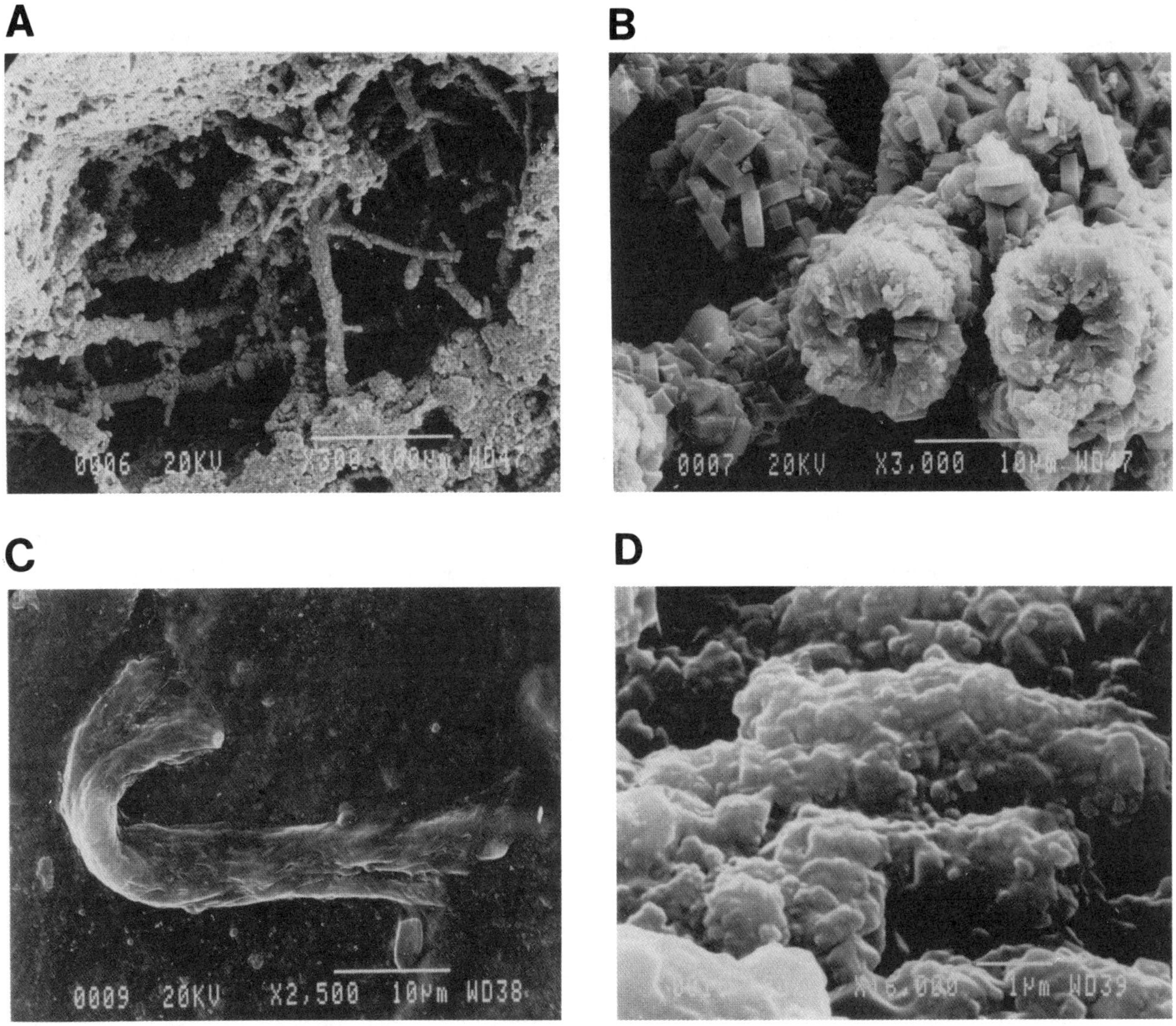

Figure 3. SEM photographs from rocks sampled from Kiribati : (A) irregularly encrusted endolithic algal filaments (7RD011-07); (B) enlarged portion of (A), showing empty filaments and well-crystallized carbonate fluorapatite adhering to filaments; (C) tubular filaments of cyanobacteria (7RD011-P); (D) straight tubes of cyanobacterial origin coated by phosphatized sheaths (7RD011-07). Energy dispersive X-ray analysis showed high calcium and phosphorus as well as minor amounts of sulfur present in these tube structures.

tremely fine grain-size of the material found in seamount manganese crusts (Burns and Burns, 1976). In addition, interference from X-ray fluorescence of the Mn-rich material may have partially obscured the diffraction peaks.

The P_2O_5 content of the Kiribati phosphorites ranged from 13 to 33% and the fluoride content varied from 2.0 to 3.8%. The F/P_2O_5 mass ratios were all in the range 0.110 to 0.129, except for one sample, which was very high at 0.15. These relatively high F contents are consistent with anticipated amounts in highly substituted carbonate apatites (McClellan and Lehr, 1969).

Phosphatic rocks from Tuvalu (stations 4 and 6)

a) Petrology

These rocks consist of thick manganese oxide crusts ranging from 3 to 60 mm, and abundant weathered clasts

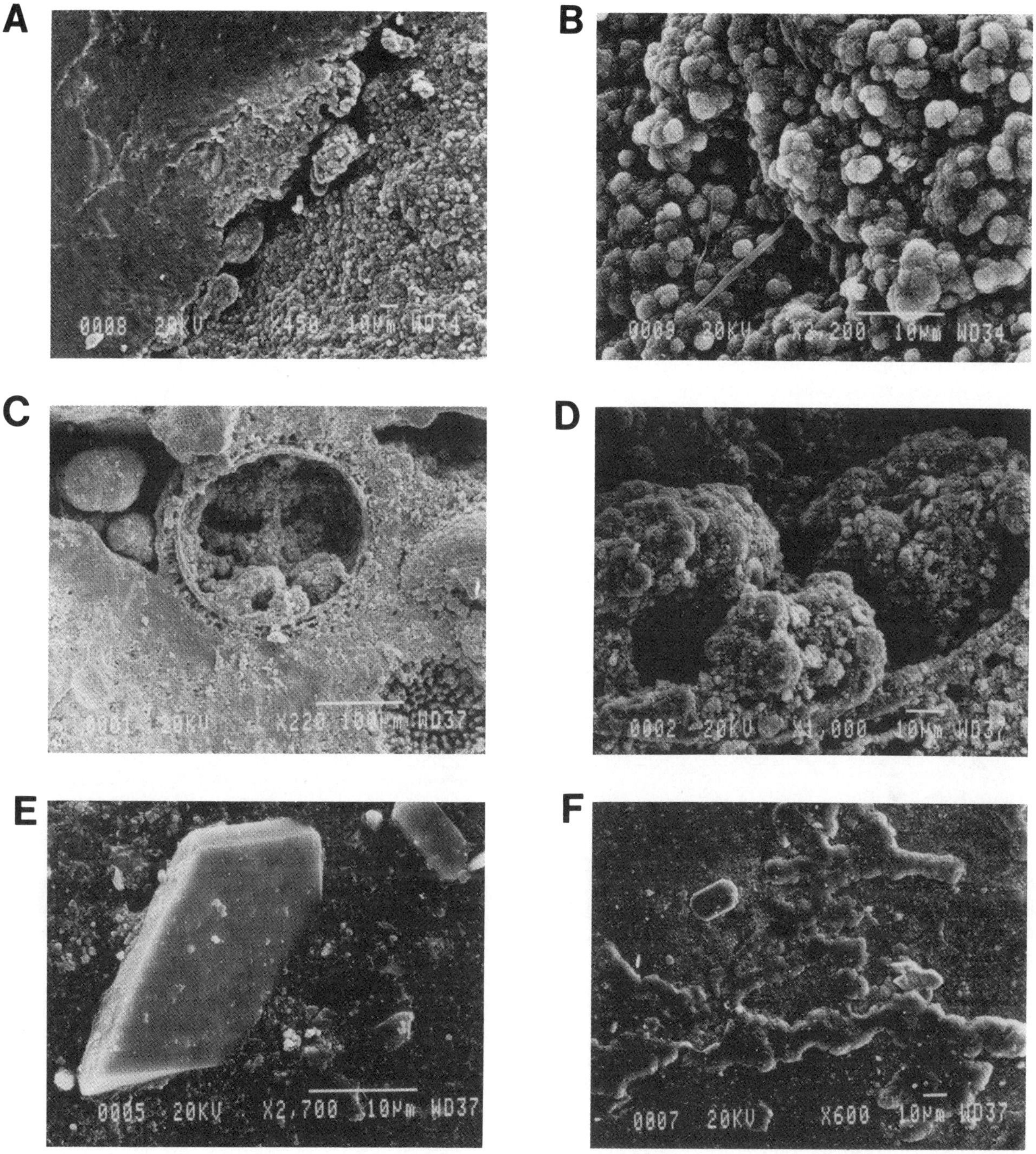

Figure 4. SEM photographs from rocks sampled from Kiribati: (A) cyanobacterial mats, coccoid unicells, and phosphate "pellets" at an interface between mat and coccoid unicells. Pelletal forms appear to be formed by the aggregation of phosphatic unicells (7RD011-P); (B) enlarged portion of (A) showing coccoid colony of cyanobacteria (X-ray analysis confirmed that the mat and unicells were phosphatic); (C) phosphate lumps within a foram shell and in cavities. Boring patterns developed within the shell chamber (lower right margin) can also be seen (7RD011); (D) enlarged portion of (C), showing that the lumps are an amalgamation of cyanobacterial unicells; (E) prismatic gypsum crystals and dendritic manganese growths (F) within a phosphate matrix (7RD011-R). X-ray analysis showed that these filaments showed high magnesium and manganese as well as calcium and phosphorus, confirming that these are algal filaments replaced by manganese and phosphate.

derived from mafic rock fragments cemented together by a carbonate-phosphate matrix (Fig. 2C). These rocks contain planktonic foraminifers as abundant components in the matrix. Clinopyroxene and plagioclase were identified in the weathered clasts. The mafic rock clasts are heavily altered and occasionally replaced by calcite, phosphate, and iron oxides.

The SEM studies revealed considerable dissolution of carbonate shell walls and borings developed within foraminiferal chambers (Fig. 5A). These borings are oriented normal to the surface (Fig. 5B). We also observed micro-dissolution features on the surfaces of the carbonate grains and subsequent formation of microgranular CFA on these surfaces (Fig. 5C). Unphosphatized coccoliths (Fig. 5D) and discosters are other components observed in the matrix.

b) Mineralogy and Geochemistry

The X-ray diffractograms show that plagioclase, apatite, hematite, goethite, and chlorite are present in these samples. The structural CO_2 content varies from 5.2 to 6.0% and the 'a' cell dimensions are about 9.3Å indicating that the CO_3:PO_4 mole ratio of the CFA is similar to the phosphorites from Kiribati (0.22). No manganese oxide mineral peaks were observed in the X-ray diffractograms of the manganese oxide crust material.

The P_2O_5 contents in these samples is similar to those of Kiribati at 18 to 30% and the fluoride content ranged from 2.3 to 3.4%. The F/P_2O_5 mass ratio varied from 0.110 to 0.124.

While the phosphorite-manganese rocks from the Kiribati seamounts displayed shallow-water fauna and thin

Table 2. Detailed mineralogy and geochemistry of the phosphatized material from the seamount samples.

Sample No.	X-ray mineralogy[1]	Structural components of CFA: CO_2 %[2]	Structural components of CFA: a(Å)	P_2O_5 %	F %	F/P_2O_5
			Kiribati			
lRD001-1	cal	--	--	--	--	--
1RD001-1	cal and CFA	6.0	9.32	13.3	2.0	0.15
7RD011-03	CFA and cal	5.2	9.325	25.9	3.25	0.125
7RD011	CFA and cal	6.0	9.32	25.7	3.25	0.115
7RD011-P	CFA and cal	6.0	9.32	32.1	3.70	0.115
7RD011-R	CFA and cal	--	--	32.8	3.70	0.113
7RD011-R[3]	CFA and cal	--	--	33.1	3.79	0.11
7RD011-07	CFA	--	--	29.3	3.74	0.127
7RD011-07[4]	opal-CT and sid	--	--	--	--	--
			Tuvalu			
4RD007-02	CFA and cal	--	--	18.6	2.4	0.129
4RD007-21A	plag, CFA, cal goet and chlo		--	--	--	--
4RD007-21B	CFA and cal	6.0	9.32	30.4	3.4	0.112
6RD008-02	CFA and cal	5.2	9.325	24.1	2.9	0.12
6RD008-03	CFA cal and plag	--	--	18.5	2.3	0.124

[1]minerals listed in approximate order of abundance: cal – calcite, CFA – carbonate fluorapatite; plag – plagioclase; goet – goethite; chlo – chlorite, and sid. – siderite.
[2]Gulbrandsen (1970) X-ray peak-pair method.
[3]phosphate layer in between metal oxide layers.
[4]rounded chert grain found embedded within algal layers.

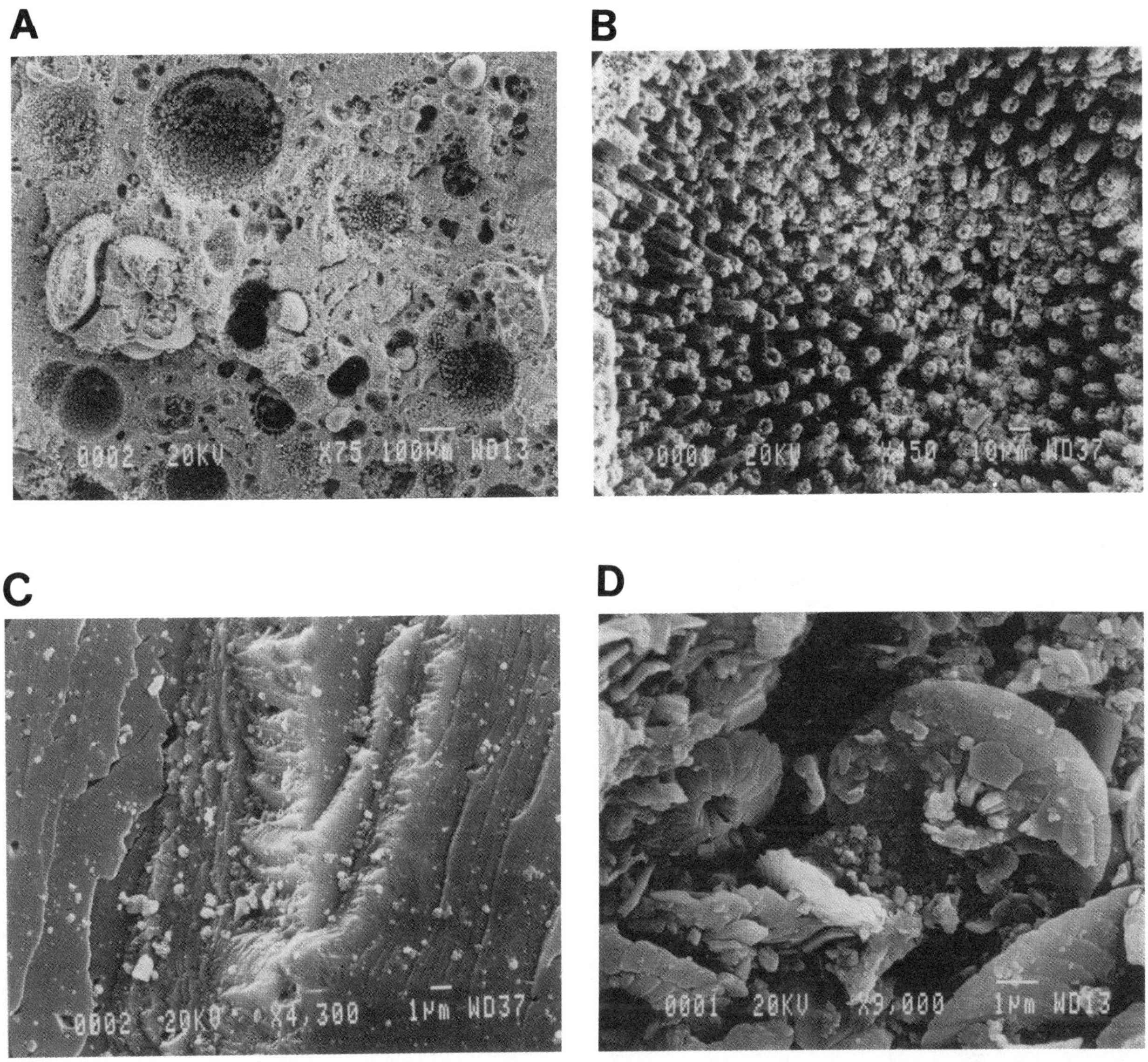

Figure 5. SEM photographs from rock sample 4RD007-02, from a seamount in Tuvalu: (A) photograph illustrating dissolution of carbonate shell walls; (B) enlarged portion of (A) showing phosphatized casts developed within foraminiferal chambers (X-ray analysis confirmed that these casts are rich in calcium and phosphorus); (C) microdissolution features on a carbonate grain surface together with small grains formed on these surfaces; and (D) unphosphatized coccoliths within the matrix material.

manganese oxide crusts, the samples from the Tuvalu seamounts showed open-ocean fauna and thick manganese crusts. However, samples from both areas displayed dissolution of grain surfaces and boring patterns developed within shell chambers. The principal geochemical and mineralogical features of samples from both areas did not show significant variation although the abundance of CFA is somewhat lower in the Tuvalu seamounts.

DISCUSSION

Thin section observations and SEM studies have revealed that endolithic algae and filamentous and unicellular cyanobacteria are widely distributed in the Kiribati phosphorites. Prismatic gypsum, larger foraminifera (Ex. *Lepidocyclina*), and gastropods are also associated with these samples. This indicates that these sediments must have accumulated when the seamount was at intertidal or subtidal depths. Cell-like structures of possible cyanobacterial origin inside the carbonate shells, marginal grain corrosion on the calcareous grains and the different boring patterns in these phosphorites (Figs. 4C,4D) appear to be caused by the action of algae. Klement and Toomey (1967) and Golubic (1973) state that certain boring and perforating algae are highly destructive of carbonate shells and coastal limestones, and their optimal distribution would be in shallow stagnant environments at depths up to about 20 m. Phosphate grains bound by cryptalgal laminations, phosphate intraclasts and grains coated by layers of micritic phosphate also suggest that the phosphatization occurred during shallow water diagenesis. X-ray energy-dispersive analysis has confirmed that the cyanobacterial cells and filaments present in these rocks are completely phosphatized.

Cyanobacteria are known to fix high concentrations of phosphorus during their metabolic activity (Azad and Borchardt, 1970). It is also known from the experimental studies of Kobluk and Risk (1977) and from studies of calcareous deposits (Krumbein, 1968; Klappa, 1979) that algae are not directly involved in calcification, and that precipitation of carbonate takes place mostly on dead filaments. We suggest that, during shallow-water diagenesis of these carbonate sediments, the decay of cyanobacterial cells and filaments of algae induced phosphate precipitation. Hence, the concentration and subsequent release of phosphate within the sediment constituted the microbial role in the formation of these phosphorites. Cyanobacterial influence in phosphogenesis has been suggested by Soudry and Champetier (1983) and Southgate (1986). The well-crystallized phosphate crystals, found on endolithic algal filaments with empty cavities (Fig. 3B), and the phosphatized sheath on cyanobacterial filaments (Fig. 3D) might suggest that phosphatization took place in the early stages of the decay of these filaments. These observations also imply that the phosphate mineralization of these filaments is a direct process rather than replacement of carbonate which preexisted on filaments. These organic filaments may not only be the source for phosphate, but may also serve as favorable sites for phosphate mineralization, as suggested by Southgate (1986).

Our studies have revealed that appreciable concentrations of phosphorus occur in the fossilized cells of possible cyanobacterial origin. High concentrations of phosphate in bacterial structures (cocco-like and bacilli forms) have been reported earlier by Cayeux (1936), Oppenheimer (1958), O'Brien et al. (1981), Riggs (1982), Prevot and Lucas (1986) and Zanin et al. (1985). To our knowledge, our observations of these structures are the first reported for seamount or "open-ocean" phosphorites. We suggest that the shallow-water phosphorites (Kiribati samples) were formed during early diagenesis of algal laminated sediments. Cyanobacteria were involved in the evolution of these phosphorites either as a source for phosphate or a suitable surface for precipitation, or both. Phosphate released to interstitial waters subsequently replaced carbonate shell materials, precipitated into cryptocrystalline aggregates in the form of pellets in rock cavities, and produced the coatings around phosphate pellets.

The structural CO_2 content of the CFA is about 6%, slightly higher than the CO_2 content of most marine phosphorites (Kolodny, 1981). This may be related to the presumed high carbonate content of the predecessor materials (Birch, 1980). The CO_3:PO_4 mole ratio of about 0.22 is close to values reported for Chatham Rise phosphorites ($\sim$0.25) where the phosphatization occurred within a carbonate-rich matrix (Von Rad and Kudrass, 1984).

Phosphorite-manganese crusts of the Tuvalu seamounts were dredged from greater depths than the Kiribati samples, and phosphate in these crusts occurs mainly as a cementing material for hyaloclastites, derived from the weathering of volcanic materials. The cementing material consists of open-ocean fauna such as planktonic foraminifera and coccoliths. Characteristic features such as the dissolution of calcareous grain surfaces, microbial borings within the carbonate grains, and micro-dissolution structures on calcareous grain surfaces (Fig. 5A, B, C and D) indicate that the destruction of carbonates has taken place by microorganisms during diagenesis. Similar observations were reported by Lamboy (1987) for North African phosphorites, which he interpreted as indicative of a microbial role in phosphogenesis. Thus, microbial activity may also have been partially responsible for the phosphatization of these deep water samples.

Ferromanganese oxide occurs both as a matrix material and as crusts on these phosphorites. Two mechanisms can be proposed for the distribution of manganese oxide within the phosphatized matrix. Empty filaments and microbial borings resulting from algal action are commonly observed in phosphorites. Manganese oxide might have replaced these borings and infilled the filaments as a secondary precipitate. Alternatively, manganese and iron released to the interstitial waters prior to sulfate reduction during early diagenesis might have precipitated in algal channels present within the sediment. If the latter was the case, manganese oxide and phosphate formation may have been contemporaneous within the matrix material. Alternatively, if the manganese oxide is a secondary replacement, the cuspate iron oxide structures within the dendritic growths of manganese oxide, and the occurrence of manganese oxide as a coating on intraclasts, cannot be explained. We conclude, therefore, that manganese was distributed during early diagenesis with some secondary replacement occurring along algal laminations. Pratt and McFarlin (1966) and Manheim et al. (1980) suggested that both replacement type and diagenetic manganese oxide were present in manganese-phosphorite crusts of the Blake Plateau. The crust-like manganese oxides that occur on these phosphorites seem similar to those reported from other open-ocean environments and are most likely of hydrogenetic origin.

SUMMARY AND CONCLUSIONS

Our studies of the phosphorite-manganese crusts from Kiribati seamounts suggest that (a) the paleodepths of the seamounts were very shallow, probably intertidal or subtidal, during phosphatization; (b) endolithic algae and cyanobacteria were influential in the early diagenesis of carbonate sediments, which took place when the seamount was at shallow depths; (c) organic filaments may have acted as both a source and a substrate for phosphate mineralization; and (d) early diagenetic as well as replacement-type manganese oxides appear to be present in the phosphatic matrix.

Study of the phosphorite-manganese crusts from Tuvalu seamounts shows that (a) weathered clasts from mafic rocks are the main constituents of these samples, whereas the phosphorite occurs as a cementing material for the clasts; (b) phosphatized pelagic fauna are the main constituents of the cementing material; (c) boring patterns and destruction of carbonate grains are similar to those in the Kiribati samples, possibly indicating microbial participation during phosphogenesis; and (d) manganese crusts are thicker than in the Kiribati samples, indicating that growth rates are faster at deeper depths, or these samples are significantly older compared to those from Kiribati.

ACKNOWLEDGMENTS

The authors are grateful to Barbara Keating and Eric De Carlo of the University of Hawaii for providing the samples upon which this study is based. Peter Southgate, Frank Manheim, and two anonymous reviewers provided many valuable suggestions on the earlier draft of this manuscript. William Landing, Koh Harada, Philip Chin, and Tom Seal of FSU (Florida State University) provided valuable laboratory assistance. We also thank Ms. Sandy Silver and Ms. Kim Riddle of the electron microscope facility at FSU for their help in the SEM studies. Financial support for W.C.B. was provided by a grant from the National Science Foundation (OCE 8520724). Mr. V.P. Rao wishes to thank the Director, National Institute of Oceanography (NIO), India for providing leave and to UNESCO for providing a travel grant to visit FSU where this study was completed. He is also thankful to Mr. P. S.N. Murty and Mr. R.R. Nair of NIO for their encouragement.

REFERENCES

Azad, H.C., and J.A. Borchardt, 1970, Variations in phosphorus uptake by algae: Environmental Science and Technology, v. 4, p. 737-743.

Baturin, G.N., 1982, Phosphorites on the Sea Floor: New York, Elsevier Science Publishing Company, 343 p.

Birch, G.F., 1980, A model of penecontemporaneous phosphatization by diagenetic and authigenic mechanisms for the western margin of southern Africa, *in* Marine Phosphorites, Y.K. Bentor ed.: Tulsa, OK, Society of Economic Paleontologists and Mineralogists, Special Publication, 29, p. 79-100.

Braithwaite, C.J.R., 1980, The petrology of the oolitic phosphorites from Esprit (Aldabra), western Indian Ocean, Philosophical Transactions of the Royal Society of London, B, v. 288, p. 511-540.

Burns, R.G., and V.M. Burns, 1976, Mineralogy, *in* G.P. Glasby, ed., Marine Manganese Deposits: Amsterdam, Elsevier, p. 185-248.

Burnett, W.C., D.J. Cullen, and G.M. McMurtry, 1987, Open-ocean phosphorites: In a class by themselves?, *in* P.G. Teleki, M.R. Dobson, J.R. Moore and U. Von Stackelberg, eds., Marine Minerals: Advances in Research and Resource Assessment: Tokyo, D. Reidel Publishing Company, p. 119-134.

Cayeux, L., 1936, Existence de nombreuses bacteries dans les phosphates sedimentaires de tout age: C. R. Acad. Sci. Paris. v. 23, p. 1198–1200.

Cullen, D.J., and W.C. Burnett, 1986, Phosphorite associations on seamounts in the tropical southwest Pacific Ocean: Marine Geology, v. 71, p. 215-236.

DeCarlo, E.H., G.M. McMurtry, and K.H. Kim, 1987, Geochemistry of ferromanganese crusts from the Hawaiian Archipelago - I. Northern survey areas. Deep-Sea Research, 34, p. 441-467.

Edmond, C.R. 1969, Direct determination of fluoride in phosphate rock samples using the specific ion electrode: Bulletin Analytical Chemistry, v. 41, p. 1327-1328.

Golubic, S., 1973, The relationship between blue-green algae and carbonate deposits, *in* N.G. Carr and B.A. Whitton, eds., The Biology of the Blue-Green Algae: University of California Press. p 434-473.

Gulbrandsen, R.A., 1970, Relation of carbon dioxide content of apatite of the Phosphoria Formation to regional facies: Washington, DC, U.S. Geological Survey. Professional Paper 700-B, p. B9-B13.

Halbach, P., F.T. Manheim, and P. Otten, 1982, Co-rich ferromanganese deposits in the marginal seamount regions of the Central Pacific Basin - results of the MidPac '81: Erzmetall, v. 35, p. 447-453.

Halbach, P., and F.T. Manheim, 1984, Potential of cobalt and other metals in ferromanganese crusts on seamounts of the Central Pacific Basin: Marine Mining, v. 4, p. 319-336.

Hamilton, E.L., 1956, Sunken Islands of the Mid-Pacific Mountains: Boulder, CO, Geological Society of America, Memoir 64, 97 p.

Heezen, B.C., B. Glass, and H.W. Menard, 1966, The Manihiki Plateau, Deep-Sea Research, v. 13, p. 445-458.

Hein, J.R., F.T. Manheim, A.S. Schwab, C.L. Daniel, R.M. Bouse, L.M. Morgenson, R.E. Sliney, D. Clague, G.B. Tate and D.A. Cacchione, 1985, Geological and geochemical data for seamounts and associated ferromanganese crusts in and near the Hawaiian, Johnston Island, and Palmyra Island Exclusive Economic Zones: Menlo Park, CA, U.S. Geological Survey File Report, 85-292, 129 p.

Keating, B., B. Bolton and Shipboard Party, 1986. Initial report of 1986 R/V MOANA WAVE cruise MW-86-02 in the Kiribati/Tuvalu region, Central Pacific Ocean, CCOP/SOPAC Cruise Report. 121, SOUTHPAC 1986.

Klappa, C.F., 1979, Calcified filaments in Quaternary calcretes: Organo-mineral interactions in the subaerial vadose environment: Journal of Sedimentary Petrology, v. 49, p. 955-965.

Klement, K.W., and D.F. Toomey, 1967, Role of the blue-green alga *Girvanella* in skeletal grain destruction and lime-mud formation in the Lower Ordovician of west Texas: Journal of Sedimentary Petrology, v. 37, p. 1045-1051.

Kobluk, D.R., and M.J. Risk, 1977, Calcification of exposed filaments of endolithic algae, micrite envelope formation and sediment production: Journal of Sedimentary Petrology, v. 47, p. 517-528.

Kolodny, Y., 1981, Phosphorites, *in* C. Emiliani, ed., The Sea, Vol. 7: New York, Wiley & Sons, New York, p. 981-1023.

Krumbein, W.E., 1968, Geomicrobiology and geochemistry of the Nari-Lime Crust (Israel), *in* G. Muller and G.M. Friedman, eds., Developments in Carbonate Sedimentology in Central Europe: Berlin, Springer-Verlag. p. 138-147.

Krumbein, W.E., 1977, Primary production, mat formation and lithification: contribution of oxygenic and facultative anoxygenic cyanobacteria, *in* E. Flugel, ed., Fossil Algae: Recent results and developments: Berlin, Springer-Verlag, p. 37-56.

Lamboy, M., 1987, Genesis of granular phosphorites, clues gathered from study of grains including foraminifers: importance and modes of precipitation: C.R. Acad. Sc. Paris, v. 304, p. 435-440.

Manheim, F.T., 1986, Marine cobalt resources: Science, v. 232, p. 600-608.

Manheim, F.T., R.M. Pratt, and P.F. McFarlin, 1980, Composition and origin of phosphorite deposits of the Blake Plateau, *in* Y.K Bentor, ed., Marine Phosphorites: Tulsa, OK, Society of Economic Paleontologists and Mineralogists, Special Publication, 29, p. 117-129.

Manheim, F.T., P. Popenoe, W. Siapno, and C. Lane, 1982, Manganese-phosphorite deposits of the Blake Plateau, *in* P. Halbach and P. Winter, eds., Marine Mineral Deposits – New Research Results and Economic Prospects: Proceedings of the Clausthaler Workshop, Republic of Germany: Verlag Gluckauf, p. 9-44.

McClellan, G.H., and J.R. Lehr, 1969, Crystal chemical investigation of natural apatites: American Mineralogist, v. 54, p. 1374-1391.

Monty, C.L.V., 1976, The origin and development of cryptalgal fabrics, *in* M. R. Walter, ed., Stromatolites, Developments in Sedimentology: Elsevier Science Publishing Company, p. 193-249.

Murphy, J., and J.P Riley, 1962, A modified single solution method for determination of soluble phosphate in natural waters: Analytica Chimica Acta, v. 27, p. 31-36.

O'Brien, G.W., J.R. Harris, A.R. Milnes, and H.H. Veeh, 1981, Bacterial origin of East Australian continental margin phosphorites: Nature, v. 294, p. 442-444.

Oppenheimer, C.H., 1958, Evidence of fossil bacteria in phosphate rocks, Institute of Marine Science Publication, v. 5, p. 156-159.

Peryt, T.M., 1983, Coated Grains: Berlin, Springer-Verlag, 660 p.

Pratt, R.M., and P.F. McFarlin, 1966, Manganese pavements on the Blake Plateau: Science, v. 151, p. 1080-1082.

Prevot, L., and J. Lucas, 1986, Microstructure of apatite-replacing carbonate in synthesized and natural samples: Journal of Sedimentary Petrology, v. 56, p. 153-160.

Riggs, S.R., 1982, Phosphatic bacteria in the Neogene phosphorites of the Atlantic coastal plain - continental shelf system: IGCP Project 156 Phosphorite Newsletter, n. 11, p. 34.

Schroeder, J.H., 1972, Calcified filaments of an endolithic alga in Recent Bermuda reefs: Neues Jahrb. Geol. Palaeont. Monatsch, v. 1, p. 16-33.

Sonnenfeld, P., 1984, Brines and Evaporites: London, Academic press, 615 p.

Soudry, D., and Y. Champetier, 1983, Microbial processes in the Negev phosphorites (Southern Israel): Sedimentology, v. 30, p. 411-423.

Southgate,P.N., 1986, Cambrian phoscrete profiles, coated grains, and microbial processes in phosphogenesis: Georgina basin, Australia: Journal of Sedimentary Petrology, v. 56, p. 429-441.

Summerhayes, C.P., 1967, Manganese nodules from the Southwest Pacific, New Zealand: Journal of Geology and Geophysics, v. 10, p. 1372-1381.

Summerhayes, C.P., 1973, Distribution, origin and economic potential of phosphate sediments from the Agulhas Bank, South Africa: Translations Geological Society of South Africa, v. 76, p. 271.

Von Rad, U., and H.R. Kudrass 1984, Phosphorite Deposits on the Chatham Rise, New Zealand, R.V.Sonne Cruise SO-17, 1981: Geologisches Jahrbuch Reihe D, Heft 65.

Zanin, Y.N., S.V. Letov, N.A.Krasil'nikova, and Y.V Mirtov, 1985. Phosphatized bacteria from Cretaceous phosphorites of East-European platform and Paleocene phosphorites of Morocco, *in* J.Lucas and L. Prevot, eds, Sciences Geologiques Memoire "phosphorites" VI International field workshop and seminar on phosphorites IGCP 156, v. 77, p. 79-81.